Radicals for Life: The Various Forms of Nitric Oxide

The cover photo shows a two-photon fluorescence image 206 × 150 μm of the carotid artery extracted from a mouse and mounted in a perfusion chamber. The tissue was stained with fluorescent dyes to visualize adventitial collagen (CNA35/OG488; green), the elastin bands (EOSIN; red), and the nuclei of the vascular smooth muscle cells in the tunica media (SYTO; blue). These nuclei appear elongated perpendicular to the direction of vascular flow. The typical wave-like appearance of the elastin bands (red) is caused by the absence of transmural pressure. The outer adventitial layer contains some purple/white nuclei from embedded fibroblasts. (Image courtesy R. Megens and M. van Zandvoort, CARIM, Maastricht University, NL.)

Radicals for Life: The Various Forms of Nitric Oxide

Edited by

ERNST VAN FAASSEN

Debye Institute, Section Interface Physics,
Ornstein Laboratory, Utrecht University,
3508 TA, Utrecht, The Netherlands

AND

ANATOLY FYODOROVICH VANIN

Semenov Institute of Chemical Physics,
Russian Academy of Sciences,
Moscow, 119991, Russian Federation

ELSEVIER

AMSTERDAM • BOSTON • HEIDELBERG • LONDON • NEW YORK • OXFORD • PARIS
SAN DIEGO • SAN FRANCISCO • SINGAPORE • SYDNEY • TOKYO

Elsevier
Radarweg 29, PO Box 211, 1000 AE Amsterdam, The Netherlands
The Boulevard, Langford Lane, Kidlington, Oxford OX5 1GB, UK

First edition 2007

Library of Congress Cataloging-in-Publication Data
A catalog record for this book is available from the Library of Congress

British Library Cataloguing in Publication Data
A catalogue record for this book is available from the British Library

ISBN-13: 978-0-444-52236-8

Transferred to Digital Printing in 2010

Foreword

Since the discovery of the physiological role of nitric oxide in the late 1980s, this simple molecule has been the focus of attention of biochemical and biomedical research. Originally, the main emphasis was on its vasodilating effect in the mammalian vasculature because it was in this context that the importance of NO had been recognized first. Only gradually it was realized that nitric oxide fulfills other important functions as well, for example, in neuronal signal transduction and in the immune system. In recent years, a fresh and powerful impulse was given by the discovery that NO acts as a modulator of gene expression and apoptosis. The list of physiological processes where a role of NO is recognized keeps growing daily and appears unlimited. It sometimes seems that NO appears wherever you take a close look. The actions of NO span a wide field covering the physiology of healthy organisms as well as many pathophysiologies in mammals and humans.

Recent years have seen a tremendous step forward: It was the discovery that a broad range of other compounds can fulfill one or more of the roles of true NO. Some but not all of these compounds are metabolites of NO, and are endogenously formed in the presence of NO. Examples are nitrite and *S*-nitrosothiols. Others like heme-nitrosyl or dinitrosyl-iron complexes are not metabolites *per se* but form by reversible binding of NO as a nitrosyl ligand to a transition metal ion. Until now, the knowledge of these alternative forms of nitric oxide was not yet collected and ordered in the form of a monograph. Instead, it was widely scattered in the specialized scientific literature and difficult to access. The purpose of the editors was to close this gap, and make the huge body of information available to researchers in a very compact form. Our aim was to make the book accessible and readable for biochemists, clinicians, biologists, and biomedical researchers from all disciplines. That is not easy as all disciplines have developed their own special jargon and ways of expression. We have tried to adopt a language common to these various disciplines, and review the current status of the field rather than dwell on technical details. Up to date reference lists will help the readers to locate more specialized literature and find the original sources of relevant reaction constants and the like.

Compiling a book is a large task, and we would not have completed the work without the cooperation and enthusiasm of our contributors and collaborators, and the patient encouragement from the staff at Elsevier. We hope that the readers will find the volume a useful tool for their research, and enjoy reading about the many surprising aspects and forms of the small diatomic molecule called NO.

Anatoly Fyodorovich Vanin and Ernst van Faassen

Contents

PART I
Introductory Overview

CHAPTER 1

Nitric oxide radicals and their reactions

Ernst van Faassen[1] and Anatoly F. Vanin[2]

[1]*Debye Institute, Section Interface Physics, Ornstein Laboratory, Utrecht University, 3508 TA, Utrecht, The Netherlands*
[2]*Semenov Institute of Chemical Physics, Russian Academy of Sciences, Moscow, 119991, Russian Federation*

Nitric oxide is a peculiar radical species in many respects. It is a radical in the sense that its electronic configuration contains a half-occupied orbital that is occupied by a single electron only. In spite of this open electronic structure, its reactivity with most other molecules is surprisingly low (notable exceptions being superoxide radicals). The half-occupied orbital provides the molecule with a nonzero total electronic angular momentum $S = 1/2$ and nonzero electronic magnetic moment from the electrons. However, the ground state of $NO^{\bullet}$ is not paramagnetic at all. This diatomic molecule has fifteen electrons in an electronic configuration $(K^2K^2) - (2s\sigma)^2(2s\sigma^*)^2(2p\pi)^4(2p\sigma)^2(2p\pi^*)^1$ and rotational symmetry around the molecular axis. The unpaired electron is located in one of the antibonding π^* orbitals. The axial symmetry of the electronic fields allows for the projections of total spin S and angular momentum L along the molecular symmetry axis as conserved quantities. As such, $NO^{\bullet}$ is a good example of a type (a) molecule in Hund's classification, where the axial projections of S, L and $J = L+S$ are given the quantum numbers Σ, Λ and $\Omega = \Lambda + \Sigma$ respectively. Type (a) molecules typically have strong internuclear fields to constrain L and strong spin-orbit coupling to constrain S to the molecular axis. Free NO has $\Sigma = 1/2$ and $\Lambda = 1$. The ground state ${}^{2\Sigma+1}\Lambda_{\Omega} = {}^2\Pi_{1/2}$ has antiparallel coupling, with the parallel coupled ${}^2\Pi_{3/2}$ state being 124.2 $cm^{-1} \sim 15$ meV higher in energy. Rotationally excited ladders of these states appear with energy spacings of about 5 cm^{-1}. The magnetic moment $\mu = g_{\Omega}\Omega$ is proportional to the gyromagnetic ratio g_{Ω}. For molecules of Hund's type (a), the gyromagnetic ratio is given by [1]

$$g_{\Omega} = \frac{(\Lambda + 2\Sigma)(\Lambda + \Sigma)}{\Omega(\Omega + 1)} = 0 \quad \text{for} \quad {}^2\Pi_{1/2}$$

$$= 4/5 \quad \text{for} \quad {}^2\Pi_{3/2}$$

We note that the ground state ${}^2\Pi_{1/2}$ of this radical has nonzero angular momentum ($\Lambda = 1$) but zero total magnetic moment. In other words, it is not paramagnetic at all when in free state. As such, the true ground state of an isolated $NO^{\bullet}$ molecule in the gas phase cannot be detected by magnetic resonance spectroscopy.

However, detection of $NO^{\bullet}$ radicals with electron paramagnetic resonance (EPR) becomes possible under certain conditions. In dilute $NO^{\bullet}$ gas at room temperature ($kT \sim$ 25 meV), ca 70% of the molecules are thermally excited to the $^{2}\Pi_{3/2}$ state with $g_{\Omega} = 4/5$. This state shows linear Zeeman splitting when brought into an external magnetic field. At X-band frequencies ~10 GHz, the EPR transitions occur at magnetic fields near 9100 G. The EPR spectrum (Fig. 1) of this dilute gas is ca 250 G wide and shows 9 well resolved lines. It appears as a triplet of triplets with 3:4:3 intensity. The larger splitting is caused by the zero-field splitting (ZFS) interaction and the smaller splitting by the hyperfine coupling to the magnetic moment of the ^{14}N nucleus ($I = 1$). At higher gas pressures, the lines are broadened by spin–spin interactions, and the resolution of the ZFS is lost.

Alternatively, $NO^{\bullet}$ molecules may be adsorbed on solid surfaces, where the interactions with atoms of the substrate lead to quenching of the orbital angular momentum. The resulting magnetic moment is then determined by the electronic spin only [3]. Finally, $NO^{\bullet}$ often appears as a ligand in paramagnetic metal complexes with Co or Fe centers. Well-known examples are the ferrous nitrosyl complexes with heme moieties or with iron–dithiocarbamate complexes as discussed in Chapters 2–5 and 18. The small difference in electronegativity between nitrogen and oxygen gives $NO^{\bullet}$ a modest electrical dipole moment of 0.159 D, i.e. more than an order of magnitude smaller than that of water.

Nitric oxide gas is colorless in the visible wavelength region, but has a prominent infrared absorption at 1878 cm^{-1} = 0.233 eV due to the fundamental vibrational band [ν(N=O) stretch mode] [4,5]. This characteristic absorption corresponds to a wavelength of 5.3 μm and is often used for detection of NO in the gas phase with optical sensors [6]. For example, the infrared absorption line has been used to detect traces of NO escaping from biological samples at rates above ca 10 pmol/s [7]. Upon binding to, for example, the metal center of a complex, the frequency of this vibrational band is changed considerably by a combination of two antagonistic effects. First, the anchoring to the heavy metal center raises the effective reduced mass of the stretch vibration and lowers the frequency. Additionally, the strength of the NO bond is affected by partial charge transfer of the unpaired electron towards the metal.

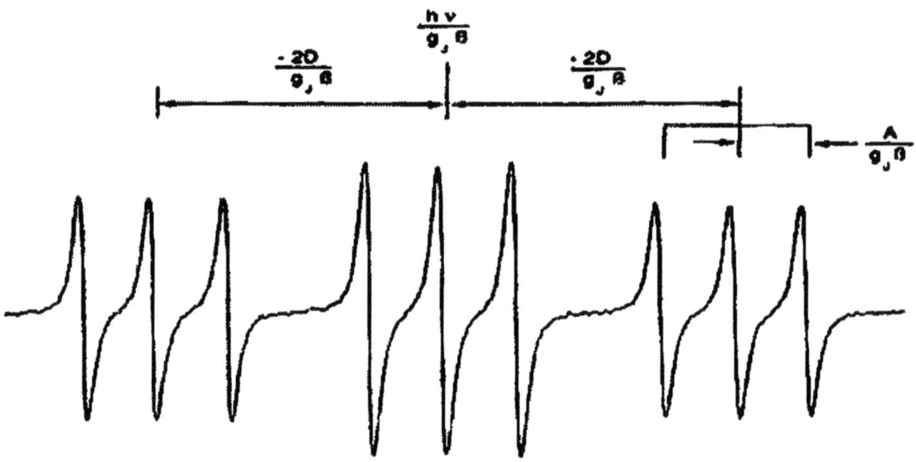

Fig. 1. X-band EPR spectrum of 1 Torr NO gas at room temperature. The scan range is 335 Gauss. (From Ref. [2].)

This transfer of the electron away from the antibonding orbital significantly stiffens the NO bond and tends to raise the vibration frequency. This stiffening is the reason that the stretch vibrations of free nitrosonium NO^+ are raised to 2200 cm^{-1} [8]. Conversely, transfer of additional electron density into the antibonding orbital lowers the frequency of the ν(N=O) stretch mode to ca 1363 cm^{-1} [9] for the free nitroxyl anion NO^-. IR and Raman spectroscopy of nitrosyl complexes have shown in accordance that the nitrosyl stretch vibrations span the full range of 1100–2000 cm^{-1}, and the stretch vibration of the NO ligand was found to be a good spectroscopic indicator of the degree of charge transfer in nitrosyl complexes with transition metal ions [10]. Prime examples are the endogenous nitrosyl–iron complexes in biological systems like tissues or blood. The extent of charge transfer is directly related to an important structural parameter, namely the orientation of the nitrosyl axis with respect to the metal ion. The M—N—O bond angle is predicted to be 120° for NO^- ligands, whereas a nearly linear alignment is found for nitrosyl cations. Such charge transfer was found to be very characteristic for nitrosyl ligands on iron atoms and the shared nature of the unpaired electron is accounted for in Enemark–Feltham notation [11] for the combined electronic configuration of the Fe—NO motif. Paramagnetic mononitrosyl iron complexes (MNICs, cf Chapter 18) have $\{FeNO\}^7$ configuration and typical frequencies of the ν(N=O) stretch mode are 1670–1720 cm^{-1}. Paramagnetic dinitrosyl iron complexes (DNICs, cf Chapter 2) have $\{Fe(NO)_2\}^7$ or $\{Fe(NO)_2\}^9$ configuration and typical ν(N=O) stretch frequencies are higher 1730–1800 cm^{-1}. Significantly, the two nitrosyl ligands are found to stretch with slightly different frequency, separated by 30–60 cm^{-1}. For example, for Cys-DNIC the two stretch frequencies are reported as 1730 and 1770 cm^{-1} [12]. Therefore, the two ligands show distinct charge transfer and net effective charge. This nonequivalence of the nitrosyl ligands is highly significant as it seems the reason for the unusual reaction chemistry of DNIC as described in Chapter 2. A selection of experimental stretching frequencies for nitrosyl–metal complexes can be found in Refs. [10,11,13,14].

For covalent nitroso compounds the ν(N=O) stretch mode [18] is a good marker as well (cf Table 1).

Table 1 Typical frequencies for the ν(N=O) stretch mode in various compounds

Class	ν(N=O) Stretch mode (cm^{-1})	Remarks	Refs.
NO radical	1878	Free molecule	[4,5]
NO^+	2200	Free molecule	[8]
NO^-	1363	Free molecule	[9]
NO^+–M	1850–1860	M—N—O is 180°	[15]
NO—Fe^{3+}–DETC	1687	1708 cm^{-1} for *N*-methyl-D, L-glucamine dithiocarbamate (MGD)	[16]
NO—Fe^{2+}–DETC	1724	M—N—O is 173° [17]	[14]
NO—Fe^{3+}–porphyrins	1850–1860	M—N—O is 175°	[10]
NO—Fe^{2+}–porphyrins	1620–1670	M—N—O is 145°	[10]
S-nitroso	1480–1530		[19]
N-nitroso aliphatic	1420–1460	ν(N—N) at 1030–1150	[20]
N-nitroso aromatic	1450–1500	ν(N—N) at 925–1025	[20]
O-nitroso	1610–1730	Two distinct modes with $\Delta\nu \sim 40$ cm^{-1}	[20]
Fe^{2+}–$(NO)_2$, DNIC	1730, 1770	Nitrosyl ligands are inequivalent with $\Delta\nu \sim 40$ cm^{-1}	[12]

The UV spectrum of nitric oxide shows absorption below 200 nm due to weak electronic transitions to unoccupied orbitals, and the onset of photoionization to nitrosonium NO^+ by ejection of the unpaired electron from the π^* orbital. As expected for an antibonding electron, the ionization threshold is fairly low with 9.26 eV. This photoionization produces a prominent photoabsorption peak near 14 eV. For still higher photon energies >20 eV, fragmentation of the molecule occurs [21].

Nitric oxide is a highly corrosive gas with a boiling point of −151.7°C at 1 atm and is prone to oxidation to nitrogen dioxide radicals when in contact with dioxygen. The reaction chemistry of $NO^\bullet$ is very complex due to several reasons: first and foremost, it is thermodynamically unstable. If kinetically allowed by conditions of high pressure and temperature, it may dismutate *via* the pathways [22]

$$3NO^\bullet \rightarrow N_2O + NO_2^\bullet \qquad \Delta H = -37.2 \text{ kcal mol}^{-1} \tag{1}$$

$$2NO^\bullet \rightarrow N_2 + O_2 \qquad \Delta H = -21.6 \text{ kcal mol}^{-1} \tag{2}$$

The prime example for the first pathway is when pure $NO^\bullet$ is compressed to high density for storage in pressure containers. Interestingly, this reaction appears as a disproportionation as it generates products with higher as well as lower oxidation state than the original $NO^\bullet$. A prominent example for the second pathway is the automotive catalytic converter where rhodium catalyzes the reductive decomposition of $NO^\bullet$ into dinitrogen and dioxygen.

The second reason is the fact that many higher oxides of nitrogen have a rather unstable electronic structure: N_2O, N_2O_4 and N_2O_3 all appear as resonance hybrids resonating between several isomeric forms with different atomic and electronic structures [23]. In addition, $NO_2^\bullet$ is a radical with tendency to dimerize and HNO dimerizes to metastable hyponitrous acid HONNOH which decomposes into N_2O and water [see below, Eq. (6)].

In addition, the analysis of the reaction chemistry and identification of the reaction products is technically challenging since many nitrogen oxides have strong broad overlapping absorptions in the ultraviolet at wavelength shorter than 280 nm. The visible bands are usually better resolved but have very small extinctions (N_2O_3 with $\varepsilon_{620} \sim 20$ Mcm^{-1}; nitrate NO_3^- with $\varepsilon_{300} \sim 7$ Mcm^{-1}; nitrite NO_2^- with $\varepsilon_{354} \sim 24$ Mcm^{-1}; nitrous acid HNO_2 with a characteristic tetrad of peaks at 347, 358, 371 and 386 nm and $\varepsilon_{371} \sim 54$ Mcm^{-1}, $pK_a = 3.37$; $NO_2^\bullet$ with $\varepsilon_{385} \sim 30$ Mcm^{-1}). This last radical, nitrogen dioxide, is a common industrial and automotive pollutant and is responsible for the orange-brown hue of smog and polluted air. Peroxynitrite anions have a stronger UV absorption with $\varepsilon_{302} \sim 1704$ Mcm^{-1}. The lifetime of peroxynitrite at physiological pH is below a second because of its reactivity towards proteins and propensity to protonate to *cis* or *trans* forms of pernitrous acid [24,25]. Far stronger absorptions in the visible region are known for the hyponitrite radical anion $N_2O_2^{-\bullet}$ ($\varepsilon_{290} \sim 6 \times 10^3$ Mcm^{-1}); its protonized form, the radical $HN_2O_2^\bullet$ ($\varepsilon_{290} \sim 3 \times 10^3$ Mcm^{-1} $pK_a = 5.5$) [32] and the unstable $N_3O_3^-$ anion ($\varepsilon_{380} \sim 3.8 \times 10^3$ Mcm^{-1}) with decay rate of 300 s^{-1} in water at room temperature [26].

The optical spectra of nitrite and nitrate are shown in Fig. 2.

Nitric oxide itself can readily participate in a wide variety of redox reactions [28]. It may be oxidized to the nitrosonium cation NO^+, which is isoelectronic to carbon monoxide CO

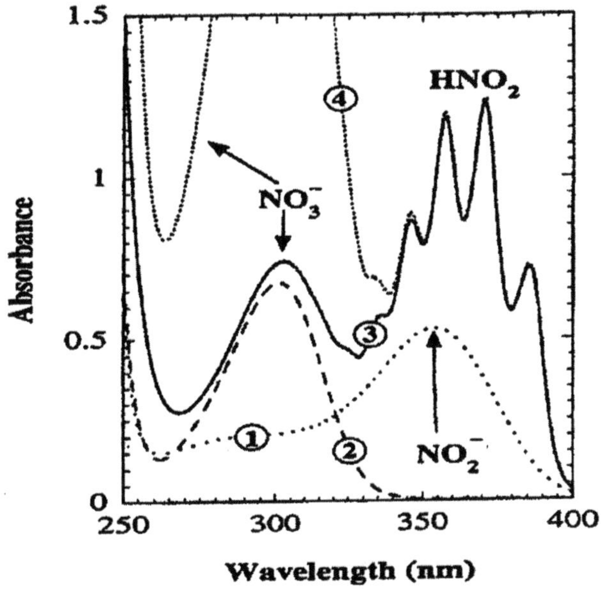

Fig. 2. Room temperature spectra of nitrite, nitrous acid and nitrate. The curves are: (1) 22.6 mM $NaNO_2$; (2) 92.8 mM $NaNO_3$; (3) 22.6 mM $NaNO_2$ plus 92.8 mM HNO_3; (4) 22.6 mM $NaNO_2$ plus 418 mM HNO_3. (From Ref. [27].)

and N_2. It has a prominent UV absorption ($\varepsilon_{220} \sim 3850\ \mathrm{Mcm}^{-1}$ [29]). The oxidation reaction is

$$NO^{\bullet} \rightarrow NO^+ + e^- \qquad E_{1/2} = -1.21\ \mathrm{V} \tag{3a}$$

The reduction potential of 1.21 V is quite high and exceeds range of the typical reduction potentials found *in vivo*. The latter range from −0.3 V (NAD^+/NADH) to +0.82 V (O_2/H_2O). Therefore, nitrosonium is not generated by simple oxidation of nitric oxide radicals. Acidic reduction of nitrite anions provides an alternative pathway [Eq. (3b)]

$$NO_2^- + 2H^+ \rightleftarrows NO^+ + H_2O \tag{3b}$$

However, in the normal physiological range pH ~ 7.0–7.4, this equilibrium is shifted to the far left side [28] with $[NO^+]/[NO_2^-] \sim 10^{-17}$ approaching infinitesimally small values. In biological systems, the formation of nitrosonium is thought to be dominated by heterolytic fission of N_2O_4 [see below Eq. (8)]. However, even when formed, nitrosonium is short lived in the presence of water. Reaction (3b) shows that the nitrosonium cation reacts rapidly with water to nitrite. It may also react with other nucleophiles, and can nitrosate proteins in biological systems. Other nitrosating pathways will be discussed below [cf Eq. (10)].

Alternatively, nitric oxide radicals may be reduced to the nitroxyl anion NO^- which is isoelectronic to molecular oxygen. Having two half-occupied orbitals, this biradical may exist [30,31] as a spin triplet ($S = 1$) or as a spin singlet ($S = 0$)

$$NO^{\bullet} + e^- \rightarrow {}^3NO^- \ (S = 1) \qquad E_{1/2} = +0.39 \text{ V} \tag{4a}$$

$$NO^{\bullet} + e^- \rightarrow {}^1NO^- \ (S = 0) \qquad E_{1/2} = -0.35 \text{ V} \tag{4b}$$

Just as with the dioxygen molecule, the spin triplet $^3NO^-$ is ground state, with the spin singlet 17–20 kcal/mol (0.74–0.87 eV) higher in energy.

The two spin states of the nitroxyl anion have markedly different reaction rates with nitric oxide to form the hyponitrite radical $N_2O_2^{\bullet 2-}$, or even the bluish trinitrogen trioxide anion $N_3O_3^{2-}$ [32]. The rates are strongly affected by the constraints imposed by spin conservation. The protonation of the ground state triplet nitroxyl is kinetically slow as it requires spin conversion between singlet HNO and triplet NO^- as well as a nuclear reorganization:

$$^1HNO \rightleftarrows {}^3NO^- + H^+ \tag{5}$$

Whereas older literature often quotes a low equilibrium constant of $pK_a \sim 4.7$, newer estimates [26,31] favor a sharply upward revision to a value of $pK_a \sim 11$ for the protonation of $^3NO^-$. This value makes the protonized HNO the predominant species at physiological pH. At higher concentrations, the chemistry of HNO is complicated by its tendency to dimerize and decompose into dinitrogen oxide N_2O and water. The irreversible decomposition has a high rate of $k = 8 \times 10^6$ $(Ms)^{-1}$:

$$2HNO \rightleftarrows [HONNOH] \rightarrow N_2O + H_2O \tag{6}$$

The nitroxyl anion NO^- may act directly on a range of biological molecules [33,34]. Angeli's salt $Na_2N_2O_3$ is a well-known water-soluble nitroxyl donor.

In water, $NO^{\bullet}$ has a solubility [35] of 1.9 mM at 25°C and against a P_{NO} of 1 atm (Fig. 3 and Table 2). The solubility is somewhat higher than that of dioxygen.

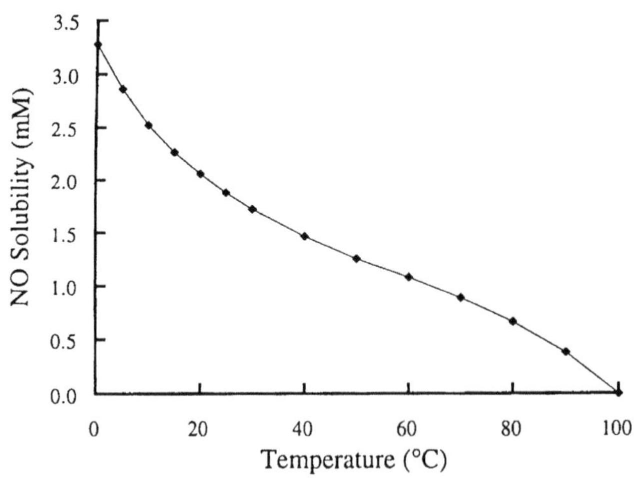

Fig. 3. Solubility in water of nitric oxide under 1 atm NO as a function of temperature. (From Ref. [35].)

Table 2 Solubility of selected small neutral molecules in pure water under a gas pressure of 1 atm. The solubilities are in millimolar concentration

Molecule	Dipole moment (*D*)	20°C (mM)	37°C (mM)
NO	0.159	2.1	1.6
N_2O	0.161	28.1	17.7
O_2	–	1.4	1.0
CO_2	–	39.2	25.9
CO	0.110	1.06	0.84

In pure deoxygenated water and at low concentrations, $NO^{\bullet}$ is indefinitely stable on a timescale of weeks. In particular, it does not undergo a hydration reaction. However, this stability is only apparent due to slowness of the irreversible disproportionation reaction Eq. (1).

At high concentrations as in compressed gases, the disproportionation reaction (1) is very noticeable and leads to rapid formation of substantial quantities of secondary radical like nitrogen dioxide $NO_2^{\bullet}$. The latter radical is a strong one-electron oxidant with a reduction potential of 1.04 V [30] for the $NO_2^{\bullet}/NO_2^-$ redox couple. Nitrogen dioxide rapidly decomposes in water. At lower concentrations of nitric oxide, the disproportionation reaction (1) is kinetically inhibited, but $NO_2^{\bullet}$ may still be formed by a complicated reaction of $NO^{\bullet}$ with dioxygen. For biological samples, the formation of this $NO_2^{\bullet}$ intermediate is highly significant due to its capacity to generate further species capable of N-nitrosation of amines and S-nitrosation of thiols and organic sulfides. Therefore, a short outline will be given here, with a comprehensive discussion being found in the excellent recent monograph of Williams [36].

The formation of $NO_2^{\bullet}$ from nitric oxide and dioxygen has a small activation energy of $\Delta H = 4.6 \pm 2.1$ kJ/mol. The precise mechanism has not been clarified and two pathways have been proposed. The first is based on the tendency of $NO^{\bullet}$ to dimerize to dinitrogen dioxide. This species may be oxidized to dimerized nitrogen dioxide.

$$\begin{gathered} 2NO^{\bullet} \rightleftarrows N_2O_2 \\ N_2O_2 + O_2 \rightarrow N_2O_4 \\ N_2O_4 \rightleftarrows 2NO_2^{\bullet} \end{gathered} \tag{7a}$$

The second proposition involves the formation of an intermediate peroxynitrite radical

$$\begin{gathered} NO^{\bullet} + O_2 \rightleftarrows ONOO^{\bullet} \\ ONOO^{\bullet} + NO \rightarrow ONOONO \\ ONOONO \rightarrow 2NO_2^{\bullet} \end{gathered} \tag{7b}$$

Both mechanisms have a net balance

$$2NO^{\bullet} + O_2 \rightarrow 2NO_2^{\bullet} \tag{7c}$$

Upon completion of reaction (7c), the appearance of nitrosating species like N_2O_3 and N_2O_4 is unavoidable. Although not radicals themselves, these species are quite reactive.

This fact is highly relevant for biological systems, where the pool of proteins provides a potential target for subsequent nitrosating reactions of amine and thiol moieties. The equilibrium reactions are

$$2NO_2^{\bullet} \rightleftarrows N_2O_4 \tag{8}$$

$$NO_2^{\bullet} + NO^{\bullet} \rightleftarrows N_2O_3 \tag{9}$$

The forward and backward reaction rates of Eq. (8) were reported as 4.5×10^8 $(Ms)^{-1}$ and 6.9×10^3 s^{-1}, respectively. The forward and backward reaction rates of Eq. (9) were reported an order of magnitude faster with 1.1×10^9 $(Ms)^{-1}$ and 8.1×10^4 s^{-1}, respectively [37].

The powerful nitrosating species dinitrogen trioxide N_2O_3 is easily identified from its weak blue color ($\varepsilon_{620} \sim 20$ Mcm^{-1}). It is a highly polar molecule with a dipole moment of 2.122 D, i.e. exceeding the moment 1.854 D of a water molecule. Experimental evidence supports the usual assumption that dinitrogen trioxide N_2O_3 is the main nitrosating species in buffered water solutions near physiological pH, in spite of the fact that it may easily react with water molecules [see below, Eq. (12)]. However, the situation is less clear for living systems, as these form a heterogeneous environment with regions of low polarity. In such regions, neutral molecules like $NO_2^{\bullet}$ or $NO^{\bullet}$ tend to accumulate to higher concentration with different equilibrium constants for the equilibria between $NO_2^{\bullet}$, $NO^{\bullet}$, N_2O_3 and N_2O_4 [38]. This phenomenon significantly accelerates the autooxidation reaction Eq. (1) [39]. The nitrosation reactions that transform thiol moieties RS—H to nitrosothiols RS—NO are

$$N_2O_4 + RS{-}H \rightarrow NO_3^- + RS{-}NO + H^+ \tag{10a}$$

$$N_2O_3 + RS{-}H \rightarrow NO_2^- + RS{-}NO + H^+ \tag{10b}$$

These nitrosation reactions are fast. For glutathione (GSH), reaction (10b) proceeds with a second order rate of $k = 6.6 \times 10^7$ $(Ms)^{-1}$ [40].

These reactions are distinguished by their primary reaction products being nitrate and nitrite, respectively. In principle, nitrosonium is a nitrosating species also. Its nitrosation reaction with thiols RS—H proceeds as

$$NO^+ + RS{-}H \rightarrow RS{-}NO + H^+ \tag{11}$$

However, in aqueous environment the concentration of free nitrosonium is negligibly small [cf Eq. (3b) and its discussion].

In aqueous solution, the nitrosation reactions (10a) and (10b) have to compete with an alternative pathway: hydrolysis to nitrite or nitrate according to the reactions

$$N_2O_4 + H_2O \rightarrow NO_2^- + NO_3^- + 2H^+ \tag{12a}$$

$$N_2O_3 + H_2O \rightarrow 2NO_2^- + 2H^+ \tag{12b}$$

The hydrolysis pathway leaves N_2O_4 with a lifetime of about 10^{-3} s in water [41]. Published rates for (12b) vary considerably [42,43], possibly because the reaction is catalyzed by the presence of phosphate. But even in absence of phosphate, reaction (12b) is fast, and the lifetime of N_2O_3 in water is shorter than ca 1 ms [37]. Therefore, the nitrosation reactions (10a,b) will be significant only at sufficiently high levels of thiol. The critical thiol concentration can be estimated from the nitrosation rate of $k = 6.6 \times 10^7$ $(Ms)^{-1}$ [40] for glutathione and the lifetime of 1 ms for N_2O_3 in water. It means that nitrosation of glutathione dominates hydrolysis if [GSH] > 15 μM. This condition is easily satisfied *in vivo* where tissue concentrations of GSH are of the order of 0.5–1 mM (cf Table 1, Chapter 9). Phrased otherwise, at physiological thiol concentrations, the hydrolysis pathway (12a,b) is not significant and thiols act as main target for the nitrosating species N_2O_4 and N_2O_3.

It should be kept in mind that N_2O_3, NO^+ and N_2O_4 are by no means the only compounds with proven capacity of S-nitrosation. Alternative pathways for S-nitrosation have been identified for organic nitrates $RONO_2$ and organic nitrites RONO. More details on these reactions are given in Chapters 9, 10 and 17.

The formation of *S*-nitrosothiols has significant implications *in vivo*. The S-nitrosated moieties in tissues reach concentrations upto ca 100 nM. Therefore, these moieties are potentially relevant as an endogenous form of nitric oxide with higher stability and longer lifetime than nitric oxide itself. The *S*-nitrosothiols are quite stable with a lifetime of several days in aqueous solutions if these solutions are kept cool in the dark and free of trace metal ions. They may release free $NO^\bullet$ radicals by hemolytic cleavage *via* thermolysis, photolysis or catalytically by trace metal ions in reduced state like Fe^{2+} or Cu^+. The possibility of heterolytic cleavage of the S—N bond under release of free NO^+ is still controversial. As will be explained in Chapters 9–11, the homolytic decomposition has high significance for the balance between the various nitrosated species in biological systems.

The balance of the above reactions makes the rate of nitrite formation third order in the constituents

$$\frac{d}{dt}[NO_2^-] = k[NO^\bullet]^2[O_2] \qquad (13)$$

The rate of $k \sim 5 \times 10^6\ M^{-2}s^{-1}$ (at 25°C) does not depend on the pH [36].

The formation of the $NO_2^\bullet$ intermediate is the rate-limiting step in this oxidation of $NO^\bullet$. Accordingly, it also determines the rate of nitrosation of thiol moieties in biological systems. The low rate guarantees that the uncatalyzed oxidation to nitrite does not significantly affect the lifetime of $NO^\bullet$ in aqueous solutions. Even at full oxygen saturation ($[O_2] \sim 1$ mM), the nanomolar concentrations of $NO^\bullet$ make that this radical would have a lifetime of many hours if kept in aqueous solution. However, in complex biological systems, the presence of a membrane fraction may strongly enhance the observed rate of autooxidation and nitrosating capacity [39,44]. The small dipole moment of $NO^\bullet$ (cf Table 2) allows this radical to concentrate in the hydrophobic phospholipid membranes and reaches a tenfold higher local concentration than in aqueous medium. This concentration effect increases the rate of autooxidation 300-fold [39]. In addition, spurious metal ions or enzymatic reaction centers could act as catalyzers and accelerate the oxidation reaction (7c). The ensuing nitrosation reactions would be accelerated as well. At the moment, it is still unknown whether such nonenzymatic autooxidation reactions are significant for the $NO^\bullet$ lifetime and nitrosation in biological systems.

In addition, enzymatic pathways exist for the oxidation of $NO^\bullet$. In biological systems, at least two important irreversible pathways for oxidation of $NO^\bullet$ have been identified: oxyhemoproteins and superoxide. Hemoglobin (Hb) is a tetrameric heme protein composed of pairs of ferrous α- and β-subunits. The four hemes may be independently nitrosylated and the ferrous forms are paramagnetic. Accordingly, blood and many biological tissues show the prominent EPR absorption as a mixture of α- and β-type heme nitrosylated ferrous state [45]. The spectral shape depends on oxygen concentration and conformation of the Hb (relaxed-R of tense-T). In fully oxygenated blood, the dominant species is $\alpha(Fe{-}O_2)_2{-}\beta(Fe{-}O_2)_2$ which rapidly oxidizes nitric oxide to nitrate

$$OxyHb^{2+} + NO^{\bullet} \rightarrow metHb^{3+} + NO_3^- \tag{14}$$

Equation (14) has a high reaction rate of $\sim 1 \times 10^7$ $(Ms)^{-1}$ and is the dominant reaction pathway for loss of nitric oxide in the normoxic vasculature. At first sight, reaction (14) might look like iron-mediated oxidation of $NO^{\bullet}$ by dioxygen. However, the rate depends crucially on whether dioxygen or nitrosyl is the first ligand to bind, since nitrosylated Hb itself is remarkably stable against transformation into metHb and nitrate by dioxygen. In oxygenated Tris buffer (50 mM, pH = 7.4) the lifetimes of $\alpha(Fe{-}NO)_2{-}\beta(Fe{-}O_2)_2$ and $\alpha(Fe{-}NO)_2{-}\beta(Fe{-}NO)_2$ were 2 h and 41 min, respectively [46]. Interestingly, the partially nitrosylated $\alpha(Fe{-}NO)_2{-}\beta(Fe{-}O_2)_2$ retained significant capacity for cooperative binding and transport of oxygen. Myoglobin is an oxygen transporter with a single heme-binding site. Oxymyoglobin also rapidly converts nitric oxide to nitrate with a comparable rate [47].

An even higher reaction rate applies to the diffusion-controlled reaction with the superoxide radical to peroxynitrite, an isomer of nitrate

$$NO^{\bullet} + O_2^{\bullet -} \rightarrow ONOO^- \tag{15}$$

Reported reaction rates in water range from 4.3×10^9 to 1.9×10^{10} $(Ms)^{-1}$ [48]. Peroxynitrite, though not itself a radical, is a powerful oxidizing agent and has been shown to be highly damaging to intracellular processes. It reacts with many proteins [48], particularly *via* the nitration of tyrosine residues and thiols. Peroxynitrite readily acts as a ligand for transition metal ions [49] and the metal-containing sites of many enzymes, hemoproteins in particular, are prone to modification by peroxynitrite leading to inhibition of catalytic activity (e.g. superoxide dismutase, cytochromes and nitric oxide synthase) or structural degradation (e.g. Zn release from zinc-finger-containing proteins). In water, the peroxynitrite anion is stable ($\varepsilon_{302} \sim 1704$ Mcm^{-1}[24]), but the protonized form (peroxynitrous acid, *cis*-ONOOH with $pK_a = 6.8$, *trans*-ONOOH with $pK_a \sim 8$) is unstable, because the *trans*-isomer is a vibrationally excited state that can rearrange to nitric acid [25]. The rearrangement shortens the lifetime of peroxynitrous acid to less than 1 s.

It was already remarked that $NO^{\bullet}$ is thermodynamically unstable and prone to dismutation. Recent investigations have shown that transition metal ions may catalyze the autooxidation of $NO^{\bullet}$ [50]. Dinitrosyl iron complexes (DNIC) with small thiol ligands possess S-nitrosating capacity reminiscent of nitrosonium, and the formation of hydroxylamines was attributed to intermediate nitroxyl NO^- anions. It suggests that the iron atom mediates a strong electronic coupling between the nitrosyl ligands and facilitates their dismutation into nitrosonium and nitroxyl. A more detailed discussion is given in Chapter 2.

Finally, a nonnegligible decay channel for $NO^{\bullet}$ may be its sequestration by ferrous iron as found in heme enzymes like hemoglobin. Although nitric oxide binds to ferrous as well as ferric form, the ferrous binding is so strong as to be considered irreversible on the relevant timescales of minutes to hours. The strong binding of $NO^{\bullet}$ ligands to ferrous heme forms a very stable paramagnetic mononitrosyl iron complex. It was reported that ca 0.004% of Hb-heme is nitrosylated in healthy human volunteers [51]. Given that blood has a heme concentration of [Hb-heme] = 4 [Hb] $\sim$ 8 mM, the quantity of nitrosylated heme in arterial blood amounts to ca 0.3 μM. This is a significant quantity indeed. The binding of nitrosyl ligands to iron strongly affects the spectroscopic properties of the complex. The fundamental intramolecular N–O vibration appears at 1878 cm^{-1} in the gas phase, but is redshifted upon binding to heme iron. The vibration is easily observable by strong IR absorption and moderate intensity of Raman scattering and may be used to discriminate the redox state, degree of coordination and conformation of the protein.

The sequestration by iron allows hemoglobin of the blood to act as a very significant sink for the nitric oxide radicals as produced by the endothelial lining of blood vessels. Numerical simulations indicate that the loss of nitric oxide in the vascular lumen should reduce the $NO^{\bullet}$ levels in the smooth muscle tissue as well [52,53] to levels below the activation threshold of the guanylate cyclase enzyme. Therefore, it might still be premature to simply equate $NO^{\bullet}$ with the endothelial relaxation factor (EDRF). We should keep an open mind to the possibility that EDRF be some reaction product of primary $NO^{\bullet}$ that is more stable against the oxidation in the vascular system and may be either reconverted to truly free $NO^{\bullet}$ or share certain $NO^{\bullet}$-like properties like vasodilation or inhibition of platelet aggregation. For fulfilling its signaling function, important parameters are the lifetime, diffusion rate and the ability to cross biological membranes. In recent years, a surprisingly large variety of such stabilizing forms of $NO^{\bullet}$ have been proposed and discussed in the literature. Some of them share only some of the properties of $NO^{\bullet}$ but lack others. For example, *S*-nitrosothiols are good inhibitors of platelet aggregation but rather inefficient activators of the guanylate cyclase enzyme. DNIC complexes are excellent vasodilators and inhibitors of platelet aggregation (cf Chapter 3). In the bloodstream of animals, they have much longer lifetime than $NO^{\bullet}$ itself. However, these complexes also may release some free iron with the risk of toxic effects on tissues or individual cells. Nitrite anions are small and show rapid diffusive transport in water. They are known to cross the membranes of red blood cells, but their entry into other cell types is uncertain. Nitrite fails to prevent platelet aggregation [54] but can induce relaxation of precontracted vessel rings [55]. The purpose of this book is to give an overview of these alternative forms of $NO^{\bullet}$ and to compile the diverse and scattered literature in the form of a monograph. Some readers may have noticed the absence of nitrate from the list of topics covered. Although dedicated nitrate reductases exist in certain strains of bacteria, they are not known as a truly mammalian enzyme. Although a modest degree of nitrate reduction is known from symbiotic bacteria in mouth and intestinal tract [56], this xenobiotic pathway does not seem to contribute significantly to endogenous nitrite or NO. However, certain mammalian enzymes are capable of reducing nitrate under hypoxia [57]. Biological systems are highly complex and often show very unexpected properties. The recent findings about reduction of nitrite anions under hypoxia are a good example of such unexpected results. The progress in the field is very rapid, and the topics discussed should be considered *in statu nascendi* with many more surprises to come in the future.

REFERENCES

1 Jevons W. Band spectra of diatomic molecules. Cambridge University Press, London, 1932.
2 Whittaker J. Molecular paramagnetic resonance of gas-phase nitric oxide. J. Chem. Edu. 1991; 68: 421–423.

3 Kasai P, Gaura R. Electron spin resonance study of nitric oxide adsorbed in Na-A zeolite. J. Phys. Chem. 1982; 86: 4257–4260.
4 Gillette R, Eyster E. The fundamental rotation-vibration band of nitric oxide. Phys. Rev. 1939; 56: 1113–1119.
5 Pope R, Wolf P. Rare gas pressure broadening of the NO fundamental vibrational band. J. Mol. Spectrosc. 2001; 208: 153–160.
6 Nelson D, Shorter J, McManus JB, Zahniser M. Sub-part-per-billion detection of nitric oxide in air using a thermoelectrically cooled mid-infrared quantum cascade laser spectrometer. Appl. Phys. B. 2002; 75, 343–350.
7 Ganser H, Horstjann M, Suschek C, Hering P, Murtz M. Online monitoring of biogenic nitric oxide with a QC laser-based Faraday modulation technique. Appl. Phys. B. 2004; 78: 513–517.
8 Westcott B, Enemark J. In Inorganic Electronic Structure and Spectroscopy, Vol. II (Solomon E, Lever A, eds.), Wiley, New York, 1999.
9 Trone M, Huetz A, Landau M, Pichou F, Reinhardt J. Resonant Vibrational excitation of the NO ground state by electron impact in the 0.1–3 eV range. J. Phys. B. 1975; 8: 1160.
10 Ford P, Lorkovic I. Mechanistic aspects of the reactions of nitric oxide with transition metal complexes. Chem. Rev. 2002; 102: 993–1017.
11 Enemark J, Feltham J. Principles of structure, bonding and reactivity for metal nitrosyl complexes. Coord. Chem. Rev. 1974; 13: 339–406.
12 Costanzo S, Menage S, Purrello R, Bonomo R, Fontecave M. Re-examination of the formation of dinitrosyl-iron complexes during reaction of S-nitrosothiols with Fe(II). Inorg. Chim. Acta 2001; 318: 1–7.
13 Khalepp B, Luchkina S, Ovchinnikov I. Valence vibration of the NO group and ESR spectra of nitrosyl complexes of iron and chromium. Rus. Chem. Bull. 1973; 22: 940–943.
14 Autreaux BD, Horner O, Oddou J-L, Jeandey C, Gambarelli S, Berthomieu C, Latour J-M, Michaud I. Spectroscopic description of the two nitrosyl-iron complexes for inhibition by nitric oxide. J. Am. Chem. Soc. 2004; 126: 6005–6016.
15 Scheidt W, Lee Y, Hatano K. Preparation and structural characterization of nitrosyl complexes of ferric porphyrinates. J. Am. Chem. Soc. 1984; 106: 3191–3198.
16 Fujii S, Yoshimura T, Kamada H. Nitric oxide trapping efficiencies of water soluble iron(III) complexes with dithiocarbamate derivatives. Chem. Lett. 1996; 785–786.
17 Goodman B, Raynor J, Simons M. Electron spin resonance of Bis (N, N-diethyldithiocarbamato) nitrosyl iron. J. Chem. Soc. (A) 1969; 2572–2575.
18 Brown CA, Pavlowsky MA, Westre TE, Yan Zhang, Hedman B, Hodgson KO, Solomon ET. Spectroscopic and theoretical description of the electronic structure of $S = 3/2$ iron-nitrosyl complexes and their relation to O_2 activation by non-heme enzyme active sites. J. Am. Chem. Soc. 1995; 117: 715–732.
19 Williams DLH. The chemistry of S-nitrosothiols. Acc. Chem. Res. 1999; 32: 869.
20 Lee J, Chen L, West A, Richter-Addo G. Interaction of organic nitrosocompounds with metals. Chem. Rev. 2002; 102: 1019–1065.
21 Iida Y, Carnovale F, Daviel S, Brion C. Absolute oscillator strengths for photoabsorption and the molecular and dissociative photoionization of nitric oxide. Chem. Phys. 1986; 105: 211–225.
22 Caulton G. Synthetic methods in transition metal nitrosyl chemistry. Coord. Chem. Rev. 1975; 14: 317–355.
23 Beckman J. In Nitric Oxide: Principles and Actions (Lancaster J, ed.), Academic Press, San Diego, 1996.
24 Bohle S, Hansert B, Paulson S, Smith B. Biomimetic synthesis of the putative cytotoxin peroxynitrite $ONOO^-$ and its characterization as a tetramethylammonium salt. J. Am. Chem. Soc. 1994; 116: 7423–7424.
25 Goldstein S, Czapski G. Direct and indirect oxidations by peroxynitrite. Inorg. Chem. 1995; 34: 4041–4048.
26 Lymar S, Shafirovitch V, Poskrebyshev G. One-electron reduction of aqueous nitric oxide: a mechanistic revision. Inorg. Chem. 2005; 44: 5212–5221.
27 Chlistunoff J, Ziegler K, Lasdon L, Johnston K. Nitric/Nitrous acid equilibria in supercritical water. J. Phys. Chem. A. 1999; 103: 1678–1688.

28 Koppenol W, Traynham J. Say NO to nitric oxide: nomenclature for nitrogen- and oxygen-containing compounds. Met. Enzym. 1996; 268: 3–11.
29 Bayliss NS, Watts DW. The spectra and equilibria of nitrosonium ion, nitroacidium ion, and nitrous acid in solutions of sulphuric,hydrochloric, and phosporic acids. Aust. J. Chem. 1955; 9: 319–332.
30 Stanbury DM. Reduction potentials involving inorganic free radicals in aqueous solution. Adv. Inorg. Chem. 1989; 33: 69–139.
31 Fukuto J, Switzer C, Miranda K, Wink D. Nitroxyl (HNO): chemistry, biochemistry and pharmacology. Annu. Rev. Pharmacol. Toxicol. 2005; 45: 335–355.
32 Poskrebyshev G, Shafirovich V, Lymar S. Hyponitrite radical, a stable adduct of nitric oxide and nitroxyl. J. Am. Chem. Soc. 2004; 126: 891–899.
33 Wink D, Mitchell J. Chemical biology of nitric oxide: insights into regulatory, cytotoxic and cytoprotective mechanisms of nitric oxide. Free Rad. Biol. Med. 1998; 25: 434–456.
34 Miranda K, Espey M, Yamada K, Krishna M, Ludwick M, Kim S, Jourd henic D, Grisham MB, Feelish M, Fukuto JM, Wink DA. Unique oxidative mechanism for the reactive nitrogen oxide species, nitroxyl anion. J. Biol. Chem. 2001; 276: 1720–1727.
35 Armor JN. Influence of pH and ionic strength upon solubility of nitric oxide in aqueous solution. J. Chem. Eng. Data 1974; 19: 82–84.
36 Williams DLH. Nitrosation reactions and the chemistry of nitric oxide. Elsevier, Amsterdam, 2004.
37 Herold S, Rock G. Mechanistic studies of S-nitrosothiol formation by NO/O_2 and by NO/methemoglobin. Arch. Biochem. Biophys. 2005; 436: 386–396.
38 Nedospasov A. Is N_2O_3 the main nitrosating species in aerated nitric oxide solutions in vivo? J. Biochem. Mol. Toxicol. 2002; 16: 109–120.
39 Liu X, Miller M, Josho M, Thomas D, Lancaster J. Accelerated reaction of nitric oxide with O_2 within the hydrophobic interior of biological membranes. Proc. Natl. Acad. Sci. USA 1998; 95: 2175–2179.
40 Fucuto JM, Cho JY, Switzer C. Nitric oxide biology and pathobiology. 1st ed. San Diego, Elsevier Science and Technology Books, 2000.
41 Keshive M, Singh S, Wishnok JS, Tannenbaum SR, Deen WM. Kinetics of S-nitrosation in nitric oxide solutions. Chem. Res.Toxicol. 1996; 9: 988–993.
42 Goldstein S, Czapski G. Mechanism of the nitrosation of thiols and amines by oxygenated NO solutions: the nature of the nitrosating intermediates. J. Am. Chem. Soc. 1996; 118: 3419–3425.
43 Lewis R, Tannenbaum S. Deen W. Kinetics of N-nitrosation in oxygenated nitric oxide solutions at physiological pH – role of nitrous anhydride and effects of phosphate and chloride. J. Am. Chem. Soc. 1995; 177: 3933–3939.
44 Gordin V, Nedospasov A. NO catastrophes *in vivo* as a result of micellar catalysis. FEBS Lett. 1998; 424: 239–242.
45 Jaszewski A, Fann Y, Chen Y, Sato K, Corbett J, Mason R. EPR spectroscopic studies on the structural transition of nitrosyl haemoglobin in the arterial-venous cycle of DEANO-treated rats. Free Rad. Biol. Med. 2003; 35: 444–451.
46 Yonetani T, Tsuneshige A, Zhou Y, Chen X. Electron Paramagnetic Resonance and oxygen binding studies of α-nitrosyl hemoglobin. J. Biol. Chem. 1998; 273: 20323–20333.
47 Doyle M, Hoekstra J. Oxidation of nitrogen oxides by bound dioxygen in hemoproteins. J. Inorg. Biochem. 1981; 14: 351–358.
48 Radi R, Denicola A, Alvarez B, Ferrer-Sueta G, Rubbo H. The biological chemistry of peroxynitrite. In Nitric Oxide (Ignarro L, ed.), Academic Press, 2000.
49 Herold S, Koppenol WH. Peroxynitritometal complexes. Coord. Chem. Rev. 2005; 249: 499–506.
50 Stojanovic S, Stanic D, Nikolic M, Spasic M, Niketic V. Iron catalyzes conversion of NO into nitrosonium and nitroxyl species. Nitric Oxide 2004; 11: 256–262.
51 Gladwin M, Ognibene F, Pannell L, Nochols J, Pease-Fye M, Shelhamer J, Schechter A. Relative role of heme nitrosylation and β-cysteine 93 nitrosation in the transport and metabolism of nitric oxide by haemoglobin in the human circulation. Proc. Natl. Acad. Sci. USA 2000; 97: 9943–9948.
52 Lancaster J. A tutorial on the diffusibility and reactivity of free nitric oxide. Nitric Oxide 1997; 1: 18–30.
53 Condorelli P, George S. Free nitric oxide diffusion in the bronchila microcirculation. Am. J. Physiol. Heart Circ. Physiol. 2002; 283: H2660–H2670.

54 Lundberg J, Weitzberg E, Cole J, Benjamin N. Nitrate, bacteria and human health. Nat. Rev. Microbiol. 2004; 2: 593–602.
55 Li H, Samouilov A, Liu X, Zweier JL. Xanthine oxidase catalyzed nitrate reduction: an important nitrite and nitric oxide generating pathway during ischemia. Circulation 2001; 104 (Suppl): 1266.
56 Mcknight GM, Smith LM, Drummond RS, Duncan CW, Golden M, Benjamin N. Chemical synthesis of nitirc oxide in the stomach from dietary nitrate in humans. Gut 1997; 10: 211–214.
57 Millar TM, Stevens CR, Benjamin N, Eisenthal R, Harrison R, Blake DR. Xanthine oxireductase catalyses the reduction of nitrates to nitrite under hypoxic conditions. FEBS Lett. 1998; 427: 225–228.

PART II

DNIC: Physico-chemical Properties and Biological Activity

Radicals for Life: The Various Forms of Nitric Oxide
E. van Faassen and A.F. Vanin (Eds.)

CHAPTER 2

DNICs: physico-chemical properties and their observations in cells and tissues

Anatoly F. Vanin[1] and Ernst van Faassen[2]

[1]*Semenov Institute of Chemical Physics, Russian Academy of Sciences, Moscow, 119991, Russian Federation*
[2]*Debye Institute, Section Interface Physics, Ornstein Laboratory, Utrecht University, 3508 TA, Utrecht, The Netherlands*

INTRODUCTION

Dinitrosyl-iron complexes (DNICs) with thiol-containing ligands were discovered in cells and tissues with EPR spectroscopy. The EPR spectrum of these paramagnetic ($S = 1/2$) complexes has a characteristic anisotropic lineshape centered at $g_{av} = 2.03$ (2.03 signal) which may be observed over a wide range of temperatures from liquid helium to room temperature. The $g = 2.03$ signal was recorded and discussed for the first time in cultured yeast cells and subsequently in the tissues of some animals [1–3] (Fig. 1A,C). Independently, Commoner and colleagues from USA (1965) observed weak signals at $g = 2.03$ in rat livers during the initial stages of chemically induced carcinogenesis [4] (Fig. 1D). Interestingly, some spectra published earlier by Mallard and Kent did show the presence of unidentified paramagnetic centers at $g = 2.03$ in rat liver in similar experiments (Fig. 1B) [5] but their nature was not discussed.

The identity of these paramagnetic centers at $g = 2.03$ was recognized when it was demonstrated that the shape and EPR spectroscopic parameters of 2.03 signal and EPR signals of low-molecular DNIC with cysteine in a frozen solution were similar (Fig. 2) [6,7]. Subsequent studies confirmed that the paramagnetic species at $g = 2.03$ in cells and tissues are really DNIC with protein or low-molecular thiol-containing ligands [3,6–12].

As a rule, at X-band the 2.03 signal is usually characterized by a g-factor with axial symmetry ($g_\perp = 2.04$, $g_{||} = 2.014$). At room temperature, the protein-bound DNICs may be distinguished easily from their low-molecular-weight analogs. At this temperature, the EPR spectrum from protein-bound DNIC retains the shape of a powder spectrum as would be recorded in frozen solution at low temperature (77 K) (Figs. 2, 3 and 5). In contrast, the low-molecular DNICs are rapidly tumbling and show a motionally narrowed isotropic line with a half-width of 0.7 mT and resolved 13-component hyperfine structure (HFS) [13–15] (Figs. 2–4).

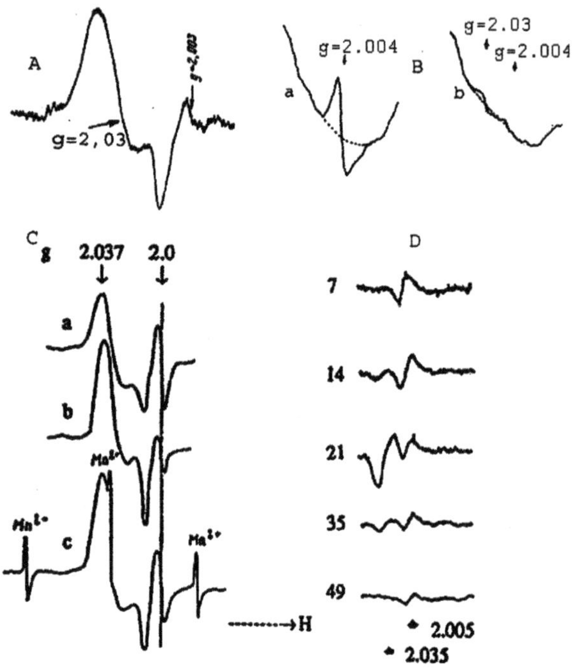

Fig. 1. The first recordings of 2.03 signal; yeast (A) [1,2]; rat liver carcinoma induced by the hepatocarcinogenic compound, *P*-dimethylamino-azobenzene (butter yellow) (panel B, spectrum b; (a) spectrum from normal liver) [5]; yeast and rabbit liver (panel C, spectra a, b and c, respectively) [3]; panel D: livers from rats maintaining 7, 14, 21, 35 and 49 days on a diet containing butter yellow [4]. Recordings were made at 77 K (A–C) or ambient temperature (D). (With permission.)

Interest in DNIC complexes sharply increased in the 1990s after the discovery of the physiological roles of endogenous NO radicals in mammals and of the L-arginine/NO pathway catalyzed by the nitric oxide synthases. The L-arginine-dependent formation of DNIC was demonstrated in various cultured cells and tissues expressing inducible, high output, NO synthase activity. These presently include macrophages [16–19], fibroblasts [20], hepatocytes [21–25], vascular smooth muscle cells [26], isolated human islets of Langerhans [27], isolated rat aorta [28], different types of tumor cells [18,29,30] (all treated with lipopolysaccharides and/or cytokines *in vitro*), as well as liver of mice treated with *Corynebacterium parvum* [31], murine tumor transplants [30,32] and rat heart allografts [33,34] *in vivo*. Formation of DNIC *via* constitutive NO synthase was also demonstrated in isolated porcine endothelial cells stimulated with bradykinin or the ionophore A23187 [35].

LOW-MOLECULAR-WEIGHT DNIC WITH THIOL-CONTAINING LIGANDS

The first description of EPR spectra of low-molecular DNIC with various anionic ligands (phosphate, pyrophosphate, arsenate, molybdate, carbonate, maleate, mercaptane, cysteine,

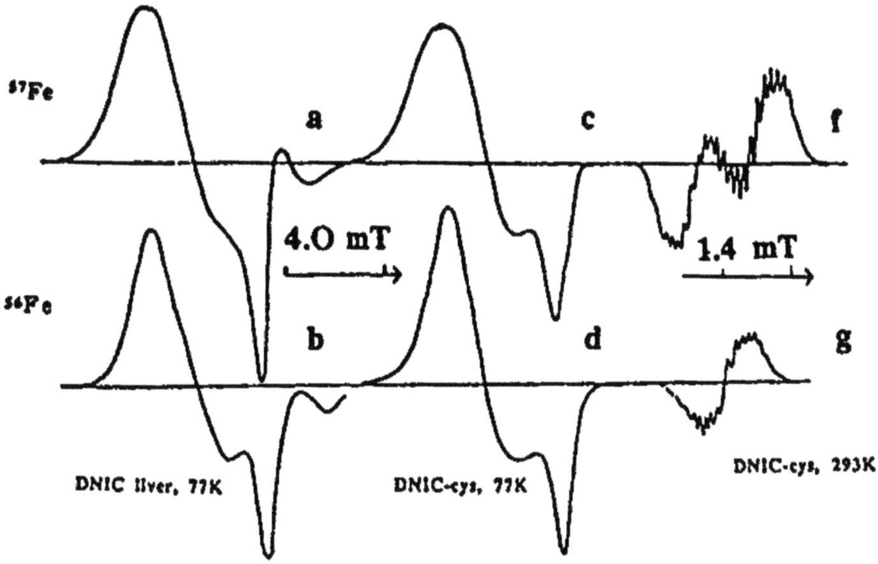

Fig. 2. 2.03 signals from the livers of mice maintained on drinking water with nitrite + ^{57}Fe–citrate complex (curve a) or nitrite + ^{56}Fe–citrate complex (curve b). EPR signals of DNIC with cysteine, containing ^{57}Fe (curves c,f) or ^{56}Fe (curves d,g). Recordings were made at 77 K (curves a–d) or ambient temperature (curves f,g). (From Ref. [15].)

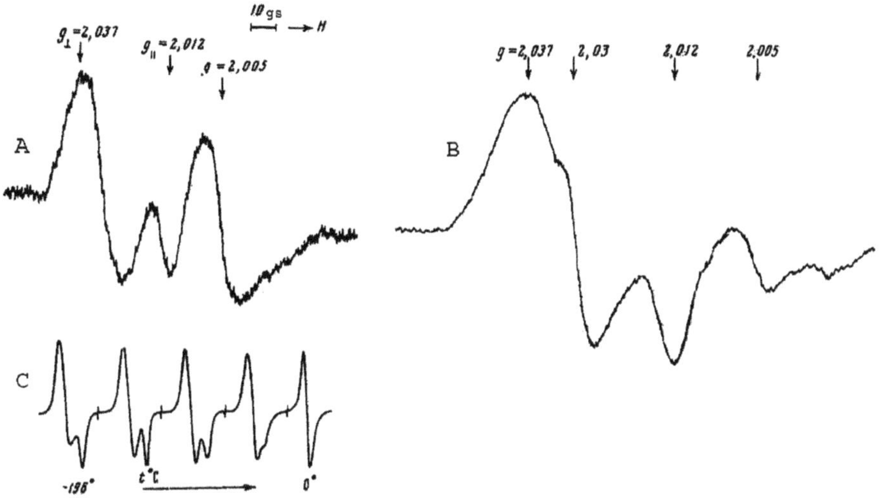

Fig. 3. 2.03 signals from rabbit liver (A) or yeast cells (B). Recordings were made at ambient temperature. A narrow signal superimposed on the 2.03 signal (B) is due to a low-molecular DNIC. (C) The change of the shape of the EPR signal of DNIC with cysteine at temperature increasing from –196 to 0°C. (From Ref. [13].)

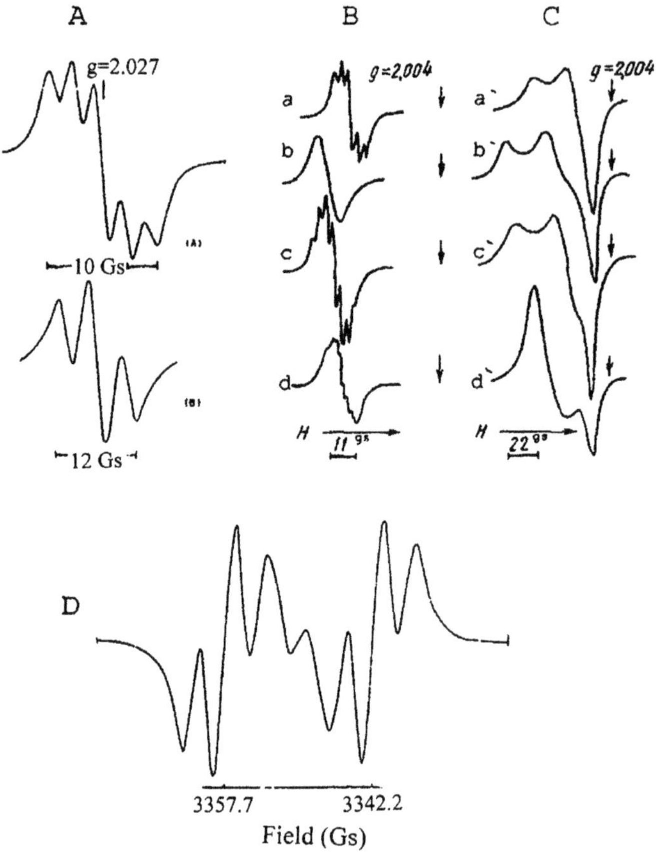

Fig. 4. (A) EPR spectrum of DNIC with OH^- (pH 11) containing ^{14}NO (a) or ^{15}NO (b) [36]. (B,C) EPR spectra of DNIC with OH^- (pH 12) (a,a`), DNIC with $H_2O$ (pH 7) (b,b`), DNIC with phosphate (c,c`) or DNIC with cysteine (d,d`) [6]. (D) EPR spectrum of DNIC with OH^- containing ^{15}NO and ^{57}Fe (pH 11) [36]. Recordings were made at ambient temperature (A,B,D) or 77 K (C). (With permission.)

thiourea, penicillamine, hydroxyl and other) was given by McDonald et al. in the 1960s. EPR spectra of the complexes in aqueous solutions were recorded at ambient temperature [36]. In dilute solution at X-band frequencies, these complexes have a characteristic EPR spectrum showing an isotropic singlet at $g = 2.02–2.04$ with resolved HFS from the nuclear magnetic moment of nitrogen atoms in NO ligands. Additional HFS may arise from protons or phosphorus atoms in anionic ligands (Fig. 4). The analysis of EPR characteristics of the complexes indicated that the paramagnetic complexes consist of one iron atom, two NO groups and two anionic ligands. EPR spectra show an additional strong doublet hyperfine (HF) coupling of 1.25 mT from the iron nucleus if the ^{57}Fe isotope is used. This magnitude shows that the unpaired electrons from the nitrosyl ligands are predominantly localized on iron atom (Fig. 4). So, d^7 electron configuration and low spin state $S = 1/2$ were proposed to explain the spectroscopic properties of the complexes [36]. Infrared studies of the intramolecular vibrations suggest that the nitrosyl ligands have donated an electron to the iron and resemble NO^+ rather than NO [37]. This nitrosonium character is further corroborated by the small HFS coupling to the nitrogen nucleus in the nitrosyl ligands in DNIC (ca 0.15 mT) [36]. Thus, the electronic structure of DNIC is better represented by the formula $\{(L^-)_2Fe^+(NO^+)_2\}^+$. The geometrical structure of DNIC is still a matter of debate and depends on the aggregation state. As will be discussed below, X-ray diffraction has shown tetrahedral surrounding of the iron in crystalline state, but in dilute solution EPR spectroscopy favors square planar configuration.

In frozen solution at X-band, the EPR spectra of DNIC with low-molecular-weight anionic ligands have the shape of a spin $S = 1/2$ center with either axial or rhombic symmetry. Axial symmetry is mainly characteristic of DNIC with thiol-containing ligands, and the complexes have g-factors with the values of $g_\perp = 2.045$ or 2.04, $g_\| = 2.014$ (Fig. 5). The spectral shape shows that the coupling tensor has $g_z < g_x = g_y$, i.e. oblate axial symmetry rather than prolate. At higher frequencies the Zeeman anisotropy is better resolved, and will be discussed

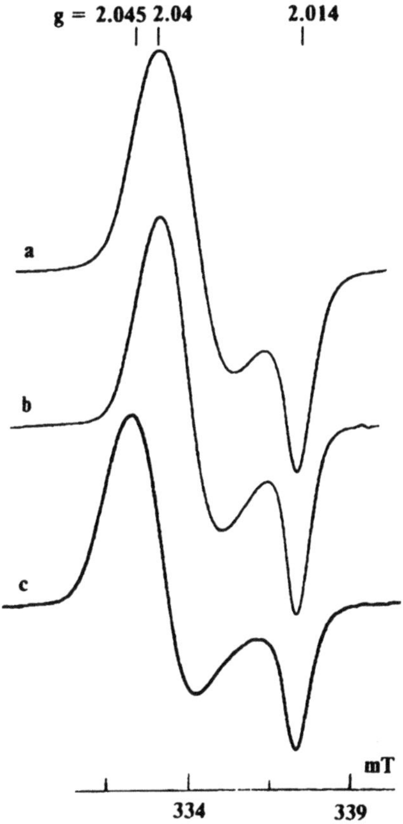

Fig. 5. EPR spectra from frozen solutions of DNIC with cysteine containing ^{14}NO (curve a) or ^{15}NO (curve b) and DNIC with thiosulphate (curve c). Recordings were made at 77 K. (From Ref. [40].)

at the end of this chapter. Lower rhombic symmetry is mainly characteristic of DNIC with various non-thiol-containing ligands (phosphate, halogens, citrate, ascorbate, histidine, etc.), the three elements of the g-tensor spanning the range 2.014–2.05. Nevertheless, some non-thiolate complexes (for example, DNIC with hydroxyl) show axial symmetry with g-factors of $g_{\perp} = 2.014$, $g_{||} = 2.05$, $g_{av} = 2.026$ [6,7]. The representative EPR spectra of DNIC with non-thiol-containing ligands are presented in Fig. 4.

DNICs with low-molecular-weight thiol-containing ligands can be divided into two groups in accordance with the shape of central part of the EPR spectrum and the value of $g_{\perp}$, $g_{||}$ and $g_{av} = (g_{||} + 2g_{\perp})/3$. The first group comprises the EPR signals from DNIC with cysteine, glutathione, N-acetylcysteine, homocysteine, N-acetylpenicillamine or dithiothreitol ($g_{\perp} = 2.04$, $g_{||} = 2.014$ and $g_{av} = 2.031$). The EPR signals from DNIC with thiosulfate and mercaptotriazole ($g_{\perp} = 2.045$, $g_{||} = 2.014$ and $g_{av} = 2.035$) constitute the second group [40]. The representative EPR spectra from both groups are presented in Fig. 5. The spectrum of frozen solution of DNIC with cysteine including ^{15}NO ligands is also shown in Fig. 5.

In dilute solution at room temperature, fast tumbling motions averages the axial anisotropy of g-factor resulting in the registration of narrow isotropic EPR signals with g_{iso} values at the range 2.029–2.031 for all the mentioned DNIC with thiol-containing ligands. The signals show resolved HFS with 5 or 13 components. The 5 component HFS has 1:2:2:2:1 intensity and is characteristic of DNIC with thiosulfate, N-acetylpenicillamine or mercaptotriazole (Fig. 6, curve c). It is caused by HF interaction with the nitrogen nuclear moments ($I = 1$) of both NO ligands. At room temperature, the DNIC with other thiol-containing ligands shows resolved 13-component HFS. This multiplicity results from HF interaction with the two nitrogen nuclei of the nitrosyl ligands plus four protons ($I = 1/2$) from two methylene groups close to the sulfur atom (Fig. 6, curve a). The HF interaction with the nitrogen may be modified by substitution with the ^{15}N isotope ($I = 1/2$): With ^{15}NO ligands, the spectrum shows 9 resolved components (Fig. 6, curve b). The description of the HFS formation in the signals is shown on the right side of Fig. 6.

Quite often, the EPR spectra show resolved HFS from four protons from two separate methylene groups. Examples are DNIC with cysteine, glutathione, N-acetylcysteine, homocysteine or dithiothreitol ligands. This property demonstrates that the complexes contain two thiol-containing ligands. Therefore, these DNICs are cationic and can be presented as $\{(RS^{-})_2Fe^{+}(NO^{+})_2\}^{+}$ as proposed in Ref. [36]. This formula was also found for DNIC with mercaptotriazole (from X-ray analysis described in Ref. [41]). Since DNIC with thiosulfate or N-acetylpenicillamine have similar EPR spectra in solution. Although not yet proven, it is plausible that they have the same formula as well.

It is noteworthy that the values of g_{iso} at room temperature differ slightly from the average $g_{av} = (g_{||} + 2g_{\perp})/3$ as obtained in frozen state. Evidently, it is due to the influence of the solvent on the electronic structure of the complexes in the solution at ambient temperature. As proposed in Refs. [38,39] besides two thiolate and two nitrosyl ligands the DNICs may also contain one or two solvent molecules (usually water). Changing the solvent to dimethylformamide, dimethylsulfoxide (DMSO), tetramethylurea or hexamethanol affects the EPR spectra significantly. In particular, it shows a marked decrease from axial to rhombic symmetry of the paramagnetic center (Fig. 7) [42]. Cysteine is highly soluble in these polar organic solvents and the DNICs undergo slow precipitation (on a timescale of several minutes) but the EPR spectra could be observed for a few minutes.

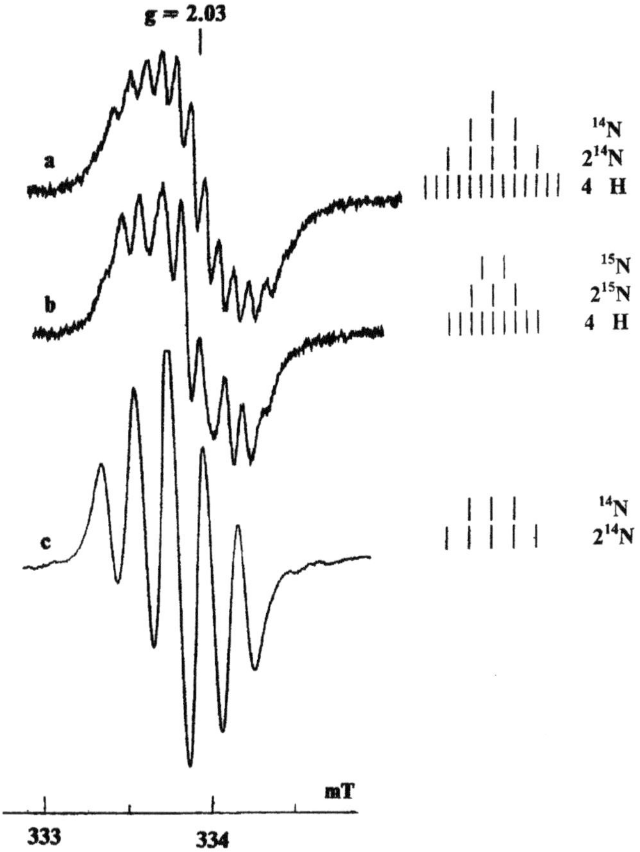

Fig. 6. EPR spectra from solutions of DNIC with cysteine containing ^{14}NO (curve a) or ^{15}NO (curve b) and DNIC (^{14}NO) with thiosulphate (curve c). Recordings were made at ambient temperature. On the right – decoding of hyperfine structure of the signals. (From Ref. [40].)

Usually, only a fraction of the available iron is included in paramagnetic DNIC with thiol-containing ligands: Often a substantial quantity of iron is included in the form of dimeric DNIC, which is favored at acid conditions and lower thiol:iron ratios [43,44]. Dimeric DNIC is diamagnetic due to antiferromagnetic coupling between the DNIC entities. Therefore, at a given iron concentration, the yield of paramagnetic monomeric DNIC is usually decreasing with pH due to dimerization. At neutral pH, only DNIC with cysteine or thiosulfate remain fully monomeric and include all iron. With other ligands, higher pH is usually required. Full incorporation of all available iron in monomeric DNIC is usually achieved at pH $\sim$ 10. The dimerization has a strong effect on the color of the solutions. The monomeric DNIC complexes have a dark green color due to d–d transitions in the wavelength region 600–850 nm (Fig. 8). A much stronger absorption band is observed at 392 nm ($\varepsilon_{392} = 3580\ (\mathrm{Mcm})^{-1}$) [45,46] (Fig. 8). Upon decreasing the pH from 10 to 6, the intensity of the EPR signals from these complexes drops reversibly 5–10 times and the color of the solutions turned

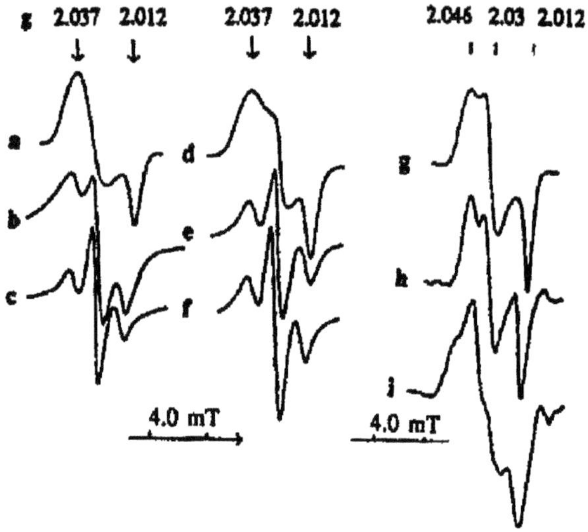

Fig. 7. Change of EPR signal shape of DNIC with cysteine (aqueous solution) (curve a) after mixing with dimethylformamide (1:100) (curves b,c), dimethylsulfoxide (1:4) (curve d), tetramethylurea (1:10) (curve e) or hexamethapol (1:10) (curve f) [42]. EPR spectra of DNIC with bovine serum albumin with ^{56}Fe (curves g,h) or ^{57}Fe (curve i) [54]. Recordings were made at 77 K (curves a–g,i) or ambient temperature (curve h).

to yellow-orange. The color is characteristic of the solutions of diamagnetic dimeric DNICs with thiol-containing ligands.

The ratio between iron and thiol decides whether the DNIC complexes appear as monomers or dimers. Except for thiosulfate, dimeric DNIC are synthesized at the stoichiometric ratio of iron:thiol of 1:2 (DNIC 1:2) [43,44,47,48]. The dimeric form does not give any EPR signal and has two optical absorption bands at 310 and 360 nm [47,48] (Fig. 8). Dimeric DNIC with cysteine is quantitatively transformed into paramagnetic monomeric DNIC by the addition of excess cysteine (cysteine:iron ratios exceeding 20:1).

In frozen state, the EPR spectra of DNIC do not show resolved HF interactions, except after isotopic substitution of iron with ^{57}Fe (Fig. 2). Still, simulation of the frozen spectrum with only axial Zeeman coupling does not lead to good fits of the EPR lineshape. In particular, the simulated intensity at $g_{||}$ remains smaller than that near the central feature, in conflict with the experimental lineshape [49] (Fig. 9). For the EPR signals from frozen solutions of DNIC with thiol-containing ligands, the amplitude at $g_{||}$ is considerably higher than that of central component. This phenomenon is attributed to unresolved anisotropic HFS from the nitrogens of the nitrosyl ligands with splitting at $g_{\perp}$ much higher than for parallel alignment of the complexes with respect to the magnetic field.

McDonald et al. also described in their article [36] the mononitrosyl iron complexes (MNICs) formed from ferrous iron, NO and dithiols. At room temperature, the complexes give rise to triplet HFS at $g = 2.027$–2.041 with component separations of 1.55 mT (Fig. 10). The EPR spectra are similar to those from MNIC with the derivatives of thiocarbonic acids, dithiocarbamates, xanthogenates, etc. [50–52]. Ferrous iron and cysteine were proved to be capable of generating MNIC also in aqueous solutions treated with nitric oxide at the amount lower than that of iron and cysteine [10,53–55]. The MNIC with cysteine is usually detected

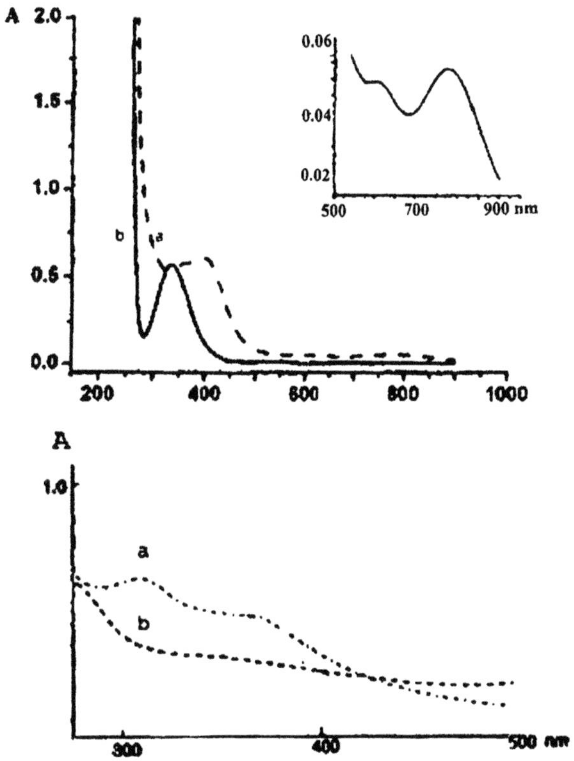

Fig. 8. Top panel: UV–Visible spectra of 0.8 mM *S*-nitrosocysteine in 100 mM HEPES buffer, pH 7.4 (curve b) and 0.15 mM monomeric DNIC-cysteine complex (curve a and curve shown in inset) [46]. Bottom panel: Optical spectra of 0.2 mM dimeric DNIC-cysteine complex in 15 mM HEPES buffer, pH 7.4 (curve a). Curve b is the spectrum of monomeric DNIC after 30 min in air. (From Ref. [47].)

at the first step of DNIC-cysteine formation when gaseous NO begins to penetrate the ferrous-cysteine solutions. Further exposure to NO results in MNIC-cysteine being transformed to DNIC-cysteine [10,53]. When recorded at ambient temperature, the MNIC-cysteine gives clearly resolved triplet HFS at $g = 2.04$ (Fig. 10). Frozen solutions show a comparatively wide asymmetrical singlet EPR signal at $g_{av} = 2.04$ ($g_{\perp} = 2.055$, $g_{||} = 2.01$) and a slightly resolved triplet HFS from the nitrogen nucleus ($I = 1$) at $g_{||}$(Fig. 10) [53].

Interestingly, the ferrous-ethylxanthogenate complex can form both MNIC and DNIC when exposed to NO in dimethylformamide or DMSO solutions [56]. The DNIC complexes dominate if some nitrogen dioxide is added together with gaseous NO. In frozen solution, DNIC-ethylxanthogenate has an anisotropic EPR spectrum at $g_{av} = 2.035$ (Fig. 11). In solution at room temperature, the isotropic EPR signal at $g = 2.03$ shows resolved quintet HFS structure (Fig. 11). Isotopic labeling with ^{57}Fe ($I = 1/2$) of DNIC-ethylxanthogenate gives signal broadening at 77 K in frozen solution. In solution at room temperature, the spectrum shows additional doublet HFS splitting of the signal from ^{57}Fe (Fig. 11).

So far we discussed the DNIC complexes with d^7 configuration, the $\{Fe(NO)_2\}^7$ structure in Enemark–Feltham notation [57], but the DNICs may undergo redox transformations to

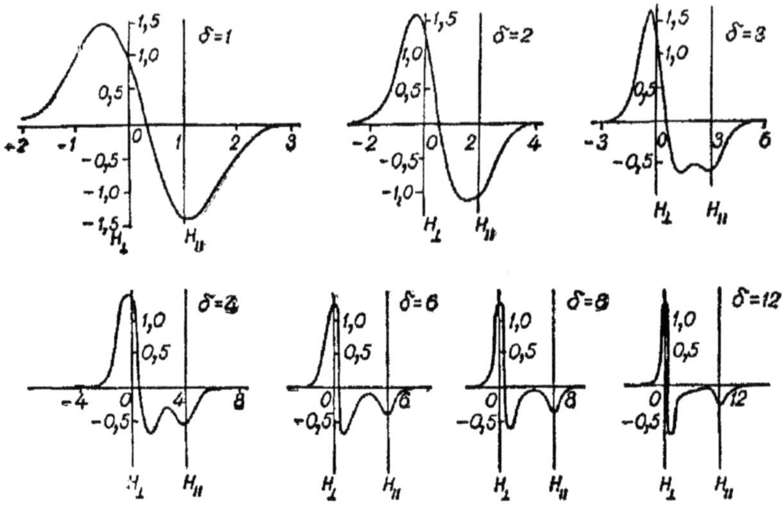

Fig. 9. Calculated shape of the EPR signals from polycrystalline samples with axial symmetry of g-factor at various values of $\delta = |H_\perp - H_\parallel|/\Delta H_{sp}$ where ΔH_{sp} is the width of a spin-packet [49]. $\Delta H_{sp} = \gamma H_1/\sqrt{T_1 T_2}$ where γ, H_1, T_1 and T_2 are electron gyromagnetic ratio, amplitude of magnetic component of microwave field, times of spin-lattice and of spin–spin relaxations, respectively.

other electronic states. DNIC with cysteine or glutathione ligands may be reduced to complexes with $\{Fe(NO)_2\}^8$ (diamagnetic) and $\{Fe(NO)_2\}^9$ (paramagnetic $S = 1/2$) centers. In frozen solution, the latter show EPR spectra with axial symmetry and $g_\perp = 2.01$ and $g_\parallel = 1.97$. In dilute solution, a single line without HFS appears at $g = 2.0$ (Fig. 12). Isotopic substitution of iron with ^{57}Fe causes broadening of the EPR spectrum and demonstrates that the unpaired electron of the $\{Fe(NO)_2\}^9$ state is localized on the iron atom (Fig. 12) [58–61].

The redox state of DNIC is clearly reflected in the optical absorption as well. The addition of dithionite to a dark-green solution of monomeric DNIC with $\{Fe(NO)_2\}^7$ changes the color to bright green and new absorption bands at 460 and 660 nm appear [60]. The redox reaction is reversible: Upon readmission to oxygen, the spectroscopic properties of the solution return to those of the $\{Fe(NO)_2\}^7$ complex.

VIBRATION SPECTROSCOPY OF NITROSYL LIGANDS IN DNIC

As mentioned in Chapter 1, the frequency ν_{NO} of the NO stretch vibration is strongly affected by the density of the unpaired electron in the half-occupied antibonding orbital of the nitrosyl ligand. Transfer of the unpaired electron from the antibonding orbital towards the iron stiffens the N—O bond and raises ν_{NO}. The DNIC represents a four-coordinate $\{Fe(NO)_2\}^n$ complex with $n = 7$, 8 or 9. The orbital diagrams were discussed in Ref. [57], which reviews the experimental data until 1974. Interestingly, the two ligands have slightly different ν_{NO} vibrations near 1760 cm^{-1}. For Cys-DNIC, the ν_{NO} vibrations appear at 1770 and 1730 cm^{-1} [45]. The separation of ca 40 cm^{-1} amounts to ca 2%, and proves that the two ligands have

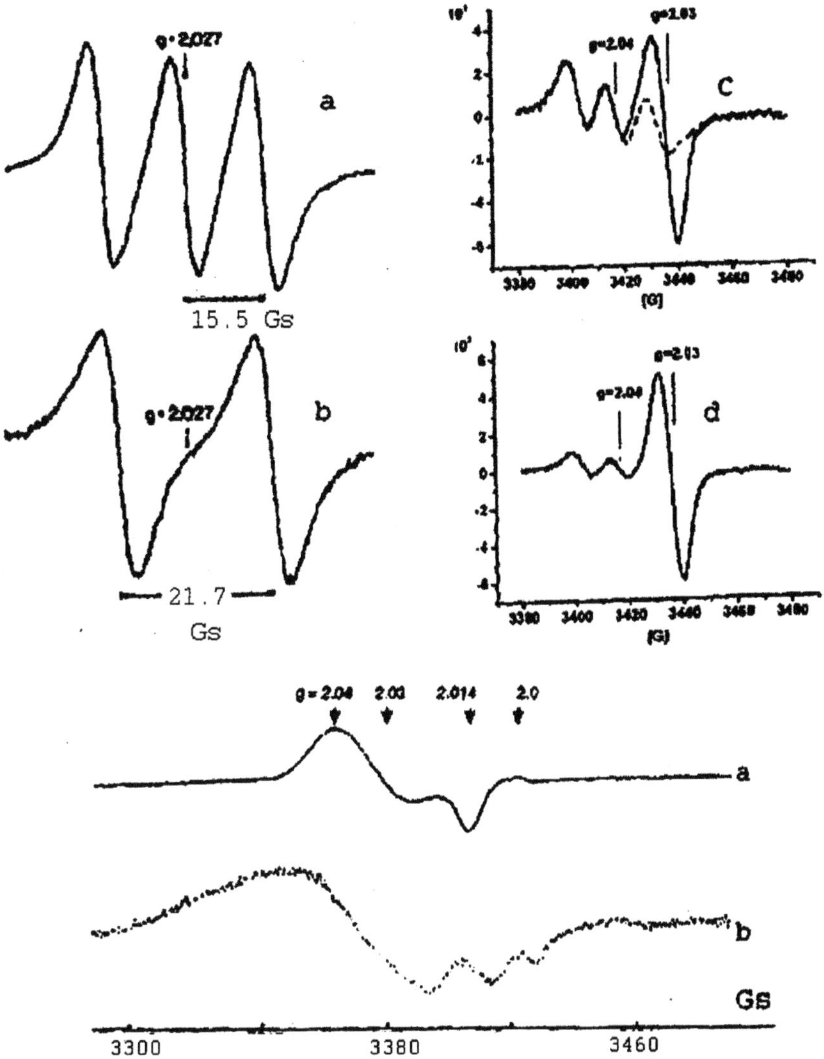

Fig. 10. Left panel: EPR spectra of mononitrosyl iron complex (MNIC) with maleonitriledithiol containing ^{14}NO (curve a) or ^{15}NO (curve b) [36]. Right panel: EPR spectra of MNIC with cysteine (triplet signal) and DNIC with cysteine (singlet signal) forming in the reaction between Fe^{2+}, *S*-nitrosocysteine and cysteine during 5 and 20 min (curves c,d, respectively) [46]. Bottom panel: EPR spectra of DNIC (a) or MNIC (b) with cysteine [47]. Recordings were made at ambient temperature (left and right panels) or 77 K (bottom panel). Reprinted with permission from McDonald CC, Phillips WD, Mower HF "An electron spin resonance study of some complexes of iron, nitric oxide and anionic ligands" (J. Am. Chem. Soc. 1965; 87: 3319–3326) © 1965, American Chemical Society.

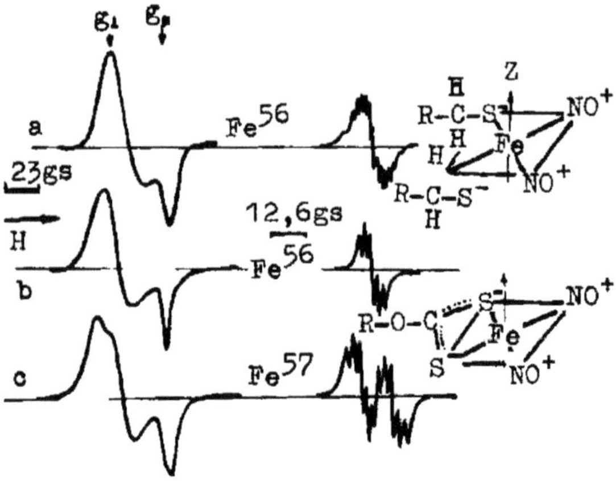

Fig. 11. EPR signals of DNIC with cysteine (curve a), DNIC with ethylxanthogenate with ^{56}Fe (curve b) or ^{57}Fe (curve c). Recordings were made at 77 K (left spectra) or ambient temperature (right spectra). The proposed geometrical structure of the complexes are also shown. (From Ref. [56].)

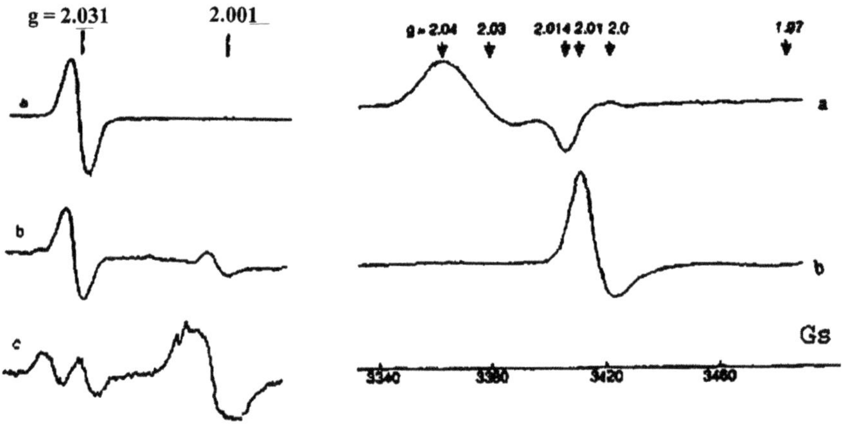

Fig. 12. Left panel: EPR spectra of monomeric DNIC with cysteine (DNIC 1:20) at $g = 2.03$ (curve a) and its reduced form (after addition of dithionite) at $g = 2.001$ with ^{56}Fe (curve b) or ^{57}Fe (curve c) [60]. Right panel: EPR spectra of monomeric DNIC with cysteine (DNIC 1:20) (curve a) and its reduced form (curve b) with ^{56}Fe [47]. Recordings were made at ambient temperature (left panel) or 77 K (right panel).

different degree of transfer of the unpaired electron towards the iron. Phrased otherwise, the effective charges of the nitrosyl ligands are different as well. This property is observed with various DNIC-containing ligands of low molecular weight, as well as with protein-bound DNIC [62]. Nevertheless, the nature of the non-nitrosyl ligands does influence the degree of electron transfer. The effect was studied with a range of ligands with increasing polarity of the Fe-ligand bond [63]. The ν_{NO} was found to increase with bond polarity. This phenomenon is explained by charge transfer: increasing polarity of the bond increases the effective (positive) charge of the iron and promotes a compensating shift of the unpaired electron from the nitrosyl towards the iron. This stiffens the N—O bond and raises ν_{NO}.

For paramagnetic $\{Fe(NO)_2\}^7$ ($S = 1/2$), the experimental ν_{NO} is found to be positively correlated with the g-factor of EPR spectroscopy [63]. High vibration frequencies (large electron transfer) bring high g-factors, and this relationship confirms that the unpaired electron of the $\{Fe(NO)_2\}^7$ resides in the ψ_z^2 orbital of the iron atom. A more detailed discussion of the electronic structure of DNIC is given later in this chapter.

MÖSSBAUER (γ-RESONANCE) PROPERTIES OF LOW-MOLECULAR DNICs WITH THIOL-CONTAINING LIGANDS

Mössbauer spectroscopic studies recoilless excitation of nuclear excited states by γ-rays. The Mössbauer effect requires a nucleus with low-lying excited states and has been detected in 43 different elements. The technique provides a particularly valuable analytical tool for iron complexes since the shape of the Mössbauer spectrum is affected by the redox state and electronic configuration of the complex as well as by the ligand field surrounding the iron. ^{57}Fe is a stable isotope with only 2.2% natural abundance. In ^{57}Fe Mössbauer spectroscopy, one observes the dipole transitions between the nuclear ground state ($I = 1/2$, nuclear g-factor $g_N = 0.181$) and the excited state at 14.4 keV ($I = 3/2$, $g_N = -0.106$). The shape of the spectrum is determined by the interaction of the nuclear quadrupole moment with the electrical field gradient (EFG) tensor of the ligand field in the complex. After enriching the iron with the ^{57}Fe isotope, the ligand field of DNIC complexes may be investigated with γ-resonance (GR) spectroscopy.

Depending on the redox state and dimerization, Cys-DNICs reveal three types of GR spectra [43,44]. Dimeric DNIC-cysteine (Fe:Cys = 1:2) is characterized by an isomeric shift (IS) of 0.14 mm/s and quadrupole splitting (QS) of 1.1 mm/s (Fig. 13a, 14A). (The IS value being relative to the metallic iron (α-Fe) at room temperature). After the addition of cysteine in a 30–40-fold excess to the complex solution, the solution changed its color from yellowish to green, characteristic of monomeric paramagnetic state of Cys-DNIC (DNIC 1:20). The corresponding GR spectrum had a complicated shape (Fig. 13e). A slightly different GR-spectrum is obtained if a solution with ^{57}Fe:Cys = 1:20 is treated with gaseous NO (Fig. 13d, 14B). Despite the non-identity of the shape, the spectra share the presence of GR absorption in a wide range of rates ("blurred" absorption). This indicates the magnetic HF interaction in these complexes, i.e. the presence of a magnetic moment of the electron shell interacting with the nucleus spin of ^{57}Fe in ground and excited states. Such absorption is typical for paramagnetic complexes with the time of electronic relaxation of 10^{-7}–10^{-8} s. This timescale is close to the characteristic time of the GR method, as defined by the lifetime of the excited state of the ^{57}Fe nucleus ($\tau \sim 10^{-7}$ s). From the GR spectra, it is estimated that the electrons produce a magnetic HF field strength of ca 180 kG at the ^{57}Fe nucleus. This field strength is compatible with electronic spin $S = 1/2$.

At higher concentrations of the monomeric DNIC-cysteine, the GR spectrum simplifies into a doublet spectrum [60] (Fig. 14C). This phenomenon is interpreted as motional narrowing caused by the increase of rapid stochastic fluctuations in the magnetic field at the iron due

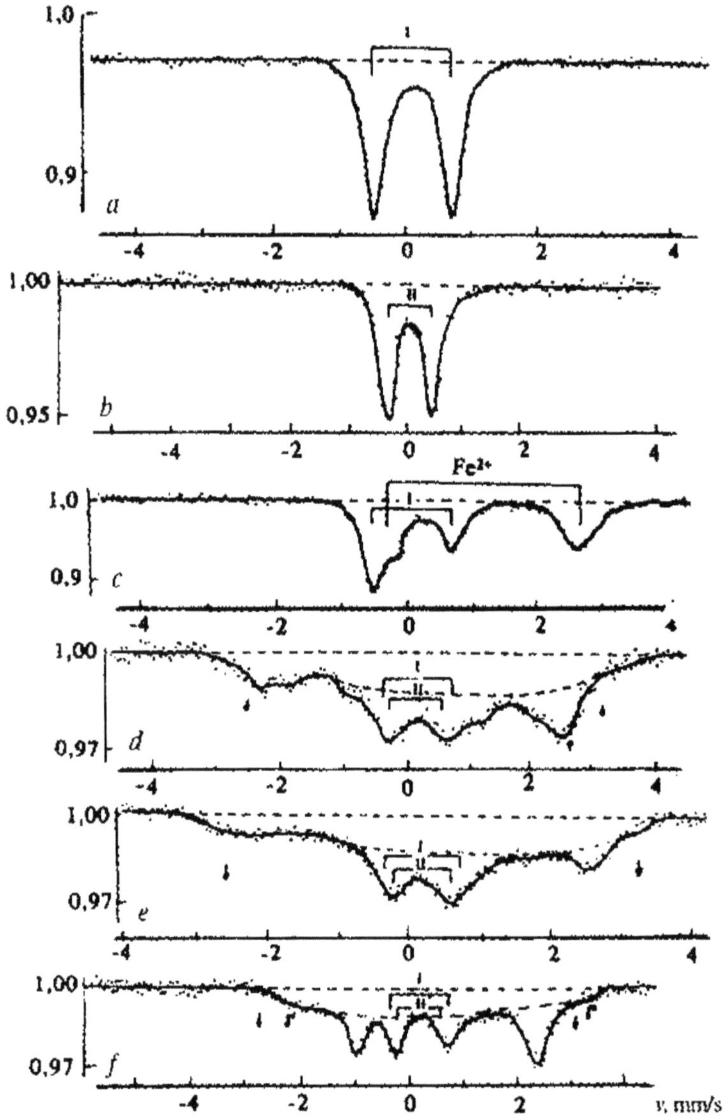

Fig. 13. Shape of Mössbauer spectra of dimeric DNIC with cysteine (DNIC 1:2) (curve a); of the same complex treated with dithionite (curve b); DNIC with cysteine synthesized for an Fe^{2+}:Cys ratio of 1:1 (curve c); monomeric DNIC with cysteine (DNIC 1:20) (curve d); dimeric DNIC with cysteine after its contact with an excess of cysteine (Fe^{2+}:Cys = 1:30) (curve e) and monomeric DNIC treated with dithionite (curve f). Recordings at 77 K. In (curves d–f), the arrows indicate the maximum values for H_{eff}. I, II and Fe^{2+} denote the positions of the doublet spectra, respectively for dimeric DNIC, reduced dimeric DNIC and the high spin Fe^{2+} complex. (From Ref. [44].)

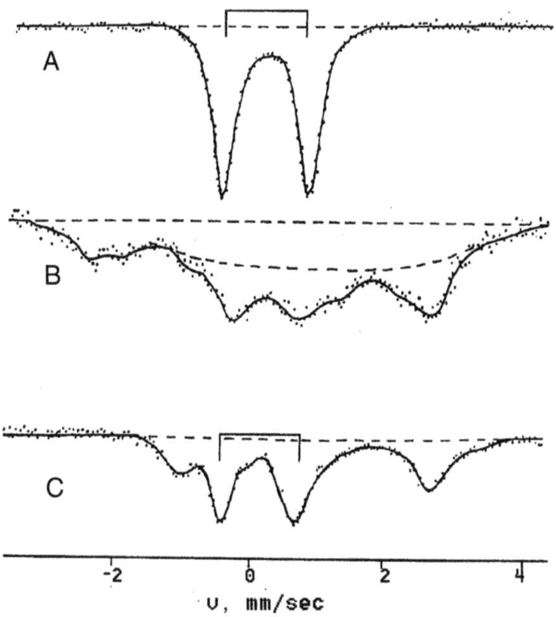

Fig. 14. Mössbauer spectra of dimeric DNIC with cysteine (A), monomeric DNIC with cysteine (B,C). The amount of ^{57}Fe was 0.5 (A,B) or 2 mg/ml (C). Recordings were made at 77 K [60].

to the spin–spin interaction between the DNIC monomers in solution. The fluctuations are sufficiently rapid to cause averaging of the HFS during the characteristic timescale of the GR method. The IS and QS of this simplified monomer spectrum coincide with those for the dimeric form (DNIC 1:2) (Fig. 14A,C). This suggests that the electronic configuration of the iron be identical in both forms. It implies that the dimeric form can be considered as a structure where two monomeric forms located in parallel planes and very weak electronic coupling between the iron centers (this structure will be considered below). The optical properties of dimeric DNIC are compatible with this interpretation.

Interestingly, the solutions of DNIC-cysteine with Fe:Cys = 1:1 show the characteristic dimeric optical absorption bands at 310 and 360 nm, except that the optical density is only half as large as expected from the iron concentration. The GR spectrum of DNIC-cysteine with ^{57}Fe:Cys = 1:1 appears as a superposition of two doublet spectra of approximately equal intensity: the first spectrum has the shape of dimeric DNIC with ^{57}Fe:Cys and the second spectrum was due to a high-spin $^{57}Fe^{2+}$ complex with IS = 1.4 and QS = 3 mm/s (Fig. 13c). Therefore, because of the cysteine shortage, only half of the total iron is sequestered in the form of Cys-DNIC. The other half remained in high-spin complex with non-thiol ligands. These data indicate unequivocally that each of the two iron atoms in dimeric DNIC-cysteine is coordinated with two cysteine ligands. These dimer comprises two monomeric DNICs and its formula in solution can be presented as $\{(RS^-)_2Fe^+(NO^+)_2\}_2^{2+}$. It is noteworthy that the proposed structure differs from the characteristic Roussin's Red Salt in crystalline state.

This dimeric iron-dinitrosyl complex has the formula $Na_2[Fe_2S_2(NO)_4]$ [64a]. In this structure, the two sulfur atoms act as bridging ligands between two iron atoms.

The proposition that the dimeric DNIC-cysteine is composed of two monomeric DNIC is corroborated by the changes in the optical and EPR spectra if the complexes are reduced with dithionite [43]. The addition of excess dithionite to dimeric Cys-DNIC rapidly changes the color from yellowish to bright green and, after 2–3 s, to stable raspberry-violet. A frozen aliquot of the intermediate bright green solution showed that the EPR spectrum is a mixture of monomeric Cys-DNIC with $\{Fe(NO)_2\}^7$ and $\{Fe(NO)_2\}^9$ configurations. The final raspberry-violet solution did not have an EPR signal, and showed that the reduced complexes may form diamagnetic dimers as well. The GR spectrum of the raspberry violet solution showed a species with doublet structure having QS = 0.8 mm/s and IS = 0.1 mm/s respectively (Fig. 13b).

The behavior was different when dithionite was added to monomeric Cys-DNIC: the color changed from green to bright green and persisted for several minutes [43]. The EPR spectrum of this bright green solution appears as a mixture of monomeric DNIC $\{Fe(NO)_2\}^7$ and $\{Fe(NO)_2\}^9$. Its GR spectrum retained the complex blurred shape characteristic of the original monomeric DNIC (Fig. 13f).

The spectroscopic changes in DNIC 1:2 and DNIC 1:20 by reduction were reversible by admission of oxygen: Upon exposure to ambient air, the raspberry-violet color of reduced DNIC 1:2 solution transformed again to yellowish characteristic of the initial DNIC 1:2 solution without EPR absorption. Similarly, the final bright green color of DNIC 1:20 solution changed to green, and the EPR signal from the solution became identical to that from initial DNIC 1:20 solution. The cycle of reduction with dithionite and reoxidation could be repeated without apparent loss of signal intensity in optical and EPR spectra.

DNIC WITH SULFIDE AND NEOCUPROINE ANIONIC LIGANDS

As well known, sulfide dianions S^{2-} often appear in biological systems as inorganic sulfur bridges between iron ions in polynuclear iron–sulfur clusters [65]. Disruption of such multinuclear clusters is one mechanism by which nitric oxide may interact with iron–sulfur proteins (ISPs) [3,66]. More details of such reactions are given in Chapters 5 and 6. The disruption can lead to the formation of iron-nitrosyl complexes, but the inorganic sulfide ions may remain bound to the iron. Incorporation of sulfide anions into non-heme nitrosyl-iron complexes was demonstrated for the first time by Goodman and Raynor by exposing an aqueous alcoholic solution of ferrous sulfate and excess sodium sulfide to gaseous NO [67]. At 77 K, EPR of frozen aliquots showed the formation of a paramagnetic MNIC containing sulfide ions (complex I). The spectrum had axial symmetry with $g_\perp = 2.01$, $g_{||} = 2.049$ and a resolved HF coupling to the nitrogen nucleus of $A_{||}(^{14}N) = 0.75$ mT (Fig. 15, curve a). At room temperature, the spectra appear as a motionally averaged triplet at $g = 2.021$ with HFS of 0.48 mT.

The parameters of the EPR absorption changed sharply when ferrous ions were added to the solution in excess over sulfide ions (Fe^{2+}:sulfide $\geq$ 10:1) [68]. The NO treatment of the solution for 10 min led to the formation of a new paramagnetic ferrous iron-nitrosyl species (complex II). In frozen solution, it had a narrow axial EPR spectrum with $g_\perp = 2.035$, $g_{||} = 2.02$, $g_{av} = 2.03$ (Fig. 15, curve b) [68]. Room temperature spectra showed a single

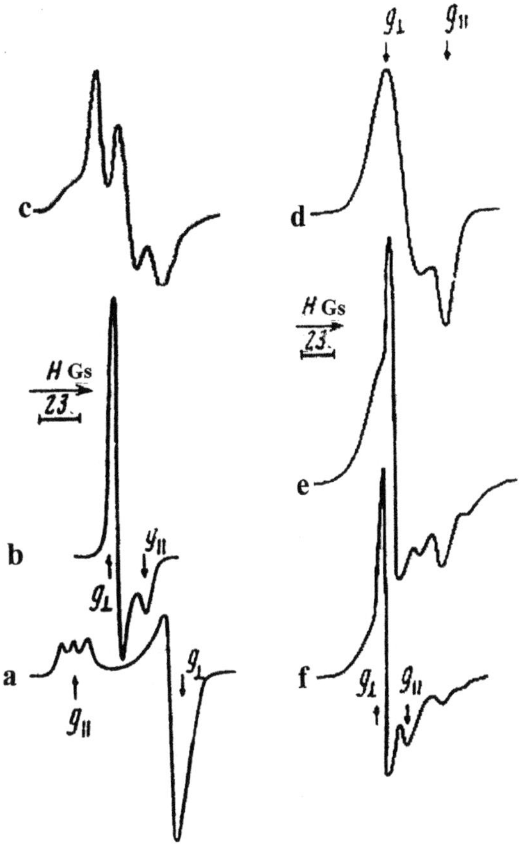

Fig. 15. EPR signals of nitrosyl-iron complexes with sulphide [66]. (Curve a): Mononitrosyl iron complex with sulphide synthesized at the ratio of [sulphide]:[Fe^{2+}] > 10 (complex I, $g_\perp = 2.01$, $g_\parallel = 2.049$); (curve b) iron-nitrosyl complex with sulphide synthesized at the ratio of [Fe^{2+}]:[sulphide] > 10 (complex II, $g_\perp = 2.035$, $g_\parallel = 2.020$); (curve c) EPR signal of complex II with ^{57}Fe; (curves d–f) respectively, EPR signals from DNIC ($g_\perp = 2.041$, $g_\parallel = 2.014$) and complexes II bound with nitrogenase enzyme. Recordings at 77 K. (From Ref. [68].)

line at $g = 2.03$ and half-width of 0.5 mT, without resolved HFS structure. Labeling of this species with the ^{57}Fe isotope produces spectra with resolved 1.2 mT doublet HFS (Fig. 15, curve c). This HFS is resolved at 77 K as well as ambient temperature, and its magnitude shows that the unpaired electron is mostly localized on the iron atom.

The balance between the complexes I and II could be reversibly shifted by the addition of competing ligands for the iron. The shifts as induced with cysteine or cyanide are given in Scheme 1:

$$\text{complex II} \overset{S^{2-}, CN^-, RS^-}{\underset{Fe^{2+}}{\longleftrightarrow}} \text{complex I}$$

Scheme 1.

It was found that anions of cyanide or cysteine do not act directly on these NO—Fe—S complexes but by binding free ferrous iron in solution. The depletion of free ferrous ions subsequently favors the shift towards complex I. Phrased otherwise, these anions act on the equilibrium indirectly by modulating the status of free iron. The addition of the reagents for persulfide, phosphites $(RO)_3P$ induced the degradation of complex II. This points to the presence of persulfide (polysulfide) ligands in the complex.

The proposition is in line with the investigation of the complex II formed from a model iron–sulfur center in serum albumin. The latter was synthesized by serum albumin treatment with ferrous iron and sodium sulfide as described elsewhere [68]. Subsequent exposure to NO gas induced the formation of complex II in the preparation. Consequent addition of triethyl phosphite or cysteine destroyed this complex II and induced the formation of protein-bound DNIC with thiol-containing ligands. This transformation was attributed to the removal of persulfide ligands from complex II, and transfer of the $Fe(NO)_2$ moiety to one of the thiols in the protein.

The formation of complex II was observed when solutions of isolated nitrogenase enzyme treated with exposure to NO gas in the presence of 8 M urea for 10 min (Fig. 15, curve f) [68]. The EPR signal of DNIC-thiol was observed in the absence of urea at NO treatment (Fig. 15, curve d). The complex II appeared at gaseous NO treatment of the enzyme solution for 20 h (Fig. 15, curve e).

It is reasonable to suggest that these observation apply to other iron–sulfur proteins also, and that exposure to NO transforms the iron–sulfur clusters of other proteins to complex II as well due to the formation of persulfide bond in the reaction between thiol groups of protein and inorganic sulfur atoms from iron–sulfur complexes. Experimental support is provided by the shape of the EPR signal of nitrosyl-iron complexes formed if the [4Fe—4S] cluster of transcription factor FNR protein is exposed to the NO-donor proline NONOate [69]. The EPR spectrum appears as the superposition of a narrow signal from complex II and that of a DNIC with thiolate ligands (Fig. 3A in Ref. [69]).

Neocuproine (2,9-dimethyl-phenanthroline) is often referred to as a selective bidentate chelator for monovalent Cu^+ ions. Recent investigations with EPR have shown that the situation with neocuproine is far less simple. In fact, neocuproine is able to bind ferrous iron as well, and it can be even incorporated as an iron ligand in DNIC complexes [70]. Stable species of paramagnetic DNIC with neocuproine were observed under the following conditions: after the addition of neocuproine to a solution of DNIC with phosphate ligands; after exposure in an aqueous solution of Fe^{2+} and neocuproine at pH 6.5–8 to NO gas; and after the addition of Fe^{2+}-citrate and neocuproine to a solution of Cys-NO [70]. Two different species of DNIC with neocuproine ligands could be distinguished from their EPR spectra. The first species had rhombic symmetry with g-factors 2.087, 2.055 and 2.025 (Fig. 16, curve d). In liquid solution at ambient temperature, the EPR spectrum simplifies to a single line at $g = 2.05$. The second species of DNIC with neocuproine can be recorded in frozen solution only ($T \sim 77$ K) and has a rhombic EPR spectrum with g-factors 2.042, 2.02 and 2.003 (Fig. 16, curve c).

The participation of neocuproine in DNIC allows us to argue against the consideration of neocuproine just as a selective bidentate chelator for Cu^+. This has important implications for the presence of S-nitrosothiols in biological systems. It is often observed that neocuproine

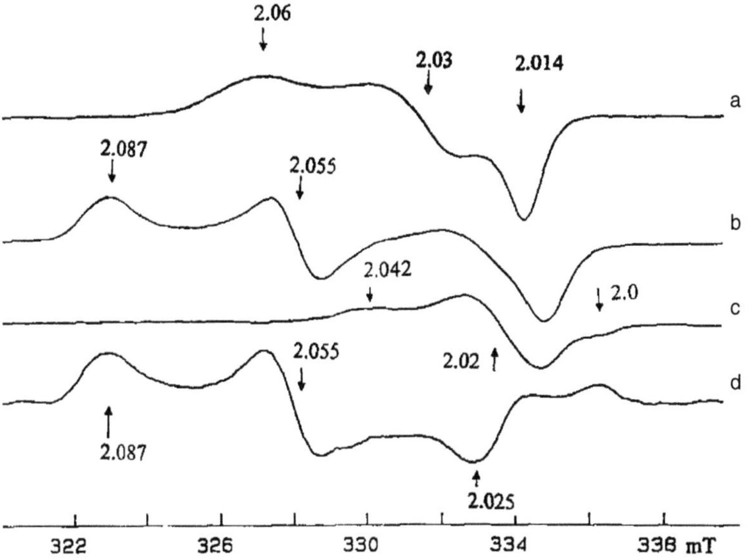

Fig. 16. EPR spectra from the solution of DNIC with neocuproine in 15 mM HEPES buffer, pH 7.0 (a) followed by pH increasing to 8.0 (b). Spectrum c was obtained by subtracting spectrum b from spectrum a. Recordings were made at 77 K [70].

inhibits the decomposition of RS—NO. This is usually interpreted as proof that Cu^+ ions dominate this decomposition. In fact, the inhibition by neocuproine may be attributable to the sequestration of ferrous iron, which is also an efficient catalyst for the decomposition of RS—NO [71]. The possibility of this alternative mechanism should be kept in mind when considering the pool of *S*-nitrosothiols in biological samples. More details of these reactions can be found in Chapters 12 and 13 of this book.

PROTEIN-BOUND DNICs

Two methods can be used for the synthesis of DNIC bound with proteins that do not contain intrinsic iron. The first is similar to the recipe for the formation of DNIC with low-molecular-weight DNIC (see above) and involves the treatment of the solution of proteins with gaseous NO in the presence of ferrous ions. The second method is the addition of low-molecular-weight DNIC to the solutions of proteins [54,72,73]. The protein-bound thiols also have high affinity for the $Fe(NO)_2$ moiety of DNIC, and replace the small thiols on a large fraction of the dinitrosyl iron moieties. The exchange results in the formation of protein-bound DNIC. This second pathway is even easier for DNIC with non-thiol ligands: The weaker binding of such DNICs facilitates the transfer of all $Fe(NO)_2$ moieties to protein. Fig. 17 demonstrate EPR spectra from DNIC-bovine serum albumin (DNIC-BSA) synthesized by these two methods.

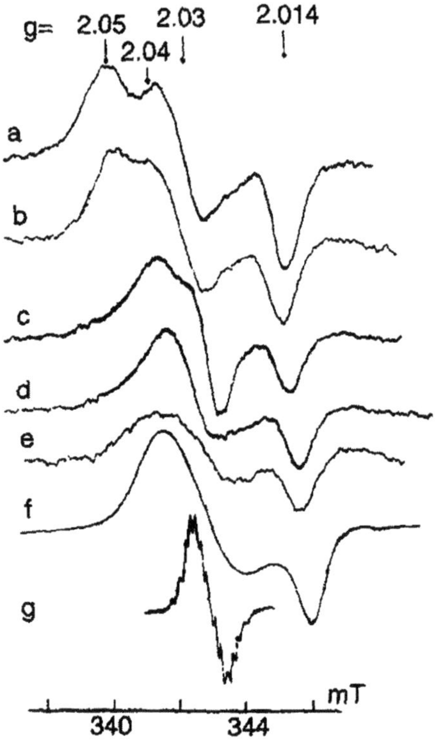

Fig. 17. EPR spectra of bovine serum albumin (BSA) (1 mM) in 15 mM HEPES buffer, pH 7.4 at room temperature. (curve a) After the addition of 1 mM DNIC with phosphate, (curve b) after the addition of 1 mM $FeSO_4$+ NO gas, (curve c) after adding 5 mM cysteine to preparation (a), (curve d) is spectrum (curve c) after subtraction of a small residual signal from Cys-DNIC, (curve e) after adding 1 mM ammonium sulfate to sample (a), (curve f) reference spectrum of 0.1 mM Cys-DNIC at 77 K and (curve g) reference spectrum of 0.1 mM Cys-DNIC at room temperature [73].

The shape of the spectra shows that the anisotropy of g-factor and HFS remain unchanged over the whole temperature range from 77 K to ambient [54,72,73]. This shows unambiguously that the DNICs are immobilized, in this case by being anchored to a large protein globule. The EPR signal from DNIC-BSA shows rhombic symmetry with g-factors 2.05, 2.03 and 2.014 [72,73].

This symmetry is lower than expected for DNIC with low-molecular-weight thiols, which have axial symmetry (see above). This low symmetry is plausibly rationalized by BSA molecule contributing only one thiol group to the DNIC. Second ligand L could be an intrinsic non-thiol amino acid residue like histidine. Thus, the different nature of the two anionic ligands could lower the symmetry of the complex from axial to rhombic. The idea is strongly supported by the observation fact that the symmetry may be upgraded to axial by the addition of cysteine to the DNIC-BSA solution (Fig. 17, curves c,d). This transformation was apparently due to the replacement of the protein-bound L ligand by cysteine in

DNIC (Scheme 2), the cysteine being free to assume its optimal orientation for binding to the iron [73].

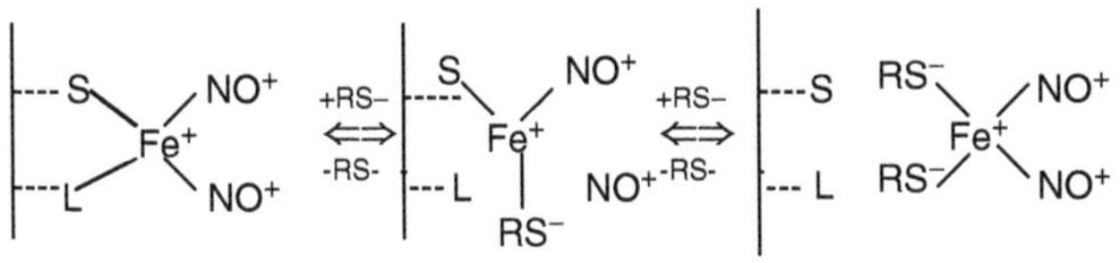

Scheme 2.

This structural change was reversible: passing the solution through the column with Sephadex G25 removes the cysteine, and the EPR spectrum of the protein fraction reverted back to rhombic symmetry. The addition of cysteine also led to the translocation of part of $Fe(NO)_2$ groups from the protein to cysteine with the formation of respective low-molecular DNIC, as demonstrated by the appearance of a narrow isotropic EPR signal superimposed on the anisotropic signal from DNIC-BSA [13,14,73] (Fig. 17, curve c). When cysteine was added in excess (at a concentration more than two order higher than the DNIC-BSA concentration) to the solution, all $Fe(NO)_2$ groups shifted from the protein to cysteine. As a result, only the motionally narrowed EPR signal from DNIC with cysteine was recorded at ambient temperature (Fig. 17, curve g). Similar transformations were observed on addition of thiosulfate ions to the solutions of DNIC-BSA. Interestingly, the shape of the EPR signals from DNIC-BSA recorded at ambient temperature in the presence of cysteine or thiosulfate is similar to that characteristic of the frozen solutions of DNIC with cysteine or thiosulfate, respectively. It can indicate that cysteine or thiosulfate molecules, while including into protein-bound DNIC (Scheme 1), can selectively influence the electronic structure of the complexes by its upgrading to axial symmetry.

Similar upgradation from rhombic to axial symmetry of DNIC-BSA was induced by the addition of dodecyl sulfate (SDS) anions at a concentration of 1% or higher (Fig. 18, curve b). The axial DNIC remained clearly anchored to the BSA because the EPR spectrum at room temperature appeared as an immobilized axial powder spectrum: It appeared similar to the EPR spectrum of a frozen solution of Cys-DNIC except that for DNIC-BSA the position of $g_{||}$ approached the central peak amplitude [73]. The same transformation of the EPR signal from DNIC-BSA was observed when the complex was incorporated into reversed micelles formed by aerosol OT, another negatively charged surfactant, regardless of the ratio between the amounts of water and aerosol ranging from 11 to 44 (Fig. 18, curves c,d).

The participation of thiol-containing ligands in DNIC-BSA was demonstrated by exposing the preparations to thiol-binding reagents. The treatment changed the shape of DNIC-BSA EPR signal to approach that of DNIC-BSA or DNIC with histidine residues as described before [72,73].

The formation of DNIC moieties in heme proteins like Hb require careful consideration. It might seem tempting to produce DNIC–Hb by directly exposing Hb to gaseous NO in the presence of iron. This method worked very well with albumin, but fails spectacularly for Hb. The reason is that the heme moiety in Hb has a high affinity for NO. Therefore, exposure to gaseous NO results in the formation of a considerable amount of ferrous nitrosyl-heme.

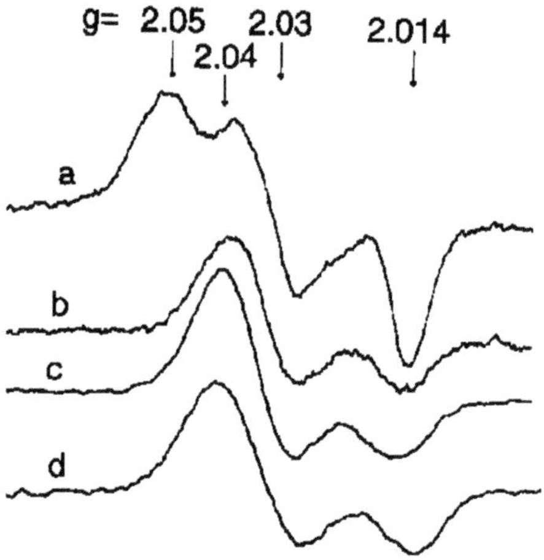

Fig. 18. EPR spectra from DNIC-bovine serum albumin (BSA) from initial preparation (curve a), after treatment with 1% dodecyl sulfate (curve b), and after its incorporation into reversed micelles formed from aerosol OT at the ratio between the amounts of water and aerosol ranging from 11 to 44 (curves c,d). Recordings were made at ambient temperature [73].

Non-heme DNIC is probably formed but remains unobservable because the EPR spectrum is dominated by the absorption from the ferrous nitrosyl-heme. But several chemical pathways exist which induce the formation of DNIC while avoiding nitrosylation of the heme: The addition of phosphate-DNIC ([phosphate] = 1 mM) to 1 mM Hb solution induces the formation of protein-bound DNIC, DNIC–Hb [73] (Fig. 19, curve a).

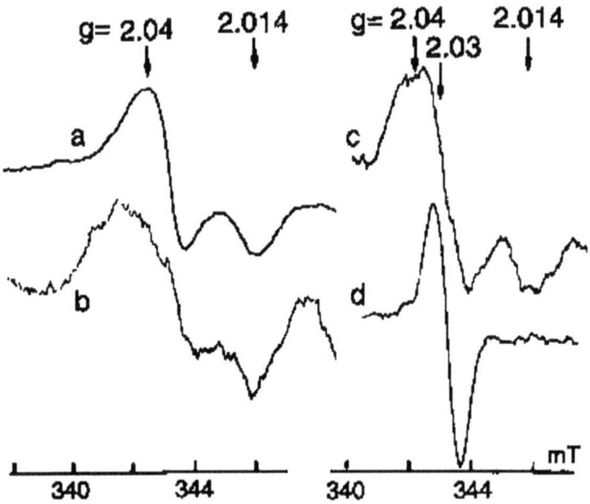

Fig. 19. EPR spectra from 1 mM solutions of horse hemoglobin after addition of 1 mM of DNIC with phosphate, including ^{56}Fe (curve a) or ^{57}Fe (curve b) or after addition of 1 mM DMIC with cysteine (curve c) or glutathione (curve d), including ^{56}Fe. Recordings were made at ambient temperature [73].

The reaction proceeds presumably *via* transfer of the Fe-$(NO)_2$ structure and does not involve the release of free NO. At room temperature, this complex shows an axially symmetric EPR spectrum, as distinct from the rhombic symmetry of DNIC-BSA. At 77 K, the EPR spectrum of DNIC–Hb coincides with that of Cys-DNIC in frozen solution. Incorporation of ^{57}Fe into DNIC–Hb induces significant broadening of the EPR spectrum (Fig. 19, curve b). The DNIC–Hb complex did include thiol ligands, since treatment with mercurate resulted in the decomposition of the complex. Exposure of DNIC–Hb to dithionite resulted in the formation of paramagnetic ferrous heme-nitrosyl complexes. Such complexes have low spin ($S = 1/2$) configuration $\{Fe(NO)\}^7$ and are unambiguously identified by EPR signal with extreme components at $g = 2.07$ and 1.98 and triplet HFS at $g = 2.01$ described earlier elsewhere [74]. Such heme-nitrosyl complexes are also slowly formed if DNIC–Hb is kept under anaerobic conditions for a long time (longer than an hour). It shows that NO has a higher affinity for ferrous heme iron than for the non-heme iron in DNIC–Hb.

DNIC–Hb could also be obtained by incubating ferric hemoglobin with Cys-DNIC. In this process, Cys-DNIC was completely consumed [72]. However, the yield of protein-bound DNIC was considerably lower than that in the hemoglobin incubation with phosphate-DNIC (Fig. 19, curve c). DNIC-glutathione were found to be stable and failed to transfer the Fe-$(NO)_2$ moiety into the hemoglobin. At room temperature, the solution retained the motionally narrowed EPR spectrum of DNIC-glutathione, and no signal from DNIC–Hb was observed [73] (Fig. 19, curve d).

DNIC-phosphate was also successfully applied to generate DNIC anchored on apo-metallothionein (apo-Mt) isolated from horse kidney [73]. The EPR signal from this complex coincided in shape and g-factor values with the EPR signal from frozen solution of DNIC with cysteine (Fig. 20, curve a). The shape of DNIC–apo-Mt signal remained unchanged

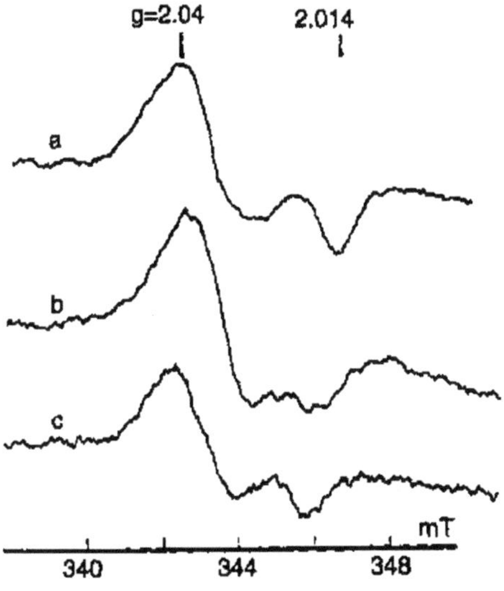

Fig. 20. EPR spectra from 1 mM solutions of apo-metallothionein (apo-Mt) from horse kidney after the addition of 1 mM DNIC with phosphate (curve a) with following addition of 1% dodecyl sulfate (SDS) (curve b) or after 1 h exposure of the solution (a) in air (c). Recordings were made at ambient temperature [73].

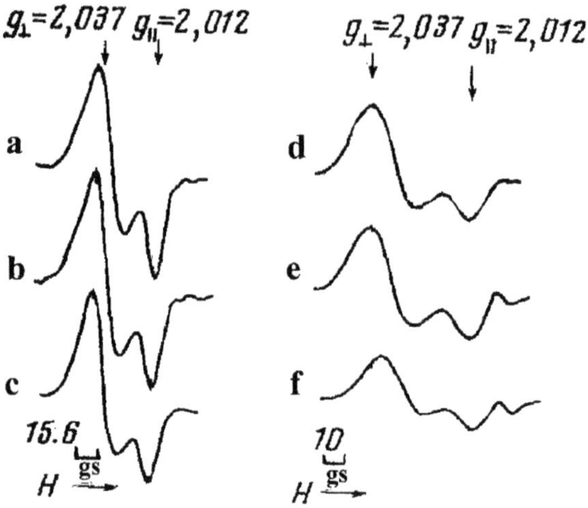

Fig. 21. Shape of the EPR signal of DNIC with undiscolored or discolored rhodopsin (curves a,d and b,e, respectively). (Curves c,f) the EPR signal of DNIC with rhodopsin which was discolored after DNIC formation. Recordings were made at 77 K (curves a–c) or at ambient temperature (curves d–f), respectively [82].

when raising the temperature from 77 K to ambient, which indicated a bond between DNIC and protein globule. Similar signal was recorded for DNIC bound with apo-Mt from rabbit liver [23,75]. On addition of 1% of DDS to the protein solution, the component at $g_{||}$ decreased and the spectrum became very similar to that of DNIC-BSA in the presence of SDS [71] (Fig. 20, curves b,c). By now, several publications document the formation of DNIC anchored on other proteins like FNR [67,76,77] flavorubredoxin [78], ferritin [79], glutathione reductase [80] or glutathione-transferase [81].

The formation of DNIC with rhodopsin was described in the preparation of the protein treated with NO gas in the presence of ferrous iron [82]. The shape of the EPR signals from rhodopsin-bound DNIC slightly differs from that of frozen solution of DNIC with cysteine and varies depending on illuminating or aging of the rhodopsin suspensions. Prolonged illumination induces bleaching and discoloration of the preparations (Fig. 21).

Protein-bound DNICs can appear in various iron–sulfur proteins treated with NO [59,66,83]. The formation of DNIC in the latter is considered in the Chapters 5 and 6 of this book.

LABORATORY SYNTHESIS OF DNIC WITH LOW-MOLECULAR ANIONIC LIGANDS

For laboratory synthesis of DNIC with low-molecular-weight thiol ligands, two facile pathways exist: First, exposure of solutions with iron and thiols to gaseous NO [69]. Second, by mixing nitrosothiols with iron [46].

As a typical example of the first pathway, a Thunberg vial with two separate containers is prepared. The upper container holds 1 ml distillate water with 25 mM ferrous sulfate, the bottom container holds 4 ml 15 mM HEPES or Tris buffer, pH 7.4 with 100 mM thiol. After evacuation to remove oxygen and addition of gaseous NO, the liquids are mixed and produce a dark-green solution with DNIC of final concentration [Fe] = 5 mM and Fe:thiol ~ 1:20. The complexes are monomeric because the thiolates are present in twentyfold excess over iron. In an alternative synthesis, dimeric DNIC is obtained in a similar way except that a smaller quantity of thiol is added. The addition of 10 mM thiol to the bottom container results in a final ratio Fe:thiol ~ 1:2. In this solution, dimeric DNIC is formed. The solution has yellow-orange color and does not give EPR spectra. The reason is that the dimers are diamagnetic due to the antiferromagnetic coupling between the two constituents. After the exposure to NO, the acidity of the mixture should have remained near pH ~ 7. In particular, acidification should have been avoided as it usually is a sign that the NO gas was contaminated with NO_2. The yield of DNIC may be verified from the optical extinction coefficient of the complexes. The monomer has strong absorption band at 392 nm with $\varepsilon_{392} = 3580\ (\text{Mcm})^{-1}$. Additionally, it has two weaker d–d absorption bands with $\varepsilon_{603} = 299\ (\text{Mcm})^{-1}$ and $\varepsilon_{772} = 312\ (\text{Mcm})^{-1}$ [45]. The dimer has $\varepsilon_{310} = 7200\ (\text{Mcm})^{-1}$ and $\varepsilon_{360} = 6800\ (\text{Mcm})^{-1}$ plus a shoulder around 450 nm. The optical spectra can be found in Refs. [45,47].

After the formation of DNIC, the solutions may be exposed to oxygen from ambient air, and will remain stable only for a limited amount of time due to oxidation of the thiol ligands by the formation of disulfide bridges. The timescale is determined by the type of thiol. Cysteine oxidizes fairly rapidly (within 10–15 min), that results in the transformation of paramagnetic monomeric DNIC to diamagnetic dimeric DNIC with cysteine. The latter can be back transformed into paramagnetic form by the addition of excess cysteine. The dimeric DNIC with cysteine remains stable in air for 1 h [47]. Monomeric glutathione complexes are stable in air for 1–1.5 h [47]. The dimeric DNIC with glutathione keep stability in air for 3–4 h [47]. Small aliquots of freshly prepared DNIC may be snap frozen and will remain stable for weeks when stored in liquid nitrogen.

DNICs with non-thiolic ligands like citrate or phosphate can also be synthesized in this way, but yields also depend on the Fe:ligand ratio and should always be verified *via* EPR spectroscopy. Additionally, such non-thiolic DNICs are considerably more susceptible to be destroyed by oxygen from ambient air. Paramagnetic DNICs with non-thiolic ligands are formed at high amount of the ligands: at the molar ratio of ligand:Fe > 50. The diamagnetic, probably dimeric DNICs appear at the lower ratio [36]. A structure similar to the structure of Roussin's Red Salt is proposed for these complexes, where non-thiolic ligands function as a bridge between two iron atoms [36].

The exposure to NO gas requires care because the NO gas should be carefully scrubbed to remove contaminating traces of higher oxides, in particular NO_2 radicals and N_2O. The use of gas from compressed pure NO is not recommended as these contaminants unavoidably form at high pressure from the dismutation reaction (cf Chapter 1).

$$3NO \rightarrow NO_2 + N_2O$$

The common procedure of bubbling the NO gas through an alkaline solution does not remove such traces with adequate precision for reproducible results. Purification by low-temperature sublimation [71] in liquid nitrogen and subsequent evaporation is much preferred.

Generation of NO gas from acidic reduction of nitrite is preferable if oxygen can be excluded from the reaction vessels. A very convenient alternative is the use of compressed inert atmospheres with low NO content. For NO contents below 1%, the above dismutation reaction is kinetically inhibited even under pressure in an inert gas like nitrogen or argon.

Recently, a very different recipe has proven useful for the synthesis of DNIC with thiol- or non-thiol ligands: The addition of ferrous iron to the solution of *S*-nitrosothiols in the presence of thiolic or non-thiolic anionic ligands, respectively [45,46,71]. The method exploits the chemical equilibrium between the various constituents in any solution containing iron, nitrosothiols and excess thiol ligands [46]. The details of this chemical equilibrium will be discussed in Chapter 11. DNIC solutions obtained in this way always contain significant quantities of free thiol ligands and nitrosothiols. This should be kept in mind when the DNIC yields are tested *via* optical absorption of the solutions. For laboratory use, this second pathway is very convenient as it completely avoids NO gas with its cumbersome requirement of the first method, namely the scrubbing and removal of oxidized contaminants. Now some groups [45,81] synthesize DNIC with thiol-containing ligands (cysteine or glutathione) from *S*-nitrosothiols and ferrous ions *via* this second pathway. Therefore, nitrosothiols will prove very useful for DNIC synthesis in the further studies of physico-chemical and biological properties of these complexes. The recipes for synthesizing DNICs from *S*-nitrosothiols or mixtures of ferrous iron and thiols are described in detail in Chapter 11 of this book.

THE MECHANISM FOR ASSEMBLY OF DNIC WITH THIOL-CONTAINING LIGANDS

The two pathways described in the previous section yield DNIC complexes with $\{Fe(NO)_2\}^7$ configuration. The mechanism for the assembly of this complex is not just simple liganding of nitrosyl and thiol ligands to the iron, but must necessarily involve some redox transformation of the constituents. The experimental evidence for redox activity is clear: The synthesis proceeds in the absence of oxygen and starts with ferrous iron in d^6 state (non-heme ferric iron is not capable of NO binding). Nitrosyl liganding should lead to a diamagnetic $\{Fe(NO)_2\}^8$ configuration, but EPR experiments show the formation of the paramagnetic $\{Fe(NO)_2\}^7$ species instead. Second, quantitative studies [77] show that three NO molecules are needed for the formation of a single DNIC. Third, the synthesis proceeds under the release of one molecule of N_2O per two DNIC complexes [85]. This N_2O represents nitrogen in a lower oxidation state than the original NO and is usually taken as a manifestation of the formation of an intermediate HNO or NO^- species. The formation of N_2O and consumption of NO remind us of the well-known dismutation reaction of NO in the gas phase under high pressure. This asymmetrical reaction has products in lower as well as higher oxidation state than the original NO (cf Chapter 1)

$$3NO \rightarrow N_2O + NO_2.$$

In dilute solutions, this dismutation reaction is kinetically forbidden and cannot play a significant role. However, the above evidence suggests that ferrous iron initiates a disproportionation reaction of the form $2NO \rightarrow NO^+ + NO^-$ [84] in the liquid phase, where the nitroxyl anion NO^- leads to subsequent formation of N_2O. Such disproportionation of NO is known to be mediated by a variety of metal complexes with iron, cobalt, copper or ruthenium [86]. The analogous disproportionation reaction leaves the dinitrosyl complex in $\{Fe(NO)_2\}^7$ state:

$2L^- + 2NO + Fe^{2+}$	$\Leftrightarrow$	$2L^-\{Fe(NO)_2\}^8$	(assembly of diamagnetic complex)
$2L^-\{Fe(NO)_2\}^8 + H^+ + NO$	$\rightarrow$	$2L^-\{Fe(NO)_2\}^7 + HNO$	(disproportionation)
$2HNO$	$\rightarrow$	$N_2O + H_2O$	(the formation of nitrous oxide)

The precise mechanism for the disproportionation step is not known, but a possible mechanism might proceed as below (Scheme 3) [46,71]:

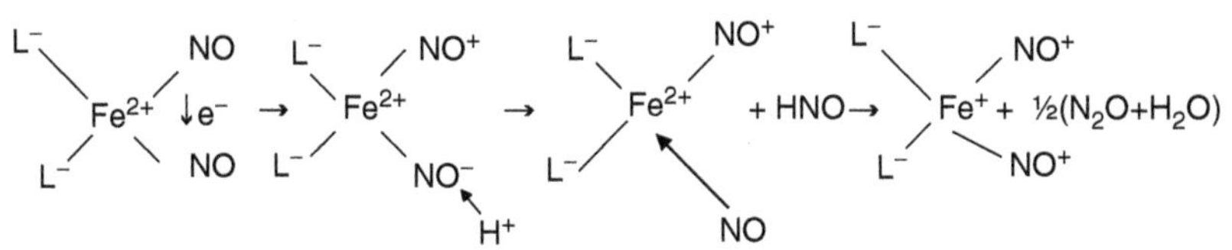

Net process: $Fe^{2+} + 3NO + 2L^- + H^+ \longrightarrow \{(L^-)_2Fe^+(NO^+)_2\} + ½(N_2O+H_2O)$

Scheme 3.

We propose that the d-orbitals of the iron mediate this disproportionation by coupling the unpaired electrons of the nitrosyls. The appearance of HNO in the above equations reflects the fact that this protonated nitroxyl rather than NO^- is the principal species at physiological pH (early reports of $pK_a \sim 4.7$ have been significantly revised upward to ca 11 [87]). NO^- is isoelectronic with dioxygen and can exist as a spin triplet ground state or spin singlet excited state. The triplet ground state has a low rate of protonation due to restrictions imposed by spin conservation [62b,87] (for properties of the nitroxyl anion cf Chapter 1).

According to Scheme 3, the formation of DNIC from ferrous iron and NO proceeds through three steps [46]. First is the formation of paramagnetic MNIC with $\{FeNO\}^7$ configuration:

$$(L^-)_2Fe^{2+} + NO \rightarrow (L^-)_2Fe^+{-}NO^+.$$

The next step involves the inclusion of a second NO ligand, the reduction of this ligand to nitroxyl and the release of this nitroxyl moiety from the complex after protonization as shown:

$$(L^-)_2Fe^+{-}NO^+ + NO \rightarrow (L^-)_2Fe^{2+}{-}(NO^+,\ NO^-) + H^+ \rightarrow (L^-)_2Fe^{2+}{-}NO^+ + HNO$$

The mononitrosyl complex at this stage is diamagnetic with $\{FeNO\}^6$ configuration. The final step is binding the third NO molecule to give DNIC with $\{Fe(NO)_2\}^7$, which is paramagnetic:

$$(L^-)_2Fe^{2+}{-}NO^+ + NO \rightarrow (L^-)_2Fe^{2+}{-}(NO^+,\ NO) \rightarrow (L^-)_2Fe^+{-}(NO^+)_2$$

It should be noted that diamagnetic mononitrosyl complex can also be formed by the inclusion of a nitrosyl ligand in a ferrous dithiocarbamate complex [46].

A similar situation may be expected if the DNIC is assembled from nitrosothiols RS—NO instead of free NO molecules, if the redox disproportionation reaction between two RS—NO molecules is proposed [46,71]:

$$2RS^{-}—NO^{+} \rightarrow RS^{-}...NO + RS...NO^{+}.$$

One-electron reduction/oxidation of RS—NO molecules leads to the appearance of instable products of the reaction. In the solution without iron, the process can be negligible. However, the disproportionation of RS—NO molecules can be sharply accelerated being catalyzed by ferrous iron. The experiments demonstrate that addition of iron to solutions of Cys-NO or GS-NO resulted in rapid formation of paramagnetic DNIC: The rate constant is comparable to that of the DNIC formation from iron and nitric oxide [46]. So, a possible mechanism might proceed as below [46,71]:

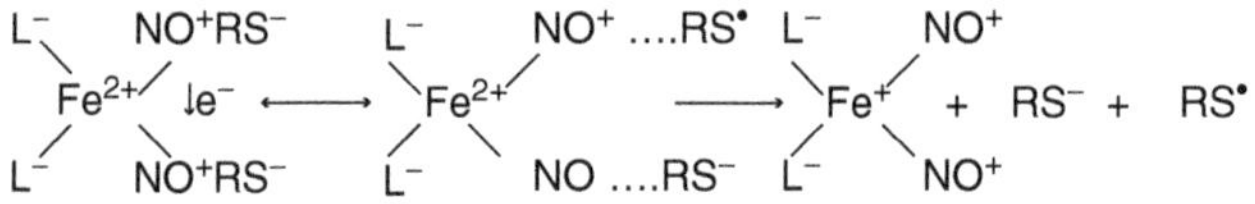

Scheme 4.

The assembly of the intermediate diamagnetic complex is reversible as it does not involve any electronic rearrangement which prevents release of the RS—NO moieties. The disproportionation with asymmetrical release of RS^- and $RS^{\bullet}$ is irreversible, however, and drives the reaction to full synthesis of DNIC in $\{Fe(NO)_2\}^7$ state. Subsequently, the thiyl radicals $RS^{\bullet}$ combine into a disulfide.

Scheme 4 predicts that two Cys-NO molecules are needed for the formation of one iron-dinitrosyl complex, in accordance with quantitative estimates from experimental data [54]. The experiments were performed on Cys-DNIC as synthesized by NO treatment of cysteine and ferrous iron solution in 10 mM HEPES buffer, pH 7.4. Upon rapid acidification of the DNIC solution to pH 1.5 with HCl, the solution turned from green to pink color and lost the EPR absorption from DNIC. The optical absorption confirmed the formation of Cys-NO, the quantity being equal to that of the initial Cys-DNIC. Subsequent rapid raising the pH back to initial value (pH 7.4) with NaOH restored the green color at reduced intensity. Optical absorption confirmed the reconstruction of around 50% of the initial Cys-DNIC. Repeating this cycle of acidification subsequently induced loss of ca 50% of the Cys-DNIC, with corresponding reductions in the amounts of Cys-NO. It means that two Cys-NO molecules were consumed for the reconstruction of one DNIC complex induced by increasing the pH from acid to neutral values.

It should be remarked that the above reactions were studied *in vitro* under strict exclusion of oxygen. Such conditions are never strictly met in actual biological systems. Therefore, it should be kept in mind that additional pathways might exist for *in vivo* formation of DNIC. In the presence of oxygen, diamagnetic DNIC with $\{Fe(NO)_2\}^8$ will be oxidized rapidly to paramagnetic $\{Fe(NO)_2\}^7$ (see previous sections).

STABILITY OF LOW-MOLECULAR-WEIGHT AND PROTEIN-BOUND DNICs IN ANAEROBIC OR AEROBIC SOLUTIONS

As mentioned earlier, low-molecular-weight DNICs with thiol-containing ligands have much higher stability against oxygen from ambient air than DNICs with non-thiolic ligands. This difference reflects the difference in binding strength of the nitrosyl ligands to the iron. This binding is enhanced significantly by the coordination of thiol ligands to the iron [39]. Low stability of DNIC with non-thiol containing ligands could be due to the hydrolysis of the nitrosonium moieties by hydroxyl anions (Scheme 5) [46,88]:

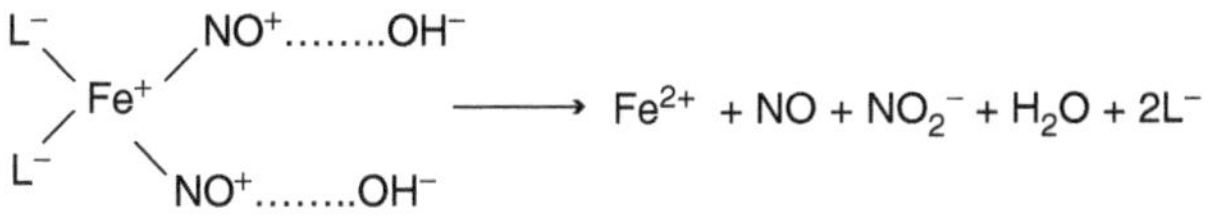

Scheme 5.

The interaction could lead to irreversible accumulation of nitrite ions. With respect to DNIC with thiol-containing ligands, due to high affinity of thiols to nitrosonium cations NO^+ [89] the interaction between them could lead to the formation of *S*-nitrosothiols (Scheme 6) by the release of ferrous iron:

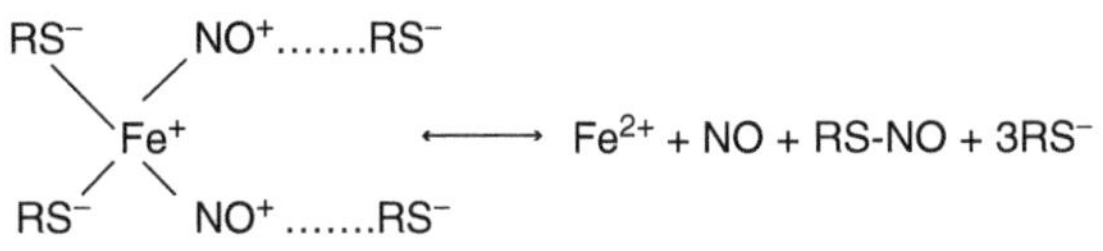

Scheme 6.

The partial decomposition of DNIC proceeds until a quasi-equilibrium between DNIC and its constituents is established. This reaction equilibrium between DNIC, NO, ferrous irons, RS^- and RS—NO is considered in detail in Chapter 11 of this book.

However, the assembly of DNIC from ferrous iron + RS—NO leads to the oxidation of thiols to thiyl radicals and formation of disulfides even at anaerobic conditions (Scheme 3). As a result, the concentration of thiols decreases with time to levels where DNIC starts to dimerize as diamagnetic DNIC (see above). Subsequent reinforcement of the thiol concentration reverts to monomeric state. These transformations are illustrated in Fig. 25 showing the formation of Cys-DNIC in an anaerobic solution of Cys-NO + ferrous iron with low cysteine:iron ratio. The sharp fall of the $g = 2.03$ signal with time reflects the loss of paramagnetic DNIC and can be reversed by the addition of excess cysteine [46]. This recovery of the 2.03 signal is even achieved after solutions of the dimeric DNIC were stored anaerobically for several hours and attest to the high stability of the complexes.

Upon exposure to oxygen, the loss of paramagnetism is significantly accelerated for weakly binding and thiol ligands. For example, the exposure of 1 mM of paramagnetic phosphate-DNIC in 100 mM phosphate buffer (pH $= 7.4$) to ambient air leads to full disappearance of the EPR signal from the solution, characteristic of the complex, within 10 min [86]. Similar behavior was characteristic of other DNIC with non-thiol-containing ligands (citrate, ascorbate, water and other; Vanin, unpublished data). With thiolate ligands like cysteine, the transformation of monomeric Cys-DNIC to dimeric Cys-DNIC is accelerated by oxygen, evidently due to oxidation of thiol compounds with oxygen. For example, exposure of 0.2 mM monomeric Cys-DNIC to ambient air leads to complete loss of paramagnetism within 40 min. The iron has been sequestered as diamagnetic dimeric MNIC, because the EPR intensity may be fully recovered by the addition of 10 mM of L-cysteine [55]. Longer exposure to oxygen subsequently leads to irreversible loss of DNIC: After storage of dimeric DNIC open to ambient air for 4 h, only 50% of the original intensity may be recovered by the addition of L-cysteine. Still, the long lifetime attests to a high stability of dimeric DNIC in particular against oxidation (Vanin, unpublished data).

Oxidation of the nitrosyl ligands is another possible pathway. French investigators have studied DNIC with bidentate NN ligands such as 2,2′-bipiridine, 4,4-dimethyl-2,2′-bipiridine, 1-10-phenanthroline, with bidentate PN ligands such as its 2-(diphenylphosphino) and 2-(diphenylphosphine oxide) derivatives, and with bidentate phosphorus ligands PP such as diphosphines 1,2-bis(diphenylphosphine)-ethane or *trans*- or *cis*-1,2-bis(diphenylphosphine)ethylene [90–93]. Exposure to dioxygen transforms the DNIC into the nitrato iron complexes. The structures of these nitrato complexes were determined as $Fe(NO_3)_2Cl(NN)$, $Fe(NO_3)_2Cl(OPN)$ or $[FeCl_4][Fe(NO_3)_2Cl(OPPPO)_2]$, respectively. Thus, it shows oxidation of the nitrosyl ligands and formation of nitrato ligands in the iron complexes. The mechanism of the process is under study now. It was noted that these complexes have sufficient electron density on the iron to transfer oxygen and allow oxidation of various organic compounds including diphenylphospine ligands [91].

Depending on the final redox state of the iron, some nitrato iron complexes would be paramagnetic, others EPR-silent state. This may explain the experimental fact that superoxide readily reacts with DNIC-cysteine to an EPR-silent product [94]. The same observation applies to DNIC bound to serum albumin. Moreover, the antioxidant features are characteristic of DNICs [94–98]. The transformation of NO to nitrato ligands by superoxide was proposed [91] to explain the observed loss of paramagnetism. The reaction would initially produce peroxynitrite ligated to the iron, followed by isomerization to nitrate. Significantly, thiosulfate was observed [99] to protect DNIC against degradation by superoxide. This protection applied to Cys-DNIC as well as DNIC bound to serum albumin, and was attributed to transfer of the $Fe(NO)_2$ moiety to the thiosulfate ligand, thereby converting the DNIC to a more stable form. Such protection is highly significant for possible therapeutical applications of thiosulfate-DNIC as vasodilator or inhibitor of platelet aggregation, where the presence of superoxide radicals is known to play an important role.

This section discussed only few of the possible reaction mechanisms of the oxidative agents with DNIC complexes. Future investigations will provide deeper insight into the complex chemistry and functions of DNIC in biosystems.

GEOMETRIC STRUCTURE AND ELECTRONIC CONFIGURATION OF DNIC IN SOLUTION AND CRYSTALLINE STATE

About forty years after their discovery, the structure of the iron center in DNIC is still a subject of controversy. Accurate structural data were obtained from diffraction experiments on crystallized DNIC, which are usually EPR silent. These diffraction studies showed that the iron was held in the center of a tetrahedral configuration [37,41,90,100–109]. In contrast, the available information on the electronic configuration was provided by EPR spectroscopy on isolated paramagnetic complexes in solution. The analysis of the EPR parameters suggested that dissolved DNIC complexes have the iron in a square plane geometry [39]. The sets of data suggest that the geometry and electronic configuration change significantly upon dissolution of the crystals in a solvent.

We will first discuss the paramagnetic DNIC with thiol-containing ligands in solution. At room temperature, the EPR spectra of DNIC show resolved HFS interactions with the ligands in coordination to the iron. The spectra of the DNIC identify four ligands: two anionic compounds and two nitrosonium ions. The DNIC with thiol-containing ligands, and thiosulfate or xanthogenate ligands all have very similar EPR spectra with axial symmetry of the g-tensor. The g-factors are $g_\perp = 2.04$–2.45 and $g_{||} = 2.014$ [39,40,56]. We noted above that DNIC in solution shows oblate axial symmetry with $g_\perp > g_{||}$. This indicates that the iron atom can be placed in plane-square structure (Scheme 7). The crystal field is dominated by four strong ligands in the plane, with small perturbation from weakly binding solvent molecules at the axial positions. In this structure, the ligand fields of the four in-plane ligands lift the ψ_{x2-y2} to a high-energy level, sufficiently high to preclude the high spin state as expected from Hund's rule. The seven electrons of the $\{Fe(NO)_2\}^7$ center distribute over the four remaining lower orbitals with a total spin $S = 1/2$. This low spin configuration is compatible with three possible orbital arrangements as indicated in Scheme 8(A–C):

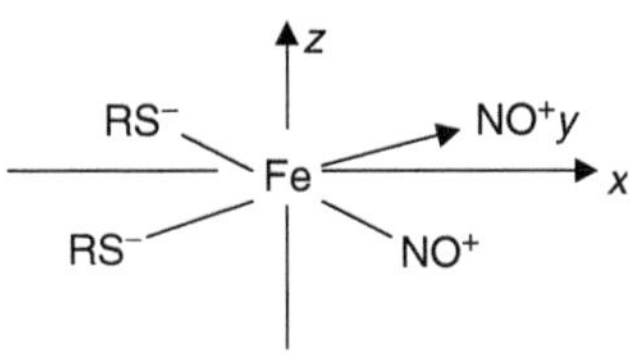

Scheme 7.

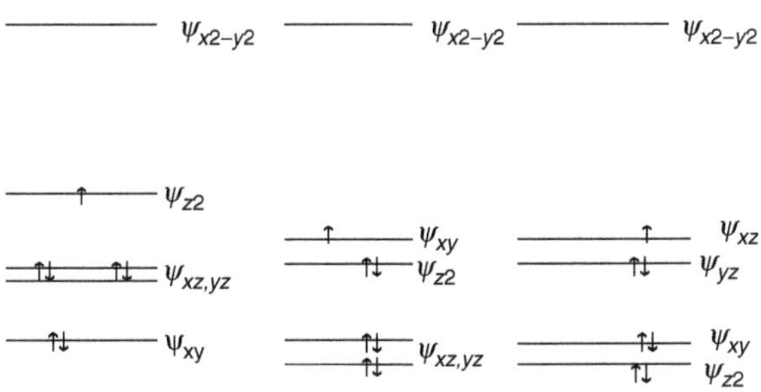

Scheme 8. Electronic configuration of iron in a square planar ligand field as expected for DNIC with $\{3d\}^7$ in solution.

The three configurations may be distinguished from the EPR properties. According to Griffith [110], the g-factor values for these orbital arrangements are

$$\text{Scheme A}: \quad g_{||} = 2.0, \qquad g_\perp = 2 + 6\alpha^2\beta^2\xi/\Delta E_{z2-(xz,yz)}$$
$$\text{Scheme B}: \quad g_{||} = 2.0 - 8\alpha^2\beta^2\xi/\Delta E_{(x2-y2)-(xz,yz)} \qquad g_\perp = 2 + 2\alpha^2\beta^2\xi/\Delta E_{xy-(xz,yz)},$$

where ξ is a one-electron constant of spin-orbital interaction, $\alpha^2\beta^2$ are covalency coefficients and $\Delta E_{z2-(xz,yz)}$ is the splitting between the respective levels.

Both the schemes A and B have oblate axiality with $g_\perp > g_{||}$ and are compatible with experimental values. For scheme C, Griffith predicts $g_{||} > g_\perp = 0$ which is incompatible with experiment. Schemes A and B are clearly distinguished by the position of the ψ_{z2} orbital of the iron. This orbital has high density in axial directions and is not strongly disturbed by the ligands in the xy plane. In contrast, the $\psi_{xz,yz,xy}$ orbitals have strong interaction with the ligands, in particular with the low-lying π^* orbitals of NO^+ ligands. This interaction can significantly decrease the energy of the $\psi_{xz,yz,xy}$ orbitals to below that of the ψ_{z2} orbital and lead to the realization of the scheme A for DNIC in solution.

Isotopic substitution of iron with the ^{57}Fe ($I = 1/2$) supports this interpretation. For ^{57}Fe-DNIC with ethylxanthogenate (Fig. 11, curve c [56]), we find experimental values for the HF couplings $|A_\perp|(^{57}\mathrm{Fe}) = 1.7 \pm 0.2$ mT, $|A_{||}|(^{57}\mathrm{Fe}) = 0.3 \pm 0.3$ mT, $|A_{\mathrm{iso}}|(^{57}\mathrm{Fe}) = 1.20 \pm 0.05$ mT [39,56]. The signs of the couplings can be deduced from the following considerations. Transition metals have negative values for $A_{\mathrm{iso}} = (2A_\perp + A_{||})/3$ [51,111]. This negative average is only compatible with negative signs for both $A_\perp$ and $A_{||}$. Therefore, $A_{||} = -0.3$ mT and $A_\perp = -1.7$ mT. The HF coupling appears as the sum of an isotropic Fermi contact term and an anisotropic dipolar interaction described by the traceless tensor $T = \mathrm{diag}(t_x, t_y, t_z)$:

$$\vec{S} \cdot A \cdot \vec{I} = A_{\mathrm{iso}}\, \vec{S} \cdot \vec{I} + \vec{S} \cdot T \cdot \vec{I}$$

Since $t_i = A_i - A_{\mathrm{iso}}$, we find a positive value for $t_z = A_{||} - A_{\mathrm{iso}} = +0.9 \pm 0.3$ mT. Since t_z is the spatial average of the dipolar interaction between the nuclear moment of iron and the unpaired electron wavefunction ψ_e, we have the usual relation

$$t_z = -g\beta g_N \beta_N \langle \psi_e | \frac{r^2 - 3z^2}{r^5} | \psi_e \rangle$$

The positive value of this quantity shows that the unpaired electron must have strong overlap with the ψ_z^2 orbital of the iron atom.

The proposed square-planar structure of thiol-DNIC is fully compatible with the formation of diamagnetic dimeric DNIC by cofacial binding as indicated in Scheme 9. Optical and GR (Mössbauer) spectroscopy [43,44] have confirmed the dimeric nature of this complex. We propose that the dimer be composed of two cofacial monomers rotated relative to each other through 180°. In this geometry, the oppositely charged RS^- and NO^+ ligands are positioned above each other, and provide electrostatic stabilization of the dimer.

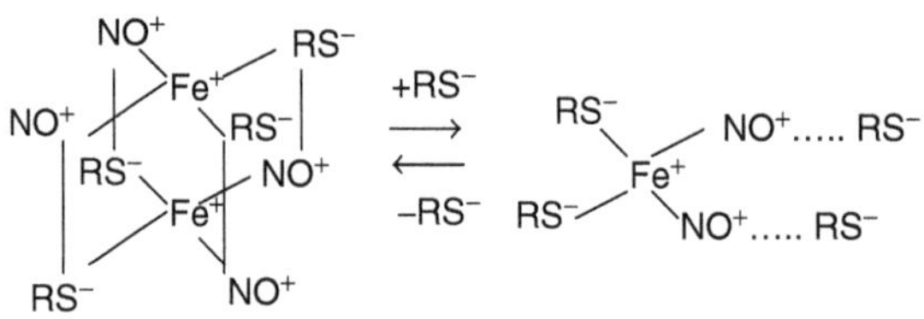

Scheme 9.

The cofacial stacking implies significant overlap between the axial d_{z2} orbitals of the adjacent iron atoms. The overlap causes antiferromagnetic spin-pairing that leaves the dimer diamagnetic [43,44].

We now consider the crystalline state of paramagnetic DNIC with thiol-containing ligands. As aforementioned diffraction experiments have revealed that the crystalline solid corresponds to iron in a tetrahedral crystal field [37,41,90,100–109]. In a tetrahedral crystal field, the degeneracy of the five d-levels of iron is increased as shown in Scheme 10. In this configuration, the three $\psi_{xz,yz,xy}$ orbitals point towards the ligands and are shifted upward in energy. The ψ_{x2-y2} and ψ_{z2} orbitals have little overlap with the ligands and remain unaffected. Scheme 10 shows the occupancy level of paramagnetic DNIC complexes in $\{Fe(NO)_2\}^9$ and $\{Fe(NO)_2\}^7$ states, respectively.

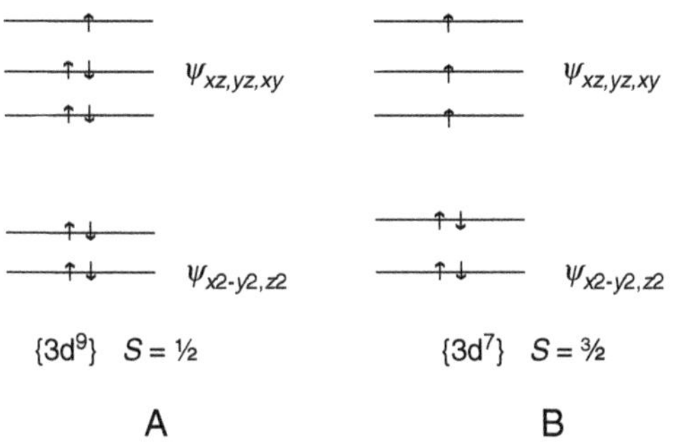

Scheme 10. Proposed electronic structures for DNIC in crystalline state with tetrahedral ligand field. In this configuration, Hund's rule makes that the $\{3d^7\}$ complex is high spin $S = 3/2$. For $\{3d^9\}$ complex $S = 1/2$.

This tetrahedral structure for the solid differs from the square-planar geometry in solution.

In principle, EPR spectroscopy should be able to distinguish between the spin state of DNIC in the molecular crystal and in solution. However, no EPR spectra seem to have been published for the crystalline state [102,105–107,109]. It could be suggested that the paramagnetism of the complexes is unobservable in the solid due to antiferromagnetic coupling or spin–spin interactions. Recently, our groups tried to perform the comparative investigation by using DNIC with 3-mercaptotriazole (1H—1,2,4-triazole-3-thiol, MT) This MT-DNIC complex can be observed with EPR both in crystalline state and in the solution. Good solvents are methanol or DMSO [40]. The complex $[Fe(SC_2H_3N_3)_2(NO)_2]1/2H_2O$ was synthesized by the reaction of 3-mercaptotriazole with $[Na_2Fe_2(S_2O_3)_2(NO)_4]2H_2O$ and $Na_2S_2O_3.5H_2O$ with molar ratio of 10:1:2 in water-alkaline solution, with subsequent re-crystallization of the complex from CH_3OH [41]. The X-ray analysis confirmed tetrahedral structure of MT-DNIC as presented in Fig. 22.

At ambient temperature, the EPR spectrum of solid MT-DNIC had the form of a single Lorentzian at $g = 2.031$ and half-width 1.7 mT (Fig. 23, curve a). The sample was polycrystalline since rotation of the sample in the magnetic field did not affect the intensity or the lineshape of the spectrum, which appears as an isotropic powder spectrum. Upon dissolving the crystals in methanol or DMSO, the spectrum narrowed to a singlet line with the center

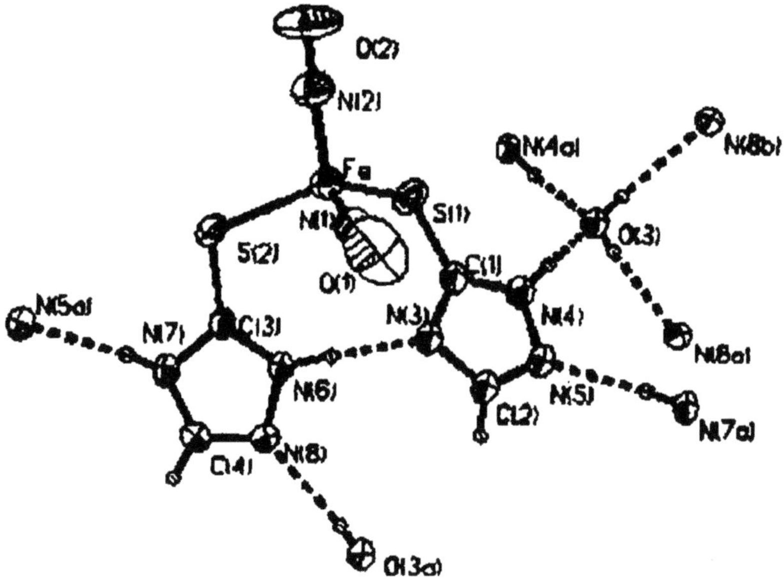

Fig. 22. A fragment of the crystal structure of DNIC with 3-mercaptotriazole (DNIC-MT) with atom indexes (from X-ray analysis [41]). (With permission.)

at $g = 2.030$ and half-width 0.7 mT (Fig. 23, curves b,c). Double integration confirmed that the number of spins in the solid and dissolved polycrystallites were equal within experimental accuracy (ca 5%). Recordings were made at modulation amplitude of 0.5 mT.

At ambient temperature in methanol, the EPR spectrum was characterized by an isotropic weakly resolved quintet HFS when recordings were made at modulation amplitude of 0.05 mT (Fig. 24, curve a). HFS resolution increased drastically when DMSO was used as a solvent (Fig. 24, curve b). At 77 K, the solution of MT-DNIC in methanol showed a $S = 1/2$ complex with axial Zeeman coupling characterized by an anisotropic EPR signal of the shape similar to that of EPR signals from frozen solutions of DNIC with various thiol-containing ligands. The difference was in lower amplitude of the peak at $g_{||}$ for MT-DNIC (Fig. 24, curve c).

It should be remarked that EPR spectra from MT-DNIC in DMSO (Fig. 24, curves b,d) were indistinguishable from EPR spectra if MT-DNIC was synthesized by exposing a mixture of Fe^{2+} and 3-mercaptotriazole (1:20) in DMSO to gaseous NO.

The spectra depend sensitively on temperature. At 77 K, the EPR signal of a polycrystalline MT-DNIC appears as a superposition of two signals. The first has an axially symmetric narrow shape with half-width 1.7 mT. The second appears as a broader powder spectrum with a peak at $g = 2.045$ (Fig. 25, curve b).

It was suggested that the latter could be attributed to a small fraction of complexes being dissolved in pockets of residual solvent (methanol) used for synthesis. When the signal characteristic of this fraction (MT-DNIC dissolved in methanol) (Fig. 24, curve c or Fig. 25, curve a) was subtracted from the signal presented in Fig. 25, curve b the isotropic EPR signal was obtained (Fig. 25, curve c).

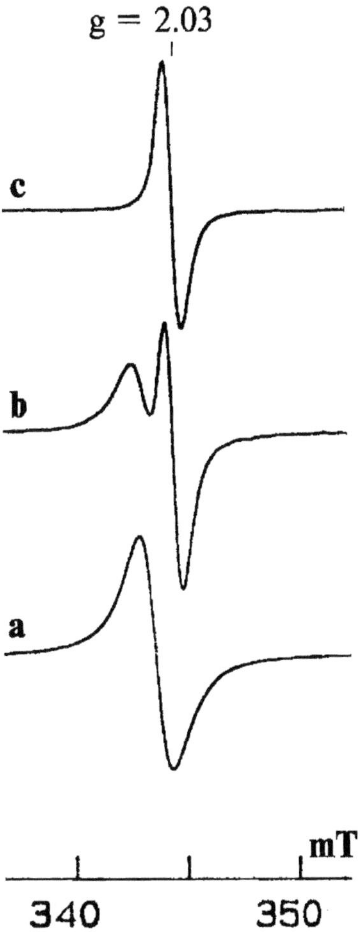

Fig. 23. The EPR signals from crystalline DNIC-MT (a) with subsequent dissolving with methanol: partial (b) and complete (c) dissolving. Recordings at ambient temperature. (From Ref. [40].)

The parameters of the latter (position in magnetic field and width) coincided with that of crystalline MT-DNIC recorded at ambient temperature (Fig. 24, curve a). The contribution of this signal to total EPR spectra (Fig. 25, curve b) estimated by double integral method exceeded 90%.

This result demonstrates that polycrystalline solids of paramagnetic MT-DNIC have a narrow symmetric singlet EPR spectrum and that this spectrum is not sensitive to changes in the temperature. It indicates the presence of strong magnetic interaction between the complexes in the solid. It is evident that tight packing of DNICs in the crystal should favor electronic coupling and exchange interactions between the adjacent centers. The latter could average the anisotropy of g-factor to g_{iso} resulting in the observation of isotropic EPR signal at $g = 2.031$ either at ambient or low temperature.

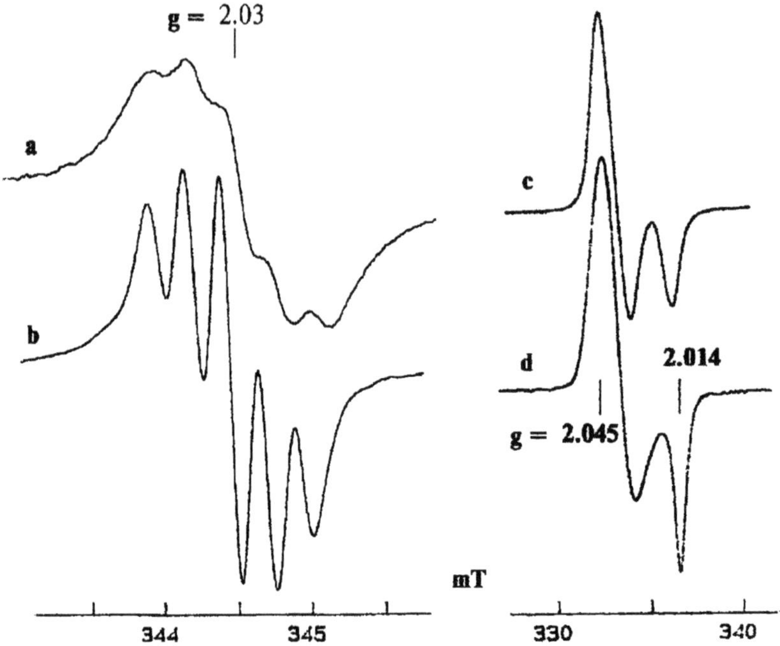

Fig. 24. The EPR signals from crystalline DNIC-MT complex dissolved in methanol (curves a,c) or dimethylsulfoxide (b,d). Recordings at ambient temperature (curves a,b) or 77 K (curves c,d). (From Ref. [40].)

In solid crystals, the EPR spectrum does not resolve the anisotropy of the Zeeman g-factor. Therefore, we cannot make a comparison with the g-values as found in solution. This does not allow us to spot any differences in the electronic configuration and ligand fields of MT-DNIC in crystalline or in dissolved states. The only subtle difference appears in the averaged g-factor being 2.031 and 2.035 for crystalline and dissolved states, respectively.

In principle, a d^9 configuration of the iron atom is possible in DNIC with thiol-containing ligands, which can arise due to transfer of two electrons on the iron from two thiol ligands (metal spin crossover effect [112]). This results in $\{(RS^{\bullet})_2Fe^{-1}(NO^+)_2\}^+$ structure ($d^7 \rightarrow d^9$ transition), i.e. the complex becomes a three-spin system, involving two thiyl radicals and iron on d^9 configuration. The total spin of this system in the ground state is $S = 1/2$. Such electronic configuration can be realized in DNICs with a tetrahedral structure (Scheme 10A), which is typical for these complexes in the crystalline state. However, such a structure is hardly typical for DNICs with thiol-containing ligands in the solution. The point is that in tetrahedral DNICs with d^9 electronic configuration, the unpaired electron should be localized on one of d_{xy}, d_{xz} or d_{zy} orbitals (Scheme 10A), while in DNICs with thiol-containing ligands it is localized on d_{z2} iron orbitals (see above). This difference suggests that the electronic structure of DNIC changes significantly upon transition from the crystalline to dissolved state. Diffraction experiments have clearly shown that the complex has tetrahedral structure in the crystalline state. In contrast, the EPR experiments favor a square-plane structure for

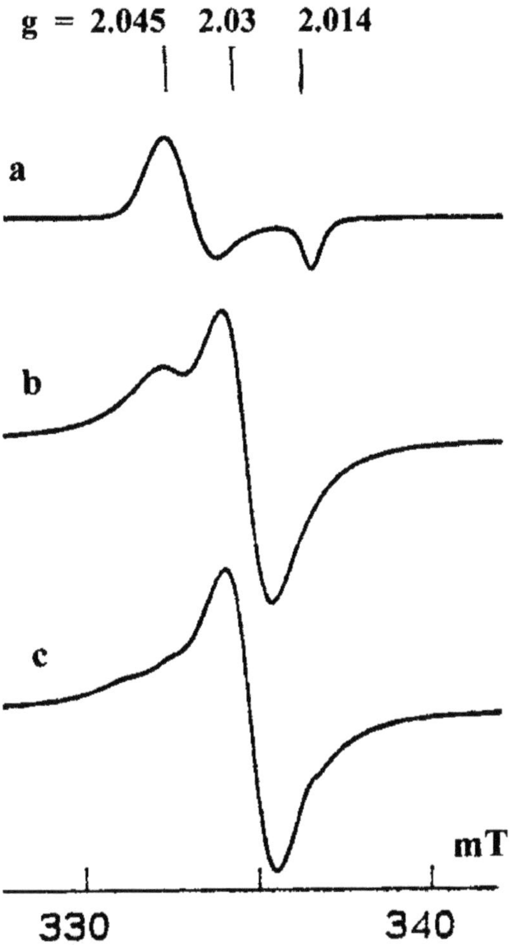

Fig. 25. The EPR signals from DNIC-MT dissolved in methanol (a) and crystalline DNIC-MT (b) recorded at 77 K. (c) The shape of the EPR signal obtained by subtracting of signal (a) from signal (b). (From Ref. [40].)

the dissolved complex. The change to square-plane structure is reflected in the backtransfer of the unpaired electrons from the iron towards the thiol sulfurs. The sulfurs change their character from thiyl to thiolate, and iron reverts to a d^7 configuration and the complex is described by $\{(RS^-)_2Fe^+(NO^+)_2\}^+$ structure. Interaction of free thiols with NO^+ can result in inclusion of these thiols into DNIC, thus changing the total charge of the complex to negative value, according to the formula $\{(RS^-)_2Fe^+(NO^+\ldots RS^-)_2\}^-$ [46,88].

It is hardly possible to keep tetrahedral structure for dissolved DNIC with thiol-containing ligands with d^7 electron configuration. In this case, the complexes should be characterized with high spin state ($S=3/2$, Scheme 10B). Such state is characteristic of various MNICs. These complexes give the EPR signal with three g-factor values (4.0, 3.95

and 2) [113–119]. Such type of signal has not been observed for DNIC with thiol-containing ligands [113].

OBSERVATIONS OF DNICs IN CELLS AND TISSUES

As aforementioned, the elucidation of the nature of paramagnetic centers giving 2.03 signal in cells and tissues (respectively called 2.03 centers) was unraveled when it was found that the EPR lineshape of the biological samples resembles spectra from DNIC with low-molecular-weight thiols in frozen solutions [6,7]. Although the coincidence was noted, the formation of dinitrosyl complexes in biological systems seems at first to be improbable because the biological role of NO was not known at the time [3]. The formation of such DNICs was subsequently proven by the appearance of the 2.03 signal in baker yeast cells that were cultured in a growth medium containing $CaNO_3$ (Reader's medium) [9,12]. Independently, the group of Commoner in USA confirmed that the elimination of nitrate from Reader's medium prevented the appearance of the $g = 2.03$ signal in anaerobically cultivated yeast cells [10]. Evidently, nitric oxide ensuring DNIC formation in yeast cells appeared in these cells due to the accumulation of nitrite. The latter was generated from nitrate which was reduced to nitrite by respective nitrate reductase. Further work demonstrated the positive correlation between the amount of the 2.03 centers in liver tissue from rats fed on a diet containing different carcinogens and the amount of nitrate anions in the ingested drinking water [10,12]. Moreover, the formation of these centers was observed in various organs of mice which were kept on a diet with a high content of KNO_3 without any addition of carcinogens (Fig. 26) [12].

Since nitrate reductase has not been found in mammals the reduction of nitrates to nitrites seemed to be caused by intestinal denitrifying bacterial microflora. Subsequent experiments on the addition of nitrite to animal diet demonstrated the formation of the 2.03 centers in animal tissues in higher amounts than that on the addition of nitrate [11,120–126] (Fig. 26). Similar result was obtained when isolated animal tissues were treated with gaseous nitric oxide [8]. These results point unequivocally to the fact that 2.03 centers are the DNICs with thiol-containing ligands.

The remaining constituents of the DNICs from yeast cells, thiols and iron, were identified by the effect of mercuric salts and chelators for ferrous iron, respectively. Exposure to competing alternative ligands caused their transformation into other types of iron-nitrosyl complexes [3]. Interesting effects were observed upon addition of xanthogenate derivatives, i.e. bidentate ligands with two binding sulfurs similar to those found in dithiocarbamate ligands. Addition of xanthogenates caused the formation of paramagnetic mononitrosyl complexes (MNICs) at the expense of DNIC. At 77 K, the EPR spectrum showed resolved triplet HFS (Fig. 27). The transformation to MNIC occurred for DNICs in biological samples, but similarly for DNIC formed *in vitro* from thiols with low molecular weight [3].

Large quantities of DNIC were formed in tissues of animals after the addition of sodium nitrite (0.3%) to drinking water of mongrel mice or rats [11,120–126]. Particularly high yields were found in liver. If the animals were maintained on the nitrite diet, the 2.03 signal in the mouse liver became observable after ca 3 days. After 7 days, the yield reached the steady state maximum of ca 8–10 μM/kg liver tissue. The intensity and narrow linewidth made

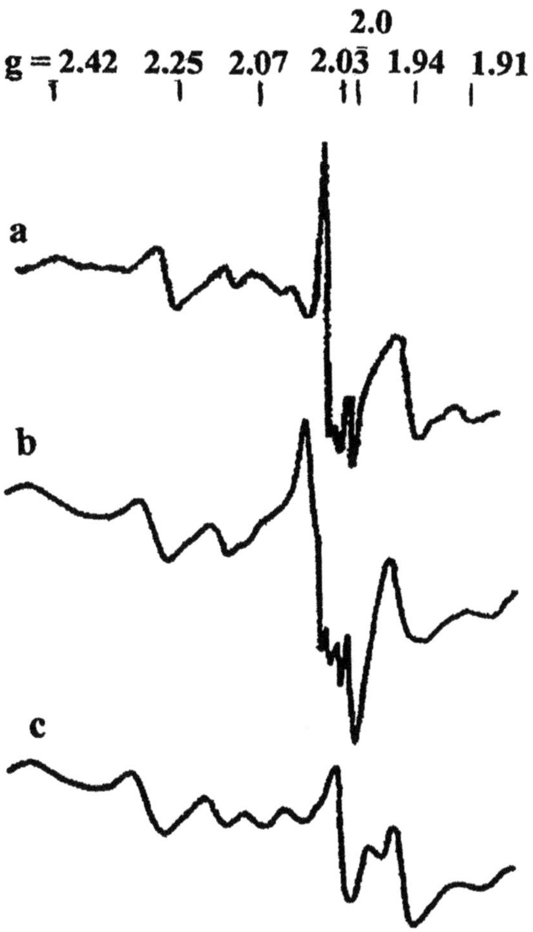

Fig. 26. EPR spectra of livers of mice maintained on a diet with nitrite [121] or nitrate (curve b) [12]. (curve c) Liver from control mice [121]. Recordings at 77 K. The EPR signals with $g = 2.42$, 2.25 and 1.91, $g = 2.07$, 2.03 or 1.94 are due to cytochrome P450, Hb–NO complexes, protein-bound DNICs or reduced iron–sulfur centers, respectively.

the signal to dominate the EPR spectra (Fig. 26). The frozen tissues showed additional EPR absorption from Hb–NO and iron–sulfur centers in the respiratory chain (cf Chapters 5 and 15 for more detail, respectively).

The yield of DNIC could be influenced by the administration of iron complexes to the nitrite-rich diet. The yields in liver increased by a factor 3–4 to ca 30 μM/kg after adding iron citrate (0.02%) to the nitrited drinking water [121–126]. Such high yields were observed only upon feeding with the combination of nitrite and iron. If the iron supplement was removed from the nitrite-containing drinking water after 20 days, the intensity of the 2.03 signal in the liver of these animals began to fall slowly on a timescale of a week. After 10 days, the amount of the DNICs had fallen to 15 μM/kg of wet tissue. The complexes

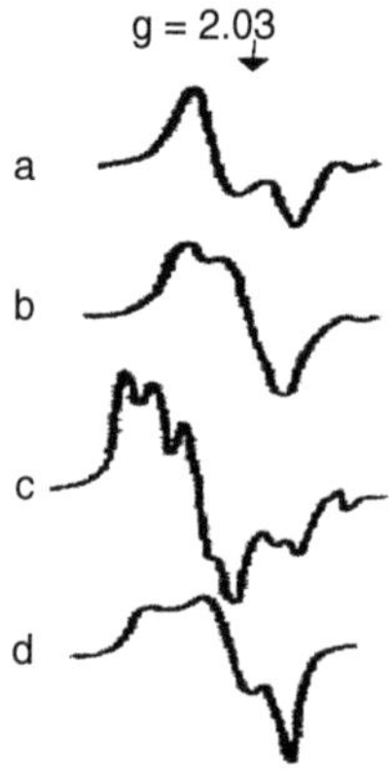

Fig. 27. Change of the shape of 2.03 signal from yeast cells (curve a) induced by treatment of yeast preparations with phenahtroline (curve b), ethylxanthogenate (curve c) or $HgCl_2$ (curve d). Recordings at 77 K [9].

were also observed in the kidney, small intestines and blood although the yield was generally 5–10 times less than in the liver. In the heart, lungs or spleen, the DNICs remained below the detection limit of ca 0.1 μM/kg due to masking of the 2.03 signal by other EPR signals.

The isotopic substitution of ^{56}Fe ($I = 0$) by ^{57}Fe ($I = 1/2$) in the drinking water caused line broadening in the DNIC spectra of tissues [122,127]. The experiments proved that a significant fraction of DNIC contained iron from the dietary intake. After prolonged dietary intake, the signal was compatible with complete replacement of ^{56}Fe by ^{57}Fe in the complexes (Fig. 28). The yields in liver tissue reached the same maximal amount of 30 μM/kg of wet liver tissue [122] observed before with unlabeled iron. Interestingly, subsequent exposure of the liver tissue to gaseous NO further increased the DNIC yield threefold [122]. At the same time, the linewidth decreased significantly and showed unequivocally that endogenous stores of ^{56}Fe iron contribute to the formation of the DNIC complexes (Fig. 28).

The predominant incorporation of exogenous iron (^{57}Fe) into DNICs allows to suggest that the Fe-NO groups forming part of the complexes appear in the blood or intercellular fluid. After being incorporated into the organ tissues, these groups link up with the thiol-containing proteins which also leads to the formation of 2.03 complexes [121–124]. We were able to prove feasibility of this formation by intraperitoneal injections of low-molecular DNIC-thiol. Such injections indeed significantly increased the yield of DNIC in all tissues [124–127].

Fig. 29 shows the evolution in time of 2.03 signal in frozen liver, kidney and blood from rats which were injected intraperitoneally with DNIC-cysteine [125]. The lineshape of the 2.03 signal from liver, kidney or blood as recorded 6 h after the injection of Cys-DNIC with ^{56}Fe or ^{57}Fe resembled that of the DNICs from the same organs extracted from rats maintained on drinking water with nitrite and iron (Fig. 30) [125]. This coincidence was also found in different species like rats, cats, guinea pigs, rabbits, mice and hamsters [125]. The lineshape remained unchanged upon thawing the samples to room temperature and proved that the DNICs were bound to protein.

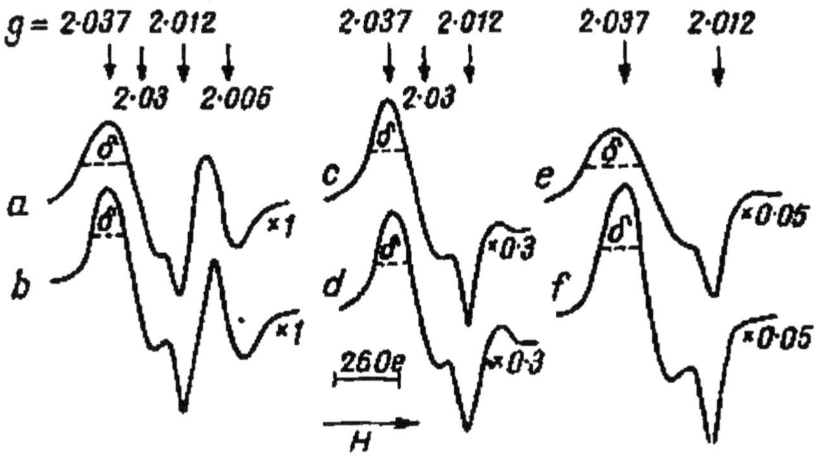

Fig. 28. 2.03 signals in livers of mice maintained on drinking water with nitrite + ^{57}Fe or ^{56}Fe during 6 days (spectra a or b, respectively) followed with gaseous NO treatment (spectra c or d, respectively). Spectra (e,f) show 2.03 signals in liver homogenate preparations from control mice (without nitrate) added respectively with ^{57}Fe or ^{56}Fe followed with gaseous NO treatment. Recordings were made at ambient temperature. On right, amplifications in relative units. δ is the width of the signal component at half amplitude ($\delta = 2.7$ or 17 mT for DNIC-cysteine EPR signal with ^{57}Fe or ^{56}Fe, respectively) [122,124].

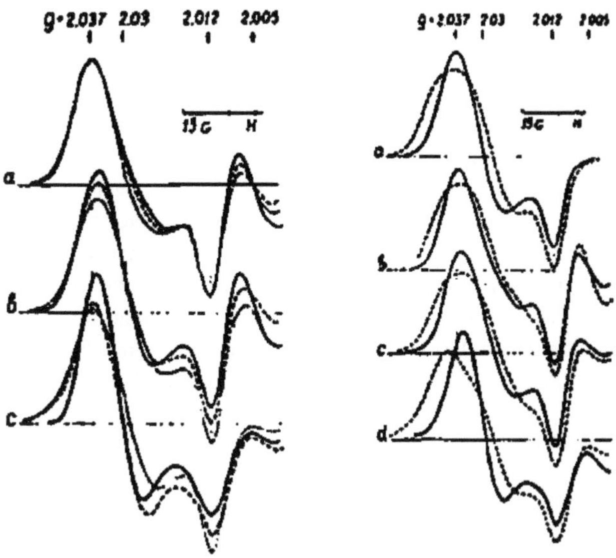

Fig. 29. Left panel: Time variation of 2.03 signals in liver (curve a), kidney (curve b) and blood (curve c) of rats injected intraperitoneally with Cys-DNIC. Recordings were conducted at 1 (– – –), 3 (------) or 6 (——) h after injection. Right panel: 2.03 signals from Cys-DNIC in frozen solution (curve a), from liver (curve b); from kidney (curve c) and from blood (curve d). The tissues were taken 6 h after the injection of Cys-DNIC. The dashed line corresponds to Cys-DNIC with ^{57}Fe, solid line to Cys-DNIC with ^{56}Fe. Spectra were recorded at 77 K. (From Ref. [125].)

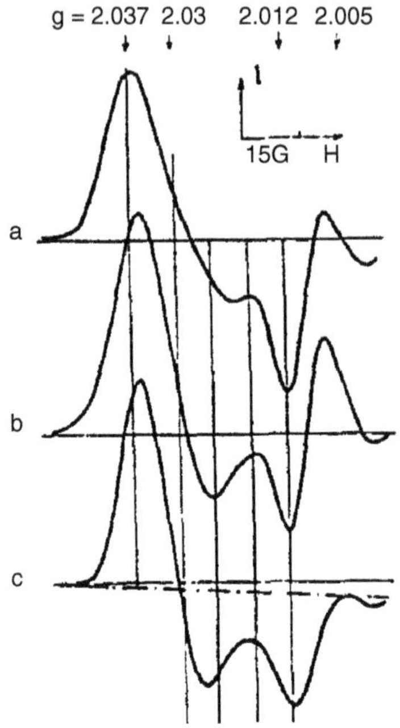

Fig. 30. 2.03 signals from liver (curve a), kidney (curve b) and blood (curve c) of rats, maintained on drinking water with nitrite and iron salts during 7 days. Recordings at 77 K [125].

At higher microwave frequencies, the anisotropy of the Zeeman interaction is better resolved. At Q-band frequencies (~30 GHz), the lineshape of the $g = 2.03$ signal in mouse liver showed a g-factor with rhombic symmetry (Fig. 31) [128].

This rhombicity is incompatible with the axial symmetry of DNIC with cysteine. Heating of the preparation to 60°C for 5 min changed the symmetry from rhombic to axial as found in DNIC with cysteine in frozen solution. The same transformation took place in samples stored at ambient temperature for 3–4 h. At X-band frequencies, the rhombicity of the Zeeman interaction is hardly noticeable. It only appears in the form of a small distortion near the central part of the 2.03 signal. Similar distortions appear in liver tissues from cat, guinea pig, rat or mouse (Fig. 32) and are attributed to rhombic distortion of the g-tensor [124,125].

Figs. 29 and 30 show that blood samples contain DNIC with a slightly different EPR lineshape: In blood, the component at $g_{||}$ has lower intensity than the common 2.03 signal [125,126]. This form is characteristic of DNIC moieties anchored to Hb that was shown before for isolated DNIC–Hb (Fig. 19). This fact is in line with the proposition about the capability of low-molecular DNICs, at least their $Fe(NO)_2$ groups penetrate through cell membranes followed by binding with thiol group of Hb or other proteins. The separation of rat blood indicated that the majority of the DNICs in blood was located in the RBC fraction rather than the plasma [125,126].

As mentioned before, Cys-DNIC shows axial symmetry in frozen aqueous solutions. Spectral distortions are observed in polar organic solvents like DMF, DMSO, tetramethylurea or

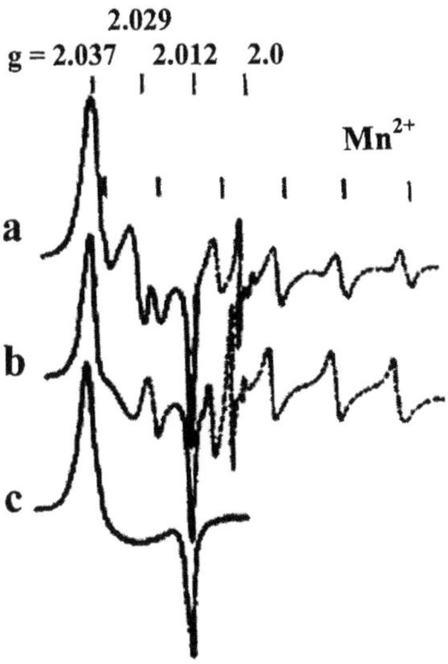

Fig. 31. Q-band spectra of DNIC at 77 K in livers of mice kept on drinking water with iron salt and nitrite (curve a). (curve b) is the same liver preparation heated at 60°C for 5 min. (curve c) is the reference spectrum of a frozen solution of Cys-DNIC. The DNIC signal in livers (spectra a,b) appears contaminated with a six-line spectrum from an unidentified Mn^{2+} complex. (From Ref. [128].)

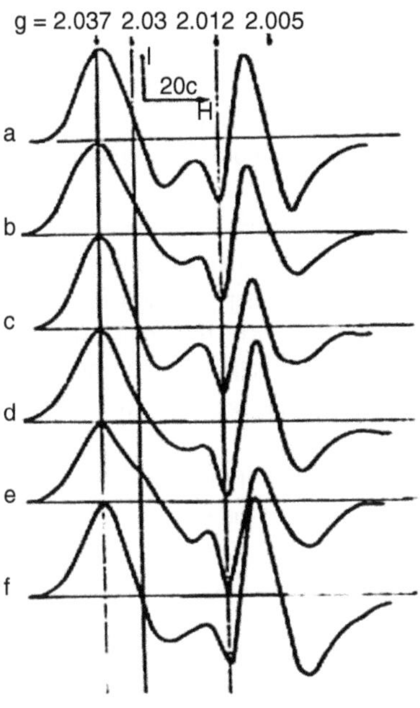

Fig. 32. 2.03 signals from livers of cat (curve a), guinea pig (curve b), rabbit (curve c), rat (curve d), mouse (curve e) and hamster (curve f). The animals were provided with nitrite + ^{56}Fe in drinking water during 7 days. Recordings at 77 K [125].

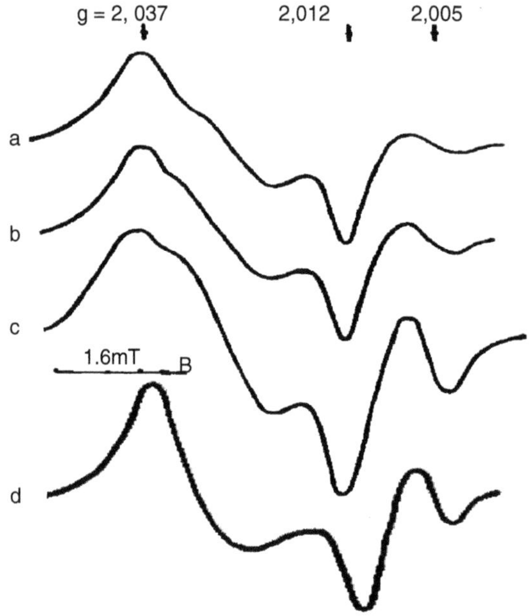

Fig. 33. EPR spectra of 2.03 complexes formed *in vivo* in mouse liver (curve a), in this tissue after homogenation (curve b), in supernatant fraction of liver homogenate (curve c) or in liver homogenate after treatment with CN^- (curve d). Recordings at 77 K [129].

hexamethanol (Fig. 7) [42]. This makes it plausible that matrices like phospholipids, fatty acids, etc. (compounds X) can also reduce the symmetry of protein-bound DNIC formed in tissues. DNIC with rhombic *g*-factors are observed in whole mouse liver tissue, liver homogenates and in supernatant fraction of the latter (Fig. 33) [129].

This points to a low-molecular nature of the compound's (compounds X) influence on the DNIC in the liver tissue. Interestingly, the dialysis of mouse blood plasma containing DNIC-BSA against mouse liver homogenates led to a sharp change in the shape of the EPR signal from plasma: it was similar to that from liver (Fig. 34) [130]. Evidently, this transformation was due to the incorporation of low-molecular compounds X into DNIC-BSA. Consequent addition of DDS to the preparation increased the symmetry of the complex to axial one (Fig. 34) [130].

It cannot be excluded that species specificity of the EPR lineshape of the DNIC from liver of mouse, rat, guinea pig or cat mentioned before (Fig. 32) is due to the specificity of these intracellular low-molecular compounds affecting the DNIC structure. The transformation to axial symmetry of DNIC from mouse liver was also achieved by the addition of surface-active compound, Na-DDS to liver homogenates [130].

In the 1970s, it was discovered that nitrite anions induce the formation of DNIC in animal tissues and microorganisms [10–12,120–122]. The appearance of DNIC was attributed to acidification of nitrite with subsequent release of free NO molecules from nitrous acid [120–122]. However, recently it was found that various heme-iron proteins and xanthine oxidase have the capacity to reduce nitrite to NO (see Chapters 14–16 of this book).

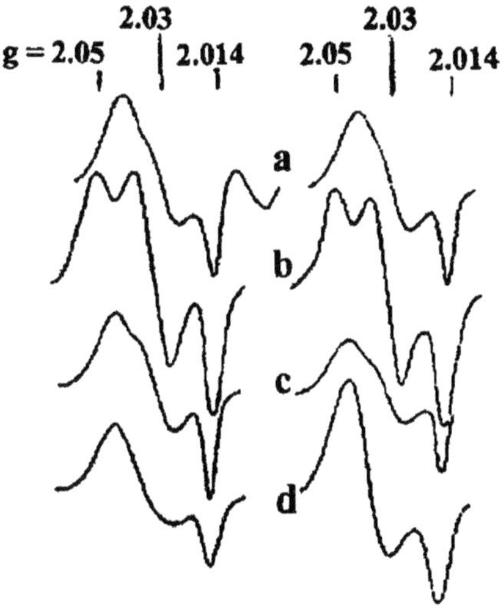

Fig. 34. Shape of the EPR signals of DNICs formed *in vivo* in mouse liver (curve a) or mouse blood plasma (curve b) after adding DNIC with phosphate to mice or plasma, respectively. (curve c) EPR signals from blood plasma after dialysis against the suspension of mouse liver homogenate preparation. (curve d) The preparation (c) was treated with dodecyl sulfate (SDS). Recordings at 77 K (left side) or ambient temperature (right side) [130].

Although the mechanisms remain controversial, hypoxia or anoxia seems a necessary condition for the reduction of nitrite at physiological pH. The new data show that nitrite should also be considered as a source of NO *in vivo*.

It is important to distinguish between acute and long-term effects of nitrite. Large quantities of DNIC complexes appear in animal tissues only if the animals are maintained on nitrite diet for days. Bolus addition of nitrite to the animals leads mainly to the formation of NO–Hb complexes with the tissue yields of DNIC remaining far lower than in animals on a long-term nitrite-rich diet [124]. The reason for this difference remains obscure. Without doubt, the elucidation of this problem will shed light on the mechanism of DNIC formation from nitrite.

The preceding discussion may have left the reader under the impression that nitrite be the dominant agent for the formation of DNIC complexes in tissues. However, it should not be forgotten that significant quantities of DNIC may also form from free NO radicals released by the enzymatic L-arginine dependent pathway. This pathway was confirmed by numerous investigators in a wide range of experiments on cultured animal and human cells [16–35]. However, the formation of the complexes in animal tissues was managed to be observed only several days after bacterial invasion artificially enhanced the NO synthesis by the inducible form of NO synthase [31]. In normal conditions or in acute experiment with activation of existing constitutive NO synthases or enhanced synthesis by inducible NO synthase (iNOS), the levels of DNIC remained below the detection limit.

This raises the question why the endogenous DNIC remains so low *in vivo*. One reason may be the continuous destruction of the DNIC by endogenous peroxynitrite and superoxide

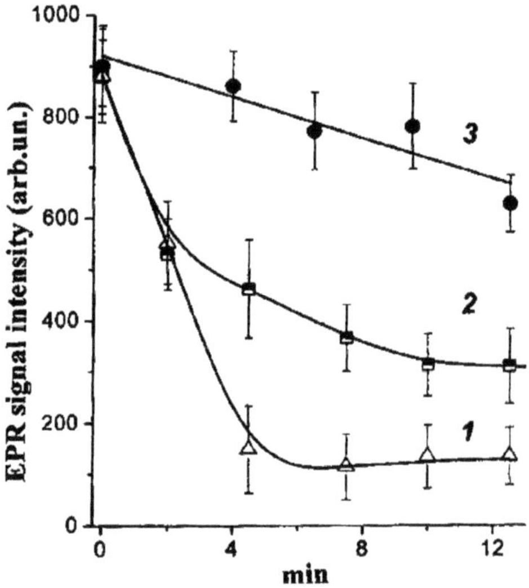

Fig. 35. Decomposition of dimeric GSH-DNIC by superoxide generated by xanthine oxidase in 0.1 mM phosphate buffer, pH 7.4. (curve 1) Kinetics of the EPR signal decrease in the solution of 0.25 mM GSH-DNIC + 0.2 U/ml xanthine oxidase + 1 mM xanthine; (curve 2) 600 U/ml catalase was added to the solution 1; and (curve 3) 600 U/ml catalase + 150 U/ml superoxide dismutase were added to the solution. Dimeric GSH-DNIC is diamagnetic and cannot be detected with EPR directly. For detection, 1.6 mM cysteine was added to transform EPR-silent dimeric GSH-DNIC into monomeric EPR-detectable Cys-DNIC monomers. (From Ref. [98].)

produced in the tissues. The destruction of DNIC by peroxynitrite was shown by our group [131]. The destruction by superoxide was demonstrated only recently [98]. Fig. 35 presents the kinetics of the decomposition of GSH-DNIC by superoxide as generated enzymatically by xanthine oxidase. With 0.2 U/ml xanthine oxidase and 1 mM xanthine, ca 80% of DNIC was lost after 4 min. The decomposition could be prevented completely by the addition of catalase (600 U/ml) + superoxide dismutase (SOD) (150 U/ml). Catalase alone attenuated the DNIC degradation for 50%. These experiments prove that superoxide and hydrogen peroxide can destroy DNIC complexes. Together with peroxynitrite, these compounds keep the basal levels of endogenous DNIC below the detection limit for EPR.

These results were corroborated by *in vitro* experiments on protein-bound DNIC with KO_2 as superoxide donor (Vanin, unpublished data). Solutions of DNIC-BSA were generated from BSA by adding low-molecular DNIC (1:20) with cysteine, phosphate or thiosulfate (forming DNIC-BSA-1, DNIC-BSA-2 or DNIC-BSA-3, respectively). The addition of equimolar KO_2 in the presence of catalase resulted in the instantaneous loss of EPR signal intensity from the solutions of DNIC-BSA-1, DNIC-BSA-2 or DNIC-BSA-3. The signal intensities were reduced by a factor 3, 2 or 1.3, respectively. In the presence of SOD, no loss of DNIC was observed.

It is important to note that the stability of DNIC-BSA depends on the type of thiol ligand. We noted that incorporation of thiosulfate anions into the DNIC greatly enhanced the stability of DNIC-BSA against destruction by superoxide or hydrogen peroxide *in vitro*. This observation was reminiscent of the high stability of low-molecular DNIC with thiosulfate ligands previously observed *in vivo*: thiosulfate-DNICs have much longer lifetime in mice than low-molecular DNICs with ligands like cysteine or glutathione [99]. The mice had been injected with bacterial lipopolysaccharide (LPS) to stimulate the synthesis of iNOS, and subsequently received a dose of preformed Cys-DNIC or GS-DNIC. The injection resulted in the formation of some protein-bound DNICs in various tissues of the animals. These animals have strong generation of NO by iNOS due to the LPS. Surprisingly, the yields of protein-bound DNICs in liver proved lower (!) than yields in controls where low-molecular DNICs were injected without LPS. This situation was reversed when thiosulfate ligands were used. With the injection of thiosulfate-DNIC, the yields of protein-bound DNIC in the LPS-treated mice were threefold higher than in controls. Clearly, the thiosulfate ligands afford far better protection against endogenous superoxide or hydrogen peroxide in inflammated animals than the usual thiols like glutathione.

Subsequent experiments [99] exploited the high stability of thiosulfate-DNIC to give the first direct proof that endogenous DNICs may form from NO released by NOS. As before, endogenous NO production was stimulated by injecting the mice with LPS. The boost of endogenous NO was confirmed by NO trapping with iron-diethyldithiocarbamate (Fe-DETC) complexes (Fig. 36, trace 1).

The endogenous NO production could be eliminated by the administration of the NOS inhibitor N^{ω}-nitro-L-arginine (NNLA). This inhibition led to a decrease in the concentration of these complexes to nearly zero (Fig. 36, curve 2). Formation of DNIC was not observed in these animals (Fig. 36, curve 3). Significant quantities of DNIC could be detected in liver

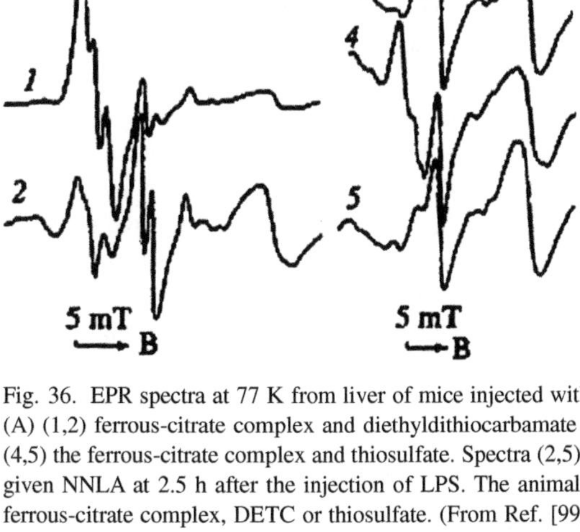

Fig. 36. EPR spectra at 77 K from liver of mice injected with bacterial lipopolysaccharide (LPS) 4 h earlier. (A) (1,2) ferrous-citrate complex and diethyldithiocarbamate (DETC); (B) (3) the ferrous-citrate complex or (4,5) the ferrous-citrate complex and thiosulfate. Spectra (2,5) were recorded in livers of mice which had been given NNLA at 2.5 h after the injection of LPS. The animals were euthanized in 15 min after injecting the ferrous-citrate complex, DETC or thiosulfate. (From Ref. [99].)

after the injection of thiosulfate ligands together with ferrous citrate (Fig. 36, curve 4). The concentration of DNIC in the tissue changed with time. The highest yields were ca 6 μM DNIC/kg of wet tissue, and were observed 10–15 min after the injection of thiosulfate. At later times, the yields diminished. We attribute the decrease of DNIC to either oxidation of the thiosulfate ligands, or the removal of the ligands from the tissue. The inhibition by LNNA showed unequivocally that the DNIC complexes were endogenously formed from endogenous NO produced enzymatically by NOS from L-arginine. We propose that a certain quantity of DNIC is formed endogenously in any organism capable of generating NO. The basal concentration of DNIC in tissues seems quite low, and remains below the detection limit of EPR (ca 10 nM tissue concentration) as a result of destruction by superoxide radicals and peroxynitrite. However, the DNIC levels may be raised above the detection threshold by supplying thiosulfate to obtain very stable DNIC complexes, or by artificially boosting the endogenous NO production, for example by the injection of LPS.

LOOSELY BOUND IRON PARTICIPATES IN THE FORMATION OF DNIC

After the discovery of endogenous DNICs in yeast cells and animal tissues, it was initially thought that the iron–sulfur clusters be the main source of the iron and that the DNIC be formed after the disruption of the clusters by free NO radicals [3]. *In vitro* studies certainly confirm that exposure of iron–sulfur clusters to free NO leads to the formation of DNIC (cf Chapters 5 and 6). Whether this disruption is the dominant pathway for DNIC formation *in vivo* is a different matter. Subsequent studies found that loosely bound iron is the main source of endogenous DNIC in animal tissues and yeast. This loosely bound iron is often referred to as "free iron" or the "labile iron pool." *In vitro* studies of the effect of gaseous NO on tissue homogenates and supernatant fractions have given further support that DNIC is formed from such loosely bound iron [132]. Independently, the same conclusion was reached by Nagata et al. [133] who observed the formation DNIC in rat liver supernatant after adding the NO donor *N*-methyl-*N*′-nitrosoguanidine (MNNG). In both cases, the formation of DNIC was inhibited if the homogenates or supernatant fractions were dialyzed against a solution with iron chelators prior to the addition of MNNG. Dialysis against Tris buffer also inhibited the formation of DNIC. Subsequent addition of some ferrous iron restored the capacity to generate DNIC. The DNIC complexes were predominantly localized in the supernatant fraction obtained by centrifugation of the liver homogenate at 100,000 *g* during 1 h. Only small amount of DNICs was detected in mitochondrial fraction. Fractionation of rat liver supernatant fraction by gel-filtration showed that DNICs were attached to protein fractions with molecular weights between 30 and 80 kD [132]. The main localization of the DNIC in supernatant fraction was found by Hutchison et al. [134] for endogenous DNIC in rabbit liver.

Many early experimental results in the literature document that exposure to NO induces the formation of DNIC in tissues with a concomitant loss of active centers in ISPs. Similar results were obtained for isolated ISP. It was widely assumed that NO induces the disintegration of the iron–sulfur clusters of the proteins immediately and that the DNIC be formed from iron released from the clusters. However, as discussed in Chapters 5 and 6, the interpretation of these data is far from clear and needs more detailed consideration. However, these centers do provide an accessible pool of loosely bound endogenous iron which may participate in the formation of DNIC. As discussed earlier in this chapter, isotopic labeling has confirmed such participation *in vitro.*

CONCLUDING REMARKS

The consideration of physico-chemical features of DNIC with thiol-containing ligands demonstrates that they can function in cells and tissues as a donor of neutral and ionized NO molecules. Besides that they can deliver $Fe(NO)_2$ groups which can influence the activity of enzymes and proteins by binding with their functionally active thiol groups. DNICs are stable enough to ensure stabilization and transport of NO and NO^+ thereby ensuring both their autocrynic and paracrynic functions. High physiological and biochemical activities of these complexes are considered below in Chapters 3–5, 7, 8, 12 and 16 as well as the formation of DNICs from endogenous NO molecules allows to propose that the complexes can be considered as a significant signaling agents in mammalian cells and tissues.

ACKNOWLEDGMENT

The work was supported by the Russian Foundation of Basic Researches (Grant 05-04-49383).

REFERENCES

1 Nalbandyan RM, Vanin AF, Blumenfeld LA. EPR signals of a new type in yeast cells.Abstracts of the Meeting "Free radical processes in biological systems," Moscow, 1964; p. 18.
2 Vanin AF, Nalbandyan RM. Free radicals of a new type in yeast cells. Biofizika (Rus.) 1965; 10: 167–168.
3 Vanin AF, Blumenfeld LA, Chetverikov AG. Investigation of non-heme iron complexes in cells and tissues by the EPR method. Biofizika (Rus.) 1967; 12: 829–841.
4 Vithaythil AJ, Ternberg JL, Commoner B. Changes in electron spin resonance signals of rat liver during chemical carcinogenesis. Nature 1965; 207: 1246–1249.
5 Mallard JR, Kent M. Difference observed between electron spin resonance signals from surviving tumour tissues and from their corresponding normal tissues. Nature 1964; 204: 1192.
6 Vanin AF. Identification of divalent iron complexes with cysteine in biological systems by the EPR method. Biokhimia (Rus.) 1967; 32: 228–232.
7 Woolum JC, Tiezzi E, Commoner B. Electron spin resonance study of iron-nitric oxide complexes with amino acids, peptides and proteins. Biochim. Biophys. Acta 1968; 160: 311–320.
8 Vanin AF, Chetverikov AG. Paramagnetic nitrosyl complexes of heme and non-heme iron. Biofizika (Rus.) 1968; 13: 608–616.
9 Vanin AF, Blumenfeld LA, Burbaev DS, Lisovskaya IL, Chetverikov AG. Investigations of some iron complexes in biological objects. Proceedings of International Jubilee Conference on Paramagnetic Resonance, 1969 June 15–20; Kazan, 1970; Part 1: 218–223.
10 Woolum JC, Commoner B. Isolation and identification of a paramagnetic complex from the livers of carcinogen-treated rats. Biochim. Biophys. Acta 1970; 201: 131–140.
11 Commoner B, Woolum JC, Senturia BH, Ternberg JL. The effects of 2-acetoaminofluorene and nitrite on free radicals and carcinogenesis in rat liver. Cancer Res. 1970; 30: 2091–2097.
12 Vanin AF, Kubrina LN, Lisovskaya IL, Malenkova IV, Chetverikov AG. Endogenous heme and non-heme nitrosyl iron complexes in cells and tissues. Biofizika (Rus.) 1971; 16: 650–658.

13 Chetverikov AG, Ruuge EK, Burbaev DSh, Vanin AF. The change of the shape of the EPR signal with g_{av} 2.03 in biological objects depending on the conditions of the registration. Biofizika (Rus.) 1969; 14: 932–935.

14 Vanin AF, Kiladze SV, Kubrina LN. On including of low molecular SH containing compounds in nitrosyl non-heme iron complexes in non-cellular or cellular preparations. Biofizika (Rus.) 1975; 20: 1068–1072.

15 Vanin AF. Nitrosyl non-heme iron complexes in animal tissues and microorganisms. D.Sc. Thesis, Institute of Chemical Physics, Moscow, 1980.

16 Lancaster JR, Hibbs JB. EPR demonstration of iron-nitrosyl complex formation by cytotoxic activated macrophages. Proc. Natl. Acad. Sci. USA 1990; 87: 1223–1227.

17 Pellat C, Henry Y, Drapier J-C. IFN-activated macrophages: detection by electronic paramagnetic resonance of complexes between L-arginine-derived nitric oxide and non-heme iron proteins. Biochem. Biophys. Res. Comm. 1990; 166: 119–125.

18 Drapier J-C, Pellat C, Henry Y. Generation of EPR-detectable nitrosyl-iron complexes in tumor target cells cocultured with activated macrophages. J. Biol. Chem. 1991; 266: 10162–10167.

19 Vanin AF, Mordvintcev PI, Hauschildt S, Mülsch A. The relationship between L-arginine-dependent nitric oxide synthesis, nitrite release and dinitrosyl-iron complex formation by activated macrophages. Biochim. Biophys. Acta 1993; 1177: 37–42.

20 Lancaster JR, Werner-Felmayer G, Wachter H. Coinduction of nitric oxide synthesis and intracellular nonheme nitrosyl-iron complexes in murine cytokine-treated fibroblasts. Free Rad. Biol. Med. 1994; 16: 869–870.

21 Stadler J, Bergonia HA, DiSilvio M, Sweetland MA, Billiar TR, Simmons RL, Lancaster JR. Nonheme nitrosyl-iron complex formation in rat hepatocytes: detection by EPR spectroscopy. Arch. Biochem. Biophys. 1993; 302: 4–11.

22 Nüssler AK, Geller DA, Sweetland MA, DiSilvio M, Billiar TR, Madariaga JB, Simmons RL, Lancaster JR. Induction of nitric oxide synthesis and its reactions in cultured human and rat hepatocytes stimulated with cytokines plus LPS. Biochem. Biophys. Res. Commun. 1993; 194: 826–835.

23 Lepoivre M, Flaman J-L, Bobe P, Lemaire G, Henry Y. Quenching of the tyrosyl free radical of ribonucleotide reductase by nitric oxide. J. Biol. Chem. 1994; 269: 21891–21897.

24 Kim Y-M, Bergonia H, Lancaster JR. Nitrogen oxide-induced autoprotection in isolated rat hepatocytes. FEBS Lett. 1995; 374: 228–232.

25 Sergent O, Griffon B, Morel I, Chevanne M, Dubos M-P, Cillard P, Cillard J. Effect of nitric oxide on iron-mediated oxidative stress in primary hepatocyte culture. Hepatology 1997; 25: 122–127.

26 Geng Y-L, Petersson A-S, Wennmalm A. Hannson G. Cytokine-induced expression of nitric oxide synthase results in nitrosylation of heme and nonheme iron proteins in vascular smooth muscle cells. Exp. Cell. Res. 1994; 214: 418–424.

27 Corbett JA, Sweetland MA, Wang JL, Lancaster JR, McDaniel ML. Nitric oxide mediates cytokine-induced inhibition of insulin secretion by human islets of Langerhans. Proc. Natl. Acad. Sci. USA 1993; 90: 1731–1735.

28 Muller B, Kleschyov AL, Stoclet J-C. Evidenvce for N-acetylcysteine-sensitive nitric oxide storage as dinitrosyl-iron complexes in lipopolysaccharide-treated rat aorta. Br. J. Pharmacol. 1996; 119: 1281–1285.

29 Lepoivre M, Flaman J-M, Henry Y. Early loss the tyrosyl radical of ribonucleotide reductase of adenocarcinoma cells producing nitric oxide. J. Biol. Chem. 1992; 267: 22994–23000.

30 Watts RN, Hawkins C, Ponka P, Richardson DR. Nitrogen monoxide (NO)-mediated iron release from cells is linked to NO-induced glutathione efflux via multidrug resistance-assotiated protein 1. Proc. Natl. Acad. Sci. USA 2006; 103: 7670–7675.

31 Chamulitrat W, Jordan SUJ, Mason RP, Litton AL, Wilson JG, Wood ER, Wolberg G, Molina Y, Vedia L. Targets of nitric oxide in a mouse model of liver inflammation by *Corynebacterium parvum*. Arch. Biochem. Biophys. 1995: 316: 30–37.

32 Doi K, Akaike T, Horie H, Noguchi Y, Fujii S, Beppu N, Ogawa M, Maeda H. Excessive production of nitric oxide in rat solid tumor and its implication in rapid tumor growth. Cancer 1996; 77: 1598–1604.

33 Lancaster JR, Langrehr JM, Bergonia HA, Murase N, Simmons RL, Hoffman RA. EPR detection of heme and nonheme iron-containing protein nitrosylation by nitric oxide during rejection of rat heart allograft. J. Biol. Chem. 1992; 267: 10994–10998.

34 Bastian NR, Xu S, Shao XL, Shelby J, Granger DL, Hibbs JB. N_{ω}-monomethyl-L-arginine inhibits nitric oxide production in murine cardiac allografts but does not affect graft rejections. Biochim. Biophys. Acta 1994; 1226: 225–231.

35 Mülsch A, Mordvintcev PI, Vanin AF, Busse R. Formation and release of dinitrosyl iron complexes by endothelial cells. Biochem. Biophys. Res. Commun. 1993; 196: 1303–1308.

36 McDonald CC, Phillips WD, Mower HF. An electron spin resonance study of some complexes of iron, nitric oxide and anionic ligands. J. Am. Chem. Soc. 1965; 87: 3319–3326.

37 Bryar M, Eaton DR. Electronic configuration and structure of paramagnetic iron dinitrosyl complexes. Can. J. Chem. 1992; 70: 1917–1926.

38 Burbaev DS. EPR investigation of the compounds modeling non-heme iron complexes from biological objects. Thesis, Physical Department, Moscow University, 1971.

39 Burbaev DS, Vanin AF, Blumenfeld LA. Electronic and spatial structures of paramagnetic dinitrosyl ferrous complexes. Zhurn. Strukt. Khimii (Rus.) 1971; 2: 252–256.

40 Vanin AF, Sanina NA, Serezhenkov VA, Burbaev DS, Lozinsky VI, Aldoshin SM. Dinitrosyl-iron complexes with thiol-containing ligands: Spatial and electronic structures. Nitric Oxide: Biol. Chem. 2007; 16: 82–93.

41 Sanina NA, Rakova OA, Aldoshin SM, Shilov GN, Shulga YM, Kulikov AV, Ovanesyan NS. Structure of the neutral monuclear dinitrosyl iron complex with 1,2,4-triazole-3-thione $[Fe(SC_2H_3N_3)_2(NO)_2]0.5H_2O$. Mendeleev Comms. 2004; (1): 1–2.

42 Mordvintcev PI, Kubrina LN, Kleschyov AL, Vanin AF. On the origin of structural difference between nitrosyl non-heme iron complexes formed in animal tissued *in vivo* and *in vitro*. Stud. Biophys. 1984; 103: 63–70

43 Vanin AF, Stukan RA, Manukhina EB. Physical properties of dinitrosyl iron complexes with thiol-containing ligands in relation with their vasodilatory activity. Biochim. Biophys. Acta 1995; 1295: 5–12.

44 Vanin AF, Stukan RA, Manukhina EB. Dimer and monomer forms of dinitrosyl iron complexes with thiol-containing ligands: physicochemical properties and vasodilatory activity. Biofizika (Rus.) 1997; 42: 7–18.

45 Constanza S, Menage S, Purello R, Bonomo RP, Fontecave M. Re-examination of the formation of dinitrosyl-iron complexes during reaction of S-nitrosothiols with Fe(II). Inorg. Chim. Acta 2001; 318: 1–7.

46 Vanin AF, Papina AA, Serezhenkov VA, Koppenol WH. The mechanism of S-nitrosothiol decomposition catalyzed by iron. Nitric Oxide 2004; 10: 60–73.

47 Vanin AF. On the stability of the dinitrosyl-iron complex, a candidate for the endothelium-derived relaxing factor. Biochemistry (Mosc.) 1995; 60: 225–230.

48 Vanin AF, Kleschyov AL. EPR detection and biological implications of nitrosyl nonheme iron complexes. In Nitric Oxide in Transplant Rejection and Anti-Tumor Defense (Lukiewicz S, Zweier JL, eds.), Kluwer Academic Publishers, New York, 1998, pp. 49–82.

49 Lebedev JaS. Computer calculations of EPR spectra. 2. Asymmetric lines. Zhurnal Strukt. Khimii (Rus.) 1963; 4: 1074–1078.

50 Butler AR, Glidewell C, Li M-H. Nitrosyl complexes of iron-sulfur clusters. Adv. Inorg. Chem. 1988; 32: 335–393.

51 Goodman BA, Raynor JB, Symons MCR. Electron spin resonance of bis(NN-diethylthiocarbamate) nitrosyliron. J. Chem. Soc. A. 1969; 2572–2575.

52 Ileperuma OA, Feltham RD. Iron-sulfur complexes of NO. 2. Synthesis and exchange studies of $Fe(NO)X[S_2CN(CH_3)_2]_2$. Inorg. Chem. 1977; 16: 1876–1883.

53 Burbaev SS, Vanin AF. On modeling of non-heme iron complexes from biological objects. Dokl. Akad. Nauk SSSR (Rus.) 1970; 190: 1348–1350.

54 Vanin AF, Malenkova IV, Mordvintcev PI, Mülsch A. Dinitrosyl iron complexes with thiol-containing ligands and their reversible conversion into nitrosothiols. Biokhimiya (Rus.) 1993; 58: 1094–1103.

55 Vanin AF. Interconversion of two forms of endothelium-derived relaxing factor, S-nitrosocysteine and dinitrosyl iron complex with cysteine. Biofizika (Rus.) 1993; 38: 751–761.

56 Vanin AF, Burbaev DS, Mardanyan SS, Nalbandyan RM, Mutuskin AA, Pshonova KV. On the coordination of iron in iron-sulphur proteins with thiol groups. Symposial papers of IY International Biophysics Congress 1973; Part 2: 678–683.

57 Enemark JH, Feltham RD. Principles of structure, bonding and reactivity for metal nitrosyl complexes. Coord. Chem. Rev. 1974; 13: 340–404.

58 Burbaev DS, Vanin AF. Reduced form of nitrosyl non-heme iron complexes. Dokl. Akad. Nauk SSSR (Rus.) 1973; 213: 860–862.

59 Kennedy MC, Antholine WE, Beinert H. An EPR investigation of the products of the reaction of cytosolic and mitochondrial aconitases with nitric oxide. J. Biol. Chem. 1997; 272: 23340–23347.

60 Lobysheva II, Serezhenkov VA, Stukan RA, Bowman MK, Vanin AF. Redox transformation and stability of dinitrosyl-iron complexes with thiolate ligands as potential donor and transporters of nitric oxide. Biokhimiya (Rus.) 1997; 62: 934–942.

61 Foster MW, Cowan JA. Chemistry of nitric oxide with protein-bound iron sulfur centers. Insights on physiological reactivity. J. Am. Chem. Soc. 1999; 121: 4093–4100.

62 D'autreaux B, Horner O, Oddou J.-L, Jeandey C, Gambarelly S, Berthomieu C, Latour J.-M, Michaud-Soret I. Spectroscopic description of the two nitrosyl-iron complexe responsible for Fur inhibition by nitric oxide. J. Am. Chem. Soc. 2004; 126: 6005–6016.

63 Khalepp B, Luchkina S, Ovchinnikov I. Valence vibrations of the NO group and ESR spectra of nitrosyl complexes of iron and chromium. Russ. Chem. Bull. 1973; 22: 940–943.

64 Williams DLH. Nitrosation reactions and the chemistry of nitric oxide. Chapter 9 (a), Chaper 13 (b). Elsevier, Cornwall, UK, 2004.

65 Beinert H. Iron-sulfur proteins. New insights and unresolved problems. Biochim. Biophys. Acta Rev. Bioenerg. 1982; 683: 246–277.

66 Ding H, Demple B. Direct nitric oxide transduction via nitrosylation of iron-sulfur centers in the SoxR transcription activation. Proc. Natl. Acad. Sci. USA 2000; 97: 5146–5150.

67 Goodman BA, Raynor IB. An electron spin resonance study of the reaction of sulfide and dithionite with some iron (I) and iron (II) complexes: the valency and stereochemistry of iron in reduced non-heme iron proteins. J. Chem. Soc. A. 1970; 2038–2043.

68 Frolov EN, Vanin AF. New type of paramagnetic nitrosyl complexes of non-heme iron. Biofizika (Rus.) 1973; 18: 605–610.

69 Cruz-Ramos H, Crack J, Wu G, Hughes MN, Scott C, Thomson AJ, Green J, Poole RK. NO sensing by FNR: regulation of the *Escherichia coli* NO-detoxifying flavophaemoglobin, Hmp. EMBO J. 2002; 21: 3235–3244.

70 Vanin AF, Serezhenkov VA, Malenkova IV. Nitric oxide initiates iron binding to neocuproine. Nitric Oxide 2001; 5:166–175.

71 Vanin AF, Malenkova IV, Serezhenkov VA. Iron catalyzes both decomposition and synthesis of S-nitrosothiols: optical and EPR studies. Nitric Oxide 1997; 1:191–203.

72 Boese M, Mordvintcev PI, Vanin AF, Busse R, Mülsch A. S-nitrosation of serum albumin by dinitrosyl-iron complex. J. Biol. Chem. 1995; 270: 29244–29249.

73 Vanin AF, Serezhenkov VA, Mikoyan VD, Genkin MV. The 2.03 signal as an indicator of dinitrosyl-iron complexes with thiol-containing ligands. Nitric Oxide 1998; 2: 224–234.

74 Henry YA, Guissani A, Ducastel B. Nitric Oxide Research from Chemistry to Biology: EPR Spectroscopy of Nitrosylated Compounds. R.G. Landes Company, Austin, Texas, USA, 1996, 61–79.

75 Kennedy MC, Gan T, Antholine WE, Petering DH. Metallothioneine reacts with Fe^{2+} and NO to form products with g = 2.039 ESR signal. Biochem. Biophys. Res. Commun. 1993; 196: 632–635.

76 D'Autreaux B, Touati D, Bersch B, Latour J-M, Michaud-Soret I. Direct inhibition by nitric oxide of the transcriptional ferric uptake regulation protein via nitrosylation of the iron. Proc. Natl. Acad. Sci. USA 2002; 99: 16619–16624.

77 D'Autreaux B, Horner O, Oddou J-L, Jeandey C, Gambarelli S, Berthomieu C, Latour J-M, Michaud-Soret. Spectroscopic description of the two nitrosyl-iron complexes responsible for Fur inhibition by nitric oxide. J. Am. Chem. Soc. 2004; 126: 6005–6016.

78 Gomes CM, Vicente JB, Wasserfallen A, Teixeira M. Spectroscopic studies and characterizatiomn of a novel electron-trasnsfer chain from *Escherichia coli* involving a flavorubredoxin and its flavoprotein reductase partner. Biochemistry 2000; 39: 16320–16327.
79 Lee M, Arosio P, Cozzi A, Chasteen ND. Identification of the EPR-active iron-nitrosyl complexes in mammalian ferredoxin. Biochemistry 1994; 33: 3679–3687.
80 Boese M, Keese MA, Becker K, Busse R, Mülsch A. Inhibition of glutathione reductase by dinitrosyl-iron-dithiolate complexes. J. Biol. Chem. 1997; 272: 21767–21773.
81 De Maria F, Pedersen JZ, Cavccuri AM, Antonini G, Turella P, Stella L, Lo Bello M, Federici G, Ricci G. The specific interaction of dinitrosyl-diglutathionyl-iron complex, a natural NO carrier, with the glutathione transferase superfamily. J. Biol. Chem. 2003; 278: 42283–42393.
82 Vanin AF, Kalamkarov GR, Ostrovskii MA. On presence of near located two SH-groups in rhodopsin molecules. Biofizika (Rus.) 1977; 22: 397–408.
83 Sellers VM, Johnson MK, Daily HA. Function of the [2Fe-2S] cluster in mammalian ferrochelatase: a possible role as a nitric oxide sensor. Biochemistry 1996; 35: 2699–2704.
84 Bonner FT, Stedman G. The chemistry of nitric oxide and redox-related species. In Methods in Nitric Oxide Research (Feelish M, Stamler JS, eds.), John Wiley & Sons Ltd, New York, 1996, pp. 3–18.
85 Stojanovich S, Stanic D, Nikolic M, Spasic M, Niketic V. Iron catalyses conversion of NO to nitrosonium (NO^+) and nitroxyl (HNO/NO^-) species. Nitric Oxide 2004; 11: 256–262.
86 Franz KJ, Lippard SJ. NO disproportionation reactivivty of Fe tropocoronal complexes. J. Am. Chem. Soc. 1999; 121: 10504–10512.
87 Fukuto J, Switzer C, Miranda K, Wink D. Nitroxyl (HNO): chemistry, biochemistry and pharmacology. Ann. Rev. Pharmacol. Toxicol. 2005; 45: 335–355.
88 Vanin AF, Muller B, Alencar JL, Lobysheva II, Nepveu F, Stoclet J-C. Evidence that intrinsic iron but not intrinsic copper determines S-nitrosocysteine decomposition in buffer solution. Nitric Oxide 2002; 7: 194–209.
89 Wink DA, Nims RW, Darbyshire JF, Christodoulou D, Handbauer I, Cox GW, Laval F, Coon JA, Krishna MC, DeGraat WG, Mithell JB. Reaction kinetics for nitrosation of cysteine and glutathione in aerobic nitric oxide solutions at neutral pH. Insights into the fate and physiological effects of intermediates generated in the NO/O_2 reaction. Chem. Res. Toxicol. 1994; 7: 5129–5525.
90 Guillaume P, Li Kam Wah H, Postel M. Coordination NO as a source of oxygen: Reactivity of the $Fe(NO)_2$ moiety in the presence of the bidentate phosphane 1,2-bis (diphenylposphino) ethane (dppen). Inorg. Chem. 1991; 30: 1828–1831.
91 Munyeiabo V, Guillaume P, Postel M. Activation of molecular oxygen by iron nitrosyls in the presence of bidentate phosphines 1,2-bis (diphenylphosphino) ethane and ethane. Inorg. Chim. Acta 1994; 221: 133–139.
92 Munyeiabo V, Damiano J-P, Postel M, Bensimon C, Roustan JL. Reactivity of (Fe-NO)-(Fe-NO) system in the presence of a ferrocene-ferricinium group tethered nearby via a ferrocenyl phosphine linkage (FcP_2): crystal structures of $\{\{Fe(NO)_2Cl\}_2\ (\mu\text{-}FcP_2)\}$ and $[Fe(NO)_2(FcP_2)]$. J. Organometal. Chem. 1995; 491: 61–69.
93 Guillaume P, Postel M. Reactivity of 2-(dihenylphospino)pyridine and 2-(dihenylphospine oxide) pyridine and its relevance to oxygen activation. Inorg. Chem. Acta 1995; 233: 109–112.
94 Zabbarova IV, Shumaev KB, Vanin AF, Gubkin AA, Petrova NE, Ruuge EK. Interaction of ferritin and myoglobin as lipid peroxidation inducers: role of reactive oxygen and nitrogen species. Biophysics (Translated from Russian) 2004; 49: 607–613.
95 Gorbunov NV, Yalowich JC, Gaddam A, Thampaaty P, Kisin ER, Elsauyed NM, Kagan VE. Nitric oxide prevents oxidative damage produced by *tert*-butyl hydroperoxide in erythroleukemia cells via nitrosylation of heme and nonheme iron. J. Biol. Chem. 1997; 272: 12328–12341.
96 Shumaev KB, Lankin VZ, Ruuge EK, Vanin AF, Belenkov YuN. The mechanism of inhibition of free-radical oxidation of β-carotene by S-nitrosoglutathione and iron dinitrosyl complexes. Dokl. Akad. Nauk SSSR (Rus.) 2001; 379: 273–275.
97 Shumaev KB, Petrova NE, Zabbarova IV, Vanin AF, Topunov AF, Lankin VZ, Ruuge EK. Interaction of oxoferrylmyoglobin and dinitrosyl-iron complexes. Biochemistry (Mosc.) 2004; 69: 569–574.

98 Shumaev KB, Gubkin AA, Gubkina SA, Gudkov LL, Timoshin AA, Topunov AF, Vanin AF, Ruuge EK. Interaction of dinitrosyl iron complexes with intermediates of oxidative stress. Biofizika (Rus.) 2006; 51: 472–477.

99 Mikoyan VD, Serezhenkov VA, Brazhnikova NV, Kubrina LN, Khachtryan GN, Vanin AF. Formation of paramagnetic nitrosyl complexes of nonheme iron in the animal organidsm with the participation of nitric oxide from exogenous and endogenous sources. Biophysics (Translated from Russian) 2004; 41: 110–116.

100 Gwost D, Caulton KG. Reductive nitrosylation of group VIIIb compounds. Inorg. Chem. 1973; 12: 2095–2099.

101 Connelly NG, Gardner C. Simple halogen nitrosyl anions of iron. J. Chem. Soc. Dalton Trans. 1976: 1525–1527.

102 Baltusis LM, Karlin KD, Rabinowitz HN, Dewan JC, Lippard SJ. Synthesis and structure of $Fe(L`H)(NO)_2$, a tetracoordinate complex having a twelve-membered chelate ring, and its conversion to pentacoordinate FeL`(NO) through formal loss of "HNO" ($L' = {}^{-}SCH_2CH_2NMECH_2CH_2CH_2NMeCH_2CH_2S^{-}$). Inorg. Chem. 1980; 19: 2627–2623.

103 Reginato N, McCrory TC, Pervitsky D, Li L. Synthesis, X-ray crystal structure, and solution behavior of $Fe(NO)_2(I\text{-}MeIm)_2$: Implications for nitrosyl non-heme-iron complexes with $g = 2.03$. J. Am. Chem. Soc. 1999; 121: 10217–10218.

104 Sanina NA, Rakova OA, Aldoshin SM, Chuev II, Atovmyasn EG, Ovanesyan NS. Synthesis and X-ray and spectral study of the compounds $[Q_4N]_2[Fe_2(S_2O_3)_2(NO)_4]$ (Q = Me, Et, n-Pr, n-Bu). Russ. J. Coord. Chem. 2001; 27: 179–183.

105 Tsai M-I, Chen C-C, Hsu I-J, Ke S-C, Hsien C-H, Chiang K-A, Lee G-H, Wang Y, Chen J-M, Lee J-F, Liaw W-F. Photochemistry of the dinitrosyl iron complex $[S_5Fe(NO)_2]^-$ leading to reversible formation of $[S_5Fe(\mu\text{-}S)_2FeS_5]^{2-}$: spectroscopic characterization of species relevant to the nitric oxide modification and repair of [2Fe-2S] ferredoxins. Inorg. Chem. 2004; 43: 5159–5167.

106 Chiang C-Y, Miller ML, Reibenspies JH, Darensbourg MY. Bismercaptoethandiazacyclooctane as a N_2S_2 chelating agent and cys-X-cys mimic for Fe(NO) and $Fe(NO)_2$. J. Am. Chem. Soc. 2004; 126: 10867–10874.

107 Chen H-W, Lin C-W, Chen C-C, Yang L-B, Chiang M-H, Liaw W-F. Homodinuclear iron thiolate nitrosyl compounds $[(ON)Fe(S,S{-}C_6H_4)Fe(NO)_2]^-$ and $[(ON)Fe(SO_2,S{-}C_6H_4)_2Fe(NO)_2]^-$ with $\{Fe(NO)\}^7{-}\{Fe(NO)_2\}^9$ electronic coupling: new members of a class of dinitrosyliron complexes. Inorg. Chem. 2005; 44: 3226–3232.

108 Wang X, Sundberg EB, Li L, Kantardjieff KA, Herron SR, Lim M, Ford P. A cyclic tetra-nuclear dinitrosyl iron complex $[Fe(NO)_2(imidazolate)_4]$: synthesis, structure and stability. Chem. Commun. 2005; 477–479.

109 Cesario E, Parker LJ, Pedersen JZ, Nuccetelli M, Mazzetti AP, Pastore A, Federici G, Caccuri AM, Ricci G, Adams JJ, Parker MW, LoBello M. Nitrosylation of human transferase P1-1 with dinitrosyl diglutathionyl iron complex in vitro and in vivo. J. Biol. Chem. 2005; 280: 42172–42180.

110 Griffith JS. The electronic structures of some first transition series metal porphyrins. Discuss. Faraday Soc. 1958; 26: 81–86.

111 McGarvey BR. The isotropic hyperfine interaction. J. Phys. Chem. 1967; 71: 51–67.

112 Van Konningsbruggen P, Maeda Y, Oshio H. Iron (III) spin crossover compounds. Top. Curr. Chem. 2004; 233: 259–324.

113 Vanin AF, Aliev DI. High spin nitrosyl iron complexes in animal tissues. Stud. Biophys. 1983; 93: 63–68.

114 Salerno JS, Siedow JH. The nature of the nitric oxide complexes of lipooxygenase. Biochim. Biophys. Acta 1979; 579: 246–251.

115 Farrar JA, Grinter R, Pountney DL, Thomson AJ. Optical and magnetic properties of iron(II)-nitrosyl complexes in model compounds. J. Chem. Soc. Dalton Trans. 1993; 2703–2709.

116 Jo D-H, Chiou Y-M, Lawrence Q. Models for extradiol cleaving catechol dioxygenases: synthesis, structures and reactivity of iron(II)-monoanionic catecholate complexes. Inorg. Chem. 2001; 40: 3181–3190.

117 Hauser C, Glaser T, Bill E, Weyhermuller T, Wieghardt K. The electronic structure of an isostructural series of octahedral nitrosyliron complexes $\{FeNO\}^{6,7,8}$ elucidated by Mössbauer spectroscopy. J. Am. Chem. Soc. 2000; 122: 4352–4365.
118 Jackson TA, Yikilmaz E, Miller A-F, Brunold TC. Spectroscopic and computational study of a non-heme iron $\{FeNO\}^7$ system: exploring the geometric and electronic structures of the nitrosyl adduct of iron nitrosyl superoxide dismutase. J. Am. Chem. Soc. 2003; 125: 8348–8363.
119 Patra AK, Rowland JM, Marlin DS, Bill E, Olmstead MM, Masharak PK. Iron nitrosyl of a pentadentate ligand containing a single carbaxamide group: synthesis, structures, electronic properties, and photolability of NO. Inorg. Chem. 2003; 42: 6812–6823.
120 Foster MA, Hutchison JMS. The origin of an EPR signal at $g = 2.03$ from normal rabbit liver and the effects of nitrites on it. Phys. Med. Biol. 1974; 19: 289–302.
121 Vanin AF, Kiladze SV, Kubrina LN. Factors influencing the formation of the dinitrosyl complexes with non-heme iron in animal organs in vivo. Biofizika (Rus.) 1977; 22: 850–857.
122 Vanin AF, Kiladze SV, Kubrina LN. Incorporation of non-heme iron into dinitrosyl complexes in the livers of mice in vivo. Biofizika (Rus.) 1978; 23: 474–480.
123 Vanin AF, Varich VJ. Formation of nitrosyl complexes of non-heme iron (complexes 2.03) in animal tissues in vivo. Biofizika (Rus.) 1979; 24: 666–670.
124 Vanin AF, Kubrina LN, Aliev DI. On the mechanism of nitrosyl non-heme iron complex formation in animal tissues. Stud. Biophys. 1980; 80: 221–230.
125 Vanin AF, Varich VJ. Nitrosyl non-heme iron complexes in animal tissues. Stud. Biophys. 1981; 86: 175–185.
126 Varich VJ, Vanin AF. Mechanism of formation of nitrosyl non-heme iron complexes in animal organisms. Biofizika (Rus.) 1983; 28: 1055–1060.
127 Tarasova NI, Kovalenko OA, Vanin AF. Mechanism of incorporation of iron into liver tissue. Biofizika (Rus.) 1981; 26: 677–682.
128 Vanin AF, Aliev DI. EPR signal shape of nitrosyl non-heme iron complexes as an indicator of prioteins components in these complexes. Stud. Biophys. 1983; 97: 223–229.
129 Mordvintcev PI, Kubrina LN, Kleschyov AL, Vanin AF. On the origin of structural differences between nitrosyl non-heme iron complexes formed in animal tissues in vivo or in vitro. Stud. Biophys. 1984; 103: 63–70.
130 Vanin AF, Kurbanov IS, Mordvintcev PI, Aliev DI. Influence of the intracellular medium on the structure of dinitrosyl complexes of non-heme iron in the liver of animals. Stud. Biophys. 1987; 120: 145–154.
131 Lobysheva II, Serezhenkov VA, Vanin AF. Interaction of peroxynitrite and hydrogen peroxide with dinitrosyl iron complexes containing thiol. Biokhimia (Rus.) 1999; 64: 194–200.
132 Vanin AF, Osipov AN, Kubrina LN, Burbaev DS, Nalbandyan RM. On the origin of paramagnetic centers with $g = 2.03$ in animal tissues and microorganisms. Stud. Biophys. 1975; 49: 13–25.
133 Nagata C, Ioki Y, Kodama M, Tagashira Y, Nakadate M. Free radical induced in rat liver by a chemical carcinogen, N-methyl-N`-nitro-N-nitrosoguanidine. Ann. NY. Acad. Sci. 1973; 222: 1031–1047.
134 Hutchison JMS, Foster MA, Mallard JR. Description of anomalous ESR signals from normal rabbit liver. Phys. Med. Biol. 1971; 16: 655–658.

Radicals for Life: The Various Forms of Nitric Oxide
E. van Faassen and A.F. Vanin (Eds.)

CHAPTER 3

Hypotensive, vasodilatory and anti-aggregative properties of dinitrosyl-iron complexes

Anatoly F. Vanin[1] and Eugenia B. Manukhina[2]

[1]*Semenov Institute of Chemical Physics, Russian Academy of Sciences, Moscow, 119991, Russian Federation*
[2]*Institute of General Pathology and Pathophysiology, Russian Academy of Medical Sciences, Moscow*

As widely accepted now, dinitrosyl-iron complexes with thiol-containing ligands $\{(RS^-)_2Fe^+(NO^+)_2\}^+$ (DNICs) as well as *S*-nitrosothiols (RS—NO) can be generated *via* the L-arginine-dependent pathway ensuring storage and transfer of NO molecules in cells and tissues [1–14]. Moreover, due to the quasi-equilibrium between DNIC and its constituents (Scheme 1):

$$\{(RS^-)_2Fe^+(NO^+)_2\}^+ \rightleftharpoons Fe^{2+} + NO + \underbrace{NO^+ + RS^-}_{\downarrow \; RS\text{—}NO} + RS^-$$

Scheme 1.

DNIC can influence a number of biochemical and physiological events acting as NO and NO^+ (RS—NO) donors. The powerful biological activity of these complexes was discovered in the 1980s by using synthetic DNIC with both thiol-containing and non-thiolic ligands. The experiments demonstrated high hypotensive [15–17] and anti-aggregative capacity of the complexes [18,19]. Later, the high vasodilatory activity of the DNIC was also found [20,21]. Now a great body of data on these properties of DNIC has been accumulated and is considered in this chapter.

HYPOTENSIVE ACTIVITY OF DNICs

The realization that DNICs have important biological potential began after the finding that a range of low-molecular DNICs (DNIC with cysteine, thiosulfate or phosphate) possesses long-lasting hypotensive activity after a bolus injection of the complexes into the portal vein of rats anesthetized with sodium ethaminal [15]. Fig. 1 and Table 1 show the data

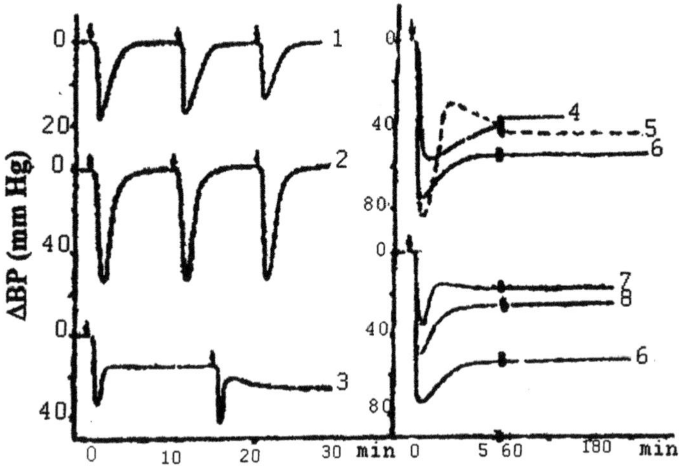

Fig. 1. Blood pressure change (ΔBP, mmHg) in Wistar rats in response to nitroglycerol (NG) 5 nM (1); sodium nitroprusside (SNP) 20 nM (2); or DNIC with thiosulfate 100 nM (2–3 injections) (3); 200 nM DNIC with phosphate (4); DNIC with cysteine 100 nM (5); or DNIC with thiosulfate (2000, 200 and 400 nM, respectively) (single injection) (6–8) [15].

Table 1 Dose dependence of blood pressure drop (ΔBP) at initial stage of DNICs, nitroprusside or nitroglycerol (NG) injection to rats [15]

Preparation	Dose (10^{-8} M)	ΔBP (mmHg)
Nitroprusside	1	40 ± 2
	2	56 ± 3
	5	77 ± 3
	10	86 ± 2
Nitroglycerol	0.44	20 ± 2
	44	50 ± 2
DNIC-cysteine	5	55 ± 4
	10	55 ± 4
	100	86 ± 2
DNIC-thiosulfate	20	35 ± 3
	40	50 ± 3
	200	85 ± 5

on hypotensive activity of these DNICs in comparison to the well-known hypotensive and vasodilatory agents, sodium nitroprusside (SNP) and nitroglycerol (NG).

All DNICs were proved to be less efficient than SNP or NG at the initial stage characterized by a rapid drop and the subsequent recovery of blood pressure (BP). The main feature of the DNICs was the prolonged character of their hypotensive action. After a bolus injection of these compounds, the animals exhibited a reduced stationary BP level retained up to awakening. For repeated injections, the plateau level was observed to decrease, each time with the proceeding rapid reversible drop of BP (Fig. 1). The injection of SNP resulted only in a short-time (2–4 min) decrease in BP, not affected by the proceeding administration of the same agent. Unlike this, in the case of NG the preliminary injection affected the subsequent reaction (the effect of tolerance to NG) as illustrated in Fig. 1.

The injection of any DNIC type into animals led to the appearance of protein-bound DNICs in blood, liver and other tissues characterized with the EPR signal at $g_{av} = 2.03$ ($g_{\perp} = 2.04$, $g_{\parallel} = 2.014$). Such type of EPR signal, recorded in liver is shown in Fig. 2, curve a.

As was demonstrated later in similar experiments, the shape of the signal did not change while the recording temperature was increased from 77 K to the ambient one [22,23]. Namely, this fact pointed unequivocally to the protein-bound nature of DNIC: The mobility of protein molecules at room temperature was not sufficiently high to ensure the averaging of anisotropy g-factor and hyperfine structure (HFS) tensors. In contrast, the low-molecular DNICs are rapidly tumbling and show a motionally narrowed isotropic line with a half-width of 0.7 mT and resolved 13- or 5-component HFS [24].

The formation of protein-bound DNICs was due to the rapid transfer of $Fe(NO)_2$ groups from low-molecular DNICs to paired RS-groups of protein. Thereby, something resembling a pool of nitric oxide (NO) appeared, which was represented by protein-bound DNICs characterized by much higher stability than that of low-molecular counterparts. As protein-bound DNICs decompose, the released nitrosyl-iron groups act as a direct source of NO ensuring the retention of decreased BP in animals for 30 min and above after a bolus injection of low-molecular DNICs [15–17].

The transition of nitrosyl-iron groups from low-molecular DNICs to proteins is confirmed by the following experimental results. The animals received injections of DNIC with thiosulfate, where ^{56}Fe was substituted by the ^{57}Fe isotope with nuclear spin $I = 1/2$. Preparations of blood, liver and other tissues showed the DNICs with a broadened EPR signal.

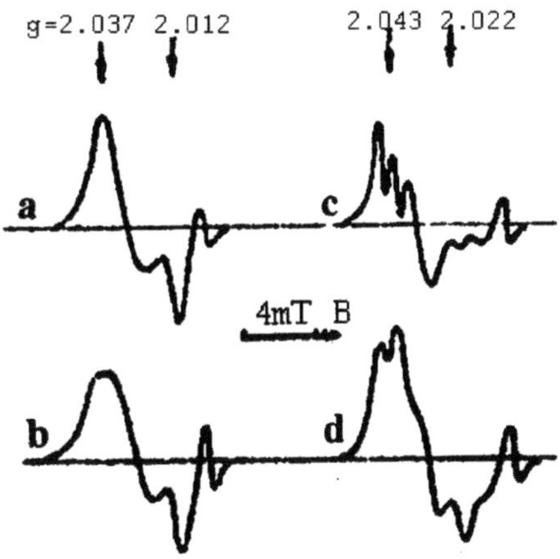

Fig. 2. The EPR spectra of DNIC with ^{56}Fe (curve a) or ^{57}Fe (curve b) formed in the rat liver after injections of DNIC with thiosulfate and mononitrosyl iron complex (MNIC) with diethyldithiocarbamate (DETC) appeared in the liver after injections of DETC into the same animals [(c) and (d), respectively]. Recordings were made at 77 K [15].

The broadening was especially pronounced at $g_\perp$ component due to the non-resolved HFS from ^{57}Fe (Fig. 2, curve b).

The hypothesis that the nitrosyl-iron group in DNIC was responsible for the hypotensive effect might be verified by adding sodium diethyldithiocarbamate (DETC) (formula: $(C_2H_5)_2{=}N{-}C_2^-$) capable of strong selective binding of Fe—NO groups that results in the formation of mononitrosyl iron complexes (MNIC) with DETC, formula: $\{(C_2H_5)_2{=}N{-}CS_2^-\}_2{-}Fe^+{-}NO^+$. These complexes are characterized by the EPR signal with $g_\perp = 2.045$, $g_{||} = 2.02$ and a well-resolved triplet HFS at $g_\perp$ [25] (Fig. 2, curves c,d). Earlier [26] it was demonstrated that DETC readily captures iron and NO radical from DNIC to form corresponding MNIC-DETC. A similar result was obtained on rats injected with DNIC and DETC [15,27]. For example, i.a. injection into the anesthetized rats of DETC 10 min after administration of the DNIC changed the shape of the EPR spectrum of the blood: instead of the EPR signal of the DNIC the MNIC-DETC was observed localized mainly (up to 90%) in the red blood cells [27]. Within 30–40 min the MNIC-DETC content in the blood fell sharply and in place of these complexes DNIC appeared, also localized in the red blood cells. The shape of the EPR signal was similar to that characteristic of the EPR signal of DNIC bound with Hb [28] (see also Chapter 2). The intensity of the signal was far lower compared with that of the initial signal of DNIC in blood before the addition of DETC. In the liver of these animals by this time, judging from the EPR spectrum of this organ, MNIC-DETC persisted [15,27].

No hypotensive effect was observed in the rats injected with "ready-to-use" MNIC-DETC in the physiological salt solution (the dose injected was the same as in the case of DNIC) [15]. Apparently, this was due to the strong binding of Fe—NO group to DETC, preventing the liberation of this group from MNIC in the organism. For this reason, it should be anticipated that the introduction of DETC to animals injected with DNIC would normalize the BP level. Indeed, experiments have confirmed this hypothesis. Irrespective of the time of DETC addition, whether prior to the introduction of DNIC or after establishing a stationary BP level induced by the injection of DNIC, the BP level either recovered or tended to the normalization (Fig. 3, curves b,c) [15,27].

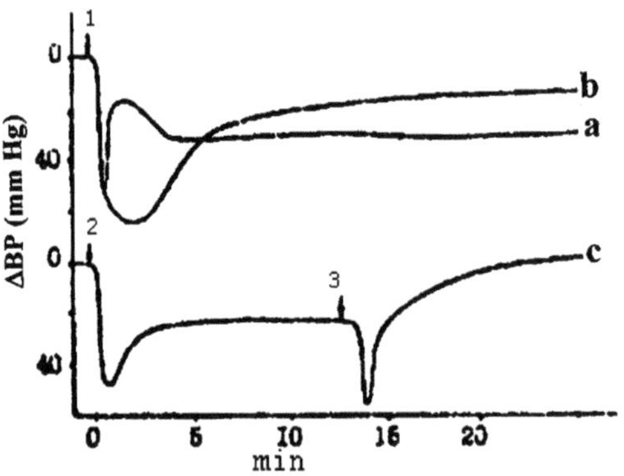

Fig. 3. Blood pressure changes (ΔBP, mmHg) in rats induced by DNIC with thiosulfate (400 nM) injected at time points 1 or 2 without DETC (curves a,b) or after DETC (5 mg) i.p. injections (40 mg) (curve b). DETC was injected i.v. at time point 3 on curve c [15].

Interestingly, the recovery of BP induced by DETC after DNIC injection was preceded as a rule by a brief drop of BP [15,17] (Fig. 3, curve c). This brief drop of BP can be attributed to different levels of NO in MNIC and DNIC (low-molecular or protein-bound DNICs). When a DETC molecule interacts with DNIC, the former takes a Fe ion and one NO molecule from DNIC to form MNIC-DETC (Scheme 2).

$$(RS^-)_2\ Fe^+(NO^+)_2\}^+ + 2\{(C_2H_5)_2{=}N{-}CS_2^-\} \longrightarrow \{(C_2H_5)_2{=}N{-}CS_2^-\}_2{-}Fe^+{-}NO^+ + RS^-{-}NO^+ + RS^-$$

Scheme 2.

The second NO^+ ligand (nitrosonium ion) becomes free and nitrosates low-molecular or protein-bound thiols forming respective *S*-nitrosothiols (RS—NOs). Due to the high vasodilatory activity [12], RS—NOs induce brief hypotension. The source of thiols recombining with nitrosonium ions remains obscure. With regard to the hypotensive effect, obviously only low-molecular RS—NOs could be responsible for it due to their free motion in the intracellular space. Rapid degradation of low-molecular RS—NO under the action of iron or copper ions in animal organisms could sharply shorten the duration of hypotensive effect by these compounds.

Interestingly, administration of DETC prior to SNP (MNIC with cyanide) to the animals completely eliminated the hypotensive effect of SNP, but without the initial brief drop of BP [15,27]. No hypotensive activity is seen in rats even after an injection of 1 μM SNP [15] (Fig. 4, curve b). For comparison, note that in the absence of DETC an injection of only 200 nM SNP was lethal due to the drop of BP below the permissible level. Apparently, the total elimination of hypotensive activity by DETC was due to the release of $Fe^+{-}NO^+$ group from SNP. This group was completely transferred to DETC without the evolution of free nitrosonium ions, in contrast to the events that could occur during the interaction of DETC with $Fe^+(NO^+)_2$ groups in DNICs.

Interestingly, Fe^{2+} ions themselves exhibited a noticeable hypotensive activity (Fig. 4, curve a). However, these ions differed in behavior from DNIC in that their action has no prolonged character and can be completely suppressed by DETC (Fig. 4, curve a).

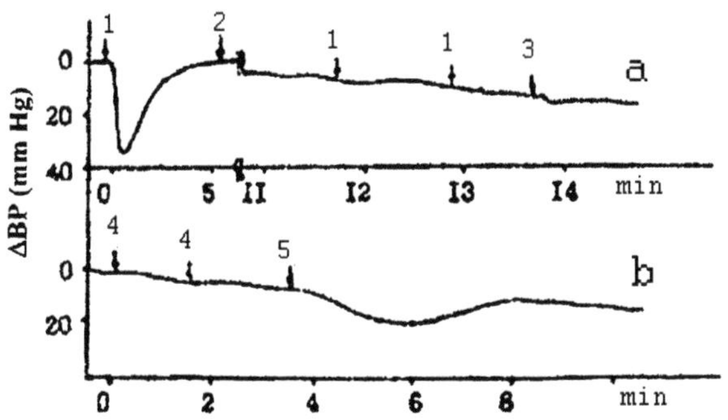

Fig. 4. Elimination of the hypotensive effects of $FeSO_4$ (curve a) or sodium nitroprusside (SNP) (curve b) by an injection of DETC (40 mg, i.p.) at time point 2 or before the nitroprusside administration. Doses of $FeSO_4$ or nitroprusside were equal to 100 nM (1), 200 nM (3), 200 nM (4) or 2000 nM (5) [15].

The idea that protein-bound DNICs as NO donors can determine long-lasting hypotension was supported by experiments with injection of "ready-to-use" protein-bound DNICs to anesthetized rats or dogs [16]. These complexes were synthesized by the administration of low-molecular DNIC with thiosulfate or phosphate to bovine serum or plasma from dogs or rats (0.4 μM DNIC/ml of serum or blood plasma). This amount of proteins in serum or plasma preparations was sufficiently high to ensure acceptance of all $Fe^{+}(NO^{+})_2$ groups from low-molecular DNICs by respective protein ligands. Prolonged hypotension induced by both types of preparations was virtually similar to that induced by the addition of low-molecular DNICs (Fig. 5). The difference was a slower time course of BP response to the injection of preparations with protein-bound DNIC. Interestingly, a bolus injection of noradrenaline to dogs or rats with reduced BP resulted in a brief recovery of BP (within 3–4 min) followed by a fall of BP to the level induced by the preceding DNIC injection (Fig. 5). Administration to animals of serum or plasma not containing protein-bound DNICs virtually left the BP level unchanged.

Hypothetically, the hypotensive activity of protein-bound DNICs could be ensured by back transfer of $Fe^{+}(NO^{+})_2$ groups from protein to low-molecular ligands with the formation of more mobile low-molecular DNICs. Specifically, the latter donated NO molecules to special enzymes such as guanylate cyclase providing a relaxation response of vascular smooth muscle to NO.

Hypotensive changes in BP evoked by low-molecular DNICs were also observed in conscious normotensive (Wistar) and spontaneously hypertensive (SHRs) rats weighing

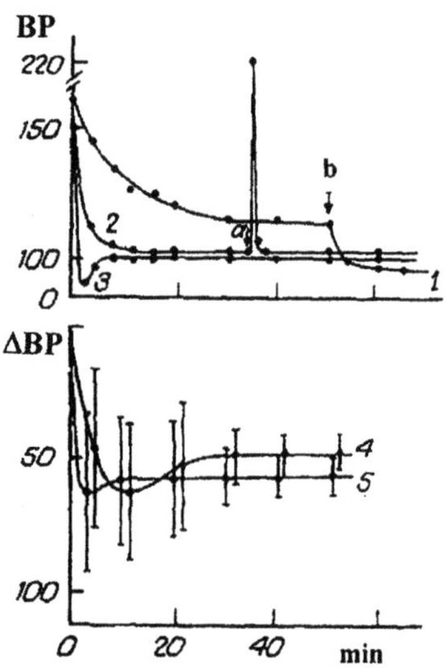

Fig. 5. Top panel: Blood pressure changes (ΔBP, mmHg) in anesthetized dogs induced by an i.v. injection of blood plasma from an animal pretreated with either DNICs (10 μM/350 ml, curve 1) or DNIC with thiosulfate in phosphate-buffered saline (PBS) (10 μM/animal, curves 2,3). Subsequently epinephrine or DNIC was added at time points a and b, respectively. Bottom panel: Blood pressure changes (ΔBP, mmHg; baseline BP = 130 ± 10 mmHg) in anesthetized Wistar rats (n = 5) induced by an i.a. injection of either rat blood plasma containing DNICs (1.5 μM/kg in 1.5 ml, curve 4) or DNIC with thiosulfate in PBS (1.5 μM/kg in 1.5 ml, curve 5) [16].

200–250 g [17]. A bolus i.v. injection of all used DNICs with glutathione, dithiothreitol or thioglycolate ligands exerted a hypotensive effect in conscious stroke-prone SHR (SHRSP) rats with initial BP value of 180 ± 10 mmHg. At DNIC doses of 1, 3 or 10 μM/kg, the decrease in BP reached 10–20, 20–30 or 70–80 mmHg. Further increment of the DNIC dose up to 20 μM/kg did not virtually intensify the effect. The dynamics of BP decrease is shown in Fig. 6, top panel. In contrast to hypotension induced by DNICs in anesthetized animals, a relatively rapid recovery of BP (within 1–1.5 h) was observed in conscious SHR rats. Normotensive Wistar rats with baseline BP of 110–120 mmHg proved more resistant to the effect of DNICs. When 5 μM/kg DNIC-thioglycolate was administered, the maximal BP decrease did not exceed 30–40 mmHg with hypotension lasting for 1–1.5 h. A bolus injection of SNP or NG into conscious animals produced only a brief hypotensive effect (1–2 min).

Prior DETC injection into SHR (150 μM/kg) sharply attenuated the hypotensive effect of DNIC with thioglycolate administered at 10 μM/kg (Fig. 6, bottom panel). Similar effect was observed when DETC was injected 5 min after DNICs. However, DETC injected 20 min after DNICs did not significantly affect intensity or duration of the hypotension induced by DNICs with the exception of a brief BP drop (Fig. 6, bottom panel). Similar effect of DETC mentioned above was characteristic of anesthetized animals (Fig. 3). Obviously, in both cases the reason was formation of short-living RS—NO due to DNIC degradation under the action of DETC as shown in Scheme 1.

The EPR assay demonstrated the formation of protein-bound DNICs in blood and other tissues from conscious animals injected with DNICs (Fig. 7) [17].

The complexes rapidly (within 1 h) disappeared from blood but remained in other tissues (liver, kidney and others) for 2–3 h despite the full recovery of BP. If DETC was administered to animals before DNIC, the latter was transformed into MNIC-DETC in all organs except for blood where DNIC disappeared from plasma but remained unchanged in red cells (Fig. 8). When DETC was injected into animals 20 min after DNIC the major part of DNIC remained unchanged in liver or kidney (Fig. 8).

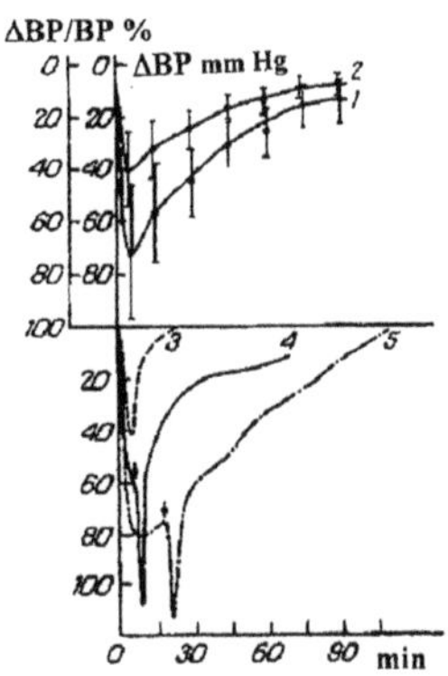

Fig. 6. Top panel: Blood pressure changes in conscious SHR rats ($n = 6$, baseline BP = 180 ± 10 mmHg) induced by an i.v. injection of DNIC with glutathione (10 μM/kg). (1) and (2) are changes in ΔBP (mmHg) and ΔBP/BP (%), respectively. Bottom panel: Blood pressure changes in conscious SHR rats induced by an i.v. injection of DNIC with thioglycolate (10 μM/kg) followed by the addition of DETC (30 mg/kg) (1 min after DNIC injection). Arrows indicate time points of subsequent additions of DETC [17].

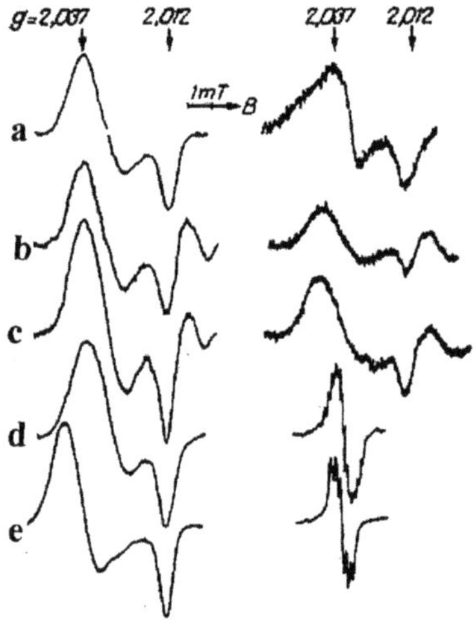

Fig. 7. EPR spectra of blood (curve a), liver (curve b) and kidney (curve c) from Wistar rats injected i.v. with DNIC with glutathione; (curves d,e) are the EPR spectra of DNIC with glutathione or thiosulfate, respectively. Recordings were made at 77 K (left side) and ambient temperature (right side) [17].

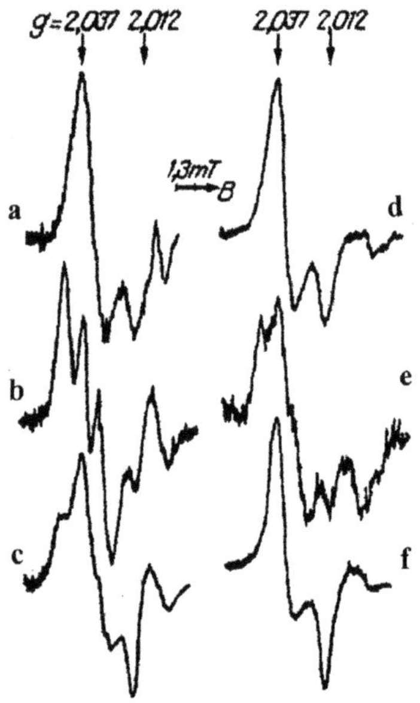

Fig. 8. EPR spectra of blood (curves a,d), liver (curves b,e) or kidney (curves c,f) from Wistar rats injected with DNIC with thiosulfate (5 μM/kg) 1 min after or 30 min before the addition of DETC (30 mg/kg) (curves a–c or d–f, respectively). Recordings were made at 77 K [17].

Thus, the attenuation of DNIC hypotension induced by the prior administration of DETC correlated with the disappearance of protein-bound DNICs as possible NO donors producing hypotension. This was not the case in experiments where DETC was added 20 min and above after a DNIC injection or DNICs were administered without DETC. In this case, BP recovered to the baseline level despite persisting protein-bound DNIC in animal tissues. The baroreflex mechanism maintaining BP could be responsible for the BP restoration. It was sufficiently strong to get over the hypotensive action of NO released from protein-bound DNICs.

The hypotensive activity of DNIC with cysteine in normotensive Wistar–Kyoto (WKY) rats was compared to DNIC distribution in organs and tissues in order to propose an optimum DNIC dosage, which would, on the one hand, provide an efficient tissue level of DNIC and, on the other hand, would not exert adverse effects on animals [29]. Blood pressure (BP) of rats injected intravenously with 0.5, 2, 4 and 6 μM/kg DNIC was monitored continuously for 3 h following the injection and measured again in 24 h. Fig. 9 shows the time course of BP after DNIC injection into conscious rats. Injection of 0.5 μM/kg DNIC did not significantly influence BP. After injection of 2 μM/kg DNIC, BP gradually decreased but remained above 90–95 mmHg in all animals. After injection of 4 μM/kg DNIC, BP significantly decreased as soon as in 10 min and reached a nadir of 80–85 mmHg in 2–2.5 h. The dose of 6 μM/kg DNIC almost immediately induced a dramatic drop of BP to 75–80 mmHg. The BP variation at the latter dose was high; in some animals BP fell to 55–60 mmHg. The decreased BP persisted for the entire period of monitoring (3 h) and was accompanied with considerable disorders of heart rhythm. Twenty-four hours after injection, BP returned to the baseline level irrespective of the DNIC dose.

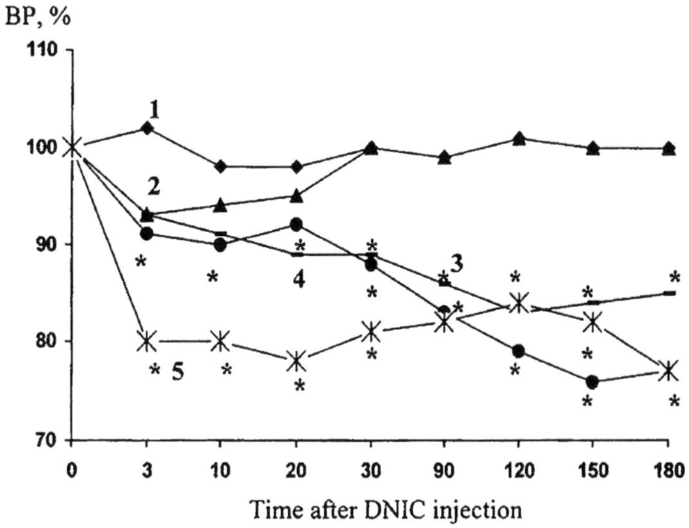

Fig. 9. Time course of BP after DNIC injection. (1) control; (2) 0.5 μM/kg; (3) 2 μM/kg; (4) 4 μM/kg; and (5) 6 μM/kg. *Significant difference from control, $p < 0.05$ [29].

Data on distribution of DNIC in tissues show that the EPR signal from DNIC was undetectable 1.5 h after the injection of 0.5 μM/kg DNIC in all analyzed organs (liver, kidneys, heart, spleen, small intestine, blood, lungs and brain). After the injection of 2 μM/kg DNIC, the EPR signal was consistently recorded in blood, kidneys and brain. After the injection of 4 and 6 μM/kg DNIC, a typical EPR signal was recorded in all the organs and tissues studied at 1.5 h and disappeared at 24 h.

Therefore, intravenous DNIC distributed to organs and tissues in a dose-dependent manner and exerted a dose- and time-related hypotensive effect, which corresponded to the content of DNIC in tissues. The dose of 4 μM/kg DNIC was efficient and did not exert adverse effects. The use of DNIC as a NO donor in diseases and conditions associated with absolute or relative NO deficiency seemed promising.

The effect of DNIC on development of hypertension was studied in SHRSP rats aged 5–6 weeks, which corresponded to the early hypertension stage. DNIC (3 μM/kg, i.p.) was injected into SHRSP and their normotensive control WKY rats every fourth day for 40 days [30]. During the development of hypertension, BP of untreated SHRSP increased from 136 ± 3 mmHg at the early hypertension stage to 216 ± 7 mmHg at the established hypertension stage ($p < 0.001$) (Fig. 10). In SHRSP treated with DNIC, BP increased only to 168 ± 17 mmHg ($p < 0.01$). Interestingly, the course of DNIC did not result in any significant changes in BP of WKY rats (124 ± 3.2 mmHg *vs.* 113 ± 2.5 mmHg before and after the DNIC course, respectively).

Therefore, DNIC exerted a selective anti-hypertensive effect in SHRs and left the BP unchanged in normotensive rats. This phenomenon may be attributed to different capacity for binding of free NO released from DNIC to NO stores in blood vessels of SHRSP and WKY rats.

NO stores primarily in the form of DNICs, which have been formed *in vivo* in vascular walls after the injection of DNIC or stimulation of endogenous NO synthesis, can be detected on isolated blood vessels [31]. The method used is based on the interaction of DETC of

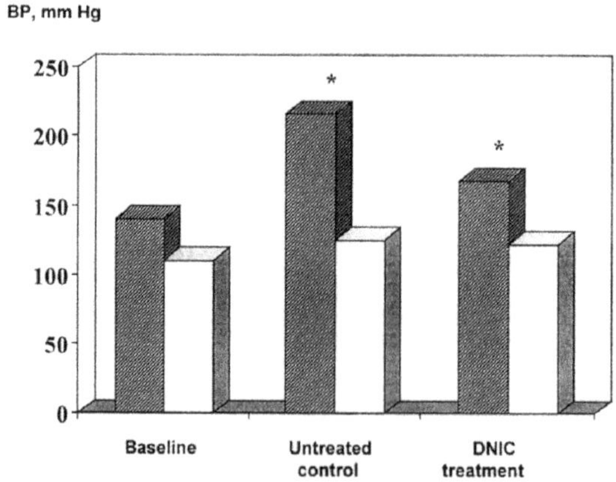

Fig. 10. Effect of DNIC with glutathione on development of hypertension in SHRSP rats. Empty bars, control; dashed bars, SHRSP. *Significant difference from baseline value, $p < 0.05$ [30]. Reprinted with permission from "Role of nitric oxide in adaptation to hypoxia and adaptive response" (Physiol. Res. 2002; 49: 89–97).

N-acetylcysteine with NO stores, which results in the release of free NO and vasorelaxation. This method was successfully modified for detecting NO stores in conscious rats pretreated with DNIC [32]. NO stores were detected 5 h after the DNIC injection (4 μM/kg, i.p.) with prior (1 h before experiment) inhibition of NO synthase with NW-nitro-L-arginine (L-NNA) to exclude contribution of *de novo* synthesized NO to the vasodilatory response. The DNIC injection induced a moderate hypotensive response which lasted for 1.5 h. Infusion of DETC (100 μM/kg, i.v.) resulted in a transient 13.0 ± 3.4% decrease in BP, which returned to the baseline level in approximately 40 min (Fig. 11). The hypotensive response was accompanied by the formation of MNIC-DETC in all analyzed organs, heart, liver, kidneys and brain as

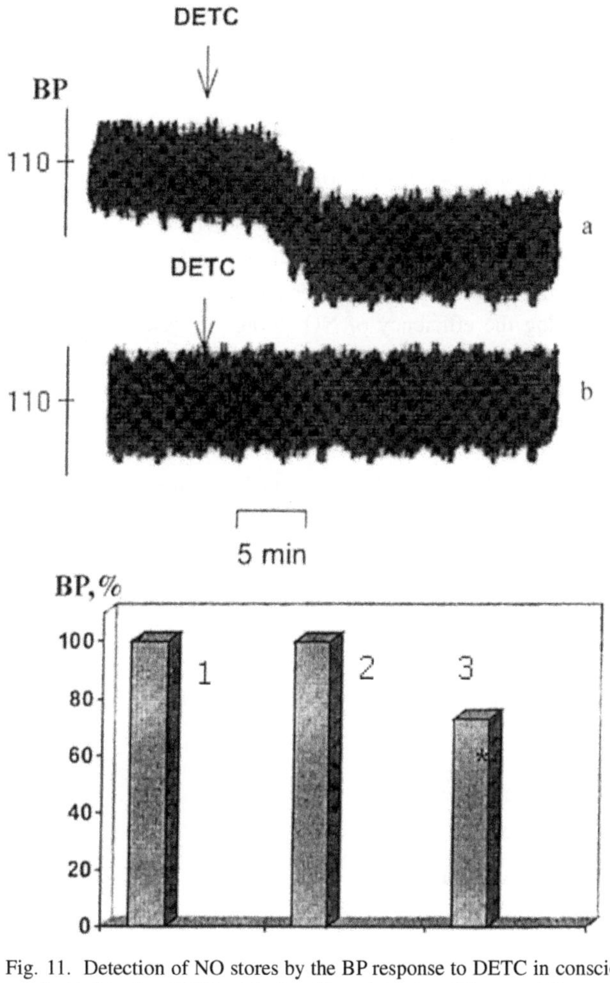

Fig. 11. Detection of NO stores by the BP response to DETC in conscious rats, % of baseline values. Upper panel: (a) 5 h after DNIC with glutathione injection; (b) control. Bottom panel: (1) baseline; (2) BP response to DETC in untreated rats; and (3) BP response to DETC in rats pretreated with DNIC. Significant difference from baseline, $p < 0.05$ [32].

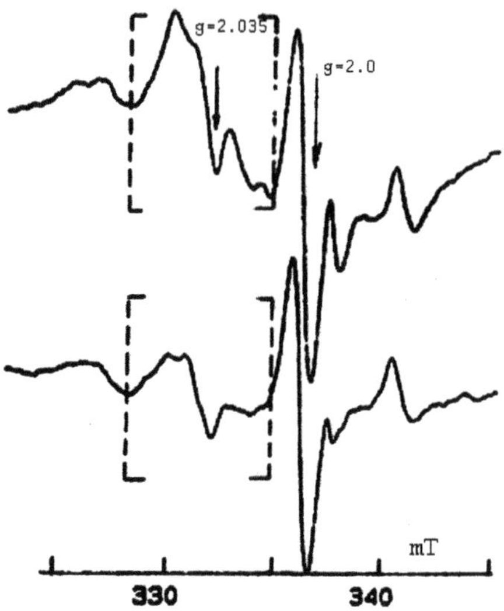

Fig. 12. EPR of MNIC-DETC after the depletion of NO stores with DETC [32].

detected by EPR signals (Fig. 12). We proposed that the DETC-induced decrease in BP 5 h after the DNIC injection was due to the depletion of NO stores induced by DETC.

We proposed a parameter describing the efficiency of NO storage in vascular wall, the maximum amount of NO stores, which can be accumulated during incubation of an isolated blood vessel with an NO donor, specifically with DNIC. Using this method, we showed that the efficiency of NO store formation in blood vessels from SHR is more than twofold lower than in their genetic control WKY [33].

A tentative relationship between the NO level, NO stores and BP in normotensive and SHRs can be represented as shown in Fig. 13.

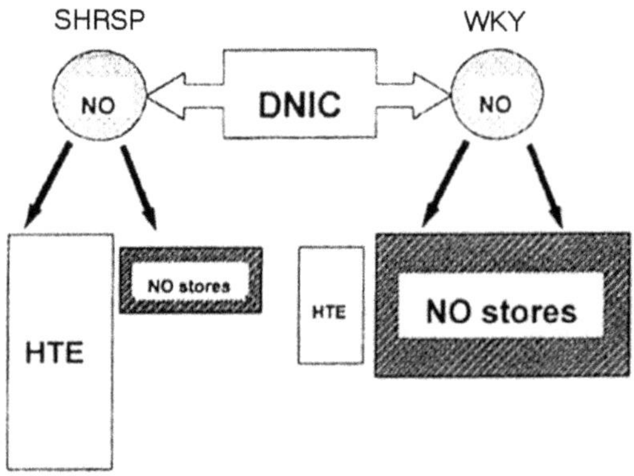

Fig. 13. A tentative relationship between the NO level, NO stores and blood pressure (BP) in normotensive and spontaneously hypertensive rats (SHRs) treated with DNIC. SHRSP, stroke-prone spontaneously hypertensive rats; WKY, Wistar–Kyoto rats; HTE, hypotensive effect; NO, nitric oxide.

After administration of the NO donor, a part of the excess NO binds to the NO store whereas the unbound part of NO exerts its biological effect, i.e. induces vasodilation and a decrease in BP. When the efficiency of NO storage is high, the volume of NO stores is large while the hypotensive effect is small. This situation apparently takes place in normotensive rats. On the contrary, when the efficiency of NO storage is low, much NO remains unbound and exerts a more pronounced hypotensive effect. This is the case in SHRs. This may explain why the hypotensive effect of DNIC is much more pronounced in SHRSP than in WKY rats.

The formation of NO stores is apparently an adaptive process. Binding of excess NO to NO stores may protect blood vessels and the body as a whole against NO cytotoxicity. Since SHRs are characterized by NO overproduction in vascular smooth muscles and macrophages resulting in vascular damage, impairment of endothelial nitric oxide synthase (eNOS) activity and endothelial dysfunction [34], the decreased capacity for NO binding observed in SHRSP may facilitate the development of these disorders.

Recently however, a possibility for modulation of NO storage capacity was demonstrated. We showed that chronic treatment of rats with DNIC increased the efficiency of NO storage and *vice versa*, chronic administration of the NO synthase inhibitor L-NNA reduced it [35]. This mechanism of vascular adaptation to chronic changes in NO level may be potentially used to improve the defense against nitrosative stress of the cardiovascular system in SHRSP.

Therefore, the DNICs can be considered as powerful long-acting hypotensive compounds which can ensure the development of new type of medicines at least with cardiovascular activity.

THE VASODILATORY ACTIVITY OF DNICs

It was reasonable to suggest that high hypotensive activity of the DNICs was caused by their capacity for inducing vasorelaxation. Organ bath experiments performed on pre-contracted denuded isolated ring segments of rat aorta supported this idea completely [20]. Original recordings of ring segment tension for monomeric and dimeric forms of DNIC with cysteine (monomeric DNIC 1:20 and dimeric DNIC 1:2, respectively; see Chapter 2) and comparison with gaseous NO and acetylcholine (Ach) are presented in Fig. 14. The experiments with Ach-induced release of endothelium-derived relaxing factor (EDRF) from endothelial cells [36] were performed on ring segments with preserved endothelium. These preparations were treated with atropine (Atr), 10^{-5} M to prevent the effect of Ach.

For all vasodilators, superoxide dismutase (SOD) potentiated their vasodilatory action. Similarity between the vasorelaxing activity of DNIC with cysteine and EDRF was observed. This similarity was seen in the slow kinetics of vessel tone restoration after the addition of DNIC or Ach at maximal concentrations (10^{-5} M) (Fig. 14). For both the agents, 10^{-5} M hemoglobin (Hb) added to the organ bath, restored the vessel tone, independent of the extent of tension recovery at the time of Hb administration. Finally, pretreatment of aortic ring segments with DETC (10^{-3} M) led to a decrease in the vasodilatory activity both of DNIC and Ach (EDRF), and prevented the vessel tone recovery (Fig. 15). In contrast, when DETC was applied after the addition of DNIC or Ach (10^{-5} M), a rapid restoration of vascular tension was observed (Fig. 15).

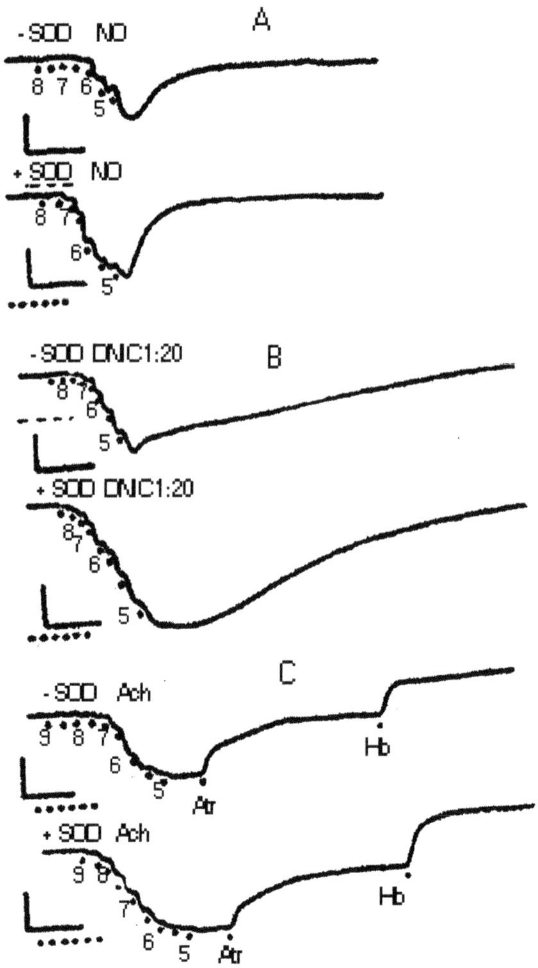

Fig. 14. Effects of nitric oxide (NO) (A), DNIC with cysteine 1:20 (B) and acetylcholine (Ach) (C) in the absence (−SOD) or in the presence (+SOD) of superoxide dismutase (SOD 30 U/ ml), on norepinephrine (NE, 10^{-7} M)-induced contractions of isolated ring segments from rat aorta. The dots show the addition of the agents (in −log M) atropine (Atr, 10^{-5} M) and hemoglobin (Hb, 10^{-5} M). Dotted line shows passive tension before the addition of NE. Vertical bars, 1 g; horizontal bars, 5 min [20].

Long-lasting relaxation induced by DNIC or Ach after pretreatment of the blood vessels with DETC was not due to an effect on the contractile ability of preparations, since after 60 min of rest and extensive washing with PBS the vessel segments responded normally to NE and endothelium-dependent and -independent relaxing agents.

We can speculate that the long-lasting vessel relaxation characteristic of the ring segments pretreated with DETC prior to DNIC resulted from the formation of high amounts of RS—NOs during degradation of DNIC induced by DETC as shown in Scheme 1. A similar

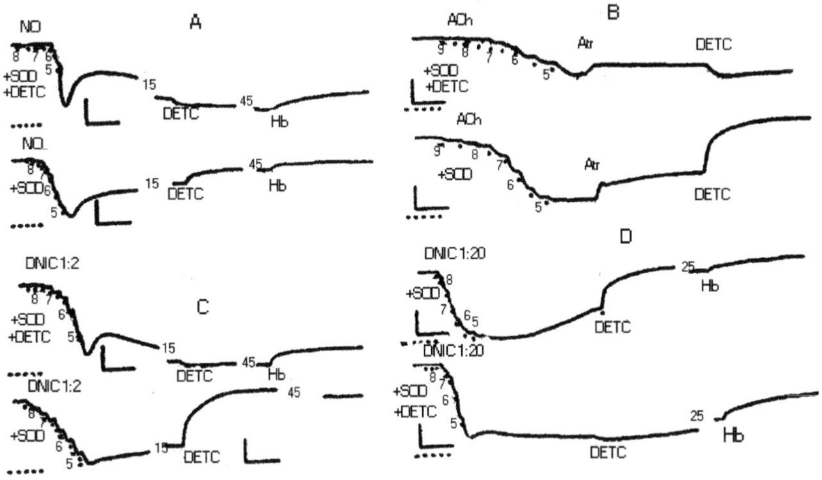

Fig. 15. Effect of diethyldithiocarbamate (DETC, 10^{-3} M) on the relaxation of rat aortic rings with (B) and without (A) endothelium pre-contracted with norepinephrine (NE, 10^{-7} M). Experiments were carried out in the presence of superoxide dismutase (SOD, 30 U/ml). Relaxation was induced by (A) nitric oxide (NO); acetylcholine (Ach); and (C) dinitrosyl iron complex with cysteine (DNIC) 1:2 and (D) DNIC 1:20. Effects of subsequent addition of DETC and hemoglobin (Hb, 10^{-5} M) are also shown. In (B), atropine (Atr, 10^{-5} M) was used to block endothelial muscarinic receptors. The dotted line indicates addition of the agents (in –log M). Vertical bars, 1 g, horizontal bars, 5 min. Numbers indicate the interval between recordings in minutes. The dotted line indicates the passive tension before the addition of norepinephrine (NE). These traces are representative of seven experiments [21].

hypothesis was made above to explain the brief drop of BP in anesthetized or conscious rats successively treated with DNIC and DETC (Figs. 3 and 6). In contrast, the relaxation induced by DNIC in isolated vessel segments in the presence of DETC had a long-lasting nature. Possibly, it was due to much more intensive accumulation of RS—NO molecules in the organ bath medium because all added DNIC molecules reacted with DETC. This was not the case when DETC was added 15 min after DNIC administration. This time DNIC decomposition led to a decrease in the amount of RS—NOs, which could have formed during the subsequent degradation of remaining DNICs by DETC. As a result, the vasorelaxation was eliminated.

A striking similarity between the effects of DETC on time course of BP on the addition of DNIC or Ach to the preparations allows to suggest that the compound accumulated in aorta segments by the time of Atr addition could be the DNIC. This is in line with the hypothesis advanced 15 years ago that EDRF was identical to DNIC [37]. Being degraded by DETC this compound produced RS—NOs. When the latter is accumulated in sufficient amounts it exerted a hypotensive effect.

Fully reversible (transient or T-type) or long-lasting (sustained or S-type) vasodilator responses were demonstrated for the pre-contracted, internally perfused rat tail artery after a bolus treatment with two iron–sulfur cluster nitrosyls (heptanitrosyltetra-μ3-thioxotetraferrate or tetranitrosyltetra-μ3-sulfidotetrahedro-tetrairon) [38]. Both the compounds produced a T-type response only at low doses below a critical threshold concentration, 10^{-4} or 10^{-5} M. The S-type response was observed at higher concentrations of both the vasodilators that comprised an initial, rapid drop of vessel tone, followed by incomplete tone restoration, resulting in a plateau of reduced tone persisting for a few hours. SNP or *S*-nitroso-*N*-acetylpenicillamine (SNAP) produced T-type responses only. DesoxyHb (NO scavenger) or methylene blue (MB), a guanylate cyclase inhibitor, initiated a prompt and complete recovery of vessel tone of all agonist-induced tone.

The T-type response was attributed to free NO released from added iron-nitrosyl cluster at the time of injection. With respect to S-type response, it was attributed to NO generated by gradual decomposition of a "store" of iron-nitrosyl complexes within the tissue. This idea was supported by the results of histochemical studies on arterial tissue treated with iron–sulfur-nitrosyl clusters. They showed high accumulation of ferrous iron, derived from these clusters. These iron deposits were predominantly located in endothelial cells.

The authors underline the similarity between vasorelaxing responses to the two used iron-nitrosyl complexes and above-considered DNIC with thiol-containing ligands, which can also induce hypotension in animals considered above or inhibit platelet aggregation [15–19,21,22]. The vasodilator and hypotensive actions of the DNICs also exhibit two-phase kinetics resembling the S-type response observed for the two used in [38] iron-nitrosyl complexes. These data also indicate the formation of iron-nitrosyl stores in tissues treated with DNICs. Interestingly, in accordance with the data described in [39] an injection of heptanitrosyltetra-μ3-thioxotetraferrate ("Roussin's Black Salt") into the rat led to the appearance of protein-bound DNICs in animal tissues *in vivo*.

A similar time course type of relaxation was also observed in isolated blood vessels for DNIC with non-thiolic ligands, DNIC with phosphate or bathocuproine disulfonate [40]. It is reasonable to suggest that long-lasting vasorelaxation induced by these DNICs was also due to the formation of protein-bound DNICs in vessel tissues acting as a depot of iron-nitrosyls. This was supported by experiments where DNIC with phosphate was injected into mice: formation of protein-bound DNIC was observed in the animal body [41].

ROLE OF DNIC/NITRIC OXIDE STORES IN PROTECTION AGAINST NITRIC OXIDE OVERPRODUCTION

Formation of NO stores in vascular wall begins after any increase in NO level induced by the stimulation of NO synthesis or administration of NO donors. An efficient method for the stimulation of NO synthesis and correspondingly, formation of NO stores is adaptation to environmental factors such as heat, exercise, mild stress or hypoxia [42]. We tested the hypothesis that DNIC/NO stores not only can provide a hypotensive effect in situations of absolute or relative NO deficiency but also can prevent NO overproduction and related injuries [43,44]. Heat stroke was used as an inductor of NO overproduction, which resulted in acute hypotension and death of some rats [45]. Adaptation to mild restraint stress was used as a means of prior increase in NO level and protection against heat stroke and NO overproduction. It was demonstrated that the protection was associated with the formation of NO stores in the vascular wall. Indeed when small doses of L-NNA were injected prior to each session of adaptation, neither protective effects developed nor NO stores formed. At the same time, injections of the NO donor DNIC with glutathione as ligand induced the formation of NO stores and completely mimicked the protection [46]. The protective effect of NO donor could not be due to the immediate action of NO because the lifetime of intravenous DNIC in the organism was about 3–5 h [29], while the effect of exogenous DNIC on both survival of rats and NO overproduction was observed 24 h after heat stroke [46]. Potential mechanisms of the protective effects of NO stores are the restriction of activity and/or expression of NOS by negative feedback [47] or the removal of excess NO by binding in NO stores. This compensatory mechanism helps in the prevention of toxic effects of increased NO formed in the process of adaptation.

Further evidence for the protective role of NO stores in NO overproduction was obtained on the rat model of Alzheimer's disease [48]. It is known that development of Alzheimer's disease is associated with toxic effects of excess NO overproduced by neurons and glia [49]. Experiments were carried out on anesthetized rats at natural ventilation. Local cerebral blood flow was continuously monitored with a laser Doppler probe. NO stores in cerebral blood vessels were detected by the vasodilatory response to *N*-acetylcysteine evident as an increase in cerebral blood flow. Since the size of NO stores is statistically significantly correlated with NO production [50], appearance and increase of NO stores may be regarded as an indirect marker of NO overproduction. Furthermore, the overproduction of NO was confirmed by the measurement of nitrite and nitrate in brain tissue.

NO stores were absent in cerebral blood vessels of control rats. In rats pretreated with DNIC, we observed an increase in cerebral blood flow, which indicated the presence of NO stores. In both hypoxia-adapted and Aβ-treated rats, *N*-acetylcysteine also revealed NO stores. In rats injected with Aβ after adaptation, the size of NO stores was significantly larger than in non-adapted rats treated with Aβ or adapted rats (Fig. 16).

We suggest that adaptation increased the ability of blood vessels for storing NO complexes. Indeed, as shown in the bottom diagram, chronic elevation of NO induced by daily injections of NO donor or by daily sessions of hypoxia, significantly increased the efficiency

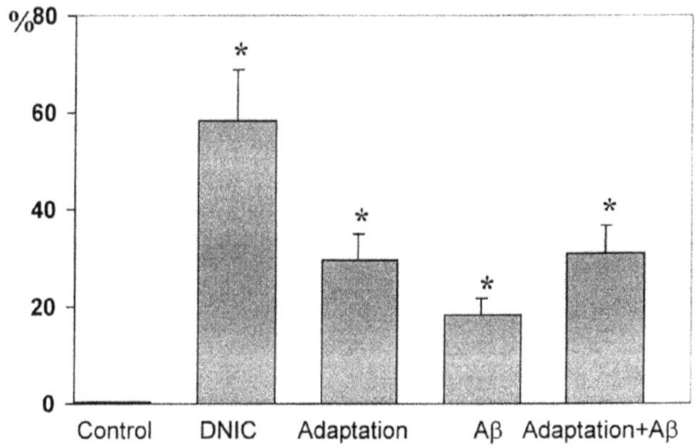

Fig. 16. Effect of DNIC, adaptation to hypoxia, Aβ(25-35) and Aβ(25-35) after pre-adaptation to hypoxia on the formation of NO stores in cerebral blood vessels. Bars reflect increases in cerebral blood flow in response to *N*-acetylcysteine in percent of baseline value. *Significant difference from control, $p < 0.05$ [50].

of binding NO to NO stores. This adaptive mechanism may enhance protection against NO overproduction.

DNIC CAPACITY OF INHIBITING PLATELET AGGREGATION

Investigations in the 1980s demonstrated that DNICs with various anion ligands (thiosulfate, cysteine or phosphate) are capable of preventing platelet aggregation induced by diverse agents (ADP, collagen, thrombin or epinephrine) [18,19]. This activity was suggested to be mediated by NO release from the complexes. According to the earlier proposed mechanisms [51], the activation of guanylate cyclase by NO radicals in platelets leads to the accumulation of guanosine 3′,5′-cyclic monophosphate (cGMP) attenuating the formation of diacylglyceride, the activator of protein kinase C, which is a factor that directly suppresses the ability of platelets to aggregate.

Comparative studies on human platelet aggregation in the presence of SNP and DNICs added 3 min before the inductor of the process (ADP) showed a much higher inhibiting activity of the latter (Fig. 17).

The effect correlated with the extensive accumulation of cGMP in platelets [19]. It cannot completely be excluded that this effect was due to the toxic action of the DNICs on platelets. However, the experiments with addition of phorbol 12-myristate 13-acetate 4-*O*-methyl ether (MPMA) did not support this hypothesis. Due to the capability for activation of protein kinase C, MPMA stimulates platelet aggregation even in the presence of cGMP [52]. It was demonstrated that MPMA efficiently induced platelet aggregation despite the addition of DNICs (Fig. 18). Moreover, MPMA stimulated the platelet aggregation when added after DNICs at the time when the process induced by ADP was blocked by DNIC addition (Fig. 18).

Interestingly, DNIC administration could induce the process of platelet disaggregation in human plasma. The effect was more notable than that characteristic of SNP (Fig. 19).

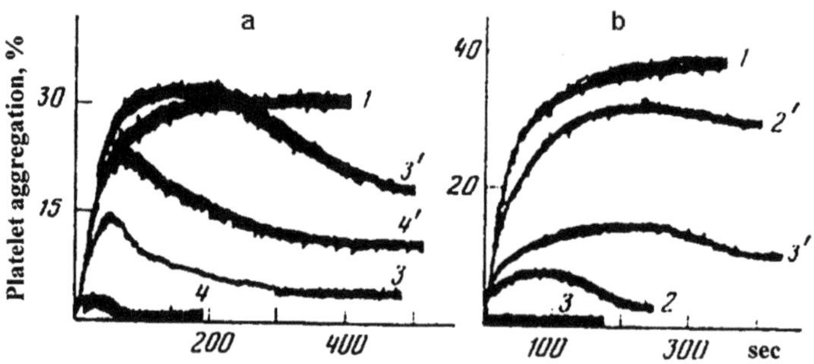

Fig. 17. Effect of DNIC with thiosulfate and nitroprusside on platelet aggregation induced by ADP (10^{-2} M). Curve 1, control preparation, curves 2–4, in the presence of 7×10^{-5}, 3.5×10^{-4} or 7×10^{-4} M DNIC (non-dashed numbers) or nitroprusside (dashed numbers), respectively. The agents were added before the inductor administration. Platelet plasma preparations are prepared from the blood of two donors (a) and (b) [18].

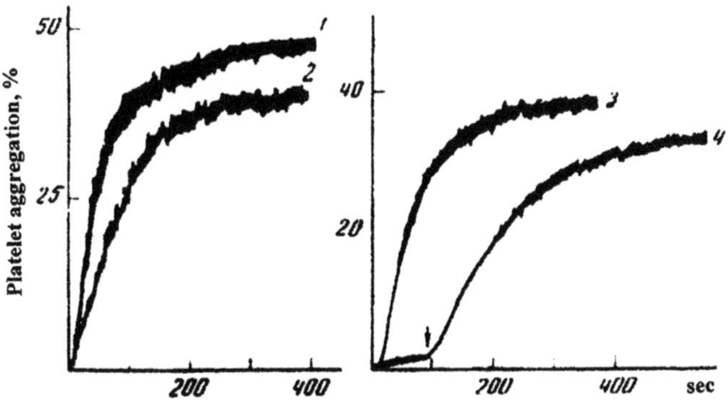

Fig. 18. Effect of phorbolic ether on platelet aggregation in the presence of DNIC with thiosulfate. Curves 1 and 3 demonstrate platelet aggregation induced by phorbolic ether (10^{-5} M) or ADP (10^{-2} M), respectively. DNIC with thiosulfate (3.5×10^{-4} M) was added 3 min before the addition of phorbolic ether or ADP (curves 2, 4, respectively). Consequently, phorbolic ether (10^{-5} M) was added at the time point shown by the arrow (curve 4). Platelet plasma preparations are prepared from blood of two donors [(a) and (c), curves 1, 2 and 3, 4, respectively] [18].

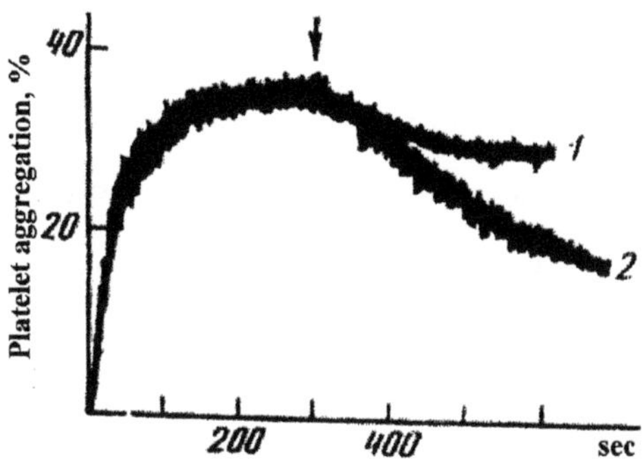

Fig. 19. Disaggregating effect of nitroprusside (curve 1) or DNIC with thiosulfate (curve 2) added to platelet plasma [donor (d)] at the concentration of 3.5×10^{-4} M after pre-aggregation of platelets induced by ADP (10^{-2} M) [18].

EPR assay of human blood plasma containing platelets revealed the formation of protein-bound DNIC in the plasma after the addition of low-molecular DNIC with thiosulfate. Evidently, these complexes function as NO donors inhibiting the process of platelet aggregation. No accumulation of protein-bound DNICs was detected immediately in platelet cells [19].

The capability of Roussin's Black Salt, another representative of iron–sulfur-nitrosyl compounds to inhibit platelet aggregation was demonstrated in the 1990s by British researchers [53]. Inhibition of platelet aggregation by this compound was eliminated by the NO scavenger Hb. Addition of a selective inhibitor of cGMP phosphodiesterase (M&B22948) enhanced the effect of Roussen's Black Salt [53]. Therefore, the data indicated that

Roussen's Black Salt affected the process of platelet aggregation as an NO donor that activated guanylate cyclase.

In conclusion, the data presented in this chapter indicate a high biological potential of DNICs with various anion ligands. Without any doubt, they can be considered as the promising basis for the development of medicines with a wide range of therapeutic effects. Moreover, the investigations on biological activities of synthetic DNICs described in this chapter can give insights into the mechanisms of endogenous DNIC function, which forms in living systems generating NO by enzymatic or non-enzymatic pathways.

ACKNOWLEDGMENT

This work was supported by the Russian Foundation of Basic Researches (Grant 05-04-49383).

REFERENCES

1 Lancaster JR, Hibbs JB. EPR demonstration of iron-nitrosyl complex formation by cytotoxic activated macrophages. Proc. Natl. Acad. Sci. USA 1990; 87: 1223–1227.
2 Drapier J-C, Pellat C, Henry Y. Generation of EPR-detectable nitrosyl-iron complexes in tumor target cells cocultured with activated macrophages. J. Biol. Chem. 1991; 266: 10162–10167.
3 Stadler J, Bergonia HA, DiSilvio M, Sweetland MA, Billiar TR, Simmons RL, Lancaster JR. Nonheme nitrosyl-iron complex formation in rat hepatocytes: detection by EPR spectroscopy. Arch. Biochem. Biophys. 1993; 302: 4–11.
4 Mülsch A, Mordvintcev PI, Vanin AF, Busse R. Formation and release of dinitrosyl iron complexes by endothelial cells. Biochem. Biophys. Res. Commun. 1993; 196: 1303–1308.
5 Geng Y-L, Petersson A-S, Wennmalm A, Hannson G. Cytokine-induced expression of nitric oxide synthase results in nitrosylation of heme and nonheme iron proteins in vascular smooth muscle cells. Exp. Cell Res. 1994; 214: 418–424.
6 Kim Y-M, Chung H-T, Simmons RL, Billiar TR. Cellular non-heme iron content is a determinant of nitric oxide mediated-apoptosis, necrosis and caspase inhibition. J. Biol. Chem. 2000; 275: 10954–10961.
7 Mikoyan VD, Serezhenkov VA, Brazhnikova NV, Kubrina LN, Khachtryan GN, Vanin AF. Formation of paramagnetic nitrosyl complexes of nonheme iron in the animal organism with the participation of nitric oxide from exogenous and endogenous sources. Biophysics (Translated from Russian) 2004; 41: 110–116.
8 Vanin AF. Dinitrosyl iron complexes are endogenous signaling agents in animal and human cells and tissues (a hypothesis). Biofizika (Rus.) 2004; 49: 581–586.
9 Stamler JS. Redox signaling: nitrosylation and related target interactions of nitric oxide. Cell 1994; 78: 931–936.
10 Hausladen A, Privalle CT, Keng T, DeAngelo J, Stamler JS. Nitrosative stress: activation of the transcription factor OxyR. Cell 1996; 86: 719–729.
11 Stamler JS, Toone EJ, Sucher NJ. S(NO) signals: translocation, regulation, and a consensus motif. Neuron 1997; 18: 691–696.
12 Butler AR, Rhodes P. Chemistry, analysis, and biological roles of S-nitrosothiols. Anal. Biochem. 1997; 249: 1–9.
13 Gaston B. Nitric oxide and thiol groups. Biochim. Biophys. Acta 1999; 1411: 323–333.
14 Vanin AF. Dinitrosyl-iron complexes and S-nitrosothiols are two possible form for stabilization and transport of nitric oxide in biological systems. Biochemistry (Mosc.) 1998; 67: 782–793.
15 Vanin AF, Mordvintcev PI, Kleschyov AL. Role of nitric oxide and iron in hypotensive action of nitrosyl iron complexes with various anion ligands. Stud. Biophys. 1985; 105: 93–102.

16 Mordvintcev PI, Putintcev MD, Galagan ME, Oranovskaya EV, Medvedev OS, Vanin AF. Hypotensive activity of dinitrosyl iron complexes with proteins in narcotized animals. Bull. Vsesoyuznogo Kardiolog. Nauchnogo Centra AMNSSSR. 1988; 1: 46–51.
17 Galagan ME, Oranovskaya EV, Mordvintcev PI, Medvedev OS, Vanin AF. Hypotensive effect of dinitrosyl iron complexes in conscious animals. Bull. Vsesoyuznogo Kardiolog. Nauchnogo Centra AMNSSSR. 1988; 2: 75–80.
18 Mordvintcev PI, Rudneva VG, Vanin AF, Shimkevich LL, Khodorov BI. The inhibition effect of low-molecular dinitrosyl iron complexes on platelet aggregation. Biokhimiya (Rus.) 1986; 51: 1851–1857.
19 Kuznetcov VA, Mordvintcev PI, Dank EK, Yurkiv VA, Vanin AF. Low-molecular and protein-bound dinitrosyl iron complexes as inhibitors of platelet aggregation. Voprosy Med. Khimii 1988; 5: 43–46.
20 Vedernikov YP, Mordvintcev PI, Malenkova IV, Vanin AF. Similarity between the vasorelaxing activity of dinitrosyl iron cysteine complexes and endothelium-derived relaxing factor. Eur. J. Pharmacol. 1992; 211: 313–317.
21 Vedernikov YP, Mordvintcev PI, Malenkova IV, Vanin AF. Effect of diethyldithiocarbamate on the activity of nitric oxide-releasing vasodilators. Eur. J. Pharmacol. 1992; 212: 125–128.
22 Chetverikov AG, Ruuge EK, Burbaev DSh, Vanin AF. The change of the shape of the EPR signal with g_{av} 2.03 in biological objects depending on the conditions of the registration. Biofizika (Rus.) 1969; 14: 932–935.
23 Vanin AF, Kiladze SV, Kubrina LN. On including of low molecular SH containing compounds in nitrosyl non-heme iron complexes in non-cellular or cellular preparations. Biofizika (Rus.) 1975; 20: 1068–1072.
24 McDonald CC, Phillips WD, Mower HF. An electron spin resonance study of some complexes of iron, nitric oxide and anionic ligands. J. Am. Chem. Soc. 1965; 87: 3319–3326.
25 Goodman BA, Raynor JB, Symons MCR. Electron spin resonance of bis(NN-diethylthiocarbamate)nitrosyliron. J. Chem. Soc. A. 1969; 2572–2575.
26 Vanin AF, Mordvintcev PI, Kleschyov AL. Appearance of nitric oxide in animal tissues in vivo. Stud. Biophys. 1984; 102: 135–146.
27 Galagan ME, Kiladze SV, Vanin AF. Dinitrosyl iron complex reactivity in a respect to diethyldithiocarbamate in the blood and narcotized rats: specifical physico-chemical and physiological properties of the products of this reaction. Biofizika (Rus.) 1997; 42: 687–692.
28 Vanin AF, Serezhenkov VA, Mikoyan VD, Genkin MV. The 2.03 signal as an indicator of dinitrosyl-iron complexes with thiol-containing ligands. Nitric Oxide: Biol. Chem. 1998; 2: 224–234.
29 Manukhina EB, Malyshev IY, Malenyuk EB, Zenina TA, Pokidyshev DA, Mikoyan VD, Kubrina LN, Vanin AF. Hypotensive effect and tissue distribution of the nitric oxide donor dinitrosyl iron complex. Bull. Exp. Biol. Med. (Rus.) 1998; 125: 30–33.
30 Manukhina EB, Mashina SY, Smirin BV, Lyamina NP, Senchikhin VN, Vanin AF, Malyshev IY. Role of nitric oxide in adaptation to hypoxia and adaptive defense. Physiol. Res. 2000; 49: 89–97.
31 Smirin BV, Vanin AF, Malyshev IY, Pokidyshev DA, Manukhina EB. Nitric oxide stores in blood vessels in vivo. Bull. Exp. Biol. Med. (Rus.) 1999; 127: 629–632.
32 Mashina SY, Vanin AF, Serezhenkov VA, Kubrina LN, Malenkova IV, Malyshev IY, Manukhina EB. Detection and evaluation of NO stores in awake rats. Bull. Exp. Biol. Med. (Rus.) 2003; 136: 26–29.
33 Manukhina EB, Malyshev IY. Role of free NO and NO stores in protective effects of adaptation to hypoxia. In Adaptation Biology and Medicine. Vol. 4: Current Concepts (Singal PK, Takeda N, eds.), Narosa Publishing House, New Delhi, 2005, pp. 82–94.
34 Wu C-C, Yen M-H. Nitric oxide synthase in spontaneously hypertensive rats. Biomed. Sci. 1997; 4: 249–255.
35 Vlasova MA, Smirin BV, Pokidyshev DA, Masina SY, Vanin AF, Malyshev IY, Manukhina EB. Mechanism of cardiovascular adaptation to chronic changes in nitric oxide (NO) level in the organism. Bull. Exp. Biol. Med. (Rus.) 2006; 142: 626–630.
36 Ignarro LJ. Endothelium-derived relaxing factor: actions and properties. FASEB J. 1989; 3: 31–49.
37 Vanin AF. Endothelium-derived relaxing factor is a nitrosyl iron complex with thiol ligands (Hypothesis). FEBS Lett. 1991; 289: 1–5.

38 Flitney FW, Megson IL, Flitney DE, Butler AR. Iron-sulfur cluster nitrosyls: a novel class of nitric oxide generator: mechanism of vasodilator action on rat isolated tail artery. Br. J. Pharmacol. 1992; 107: 842–848.

39 Chiang RW, Woolum JC, Commoner B. Further study of the properties of the rat liver protein involved in a paramagnetic complex in the liver of carcinogen-treated rats. Biochim. Biophys. Acta 1972; 257: 452–460.

40 Vanin AF, Muller B, Alencar JL, Lobysheva II, Nepveu F, Stoclet J-C. Evidence that intrinsic iron but not intrinsic copper determines S-nitrosocysteine decomposition in buffer solution. Nitric Oxide: Biol. Chem. 2002; 7: 194–209.

41 Vanin AF, Varich VJ. Formation of nitrosyl complexes of non-heme iron (complexes 2.03) in animal tissues in vivo. Biofizika (Rus.) 1979; 24: 666–670.

42 Manukhina EB, Malyshev IY. Role of nitric oxide in protective effects of adaptation. In Adaptation Biology and Medicine. Vol. 3. (Moravec J, Takeda N, Singal PK, eds.), Narosa Publishing House, New Delhi, 2002, pp. 312–327.

43 Manukhina EB, Pokidyshev DA, Malenyuk EB, Malyshev IY, Vanin AF. The protective effect of nitric oxide in heat shock. Izvestia Akad. Nauk. Ser. Biol. (Rus.) 1997; 1: 54–58.

44 Manukhina EB, Pokidyshev DA, Malyshev IY. Prevention of acute hypertension and endothelial over-activation in heat shock by adaptation to stress exposure. Bull. Exp. Biol. Med. (Rus.) 1997; 124: 380–384.

45 Manukhina EB, Malyshev IY, Mikoyan VD, Kubrina LN, Vanin AF. Increase of nitric oxide production in rat organs in heat shock. Bull. Exp. Biol. Med. (Rus) 1996; 121: 520–523.

46 Smirin BV, Pokidyshev DA, Malyshev IY, Vanin AF, Manukhina EB. Nitric oxide storage as a factor of adaptive defense. Ross. Fiziol. Zh. Im. I. M. Sechenova (Rus.) 2000; 86: 447–454.

47 Assreuy J, Cunha FQ, Liew FY. Feedback inhibition of nitric oxide synthase activity by nitric oxide. Br. J. Pharmacol. 1993; 108: 833–887.

48 Mashina SY, Alexandrin VV, Goryacheva AV, Vlasova MA, Vanin AF, Malyshev IY, Manukhina EB. Adaptation to hypoxia prevents disorders of cerebral circulation in neurodegenerative damage: role of nitric oxide. Bull. Exp. Biol. Med. (Rus.) 2006; 142: 132–135.

49 de la Torre JC, Stefano GB. Evidence that Alzheimer's disease is a microvascular disorder: the role of constitutive nitric oxide. Brain Res. Rev. 2000; 34: 119–136.

50 Manukhina EB, Malyshev IY, Smirin BV, Mashina SY, Saltykova VA, Vanin AF. Production and storage of nitric oxide in adaptation to hypoxia. Nitric Oxide: Biol. & Chem. 1999; 5: 393–401.

51 Mellion BT, Ignarro LJ, Meyers CB, Ohlstein EH, Ballot B, Ryman AL, Kadowitz PJ. Inhibition of platelet aggregation by S-nitrosothiols. Heme-dependent activation of soluble guanylate cyclase and activation of cyclic GMP accumulation. Mol. Pharmacol. 1983; 23: 653–664.

52 Castagna M, Takaio Y, Kaibuchi K, Sano K, Kikkawa U, Nishizuka Y. Direct activation of calcium-activated, phospholipid-dependent protein kinase by tumor-promoting phorbol esters. J. Biol. Chem. 1982; 257: 7847–7851.

53 Ludbrook SB, Scrutton MC, Joannou CL, Cammack R, Hughes MN. Inhibition of platelet aggregation by Roussin's Black Salt, sodium nitroprusside and other metal nitrosyl complexes. Platelets 1995; 6: 209–212.

CHAPTER 4

DNICs and intracellular iron: nitrogen monoxide (NO)-mediated iron release from cells is linked to NO-mediated glutathione efflux via *MRP1*

Des R. Richardson*

Department of Pathology, Iron Metabolism and Chelation Program, Blackburn Building, University of Sydney, Sydney, New South Wales, 2031, Australia

GENERAL INTRODUCTION: NITROGEN MONOXIDE IS A VITAL MESSENGER MOLECULE AND CYTOTOXIC EFFECTOR

Nitrogen monoxide (NO) is a crucial effector and messenger molecule that plays roles in a wide variety of biological processes in animals and plants [1,2]. In fact, NO has roles in neurotransmission, smooth muscle relaxation, blood clotting, iron (Fe) metabolism and contributes to the cytotoxicity of activated macrophages against tumor cells and intracellular parasites [2–7]. The production of NO in living organisms is mediated *via* the nitric oxide synthase (NOS) family of enzymes [1,8,9]. Many of the regulatory functions of NO are due to its ability to bind Fe within the haem prosthetic group of guanylate cyclase [2,10,11]. In fact, the high affinity of NO for Fe is a well-known branch of coordination chemistry [10].

NITROGEN MONOXIDE FORMS INTRACELLULAR COMPLEXES WITH IRON

There is strong evidence that NO forms intracellular complexes that play important roles in biological processes. Apart from the formation of haem–NO complexes in guanylate cyclase, it is well-known that the cytotoxic effect of NO is, at least in part, due to its interaction

* Author for Correspondence. E-mail: d.richardson@pathology.usyd.edu.au

with Fe in a number of proteins, including: (i) the rate-limiting enzyme in DNA synthesis, ribonucleotide reductase [12]; (ii) the Fe storage protein, ferritin [13]; (iii) haem-containing proteins [14–16]; (iv) the [Fe—S] cluster enzyme catalysing the final step in the haem synthesis pathway, ferrochelatase [17]; (v) other [Fe—S] cluster proteins involved in energy metabolism such as mitochondrial aconitase and those in complex I and II of the electron transport chain [18,19]; and (vi) the [Fe—S] cluster-containing molecule, iron-regulatory protein-1 (IRP1), that plays a role in regulating Fe homeostasis [20–25]. In fact, tumor cells co-cultured with activated macrophages exhibit a decrease in DNA synthesis with a concomitant loss of 64% of tumor cell Fe over 24 h [3,4]. The Fe released was postulated to be derived from [Fe—S]-containing proteins and enzymes [4,18,26,27] and has been hypothesized to be a complex of NO and Fe with thiol-containing ligands, for example glutathione (GSH) or cysteine (Cys) [28–30]. In fact, a wide variety of studies using electron paramagnetic resonance (EPR) spectroscopy demonstrated that NO forms dinitrosyl-dithiolato-Fe complexes (DNICs) in activated macrophages, tumor target cells and other tissues [26,27,29,31–34]. The above just mentions a range of subjects relevant for the fate of NO in the presence of iron. More details can be found in other chapters of this book. Additional information on the structure, properties and effects of DNICs are given in Chapters 2, 3 and 5. The effect of NO on iron–sulfur clusters is discussed in Chapters 5 and 6. The role of iron in apoptosis is the subject of Chapter 12. The class of nitrosyl-iron complexes with dithiocarbamate ligands is discussed in Chapters 18 and 19.

CELLULAR IRON METABOLISM

Before describing in detail about the effects of NO on intracellular Fe pools, it is important to discuss the mechanisms involved in Fe metabolism. Iron is fundamental for life as it is a cofactor of enzymes such as cytochrome c and ribonucleotide reductase, which are essential for ATP production and DNA synthesis, respectively (for reviews see [11,35]).

The maintenance of mammalian iron (Fe) homeostasis begins with the ability to absorb and control dietary Fe from the gut *via* enterocytes (for review see [36]). Most Fe in food is ingested as the ferric form (Fe^{3+}) and is fairly insoluble. Hence, the absorption of dietary Fe requires the reduction of the ferric state to its ferrous form and this was thought to be achieved by the Fe-regulated duodenal cytochrome b enzyme (Dcytb) [37]. However, more recent studies using Dcytb knockout mice have demonstrated that this molecule is not essential for normal Fe metabolism in the whole animal [38]. Once Fe^{3+} is reduced to Fe^{2+}, it is transported into the cell by the divalent metal ion transporter 1 (DMT1; also known as the divalent cation transporter or the natural resistance associated macrophage protein 2) [39–41]. Haem can also be transported into enterocytes and recently a candidate molecule that possesses this activity [haem carrier protein 1 (HCP1)] has been described [42]. However, while this protein appears to transport haem, there is no strong evidence as yet that it is the physiologically relevant mechanism. Intriguingly, another haem transporter known as the human feline leukemia virus subgroup C receptor (FLVCR) has also been identified [43]. However, its role in haem trafficking in the intestine remains unclear.

The trafficking of Fe in the enterocyte and its subsequent release into the bloodstream remains the subject of intensive research and is only briefly discussed herein (Fig. 1).

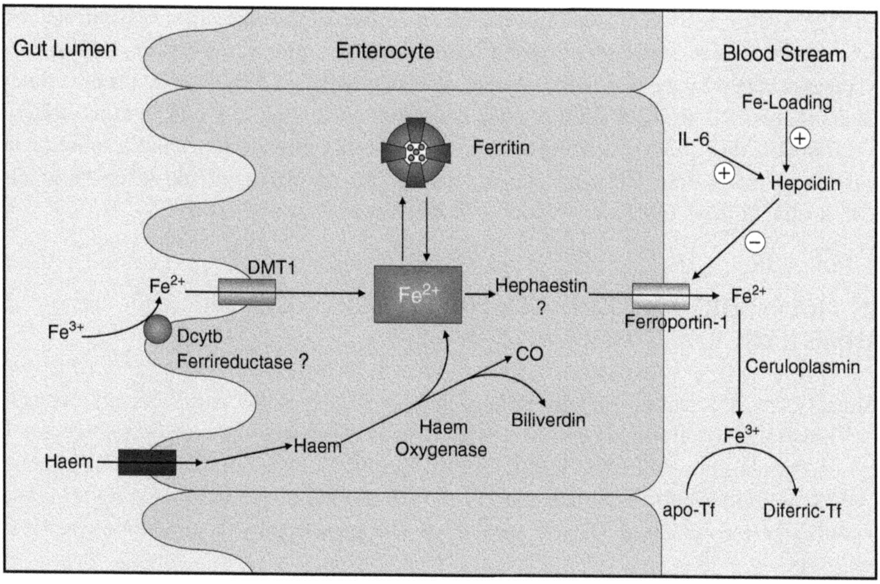

Fig. 1. Schematic illustration of the mechanism of Fe uptake by the enterocyte. Iron is internalized after inorganic Fe^{3+} in the diet is reduced to Fe^{2+} possibly by a ferrireductase known as Dcytb. However, the identity of this ferrireductase remains controversial [38]. Haem may be taken up from the gut lumen *via* specific transporters, e.g. haem carrier protein 1 (HCP1) [42] and/or feline leukemia virus subgroup C receptor (FLVCR) [43]. Internalized haem is then metabolized by haem oxygenase 1 to Fe^{2+}, bilirubin and carbon monoxide (CO). The Fe^{2+} probably then enters a compartment known as the intracellular Fe pool or is stored in ferritin. In a sequence of events that remains unclear, the ferroxidase hephaestin may be involved in the conversion of Fe^{2+} to Fe^{3+} and the subsequent release from the cell *via* ferroportin-1. The Fe efflux mediated by ferroportin-1 is thought to be Fe^{2+} which may then be oxidized by the ferroxidase activity of ceruloplasmin or apo-transferrin in the serum. The Fe^{3+} is then subsequently bound by apo-Tf to form diferric Tf. The release of Fe from the enterocyte is regulated at least to some degree by the peptide, hepcidin, which is synthesized by the liver. High levels of storage Fe in the liver or the inflammatory cytokine, interleukin-6 (IL-6), increase hepcidin expression that decreases Fe release from the enterocyte, probably by down-regulation of ferroportin-1 (see text for details). (Taken with permission from: Richardson DR. Curr. Med. Chem. 2005; 12: 2711–2729.)

The process involves a series of molecular events that include the function of the proteins, hephaestin and ferroportin-1 [44]. The latter molecule is also known as metal transporter protein 1 (MTP1; [45]) or Ireg1 [46]. Hephaestin is a transmembrane, multi-copper ferroxidase that has homology to the serum protein, ceruloplasmin [11,47]. By way of its ferroxidase activity, hephaestin may facilitate Fe export from intestinal enterocytes, perhaps in cooperation with the basolateral Fe transporter, ferroportin-1 [47] (Fig. 1). Ferroportin-1 has been suggested to be down-regulated by the direct binding of the iron-regulatory hormone, hepcidin [48] (see below). This physical interaction then results in the internalization and degradation of ferroportin-1, leading to decreased cellular Fe efflux [48]. However, the mechanism of Fe release does not only appear to be dependent upon these molecules. Indeed, it is likely that the serum ferroxidase ceruloplasmin also plays a role in Fe efflux [11,49]. Iron released from the enterocyte *via* ferroportin-1 is subsequently bound to the serum Fe-binding and transport protein, transferrin (Tf) (Fig. 1).

The absorption of Fe from the gut is under the control of a number of molecules, including DMT1 [39,40], hepcidin [50], hemojuvelin [51] and the hemochromatosis gene product

(HFE) [52]. The expression of DMT1 at the apical surface of the enterocyte is controlled by the presence of an iron-responsive element (IRE) in the 3′-untranslated region (UTR) of its mRNA that is bound by the two IRPs, namely IRP1 and IRP2 [11,53]. High levels of cellular Fe results in low IRP–RNA-binding activity that prevents binding of the IRPs to the IRE in the 3′-UTR of DMT1 [53,54], leading to decreased stability of the mRNA and a subsequent decrease in its translation. The opposite response occurs during the period when Fe levels are low, leading to high DMT1 expression and increased dietary Fe uptake.

IRON TRANSPORT AND UPTAKE: TRANSFERRIN AND THE TRANSFERRIN RECEPTOR 1

Transferrin (Tf) is a plasma protein which binds two Fe^{3+} atoms with high affinity (for reviews see [11,55]). Diferric Tf can be bound by cells expressing the transferrin receptor 1 (TfR1) on the plasma membrane. The uptake of Fe from Tf is controlled by TfR1 expression which is modulated by intracellular Fe levels *via* IRP1 and IRP2 (for reviews see [11,55,56]). The interaction of Tf with the TfR1 is also regulated by the binding of the HFE protein [57].

Once Tf is bound to TfR1, the complex is internalized *via* receptor-mediated endocytosis [11,55]. An ATP-dependent proton pump in the endosomal membrane allows the release of Fe^{3+} from Tf by mediating a decrease in endosomal pH [11]. Once released, ferric iron is probably reduced to the ferrous state by an endosomal ferrireductase. A candidate for this molecule, namely the six-transmembrane epithelial antigen of the prostate 3 (Steap3), has recently been discovered [58]. Transfer of the so-formed Fe^{2+} through the endosomal membrane is mediated by DMT1 [39,40,54,59]. After the release of Fe from Tf, and its transport through the endosomal membrane, Fe^{2+} becomes part of the elusive intracellular Fe pool [60], where it can be incorporated into haem and non-haem Fe-containing proteins. Inside the cell, Fe^{2+} can be either stored in ferritin where it is converted into Fe^{3+}, or used for metabolic functioning (e.g. haem synthesis; [11]). The nature of the labile intracellular pool (LIP) of Fe remains controversial. For instance, it may exist as compounds in the Fe^{2+} or Fe^{3+} state and the mechanism of how Fe is transported within cells remains a long unsolved question. Initially, Fe was suggested to be bound to low-molecular weight Fe complexes [60], while other studies have found no evidence of such intermediates [61,62]. Some evidence for the trafficking of Fe has been presented *via* organelle interactions and chaperone proteins [63]. After Fe delivery, the Tf–TfR1 complex returns to the plasma membrane *via* exocytosis and Tf is then released to the circulation for re-utilization [11].

EFFECT OF NITROGEN MONOXIDE ON INTRACELLULAR IRON METABOLISM

The high affinity of NO for Fe means that it will affect the activity of many Fe-containing proteins [7,11]. One of the most important examples of NO interacting with Fe-containing molecules is its effect on IRP1 that is known to regulate Fe metabolism [56]. In fact, NO can activate the RNA-binding activity of IRP1 that plays a role in the homeostatic regulation of cellular Fe homeostasis [20–22]. The mechanism by which NO exerts its effects on IRP1 is probably by both depleting intracellular Fe and by interacting with its [Fe—S] cluster [20,21,25,64]. As will be described in detail below, NO can enter cells and act to some degree like a chelator by binding Fe and inducing its release [25,65,66]. The NO-mediated Fe release is physiologically relevant, as it is observed upon the interaction of activated macrophages with tumor cells and other targets [3,4,67]. In addition, it is of interest to note that the effect of NO at inducing Fe release is quite specific for this molecule, as it is not observed with another closely related diatomic effector, namely carbon monoxide (CO) [66].

The relative roles of NO-mediated intracellular Fe release or direct interaction with the [Fe—S] cluster in activating IRP1–RNA-binding activity depend on the source of NO, the redox-related state of NO generated and the amount produced [25]. In contrast to IRP1, NO and $ONOO^-$ have been reported to decrease IRP2–RNA-binding activity [68,69], perhaps by a mechanism involving NO^+-mediated degradation [69]. Direct measurement of S-nitrosylation of critical sulfhydryl groups of IRP2 has been shown [70], suggesting this mechanism could be a physiological regulator of this molecule.

Considering that inducible NOS (iNOS) can be regulated by intracellular Fe, this could lead to an auto-regulatory-loop whereby low Fe levels induce iNOS expression [71]. This enzyme subsequently generates NO which activates IRP1–RNA-binding activity [71]. Increased IRP-1 binding to the 3′-IRE of *TfR1* mRNA would increase TfR1 expression that theoretically leads to elevated Fe uptake from Tf [56]. The enhanced cellular Fe levels would then induce the opposite effect by reducing iNOS transcription, resulting in decreased NO, and thus, IRP1–RNA-binding activity [71]. Upon this complex level of control, other studies using murine macrophages have shown that IRP1 gene expression can be down-regulated by NO [72]. Moreover, while NO can induce an increase in *TfR1* mRNA and protein expressions [20–22], the increase in Fe uptake from Tf is only minimal [22]. This is probably because NO can also inhibit Fe uptake from Tf and mobilize Fe from cells [65,73]. Hence, the effect of NO on cellular Fe metabolism is complex.

NO can decrease Fe uptake from Tf by cells *via* interfering with intracellular Fe trafficking and inducing cellular Fe mobilization [25,65,66]. The effect of NO in reducing total cellular Fe uptake and also Fe incorporation into ferritin is seen with CO also, that, like NO has a high affinity for Fe (Fig. 2A,B) [66]. Since NO was reported to form a complex with Fe in lactoferrin [74], it could be expected that based upon their high homology, a comparable reaction could occur with Tf. However, using EPR spectroscopy, it was demonstrated that the ability of NO to reduce Fe uptake from Tf [73] was neither due to direct removal of Fe from the protein nor the formation of an NO–Fe complex within this molecule [73,75]. In fact, the effect of both NO and CO in reducing Tf-bound Fe uptake appeared to be mediated by its ability to inhibit cellular ATP production [66,75], which is essential for cellular Fe internalization [76]. Considering that Tf binds high-spin Fe^{3+} and that its Fe-binding site does not interact with NO [75], it is questionable whether NO can form an NO–Fe complex with the lactoferrin–Fe complex as reported by others [74]. This is because the Fe-binding sites of these molecules are very similar [75].

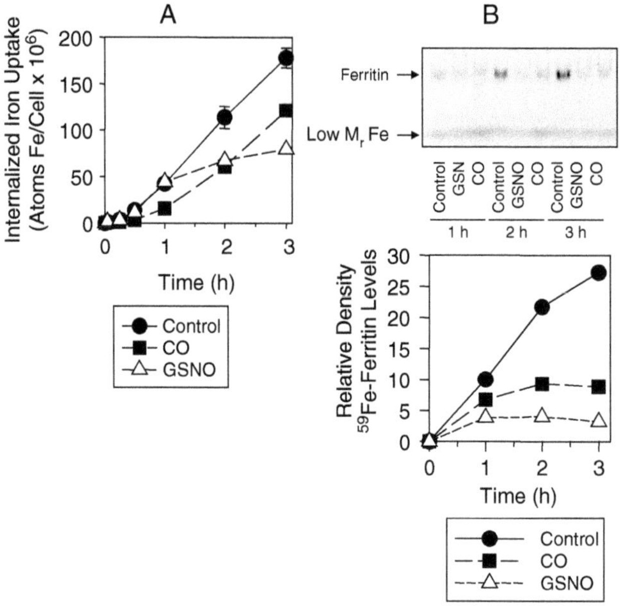

Fig. 2. Both CO gas and GSNO reduce: (A) ^{59}Fe uptake from ^{59}Fe–transferrin (^{59}Fe–Tf) and (B) ^{59}Fe incorporation into ferritin by LMTK$^-$ fibroblasts. (A) Cells were labelled with ^{59}Fe–Tf (0.75 μM) for 5–180 min at 37°C in the presence or absence of 2% CO gas or GSNO (0.5 mM). The cells were then washed four times on ice and incubated for 30 min with Pronase (1 mg/ml) at 4°C to separate internalized from membrane-bound ^{59}Fe. (B) The LMTK$^-$ cells were labelled with ^{59}Fe–Tf (0.75 μM) for 60–180 min at 37°C, washed, and native-PAGE ^{59}Fe-autoradiography performed. Densitometric analysis is presented below the autoradiograph. The results in (A) are expressed as mean of duplicate determinations in a typical experiment of 5 performed, while the results in (B) are representative of 3 separate experiments. (Taken with permission from Ref. [66].)

THE MECHANISM OF NITROGEN MONOXIDE-MEDIATED Fe RELEASE FROM CELLS

Regarding the effect of NO on cellular Fe release, initial studies using NO^+ generators (e.g. sodium nitroprusside) or $ONOO^-$ donors (e.g. SIN-1) [77] demonstrated that these agents did not result in appreciable Fe mobilization from cells [73]. In contrast, $NO^•$-releasing agents [e.g. *S*-nitroso-*N*-acetylpenicillamine (SNAP), *S*-nitrosoglutathione (GSNO) or spermine-NONOate (SperNO)] showed high efficacy [25,65,78]. The effect of these agents in inducing Fe release was due to their ability to generate NO, as their precursor compounds which do not bear the NO-moiety had no effect [25,65]. Hence, it was the $NO^•$ redox state which was important for forming intracellular Fe complexes, probably because this form of NO is capable of generating coordination complexes [10]. While NO could markedly induce intracellular release, other similar diatomic molecules such as CO had little effect (Fig. 3A) [66]. Further, the ability of NO to mobilize cellular Fe could not be augmented by CO, despite their similar chemistry and the potential of CO to directly ligate Fe pools (Fig. 3B) [10,66,79]. Like synthetic high-affinity Fe chelators [80], the efficacy of NO in mobilizing Fe from cells decreased as the labelling time with ^{59}Fe–transferrin (^{59}Fe–Tf)

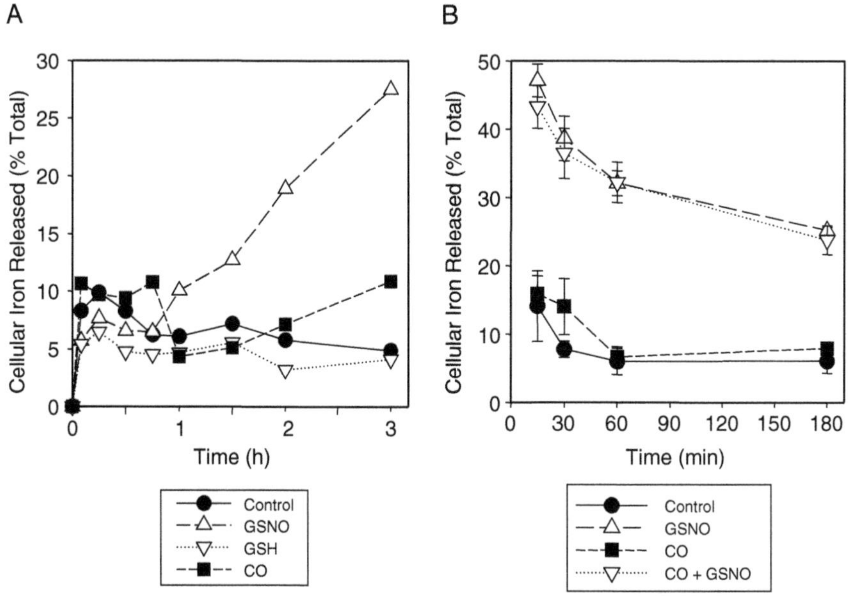

Fig. 3. (A) The NO-donor GSNO but not CO gas results in ^{59}Fe mobilization from prelabelled LMTK⁻ fibroblasts. (B) Iron mobilization from LMTK⁻ fibroblasts prelabelled for various times in the presence of GSNO, CO gas or a combination of both. (A) Cells were prelabelled with ^{59}Fe–Tf (0.75 μM) for 180 min at 37°C, washed, and then reincubated for up to 180 min at 37°C with GSNO (0.5 mM), GSH (0.5 mM) or 2% CO gas. Results are means of duplicate determinations in a typical experiment from 3 performed. (B) Cells were prelabelled with ^{59}Fe–Tf (0.75 μM) for 15–180 min at 37°C, washed and then reincubated for 180 min with GSNO (0.5 mM), 2% CO gas or GSNO (0.5 mM) and 2% CO gas. Results are expressed as mean ± SD of triplicate determinations in a typical experiment of 3 experiments performed. (Taken with permission from Ref. [66].)

increased (Fig. 3B) [66]. This can be explained by the entry of Fe into less accessible cellular pools (e.g. ferritin) as the incubation time increased [80].

Recently, my laboratory has extensively examined the mechanism of NO-mediated Fe release from cells. These studies have demonstrated that it is GSH- and is energy-dependent and relies on the uptake and metabolism of D-glucose (D-Glc) [65,78]. In fact, only sugars that can be taken up and metabolized by cells were effective in increasing NO-mediated Fe release [65]. Fig. 4 is a schematic illustration summarizing a model of NO-mediated Fe release based upon these investigations [65,78]. Glucose enters the cell by the well-characterized family of glucose transporters [81] and is subsequently phosphorylated to glucose-6-phosphate (G-6-P; Fig. 4) [79]. Glucose-6-phosphate (G-6-P) is metabolized by two major pathways, either through glycolysis and/or the tricarboxylic acid cycle (TCA) to form ATP, or through the pentose phosphate pathway (PPP; that is also known as the hexose monophosphate shunt) to form reduced NADPH (e.g. for GSH synthesis) and pentose sugars (Fig. 4) [66].

Our studies demonstrated that D-Glc uptake and metabolism by the PPP was essential for NO-mediated Fe release [65]. Significantly, depletion of GSH using the specific

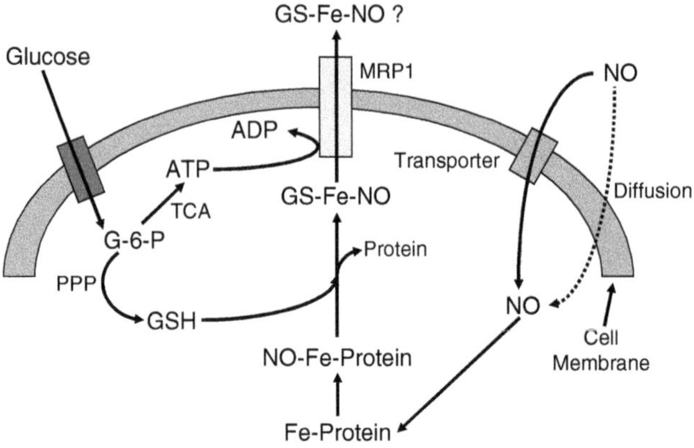

Fig. 4. Hypothetical model of D-Glucose-dependent NO-mediated Fe mobilization from cells. D-Glucose is transported into cells and is used by the tricarboxylic acid cycle (TCA) for the production of ATP and by the pentose phosphate pathway (PPP) for the generation of pentose sugars and NADPH. This reductant is involved in the production of reduced glutathione (GSH). Nitrogen monoxide (NO) either diffuses or is transported into cells where it intercepts and binds Fe bound to proteins or Fe in route to ferritin. The high affinity of NO for Fe results in the formation of an NO–Fe complex and GSH may either be involved as a reductant to remove Fe from endogenous ligands or may complete the Fe coordination shell along with NO. This complex may then be released from the cell by an active process requiring a transporter that has recently been identified as multi-drug resistance associated protein 1 (MRP1). (Modified from Ref. [78].)

GSH synthesis inhibitor, buthionine sulfoximine (BSO) [82], prevented NO-mediated Fe release from cells [65,66,78]. In addition, Fe mobilization after GSH depletion could be reconstituted by incubation of cells with *N*-acetylcysteine that increased cellular GSH levels [65].

It is probable that the effect of D-Glc on stimulating NO-mediated Fe mobilization from cells was not just due to its effect on GSH metabolism. Indeed, our experiments showed that NO-mediated ^{59}Fe release was temperature- and energy-dependent, suggesting a membrane transport mechanism could be involved [65]. As shown in Fig. 4, NO can enter cells through diffusion and there has been some evidence that this can occur by a transport molecule such as the protein disulfide isomerase that catalyses transnitrosylation [83]. A major intracellular target of NO appeared to be the Fe storage molecule ferritin [78]. NO prevented the uptake of Fe into ferritin and also appeared to indirectly mobilize Fe from this protein [78]. An indirect mechanism of Fe release from this protein was postulated, as NO-generating agents added to cellular lysates had no effect on ferritin Fe mobilization [78]. The efflux of Fe from cells and its removal from ferritin was GSH dependent, and could be inhibited using BSO [65,78]. It was speculated that GSH may assist in the removal of Fe from cells by either acting as a reducing agent or by filling the coordination shell of an Fe complex composed of NO and GSH ligands. This complex could be lipophilic enough to pass through the plasma membrane to exit the cell. However, our experiments using a variety of metabolic inhibitors showed that NO-mediated Fe removal from cells was an ATP-dependent event [65,78]. While NO appeared to act like a typical synthetic chelator (e.g. DFO or dipyridyl) in terms of its ability to mobilize cellular Fe, the mechanism was quite different, as chelator-mediated Fe release was not dependent on cellular GSH levels [65].

NITROGEN MONOXIDE MEDIATES IRON EXPORT FROM CELLS BY THE GSH TRANSPORTER, MRP1

As discussed above, NO-mediated Fe release from cells was a temperature- and ATP-dependent event suggesting the involvement of a transport system. Considering possible transport molecules responsible for NO-mediated Fe release, recently our attention became focused on the multi-drug resistance-associated protein 1 (MRP1 or ABCC1) [84]. MRP1 is an ABC transporter that is expressed ubiquitously in tissues [84]. Apart from the role of MRP1 as a detoxifying mechanism for the efflux of drugs from cells, this transporter also plays physiological roles where it is involved in the export of GSH and leukotriene C4 [84–88]. Earlier studies demonstrated that MRP1 can transport GSH coordinated to heavy metals such as As and Sb [85–88], but its role in Fe transport has not been previously proposed.

We examined NO-mediated ^{59}Fe mobilization from several well-characterized cell models expressing high MRP1 levels, namely MCF7-VP cells [89]. These cells were compared to their wild-type parental counterparts (MCF7-WT) which do not express high MRP1 levels. To confirm functionality and expression of MRP1 in MCF7-VP cells we examined efflux of the classical substrate, tritiated-vincristine (^{3}H-VCR) [90], and also MRP1 expression by RT-PCR and Western analysis. Fig. 5A shows that ^{3}H-VCR efflux from MCF7-VP cells was significantly increased ($p < 0.05$–0.0001) compared to MCF7-WT parental cells. This suggested that MRP1 was expressed on the plasma membrane and was functional. These results were further confirmed in studies showing the higher levels of *MRP1* mRNA and protein in MCF-VP cells compared to their WT counterparts (Fig. 5B). In addition, expression of several other potential GSH transporters, namely MRP2-4 or cystic fibrosis transmembrane conductance regulator (CFTR) [84] were not up-regulated in MCF7-VP cells compared to MCF7-WT. MCF7-VP cells are well-known to hyper-express MRP1 but not other drug transporters such as P-glycoprotein (multi-drug resistance protein 1; MDR1) [91,92]. Hence, these cells were implemented as an appropriate model to characterize ^{59}Fe and GSH efflux *via* MRP1 after incubation with NO. They were also used in preference to several types of MRP1-transfected cells that do not express substantial functional MRP1 on the plasma membrane (data not shown).

Using the MCF7-VP and MCF7-WT cell lines, we showed [89] that MRP1 was involved in NO-mediated ^{59}Fe and GSH efflux from our studies demonstrating that: (1) NO-mediated ^{59}Fe release (Fig. 6A,B) and GSH efflux (Fig. 6D,E) were greater in MCF7-VP cells hyper-expressing MRP1 compared to MCF7-WT; (2) cellular ^{59}Fe release (Fig. 7A,C) and GSH efflux (Fig. 7B) occurred by temperature- and metabolic energy-dependent mechanisms consistent with active transport; (3) the specific GSH inhibitor, BSO, that inhibits MRP1 transport activity [84] markedly prevented both NO-mediated ^{59}Fe and GSH efflux from cells (Fig. 6A,D); (4) Well-characterized inhibitors of MRP1 such as MK571, difloxacin, verapamil and probenecid prevent NO-mediated ^{59}Fe efflux (Fig. 8A,B); and (5) Potent inhibitors of MRP1 transport activity such as MK571 and probenecid result in an intracellular build-up of EPR-detectable DNICs (Fig. 9). Moreover, the extent of accumulation of

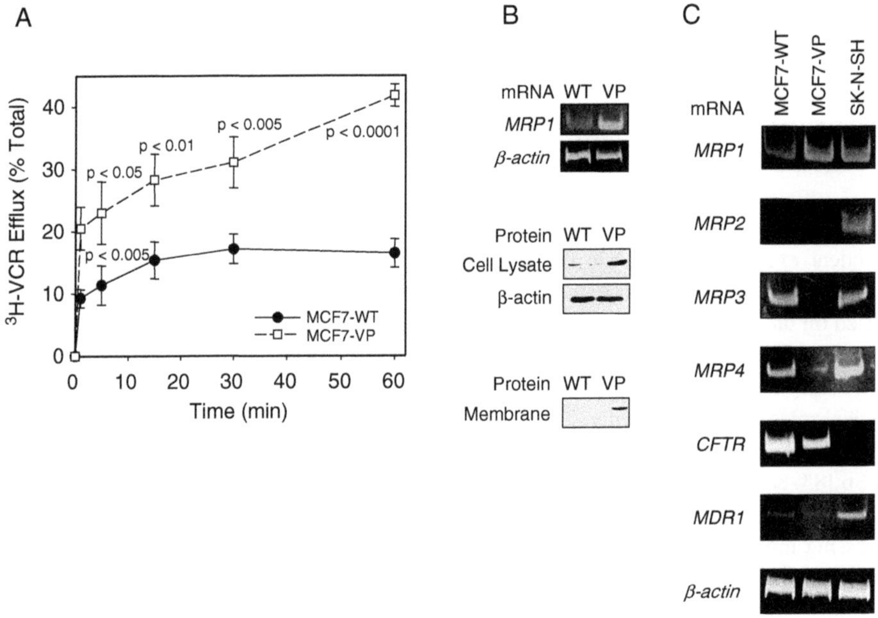

Fig. 5. The efflux of the MRP1 substrate, ^{3}H-vincristine (^{3}H-VCR), is greater from the MRP1-hyper-expressing cells, MCF7-VP, compared to control cells (MCF7-WT). (B,C) In contrast to other ABC transporters, only MRP1 is hyper-expressed in MCF7-VP cells relative to MCF7-WT. (A) ^{3}H-VCR efflux from MCF7-WT and MCF7-VP cells. Cells were prelabelled with ^{3}H-VCR (20 μM) for 30 min at 37°C, washed, and reincubated in the presence of non-radioactive VCR (20 μM). Released ^{3}H-VCR was expressed as a percentage of the total. Results are mean ± SD (4 determinations) in a typical experiment from 4. (B) The expression of *MRP1* mRNA and MRP1 protein in total cell lysates of MCF7-WT and MCF7-VP cells and MRP1 protein in the membrane fraction. (C) *MRP1* mRNA expression compared to *MRP2*, *MRP3*, *MRP4*, *CFTR* and *P-glycoprotein* (*MDR1*) in MCF7-WT, MCF7-VP and SK-N-SH (childhood neuroblastoma) cells. The results shown in (B) and (C) are representative from 3 experiments.

these later species correlates with the ability of these inhibitors to prevent NO-mediated ^{59}Fe efflux from MRP1-hyper-expressing cells (Fig. 8B) [89].

BIOLOGICAL RELEVANCE OF NITROGEN MONOXIDE-MEDIATED TRANSPORT *VIA* MRP1

The transport of NO into and out of cells is of great importance, particularly as it relates to its messenger and cytotoxic effector functions. We have demonstrated that NO can stimulate cellular ^{59}Fe and GSH release by a mechanism that is mediated by MRP1 (Fig. 6) [89]. This may be important for a number of reasons. For instance, the release of DNICs from the cell by a specific transport process could be important for intercellular responses to this messenger molecule and may impart NO with a greater half-life. In fact, a number of investigations have reported that DNICs (e.g. dinitrosyl–diglutathionyl–Fe complex) are natural carriers and storage forms of NO that possess a greater half-life than NO alone [93–97]. Considering this, it is relevant to note that DNICs have been found in animal tissues, human

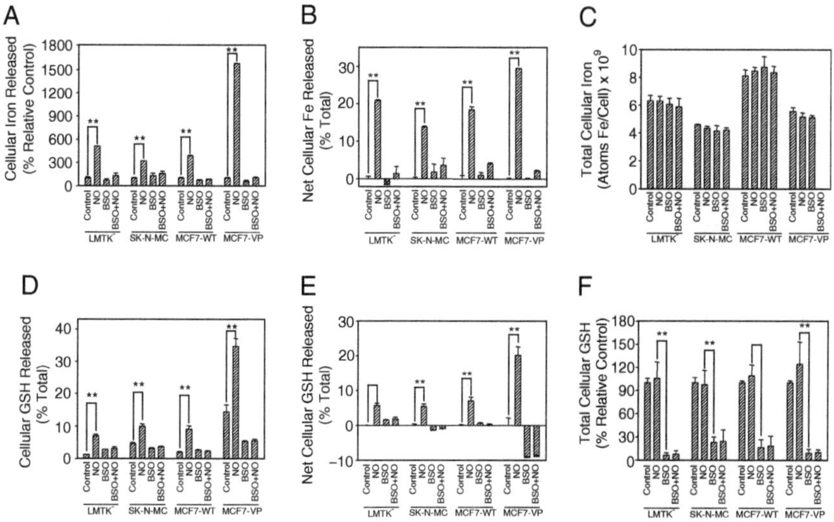

Fig. 6. Iron (Fe) and glutathione (GSH) efflux increase concurrently after incubation with the NO generator, spermine-NONOate (SperNO), and are more marked in MRP1 hyper-expressing MCF7-VP cells than MCF7-WT or SK-N-MC (childhood neuroblastoma). Both NO-mediated ^{59}Fe and GSH efflux were inhibited by the GSH synthesis inhibitor, buthionine sulfoximine (BSO). NO-mediated ^{59}Fe and GSH efflux was examined by preincubating cells with or without 0.1 mM BSO for 20 h at 37°C before labelling for 3 h at 37°C with ^{59}Fe–transferrin (^{59}Fe–Tf) (0.75 μM), washing, and subsequent incubation with the NO generator, SperNO (0.5 mM) for 3 h at 37°C. (A) NO-mediated ^{59}Fe efflux results are expressed as a percentage of the control. (B) Results calculated as a percentage of total cell ^{59}Fe. These data are expressed as net ^{59}Fe efflux over that found for the control. (C) Total cellular ^{59}Fe (i.e. cell ^{59}Fe + efflux ^{59}Fe). (D) NO-mediated GSH efflux expressed as a percentage of total GSH. (E) GSH efflux data expressed as net GSH efflux over that found for the control. (F) Total GSH measured showing the GSH depletion in the presence of BSO. Results are mean ± SD (4 determinations) in a typical experiment of 4. **Denotes $p < 0.0001$.

sera, activated macrophages and a range of different cell types [26,27,29,31–34,93]. A recent study has suggested that the dinitrosyl–diglutathionyl–Fe complex can associate with glutathione *S*-transferase enzymes to stabilize NO for many hours ($t_{1/2} = 4.5$–8 h; [95]) which markedly exceeds the $t_{1/2}$ of “free NO” which is 2 ms–2 s [98]. Furthermore, DNICs (e.g. $(NO)_2(GS)_2Fe$ complexes) can transverse cell membranes to donate Fe to tissues [97] and can transnitrosylate acceptor targets *in vitro* and *in vivo* [96,99,100] demonstrating their bioavailability and potential role as NO-carrier molecules.

The efficient efflux of DNICs by an active transport mechanism could be crucial at sites where NO is produced in small physiological quantities as a messenger molecule [89]. For example, in blood vessels where the small quantities of DNICs released from endothelial cells could be important for regulating smooth muscle tone [15] (Fig. 10A). The ability of cells to actively transport and traffic NO overcomes the random process of diffusion that would be inefficient and non-targeted.

In converse to the situation when NO acts as a messenger molecule, under conditions where NO is used as a cytotoxic effector, the substantial quantities generated by iNOS of activated macrophages could lead to the efflux of a relatively large proportion of Fe and

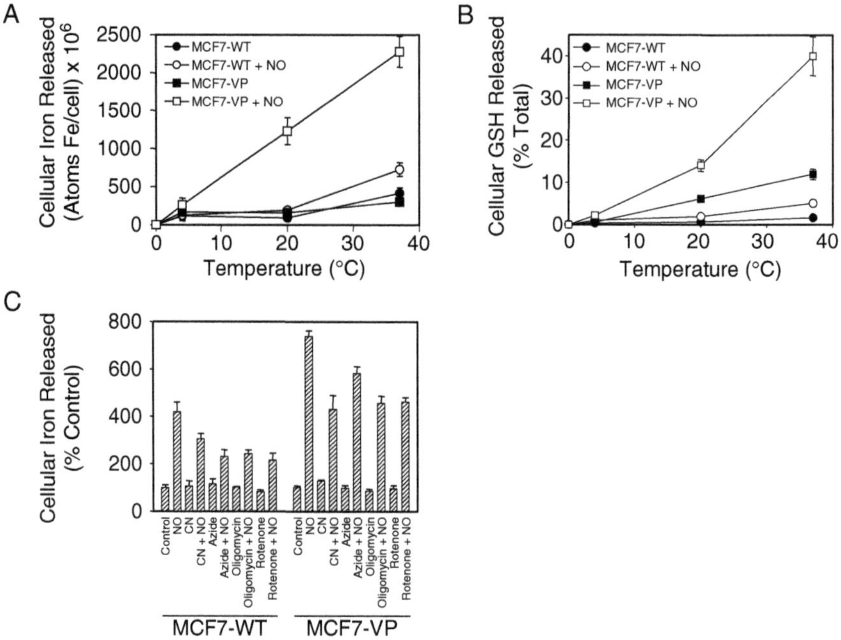

Fig. 7. NO-mediated ^{59}Fe and GSH efflux from parental (MCF7-WT) and MRP1 hyper-expressing MCF7-VP cells are temperature dependent (A,B) and NO-mediated ^{59}Fe efflux is decreased by metabolic inhibitors (C). (A,B) Cells were prelabelled for 3 h at 37°C with ^{59}Fe–transferrin (^{59}Fe–Tf; 0.75 μM), washed 4 times, and reincubated with or without SperNO (0.5 mM) at 4, 20 or 37°C. (C) Cells were prelabelled with ^{59}Fe–Tf as described above, and then pre-treated for 30 min with or without cyanide (CN; 5 mM), azide (30 mM), oligomycin (15 μM) or rotenone (20 μM). Cells were then incubated for 3 h at 37°C with or without SperNO (0.5 mM) in the presence or absence of the inhibitors. Results are mean ± SD (4 determinations) from a typical experiment of 4. (Taken with permission from Ref. [89].)

GSH from tumor target cells (Fig. 10B) [89]. Since Fe and GSH are critical for proliferation [11,84], their release from tumor cells in large amounts would be cytotoxic. In addition to our current results, this hypothesis is strongly supported by previous studies where cytotoxic macrophages induced the release of a large proportion (64%) of intracellular Fe from tumor target cells [3], an effect described to be mediated by NO [4]. Previous investigations have shown that increased GSH efflux from cells is a key signal that mediates apoptosis [101], and it is well-known that Fe mobilization from cells using chelators results in marked anti-tumor activity [102]. Hence, the dual action of NO resulting in both Fe and GSH mobilization may play a vital role in the cytotoxic activity of activated macrophages against tumor cells. We also showed that under conditions that lead to Fe and GSH efflux, proliferation of MCF7-VP cells hyper-expressing MRP1 was more sensitive to the effects of NO than their wild-type counterparts (Fig. 11) [89]. This supports the hypothesis that enhanced GSH and Fe efflux from cells hyper-expressing MRP1 leads to greater anti-proliferative activity. In addition, GSH depletion increased the cytotoxicity of NO particularly in the MRP1 hyper-expressing MCF7-VP cell type, suggesting the critical role played by GSH in the anti-proliferative activity of NO (Fig. 11).

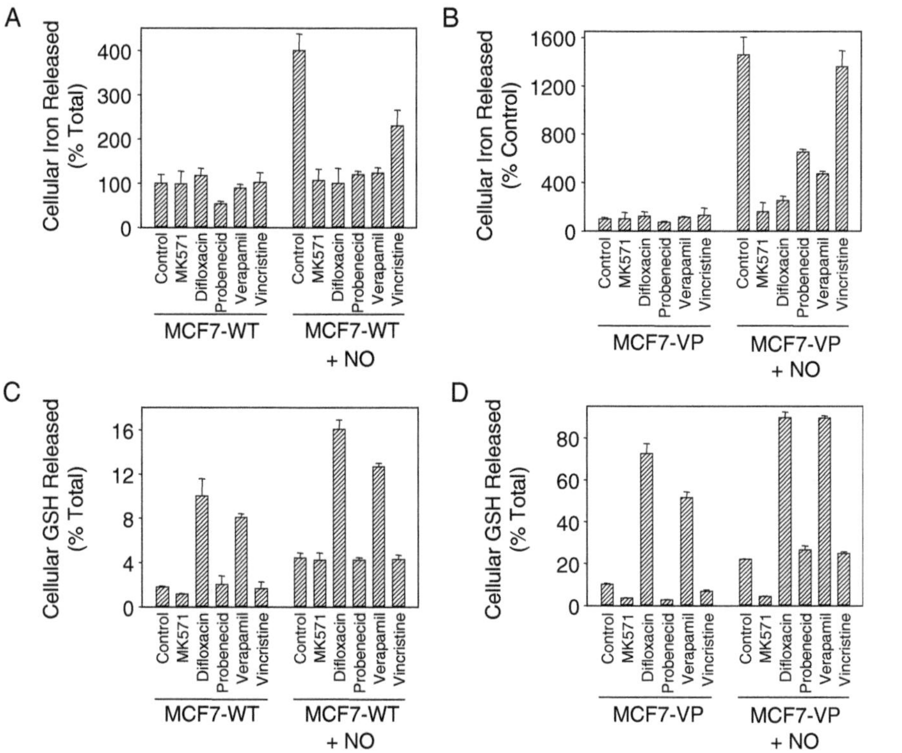

Fig. 8. Effect of MRP1 inhibitors on NO-mediated ^{59}Fe (A,B) and GSH release (C,D) from MCF7-WT and MRP1 hyper-expressing MCF7-VP cells. MRP1 inhibitors used were: MK571 (20 μM), difloxacin (20 μM), probenecid (0.5 mM), verapamil (20 μM) or vincristine (VCR; 20 μM). (A,B) Cells were labelled for 3 h at 37°C with ^{59}Fe–Tf (0.75 μM), washed and reincubated for 30 min at 37°C with the inhibitors. A further 3-h incubation at 37°C was performed with the inhibitors in the presence or absence of SperNO (0.5 mM) and ^{59}Fe efflux assessed. (C,D) Cells were incubated as in (A,B) and GSH release examined. The results are mean ± SD (4 determinations) from a typical experiment of 4. (Taken with permission from Ref. [89].)

The role of GSH in NO-mediated Fe release may not only be important for its transport out of the cell *via* the GSH transporter, MRP1. In fact, we hypothesize that it is also essential for Fe release from cellular proteins targeted by NO such as those with [Fe—S] clusters, e.g. aconitase [18,103]. Thus, we suggest that there is an intracellular equilibrium between protein-bound DNICs and low-M_r DNICs and that GSH is necessary for the conversion to the low-M_r form that is then transported out of the cell by MRP1 [89]. This idea is supported by our studies showing that incubation of cells with BSO markedly prevented both GSH and ^{59}Fe release (Fig. 6A,D) and the potent MRP1 transport inhibitors, MK571 or probenecid, resulted in an accumulation of DNICs (Fig. 9). Our hypothesis is also consistent with the studies of Ding and colleagues that postulate that GSH and/or Cys are necessary for the release of DNICs from [Fe—S] clusters [104,105].

Considering how Fe and GSH are transported by MRP1, five potential mechanisms have been proposed to account for the interaction of the transporter with GSH and its

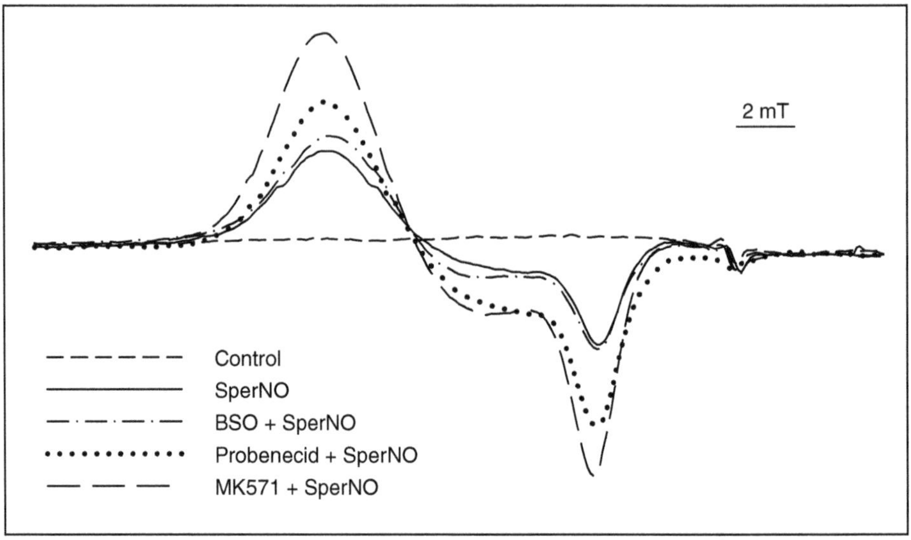

Fig. 9. Electron paramagnetic resonance (EPR) spectroscopy demonstrates that SperNO results in dinitrosyl Fe complex (DNIC) formation and incubation with MRP1 inhibitors increases intracellular DNICs. Low-temperature EPR spectra (77 K) of MCF7-VP cells (10^{10} cells) preincubated with media alone (control) or the GSH synthesis inhibitor, BSO (0.1 mM), for 20 h at 37°C. The cells were then washed and incubated for 3 h at 37°C in the presence or absence of SperNO (0.5 mM). Alternatively, cells were incubated for 30 min with medium alone (control) or the MRP1 inhibitors, probenecid (0.5 mM) or MK571 (20 μM), and then SperNO (0.5 mM) added and the incubation continued for 3 h at 37°C. Results are typical spectra from 6 experiments. (Taken with permission from Ref. [89].)

substrates [84]. First, GSH itself can directly act as a MRP1 substrate and be transported out of the cell. Second, GSH can form a complex or conjugate with a metal or compound that is then transported by MRP1, e.g. leukotriene C4 [106] or the $As(GS)_3$ complex [107]. Third, GSH can be co-transported with some MRP1 substrates, e.g. vincristine. Fourth, GSH can stimulate the transport of certain compounds (e.g. estrone-3-sulfate) by MRP1 without itself being transported out of the cell, and fifth, GSH transport by MRP1 can be enhanced by interaction with certain compounds (e.g. verapamil) that are not translocated themselves across the membrane [84,108,109].

We showed that the GSH synthesis inhibitor, BSO, markedly inhibited both NO-mediated ^{59}Fe and GSH efflux suggesting a dependence on GSH for transport [65] (Fig. 6). Considering this, it can be suggested that either a complex containing both ^{59}Fe and GSH are effluxed together or ^{59}Fe and GSH are separately transported across the membrane by MRP1. We favor a hypothesis necessitating the formation of a complex intracellularly. This is because we can detect intracellular DNICs directly by EPR and the fact that these accumulate after incubation with the MRP1 inhibitors MK571 and probenecid (Fig. 9) [89]. Relevant to the molecular mechanism of GSH and Fe transport by MRP1, it is of interest that verapamil and difloxacin markedly prevent NO-mediated ^{59}Fe efflux from MCF7-VP and MCF7-WT cells (Fig. 8A,B), while both of these MRP1 inhibitors stimulate GSH release (Fig. 8C,D).

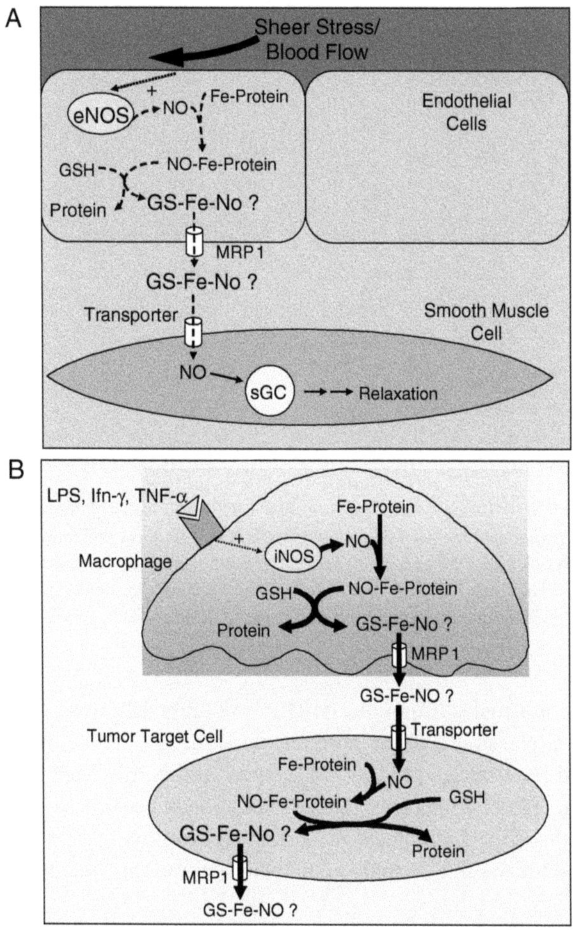

Fig. 10. Schematic illustration of hypotheses proposing the consequences of MRP1-mediated DNIC efflux from cells. (A) The efficient efflux of DNICs by an active transport mechanism could be crucial where NO is produced in small quantities as a messenger molecule, e.g. in blood vessels where endothelial NOS (eNOS) generates NO. The ability of cells to actively transport and traffic NO overcomes the random process of diffusion that would be inefficient and non-targeted. The small quantities of GS–Fe–NO complexes released from endothelial cells may be taken up by smooth muscle cells through diffusion or active transport, e.g. *via* protein sulfide isomerase for instance [83]. (B) Where NO is used as a cytotoxic effector, the substantial quantities generated by inducible NOS (iNOS) of activated macrophages could lead to the efflux of a large proportion of Fe and GSH from tumor cell targets that would be cytotoxic. (Taken with permission from Ref. [89].)

Hence, at least in the presence of these inhibitors, GSH and ^{59}Fe appear to be separate transportable entities [89].

Iron efflux from cells has been described for many years [11], and recently, ferroportin-1 has been shown to be a physiologically relevant Fe exporter [44–46]. It is probable that the role of MRP1 is quite different from that of ferroportin-1 and this is suggested by differences

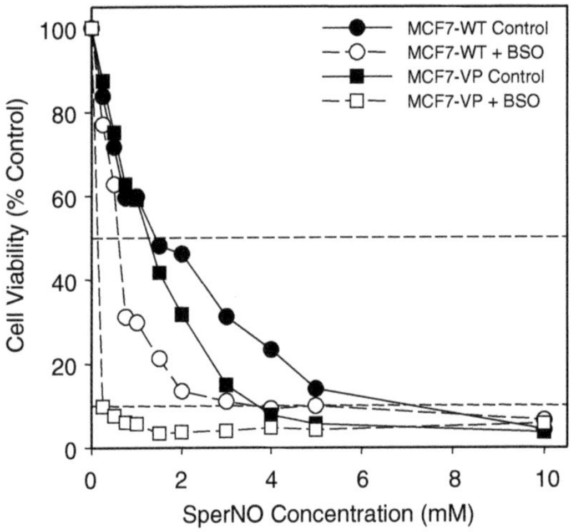

Fig. 11. Cells hyper-expressing MRP1 (MCF7-VP) are more sensitive than MCF7-WT cells to the anti-proliferative effects of SperNO and this is potentiated by GSH depletion using BSO. Cells were preincubated with media alone or media containing BSO (0.1 mM), for 20 h at 37°C. The cells were then incubated in the presence of SperNO for 24 h at 37°C. Horizontal broken lines indicate the concentration that inhibits proliferation by 50% (IC_{50}) and 90% (IC_{90}). Results are means of 4 experiments. (Taken with permission from Ref. [89].)

in the pattern of expression of these molecules. Certainly, MRP1 is ubiquitously expressed in tissues [110], while ferroportin-1 expression appears more restricted, being found at high levels in macrophages, placenta and hepatocytes [44–46]. Moreover, ferroportin-1 has not been reported to transport GSH, and its function appears to be the regulation of Fe homeostasis [48]. Finally, while we have shown that MRP1 facilitates NO-mediated Fe efflux, we cannot rule out that other transporters not examined in this study could also perform this function.

CONCLUSIONS

We have demonstrated that NO mediates the efflux of Fe and GSH from cells by an active mechanism mediated by the GSH transporter, MRP1 [89]. The ability of the cell to actively transport and traffic NO overcomes the random process of diffusion that would be inefficient and non-targeted. Hence, our results are relevant to NO transport, intra- and extra-cellular NO-signaling, NO-mediated apoptosis and the cytotoxicity of activated macrophages that is due, in part, to Fe release from tumor cell targets.

ACKNOWLEDGMENTS

I very much appreciate the help of Mr. Yohan Suryo Rahmanto in helping with the referencing of this article. This work was supported by an Australian Research Council Discovery Grant and an NHMRC Fellowship and Project grants to D.R.R.

REFERENCES

1 Marletta MA. Nitric oxide synthase: aspects concerning structure and catalysis. Cell 1994; 78: 927–930.

2 Moncada S, Palmer RMJ, Higgs EA. Nitric oxide: physiology, pathophysiology and pharmacology. Pharmacol. Rev. 1991; 43: 109–142.

3 Hibbs Jr. JB, Taintor RR, Vavrin Z. Iron depletion: possible cause of tumor cell cytotoxicity induced by activated macrophages. Biochem. Biophys. Res. Commun. 1984; 123: 716–723.

4 Hibbs Jr. JB, Taintor RR, Vavrin Z, Rachlin EM. Nitric oxide: a cytotoxic activated macrophage effector molecule. Biochem. Biophys. Res. Commun. 1988; 157: 87–94.

5 Nathan CF, Hibbs Jr. JB. Role of nitric oxide synthases in macrophage antimicrobial activity. Curr. Opin. Immunol. 1991; 3: 65–70.

6 Stuehr DJ, Gross SS, Sakuma I, Levi R, Nathan CF. Activated murine macrophages secrete a metabolite of arginine with the bioactivity of epithelium-derived relaxing factor and the chemical reactivity of nitric oxide. J. Exp. Med. 1989; 169: 1011–1020.

7 Richardson DR, Ponka P. Effects of nitrogen monoxide on cellular iron metabolism. In Nitric Oxide Synthase: Characterization & Functional Analysis (Maines MD, ed.), Academic Press, San Diego, 1996, pp. 329–345.

8 Nathan C, Xie Q-W. Nitric oxide synthases: roles, tolls, and controls. Cell 1994; 78: 915–918.

9 Alderton WK, Cooper CE, Knowles RG. Nitric oxide synthases: structure, function and inhibition. Biochem. J. 2001; 357: 593–615.

10 Stamler JS, Singel DJ, Loscalzo J. Biochemistry of nitric oxide and its redox-activated forms. Science 1992; 258: 1898–1902.

11 Richardson DR, Ponka P. The molecular mechanisms of the metabolism and transport of iron in normal and neoplastic cells. Biochim. Biophys. Acta 1997; 1331: 1–40.

12 Lepoivre M, Fieschi F, Coves J, Thelander L, Fontecave M. Inactivation of ribonucleotide reductase by nitric oxide. Biochem. Biophys. Res. Commun. 1991; 179: 442–448.

13 Lee M, Arosio P, Cozzi A, Chasteen ND. Identification of the EPR-active iron-nitrosyl complexes in mammalian ferritins. Biochemistry 1994; 33: 3679–3687.

14 Griscavage JM, Fukuto JM, Komori Y, Ignarro LJ. Nitric oxide inhibits neuronal nitric oxide synthase by interacting with the heme prosthetic group. Role of tetrahydrobiopterin in modulating the inhibitory action of nitric oxide. J. Biol. Chem. 1994; 269: 21644–21649.

15 Ignarro LJ. Heme-dependent activation of guanylate cyclase by nitric oxide: a novel signal transduction mechanism. Blood Vessels 1991; 28: 67–73.

16 Khatsenko OG, Gross SS, Rifkind RR, Vane JR. Nitric oxide is a mediator of the decrease in cytochrome P450-dependent metabolism caused by immunostimulants. Proc. Natl. Acad. Sci. USA 1993; 90: 11147–11151.

17 Kim Y-M, Bergonia HA, Muller C, Pitt BR, Watkins WD, Lancaster Jr. JR. Loss and degradation of enzyme-bound heme induced by cellular nitric oxide synthesis. J. Biol. Chem. 1995; 270: 5710–5713.

18 Drapier J-C, Hibbs Jr. JB. Murine cytotoxic activated macrophages inhibit aconitase in tumor cells. J. Clin. Invest. 1986; 78: 790–797.

19 Drapier J-C, Hibbs Jr. JB. Differentiation of murine macrophages to express nonspecific cytotoxicity for tumor cells results in *L*-arginine-dependent inhibition of mitochondrial iron-sulfur enzymes in the macrophage effector cells. J. Immunol. 1988; 140: 2829–2838.

20 Drapier J-C, Hirling H, Weitzerbin J, Kaldy P, Kuhn LC. Biosynthesis of nitric oxide activates iron regulatory factor in macrophages. EMBO J. 1993; 12: 3643–3649.

21 Weiss G, Goossen B, Doppler W, Fuchs D, Pantopoulos K, Werner-Felmayer G, Wachter H, Hentze MW. Translational regulation via iron-responsive elements by the nitric oxide/NO-synthase pathway. EMBO J. 1993; 12: 3651–3657.

22 Richardson DR, Neumannova V, Nagy E, Ponka P. The effect of redox-related species of nitrogen monoxide on transferrin and iron uptake and cellular proliferation of erythroleukemia (K562) cells. Blood 1995; 86: 3211–3219.

23 Cairo G, Ronchi R, Recalcati S, Campanella A, Minotti G. Nitric oxide and peroxynitrite activate the iron regulatory protein-1 of J774A.1 macrophages by direct dissembly of the Fe-S cluster of cytoplasmic aconitase. Biochemistry 2002; 41: 7435–7442.

24 Soum E, Brazzolotto X, Goussias C, Bouton C, Moulis J-M, Mattioli TA, Drapier J-C. Peroxynitrite and nitric oxide differently target the iron-sulfur cluster and amino acid residues of human iron regulatory protein 1. Biochemistry 2003; 42: 7648–7654.
25 Wardrop SL, Watts RN, Richardson DR. Nitrogen monoxide (NO) activates iron regulatory protein 1-RNA-binding activity by two possible mechanisms: Effect on the [4Fe-4S] cluster and iron mobilization from cells. Biochemistry 2000; 39: 2748–2758.
26 Lancaster Jr. JR, Hibbs Jr. JB. EPR demonstration of iron-nitrosyl complex formation by cytotoxic activated macrophages. Proc. Natl. Acad. Sci. USA 1990; 87: 1223–1227.
27 Pellat C, Henry Y, Drapier J-C. IFN-gamma-activated macrophages: Detection by electron paramagnetic resonance of complexes between L-arginine-derived nitric oxide and non-heme iron proteins. Biochem. Biophys. Res. Commun. 1990; 166: 119–125.
28 Vanin AF, Bliumenfeld LA, Chetverikov AG. EPR study of non-heme iron complexes in cells and tissues. Biofizika (Rus.) 1967; 12: 829–841.
29 Vanin AF. Endothelium-derived relaxing factor is a nitrosyl iron complex with thiol ligands. FEBS Lett. 1991; 289: 1–3.
30 Woolum JC, Tiezzi E, Commoner B. Electron spin resonance of iron-nitric oxide complexes with amino acids, peptides and proteins. Biochim. Biophys. Acta 1968; 160: 311–320.
31 Bastian NR, Yim C-Y, Hibbs Jr. JB, Samlovski WE. Induction of iron-derived EPR signals in murine cancers by nitric oxide. J. Biol. Chem. 1994; 269: 5127–5131.
32 Commoner B, Woolum JC, Senturia Jr. BH, Ternberg JL. The effects of 2-acetylaminofluorene and nitrite on free radicals and carcinogenesis in rat liver. Cancer Res. 1970; 30: 2091–2097.
33 Drapier J-C, Pellat C, Henry Y. Generation of EPR-detectable nitrosyl-iron complexes in tumor target cells cocultured with activated macrophages. J. Biol. Chem. 1991; 266: 10162–10167.
34 Vanin AF, Men'shikov GB, Moroz IA, Mordvintcev PI, Serezhenkov VA, Burbaev DS. The source of non-heme iron that binds nitric oxide in cultivated macrophages. Biochim. Biophys. Acta 1992; 1135: 275–279.
35 Andrews NC. Disorders of iron metabolism. New Engl. J. Med. 1999; 341: 1986–1995.
36 Richardson DR. Molecular mechanisms of iron uptake by cells and the use of iron chelators for the treatment of cancer. Curr. Med. Chem. 2005; 12: 2711–2729.
37 McKie AT, Barrow D, Latunde-Dada GO, Rolfs A, Sager G, Mudaly E, Mudaly M, Richardson C, Barlow D, Bomford A, Peters TJ, Raja KB, Shirali S, Hediger MA, Farzaneh F, Simpson RJ. An iron-regulated ferric reductase associated with the absorption of dietary iron. Science 2001; 291: 1755–1759.
38 Gunshin H, Starr CN, Direnzo C, Fleming MD, Jin J, Greer EL, Sellers VM, Galica SM, Andrews NC. Cybrd1 (duodenal cytochrome b) is not necessary for dietary iron absorption in mice. Blood 2005; 106: 2879–2883.
39 Gunshin H, Mackenzie B, Berger UV, Gunshin Y, Romero MF, Boron WF, Nussberger S, Gollan JL, Hediger MA. Cloning and characterization of a mammalian proton coupled metal ion transporter. Nature 1997; 388: 482–488.
40 Fleming MD, Trenor CC, Su MA, Foernzler D, Beier DH, Dietrich WF, Andrews NC. Microcytic anaemia mice have a mutation in Nramp2, a candidate iron transporter gene. Nat. Genet. 1997; 16: 383–386.
41 Fleming MD, Romano MA, Su MA, Garrick LM, Garrick MD, Andrews NC. *Nramp2* is mutated in the anemic Belgrade (b) rat: evidence of a role for Nramp2 in endosomal iron transport. Proc. Natl. Acad. Sci. USA 1998; 95: 1148–1153.
42 Shayeghi M, Latunde-Dada GO, Oakhill JS, Laftah AH, Takeuchi K, Halliday N, Khan Y, Warley A, McCann FE, Hider RC, Frazer DM, Anderson GJ, Vulpe CD, Simpson RJ, McKie AT. Identification of an intestinal heme transporter. Cell 2005; 122: 789–801.
43 Quigley JG, Yang Z, Worthington MT, Phillips JD, Sabo KM, Sabath DE, Berg CL, Sassa S, Wood BL, Abkowitz JL. Identification of a human heme exporter that is essential for erythropoiesis. Cell 2004; 118: 757–766.
44 Donovan A, Alison AYZ, Shepard J, Pratt S, Moynihan J, Paw BH, Drejer A, Barut B, Zapata A, Law TC, Brugnara C, Lux SE, Pinkus GS, Pinkus JL, Kingsley PD, Palis J, Fleming MD, Andrews NC, Zon LI.

Positional cloning of zebrafish ferroportin 1 identifies a conserved vertebrate iron transporter. Nature 2000; 403: 776–781.
45 Abboud S, Haile DJ. A novel mammalian iron-regulated protein involved in intracellular iron metabolism. J. Biol. Chem. 2000; 275: 19906–19912.
46 McKie AT, Marciani P, Rolfs A, Brennan K, Wehr K, Barrow D, Miret S, Bomford A, Peters TJ, Farzaneh F, Hediger MA, Hentze MW, Simpson RJ. A novel duodenal iron-regulated transporter, IREG1, implicated in the basolateral transfer of iron to the circulation. Mol. Cell 2000; 5: 299–309.
47 Petrak J, Vyoral D. Hephaestin—a ferroxidase of cellular iron export. Int. J. Biochem. Cell Biol. 2005; 37: 1173–1178.
48 Nemeth E, Tuttle MS, Powelson J, Vaughn MB, Donovan A, Ward DM, Ganz T, Kaplan J. Hepcidin regulates cellular iron efflux by binding to ferroportin and inducing its internalization. Science 2004; 306: 2090–2093.
49 Richardson DR. The role of ceruloplasmin and ascorbate in cellular iron release. J. Lab. Clin. Med. 1999; 134: 454–465.
50 Vyoral D, Petrak J. Hepcidin: a direct link between iron metabolism and immunity. Int. J. Biochem. Cell Biol. 2005; 37: 1768–1773.
51 Celec P. Hemojuvelin: a supposed role in iron metabolism one year after its discovery. J. Mol. Med. 2005; 83: 521–525.
52 Feder JN, Gnirke A, Thomas W, Tsuchihashi Z, Ruddy DA, Basava A, Dormishian F, Domingo Jr. R, Ellis MC, Fullan A, Hinton LM, Jones NL, Kimmel BE, Kronmal GS, Lauer P, Lee VK, Loeb DB, Mapa FA, McClelland E, Meyer NC, Mintier GA, Moeller N, Moore T, Morikang E, Prass CE, Quintana L, Starnes SM, Schatzman RC, Brunke KJ, Drayna DT, Risch NJ, Bacon BR, Wolff RK. A novel MHC class I-like gene is mutated in patients with hereditary haemochromatosis. Nat. Genet. 1996; 13: 399–408.
53 Hentze MW, Kuhn LC. Molecular control of vertebrate iron metabolism: mRNA-based regulatory circuits operated by iron, nitric oxide, and oxidative stress. Proc. Natl. Acad. Sci. USA 1996; 93: 8175–8182.
54 Gunshin H, Mackenzie B, Berger UV, Gunshin Y, Romero MF, Boron WF, Nussberger S, Gollan JL, Hediger MA. Cloning and characterization of a mammalian proton coupled metal ion transporter. Nature 1997; 388: 482–488.
55 Ponka P, Beaumont C, Richardson DR. Function and regulation of transferrin and ferritin. Semin. Haematol. 1998; 35: 35–54.
56 Hentze MW, Kuhn LC. Molecular control of vertebrate iron metabolism: mRNA-based regulatory circuits operated by iron, nitric oxide, and oxidative stress. Proc. Natl. Acad. Sci. USA 1996; 93: 8175–8182.
57 Giannetti AM, Bjorkman PJ. HFE and transferrin directly compete for transferrin receptor in solution and at the cell surface. J. Biol. Chem. 2004; 279: 25866–25875.
58 Ohgami RS, Campagna DR, Greer EL, Antiochos B, McDonald A, Chen J, Sharp JJ, Fujiwara Y, Barker JE, Fleming MD. Identification of a ferrireductase required for efficient transferrin-dependent iron uptake in erythroid cells. Nat. Genet. 2005; 37: 1264–1269.
59 Fleming MD, Romano MA, Su MA, Garrick LM, Garrick MD, Andrews NC. *Nramp2* is mutated in the anemic Belgrade (b) rat: evidence of a role for Nramp2 in endosomal iron transport. Proc. Natl. Acad. Sci. USA 1998; 95: 1148–1153.
60 Jacobs A. Low molecular weight intracellular Fe transport compounds. Blood 1977; 50: 433–439.
61 Richardson DR, Ponka P, Vyoral D. Distribution of iron in reticulocytes after inhibition of heme synthesis with succinylacetate: examination of the intermediates involved in iron metabolism. Blood 1996; 87: 3477–3488.
62 Vyoral D, Petrak J. Iron transport in K562 cells: a kinetic study using native gel electrophoresis and ^{59}Fe autoradiography. Biochim. Biophys. Acta 1998; 1403: 179–188.
63 Ponka P. Tissue-specific regulation of iron metabolism and heme synthesis: Distinct control mechanisms in erythroid cells. Blood 1997; 89: 1–25.
64 Pantopoulos K, Weiss G, Hentze MW. Nitric oxide and oxidative stress (H_2O_2) control mammalian iron metabolism by different pathways. Mol. Cell. Biol. 1996; 16: 3781–3788.
65 Watts RN, Richardson DR. Nitrogen monoxide and glucose: unexpected links between energy metabolism and NO-mediated iron mobilisation from cells. J. Biol. Chem. 2001; 276: 4724–4732.

66 Watts RN, Richardson DR. Differential effects on cellular iron metabolism of the physiologically relevant diatomic effector molecules, NO and CO, that bind iron. Biochim. Biophys. Acta 2004; 1692: 1–15.

67 Nestel FP, Greene RN, Kichian K, Ponka P, Lapp WS. Activation of macrophage cytostatic effector mechanisms during acute graft-versus-host disease: release of intracellular iron and nitric oxide-mediated cytostasis. Blood 2000; 50: 433–439.

68 Kim S, Ponka P. Control of transferrin receptor expression via nitric oxide-mediated modulation of iron-regulatory protein 2. J. Biol. Chem. 1999; 274: 33035–33042.

69 Kim S, Ponka P. Effects of interferon-γ and lipopolysaccharide on macrophage iron metabolism are mediated by nitric oxide-induced degradation of iron regulatory protein 2. J. Biol. Chem. 2000; 275: 6220–6226.

70 Kim S, Wing SS, Ponka P. S-nitrosylation of IRP2 regulates its stability via the ubiquitin-proteasome pathway. Mol. Cell Biol. 2004; 24: 330–337.

71 Weiss G, Werner-Felmayer G, Werner ER, Gunewald K, Wachter H, Hentze MW. Iron regulates nitric oxide synthase activity by controlling nuclear transcription. J. Exp. Med. 1994; 180: 969–976.

72 Oliveira L, Drapier J-C. Down-regulation of iron regulatory protein 1 gene expression by nitric oxide. Proc. Natl. Acad. Sci. USA 2000; 97: 6550–6555.

73 Richardson DR, Neumannova V, Ponka P. Nitrogen monoxide decreases iron uptake from transferrin but does not mobilize iron from prelabelled neoplastic cells. Biochim. Biophys. Acta 1995; 1266: 250–260.

74 Carmichael AJ, Steel-Goodwin L, Gray B, Arroyo CM. Nitric oxide interaction with lactoferrin and its production by macrophage cells studied by EPR and spin trapping. Free Radic. Res. Comm. 1993; 19: S201–S209.

75 Watts RN, Richardson DR. Examination of the mechanism of action of nitrogen monoxide on iron uptake from transferrin. J. Lab. Clin. Med. 2000; 136: 149–156.

76 Morgan EH, Baker E. The effect of metabolic inhibitors on transferrin and iron uptake and transferrin release from reticulocytes. Biochim. Biophys. Acta 1969; 184: 442–454.

77 Lipton SA, Choi YB, Pan SH, Lei SZ, Chen HSV, Sucer NJ, Loscalzo J, Singel DJ, Stamler JS. A redox-based mechanism for the neuroprotective and neurodestructive effects of nitric oxide and related nitroso-compounds. Nature 1993; 364: 626–631.

78 Watts RN, Richardson DR. The mechanism of nitrogen monoxide (NO)-mediated iron mobilization from cells. Eur. J. Biochem. 2002; 269: 3383–3392.

79 Watts RN, Ponka P, Richardson DR. Effects of nitrogen monoxide and carbon monoxide on molecular and cellular iron metabolism: mirror-image effector molecules that target iron. Biochem. J. 2003; 369: 429–440.

80 Richardson DR, Milnes K. The potential of iron chelators of the pyridoxal isonicotinoyl hydrazone class as effective antiproliferative agents II: the mechanism of action of ligands derived from salicylaldehyde benzoyl hydrazone and 2-hydroxy-1-naphthaldehyde benzoyl hydrazide. Blood 1997; 89: 3025–3038.

81 Joost HG, Thorens B. The extended GLUT-family of sugar/polyol transport facilitators: nomenclature, sequence characteristics, and potential function of its novel members (review). Mol. Membr. Biol. 2001; 18: 247–256.

82 Griffith OW, Meister A. Potent and specific inhibition of glutathione synthesis by buthionine sulfoximine (*S*-n-butyl homocysteine sulfoximine). J. Biol. Chem. 1979; 254: 7558–7560.

83 Zai A, Rudd MA, Scribner AW, Loscalzo J. Cell-surface protein disulfide isomerase catalyzes transnitrosation and regulates intracellular transfer of nitric oxide. J. Clin. Invest. 1999; 103: 393–399.

84 Ballatori N, Hammond CL, Cunningham JB, Krance SM, Marchan R. Molecular mechanisms of reduced glutathione transport: role of the MRP/CFTR/ABCC and OATP/SLC21A families of membrane proteins. Toxicol. Appl. Pharmacol. 2005; 204: 238–255.

85 Liu J, Chen H, Miller DS, Saavedra JE, Keefer LK, Johnson DR, Klaassen CD, Waalkes MP. Overexpression of glutathione *S*-transferase II and multidrug resistance transport proteins is associated with acquired tolerance to inorganic arsenic. Mol. Pharmacol. 2001; 60: 302–309.

86 Liu J, Liu Y, Powell DA, Waalkes MP, Klaassen CD. Multidrug-resistance mdr1a/1b double knockout mice are more sensitive than wild type mice to acute arsenic toxicity, with higher arsenic accumulation in tissues. Toxicology 2002; 170: 55–62.

87 Salerno M, Petroutsa M, Garnier-Suillerot A. The MRP1-mediated effluxes of arsenic and antimony do not require arsenic-glutathione and antimony-glutathione complex formation. J. Bioenerg. Biomembr. 2002; 34: 135–145.

88 Vernhet L, Allain N, Payen L, Anger J-P, Guillouzo A, Fardel O. Resistance of human multidrug resistance-associated protein 1-overexpressing lung tumor cells to the anticancer drug arsenic trioxide. Biochem. Pharmacol. 2001; 61: 1387–1391.

89 Watts RN, Hawkins C, Ponka P, Richardson DR. Nitrogen monoxide (NO)-mediated iron release from cells is linked to NO-induced glutathione efflux via multidrug resistance-associated protein 1. Proc. Natl. Acad. Sci. USA 2006; 103: 7670–7675.

90 Dallas S, Zhu X, Baruchel S, Schlichter L, Bendayan R. Functional expression of the multidrug resistance protein 1 in microglia. J. Pharmacol. Exp. Therapeut. 2003; 307: 282–290.

91 Schneider E, Horton JK, Yang CH, Nakagawa M, Cowan KH. Multidrug resistance-associated protein gene overexpression and reduced drug sensitivity of topoisomerase II in a human breast carcinoma MCF7 cell line selected for etoposide resistance. Cancer Res. 1994; 54: 152–158.

92 Benderra Z, Trussardi A, Morjani H, Villa AM, Doglia SM, Manfait M. Regulation of cellular glutathione modulates nuclear accumulation of daunorubicin in human MCF7 cells overexpressing multidrug resistance associated protein. Eur. J. Cancer 2000; 36: 428–434.

93 de Maria F, Pedersen JZ, Caccuri AM, Antonini G, Turella P, Stella L, lo Bello M, Federici G, Ricci G. The specific interaction of dinitrosyl-diglutathionyl-iron complex, a natural NO-carrier, with the glutathione transferase superfamily. J. Biol. Chem. 2003; 278: 42283–42293.

94 Mulsch A, Mordvintcev P, Vanin AF, Busse R. The potent vasodilating and guanylyl cyclase activating dinitrosyl-iron(II) complex is stored in a protein-bound form in vascular tissue and is released by thiols. FEBS Lett. 1991; 294: 252–256.

95 Turella P, Pedersen JZ, Caccuri AM, de Maria F, Mastroberardino P, lo Bello M, Federici G, Ricci G. Glutathione transferase superfamily behaves like storage proteins for dinitrosyl-diglutathionyl-iron complex in heterogeneous systems. J. Biol. Chem. 2003; 278: 42294–42299.

96 Ueno T, Suzuki Y, Fujii S, Vanin AF, Yoshimaru T. *In vivo* nitric oxide transfer of a physiological NO carrier, dinitrosyl dithiolato iron complex, to target complex. Biochem. Pharmacol. 2002; 63: 485–493.

97 Ueno T, Suzuki Y, Fujii S, Vanin AF, Yoshimura T. *In vivo* distribution and behaviour of paramagnetic dinitrosyl dithiolato iron complex in the abdomen of mouse. Free Radic. Res. 1999; 31: 525–534.

98 Thomas DD, Liu X, Kantrow SP, Lancaster Jr. JR. The biological lifetime of nitric oxide: implications for the perivascular dynamics of NO and O_2. Proc. Natl. Acad. Sci. USA 2001; 98: 355–360.

99 Boese M, Mordvintcev PI, Vanin AF, Busse R, Mulsch A. *S*-nitrosation of serum albumin by dinitrosyl-iron complex. J. Biol. Chem. 1995; 270: 29244–29249.

100 Vanin AF, Malenkova IV, Serenzhenkov VA. Iron catalyses both decomposition and synthesis of S-nitrosothiols: optical and electron paramagnetic resonance studies. Nitric Oxide 1997; 1: 191–203.

101 He Y-Y, Huang JL, Ramirez DC, Chignell CF. Role of reduced glutathione efflux in apoptosis of immortalised human keratinocytes induced by UVA. J. Biol. Chem. 2003; 278: 8058–8064.

102 Richardson DR, Tran EH, Ponka P. The potential of iron chelators of the pyridoxal isonicotinoyl hydrazone class as effective antiproliferative agents. Blood 1995; 86: 4295–4306.

103 Drapier J-C, Hibbs Jr. JB. Aconitases: a class of metalloproteins highly sensitive to nitric oxide synthesis. Methods 1996; 269: 26–36.

104 Ding H, Demple B. Glutathione-mediated destabilization *in vitro* of [2Fe-2S] centers in the SoxR regulatory protein. Proc. Natl. Acad. Sci. USA 1996; 93: 9449–9453.

105 Rogers PA, Ding H. L-Cysteine-mediated destabilization of dinitrosyl iron complexes in proteins. J. Biol. Chem. 2001; 276: 30980–30986.

106 Leier I, Jedlitschky G, Buchholz U, Cole SPC, Deeley RG, Keppler D. The MRP gene encodes an ATP-dependent export pump for leukotriene C_4 and structurally related conjugates. J. Biol. Chem. 1994; 269: 27807–27810.

107 Leslie EM, Haimeur A, Waalkes MP. Arsenic transport by the human multidrug resistance protein 1 (MRP1/ABCC1). Evidence that tri-glutathione conjugate is required. J. Biol. Chem. 2004; 279: 32700–32708.

108 Cullen KV, Davey RA, Davey MW. Verapamil-stimulated glutathione transport by the multidrug resistance-associated protein (MRP1) in leukaemia cells. Biochem. Pharmacol. 2001; 62: 417–424.
109 Trompier D, Chang XB, Barattin R, du Moulinet D'Hardemare A, Di Pietro A, Baubichon-Cortay H. Verapamil and its derivative trigger apoptosis through glutathione extrusion by multidrug resistance protein MRP1. Cancer Res. 2004; 64: 4950–4956.
110 Flens MJ, Zaman GJ, van der Valk P, Izquierdo MA, Schroeijers AB, Scheffer GL, van der Groep P, de Haas M, Meijer CJ, Scheper RJ. Tissue distribution of the multidrug resistance protein. Am. J. Pathol. 1996; 148: 1237–1247.

CHAPTER 5

Low-molecular dinitrosyl iron complexes can catalyze the degradation of active centers of iron–sulfur proteins

Anatoly F. Vanin[1] and Ernst van Faassen[2]

[1]*Semenov Institute of Chemical Physics, Russian Academy, of Sciences, Moscow, 119991, Russian Federation*
[2]*Debye Institute, Section Interface Physics, Ornstein Laboratory, Utrecht University, 3508 TA, Utrecht, The Netherlands*

Many recent studies have shown that iron–sulfur proteins (ISPs) are targets for endogenous nitric oxide (NO) radicals. The action of NO on the iron–sulfur centers (ISCs) of ISP has been recognized as one of the signaling pathways of NO in cells and tissues [1–15]. The interaction leads to the deactivation of ISPs due to degradation of the ISCs. The degradation is often concomitant with the formation of paramagnetic dinitrosyl iron complexes (DNICs) with thiolate ligands. These complexes have the generic structure $\{(RS^-)_2Fe^+(NO^+)_2\}^+$, where the unpaired electrons of the nitrosyl ligands are largely transferred to the iron atom. In Enemark–Feltham notation, the shared nature of these electrons is expressed as the complex having $\{3d^7\}$ electron configuration [1,2,9–17]. The complexes are paramagnetic and their EPR absorption near $g_{av} = 2.03$ is often considered as a marker of NO-mediated degradation of the ISCs. The spectroscopic and physico-chemical properties of DNIC and their experimental observation in cells and tissues were discussed in Chapter 2.

ISPs appear in a wide variety of organisms, from plants to bacteria and vertebrates. The ISPs are usually classified according to the number of iron atoms in the ISC clusters: one, two, three or four, respectively [18,19]. In most of these proteins, the iron atoms are held by sulfur atoms of cysteine side chains, with a few exceptions having coordination of nitrogen atoms from histidines. In polynuclear clusters, the iron atoms are held together by additional sulfide ions (labile inorganic sulfur). Mononuclear clusters appear in rubredoxin and desulforedoxin where the iron is held by four thiolate sulfurs [18,20]. The binuclear $[Fe_2S_2]$ cluster appears in many different ferredoxins found in plants, microorganisms and vertebrates. Typical binuclear centers are found in spinach ferredoxin and bovine adrenodoxin (Adr). Usually, the cluster is held by a total of four thiolate sulfurs, pairwise coordinated to each of the iron atoms coupled through the bridged inorganic sulfur atoms [18,19]. Closely related to the ferredoxins is the $[Fe_2S_2]$ cluster of the Rieske center, as found in mitochondrial electron transport chain. In the

Rieske center, the iron atoms are held by a pair of cysteinate sulfurs and a pair of histidine nitrogens, respectively.

A possible mechanism of disruption of a reduced binuclear [Fe_2S_2] cluster by NO is shown in Scheme 1. DNICs are formed only by the interaction between NO and ferrous iron (Fe^{2+}) [16,17]. Therefore, the formation of DNIC involves the reaction of NO with reduced ISPs only. The proposed [15,21] mechanism proceeds by ligand replacement, where the inorganic sulfur atoms (S*) are replaced by NO moieties:

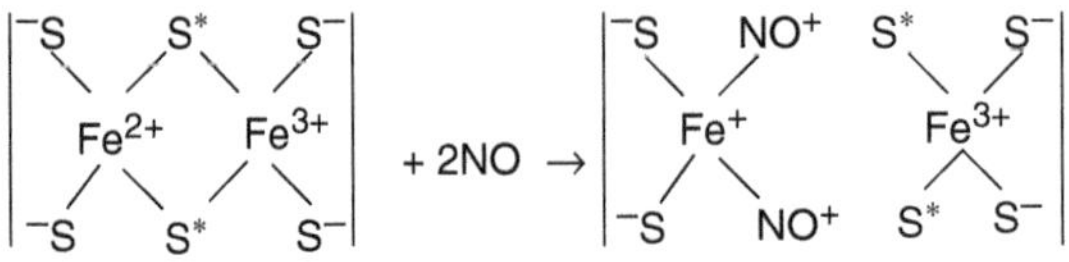

Scheme 1. The supposed mechanism of NO-catalyzed destruction of reduced binuclear iron–sulfur center (ISC). The inorganic sulfurs are marked with an asterisk. The disruption leads to the formation of a paramagnetic DNIC.

An analogous mechanism is often assumed for the disruption of tetranuclear cubane clusters [Fe_4S_4]. Such disruption was observed in many investigations of bi- or tetranuclear ICSs in isolated ISPs or in whole cells or cell preparations. When exposed to NO, substantial quantities of DNIC were formed. At the same time, usually a significant decrease of the ISC was observed *via* EPR or optical absorption [1,11,14,15].

However, certain types of ISC seem to be resistant to the disruptive action of NO. They comprise isolated Adr [21,22], and the ISCs in extracted mouse liver [23], cultured macrophages [24] or anaerobic bacteria *Clostridium sporogenus* [25]. Upon exposure to gaseous NO, or NO donors, the formation of some DNIC was observed, but it did not lead to a significant change in quantity of ISCs as detected with EPR. At the time, it could not be confirmed or refuted that the formation of DNIC involved the destruction of a relatively small part of the ISCs under the action of NO. To verify this proposition, we used EPR to quantify the formation of DNIC in cultured macrophages treated with NO. In addition, the effect of NO on the concentration of ISCs in these cells was studied. The results of the investigation are illustrated in Fig. 1.

The EPR spectra from control preparations of macrophages show a small quantity of paramagnetic reduced ISC in native state. The intensity of the EPR signal from reduced ISC at $g = 1.92$–1.94 is insignificant, but the EPR spectrum is dominated by signals at $g \sim 2.02$. This line is characteristic of trinuclear [Fe_3S_4] clusters in oxidized state [26,27]. The quantity of paramagnetic ISC at $g \sim 1.92$–1.94 increased sharply after reduction with dithionite and methyl viologen as a redox mediator. It shows that the ISCs of the macrophage mitochondrial respiratory chain are nearly fully oxidized in native state. The contribution of the EPR signals from various ISPs in mitochondrial chain is shown in Fig. 1. The EPR spectrum of the N-2 ISC at 23 K has a minimum at $g = 1.92$ and maximum at $g = 1.935$ and was identified from the lineshape and microwave power saturation similar to that of [Fe_4S_4] N-2 ISC in animal tissues [28,29]. The [Fe_2S_2] N-1b and N-1a ISCs have EPR signals with $g \sim 1.94$ or 1.93 and were also detected at 23 K. The quantity of the [Fe_4S_4] N-3 and N-4 ISCs was very small as the EPR intensity barely exceeded the noise level of the spectrometer. The EPR

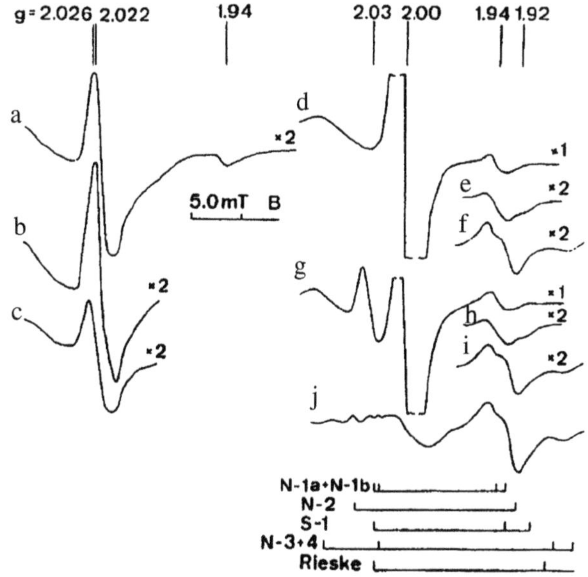

Fig. 1. EPR spectra of control macrophage preparations [24]. Left panel: unreduced preparations at 25 K (curve a), 16 K (curve b) and 30 K (curve c). Right panel: preparations treated by dithionite (1 mM) + methyl viologen (5 mM), recorded at 40 K (curves d,e) and 23 K (curve f). Preparations treated with dithionite–methyl viologen after 5 min incubation with 10^{-7} M NO, recorded at 40 K (curves g,h) and 23 K (curve i). (Curve j) shows the EPR spectrum of mouse liver at 20 K.

absorption near $g \sim 1.94$ as recorded at 40 K was attributed to a superposition of signals from reduced $[Fe_2S_2]$ centers like N-1b, N-1a, S-1 and Rieske ISCs [26,28,30]. Thus, the EPR spectra show that the ISC content of the respiratory chain in macrophages is similar to that in animal tissues. Further support for this similarity is seen in the EPR spectrum attributed to the various ISCs in mouse liver (Fig. 1, curve j).

Subsequently, we investigated the disruption by NO. The bolus injection of 100 nM NO to 10^7 macrophages in 0.3 ml buffer did not significantly affect the EPR spectra from reduced ISCs. In particular, there was no significant loss of intensity of their EPR signals. However, the EPR spectra showed the formation of a small quantity of paramagnetic DNICs centered at $g = 2.03$ (Fig. 1, curve g). It is noteworthy that NO induced the formation of significant quantities of DNIC only in macrophages treated with dithionite–methyl viologen. It did not matter whether the reduction was before or after NO addition. The EPR spectrum of DNIC is much narrower and sharper than that of reduced ISCs. Therefore, the derivative EPR spectra are much more sensitive to a gain in DNIC than for a loss of reduced ISC. In principle, double integration of the EPR spectrum should reveal the iron content in ISCs and DNICs, respectively. In practice, the strong EPR signal from methyl viologen radicals at $g = 2.0$ prevented the reliable quantification of the iron of the various ISCs. Therefore, the double integration was applied to a selection of the ISCs only, namely N-1a, N-1b and S-1. The EPR signals from these ISCs as well as the (small) Rieske $[Fe_2S_2]$ were recorded at 40–80 K [24,31]. The amount of iron in the remaining ISCs was estimated from the known

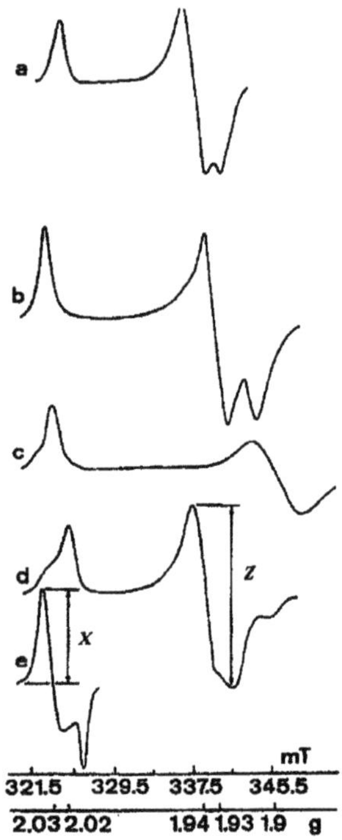

Fig. 2. EPR spectra of various ISCs from the electron transport chain. (Curve a) N-1a + N-1b [31], (curve b) S-1 [32], (curve c) Rieske [33] and (curve d) the sum with equal weight of N-1a, N-1b and S-1 clusters. (Curve e) reference spectrum of DNIC with thiol ligands.

stoichiometry of ISCs in electron transport chains in animal mitochondria [32]. We explicitly assumed that the relative contributions of the various ISCs in macrophages are similar to that in the mitochondrial chains of animal tissues.

Quantification of the spin density in these tissues poses a technical challenge since the EPR spectra appear as a superposition of broad signals from the various ISCs in the electron transport chain. We used the following protocol to determine the ratio of iron in ISPs *vs.* DNIC with EPR in the temperature range 40–80 K. First, we measured the double integrals S_{DNIC} and S_{ISC} of DNIC and sum of N-1a + N-1b + S-1, respectively. The absorption peaks are indicated in Fig. 2 as X (for DNIC) or Z (for ISC).

The small contribution from Rieske ISP was considered negligible. The ratio $S_{\mathrm{DNIC}}/S_{\mathrm{ISC}}$ is proportional to X/Z, and the proportionality factor was calibrated by numerical simulation of the experimental EPR spectra as a mixture of DNIC and a synthetic ISC spectrum. The synthetic spectrum was generated as the equimolar mixture of experimental EPR signals of N-1a, N-1b and S-1 ISC, described in Refs. [31–33]. Technical details of the procedure are given in Ref. [24]. From the numerical calculations [24], we found that the two paramagnetic species are related as $S_{\mathrm{ISC}}/S_{\mathrm{DNIC}} = 5.0 \pm 0.1$ when $X/Z = 1$. Having in mind that the ISPs in S_{ISC} are binuclear clusters [19,30–32] we find an iron ratio of $[\mathrm{Fe_{ISC}}]/[\mathrm{Fe_{DNIC}}] = 10$. In other words, we find that the paramagnetic population S_{ISC} (N-1a + N-1b + S-1) contains 10-fold more iron than the DNICs when $X/Z = 1$ [24]. However, these ratios only account for the EPR visible subfraction of iron in the electron transport chain, and the total iron content of the chain is significantly higher still. It was reported [19,30–33] that the full electron transport chain in animal mitochondria contains a total of 28 iron atoms in ICSs: 16 {$[Fe_2S_2]$ N-1a, $[Fe_2S_2]$N-1b, $[Fe_4S_4]$ N-2, $[Fe_4S_4]$ N-3, $[Fe_4S_4]$ N-4}, 8 {$[Fe_2S_2]$S-1, $[Fe_2S_2]$S-2, $[Fe_4S_4]$S-3} and 4 {$[Fe_2S_2]$-1 and $[Fe_2S_2]$-Rieske ISC} in the 1st, 2nd and 3rd Green's complexes, respectively. Therefore, the paramagnetic population S_{ISC} (N-1a + N-1b + S-1) represents ca 20% of total "iron–sulfur" iron [total Fe_{ISC}] in the transport chain. Therefore, the total iron ratio is found [24] as

$$[\text{total Fe}_{ISC}]/[\text{Fe}_{DNIC}] = 50 \text{ when } X/Z = 1$$

In this relation, [Fe_{DNIC}] represents the iron sequestered in the form of the DNIC. All of this iron is EPR visible. The [total Fe_{ISC}] represents the cumulative iron from all ISC of the electron transport chain, and only 20% of this iron is included in S_{ISC}.

After exposure to NO, prominent DNIC signal was visible in the preparation pre-treated with dithionite–methyl viologen (Fig. 1, curve g). At 40 K and 50 mW microwave power, the ratio of X/Z was equal to ~10 (Fig. 1, curves g,h). However, under these conditions, the DNIC signal at $g = 2.03$ is compressed by significant power saturation. The unsaturated X/Z ratio should be higher. Fig. 3 illustrates the effect of power saturation on the EPR intensity of DNIC in macrophage preparations.

As followed from the figure, at 60 K the EPR intensity is independent of microwave power if the latter exceeds 3 mW. Such behavior is characteristic of inhomogenous saturation of the DNIC complexes [34,35]. It was shown earlier [35] that the EPR signal from the DNIC begins to saturate namely at 3 mW of microwave power when the signal is recorded at higher temperature (77 K). So it is reasonable to suggest that the signal begins to saturate at microwave power of ca 2 mW when recorded at 40 K (Fig. 1, curve g). Therefore, at this

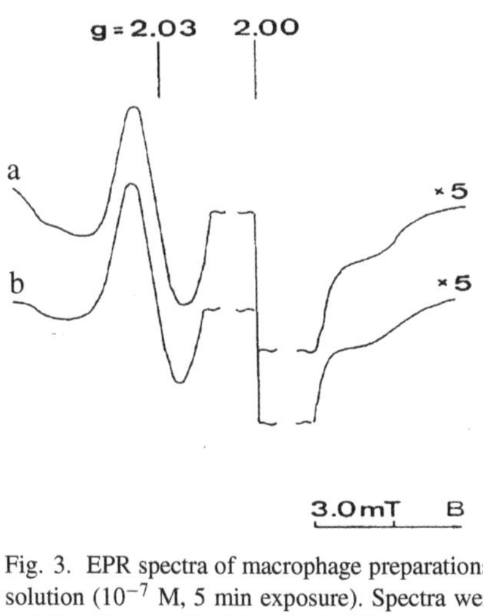

Fig. 3. EPR spectra of macrophage preparations treated with dithionite–methyl viologen and by aqueous NO solution (10^{-7} M, 5 min exposure). Spectra were taken at 60 K with microwave power of 50 mW (curve a) and 3 mW (curve b). (From Ref. [24].)

value of microwave power (in the absence of saturation) the amplitude of the EPR signal of DNIC should be larger by a factor of ca $(50/2)^{1/2} = 5$ than that at 5 mW microwave power saturation. It means that at low microwave powers below saturation, the value of X/Z estimated from Fig. 1, curves g,h (~10) increase by a factor of $(50/2)^{1/2}$ to ca *50!*

This high X/Z ratio shows that within experimental error of ca 10%, the quantity of iron in DNIC was equal to the total iron in ISCs. It proves that the mechanism of DNIC formation cannot be simple disruption of the ISC clusters by free NO molecules, since there was no notable change in EPR signal intensities for ISCs treated with NO solution (Fig. 1, curves d–f and g–i, respectively). It is likely that iron sequestered in DNIC mainly originated from sources other than the tight-binding ISC. As discussed in Chapter 2, loosely bound iron from the so-called labile iron pool (LIP) [36] can contribute to the formation of DNIC (see Chapter 2).

The existence of such loosely bound "non-heme non-FeS" iron in rat liver mitochondria was demonstrated by optical spectroscopy in Ref. [37]. The optical assay was based on the formation of a chelate of Fe^{2+} with bathophenanthroline sulfonate (BPS) in osmotically swollen mitochondria. Fig. 4 shows that the quantity of paramagnetic ISCs was not affected by the addition of BPS and shows that the clusters did not lose their iron to this particular chelator. The EPR signal from the ISC was lost under the action of BPS only when the mitochondria were exposed to sodium dodecyl sulfate and acid pH.

The quantity of "non-heme non-FeS iron" was estimated from the optical density of the Fe–PBS complexes. Its value was nearly double the quantity of iron in ISCs as estimated from EPR (3.1 ± 0.6 nM Fe–PBS/mg protein *vs.* 1.7 ± 0.3 nM Fe-ISC/mg protein). The major part of BPS-chelated iron was confined to the mitochondrial inner compartment, i.e. matrix and inner membrane. More than half of the "non-heme non-FeS iron" of the "inner" pool was in the ferrous state in mitochondria as isolated. The biological significance of this iron

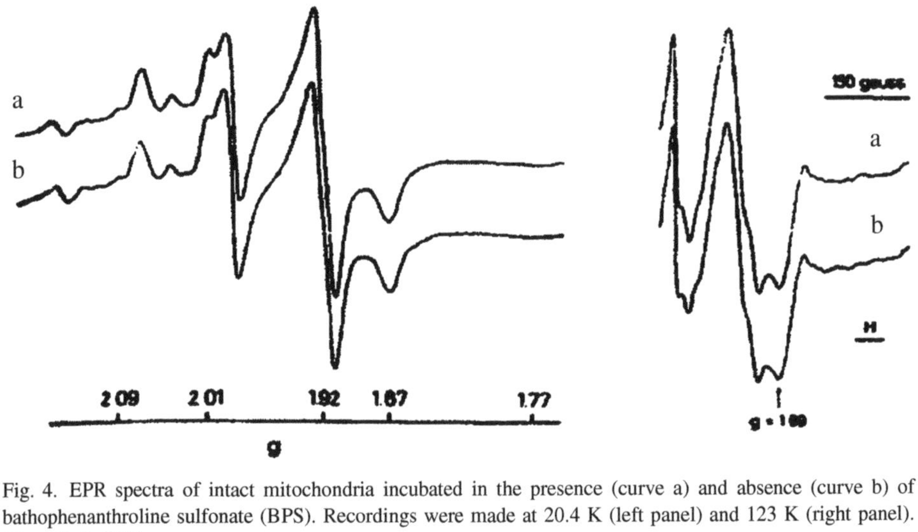

Fig. 4. EPR spectra of intact mitochondria incubated in the presence (curve a) and absence (curve b) of bathophenanthroline sulfonate (BPS). Recordings were made at 20.4 K (left panel) and 123 K (right panel). (From Ref. [37].)

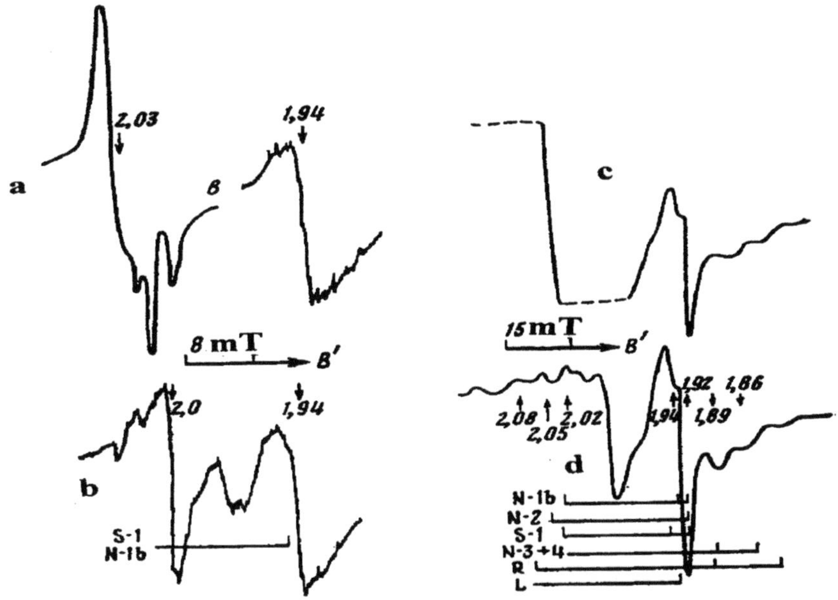

Fig. 5. EPR spectra of perfused mouse liver recorded at 77 K (curves a,b) or 20 K (curves c,d). (Curves b,d) initial preparations; (curves a,c) preparations after exposure to gaseous NO at the pressure of 200 mmHg for 30 min at ambient temperature. The amplifications of spectra (curves c,d) are equal to 1.0 and 1.5, respectively. (From Ref. [23].)

pool remained obscure. This pool was discussed as the source of iron for the synthesis of heme by ferrochelatase. It is reasonable to propose that this pool of "non-heme non-FeS iron" in mitochondria provides a significant fraction of the iron needed for the formation of some DNIC upon exposure to free NO or NO donors. This proposition would explain the formation of DNIC in preparations of rat liver mitochondria when treated with nitrite [38].

Fig. 5 shows independent experiments on perfused mouse liver preparations [23]. In this case, the presence of endogenous substrates was sufficiently high to bring all ISPs to reduced state without the addition of exogenous reductants. EPR showed that the ISCs together contained ca 1.5–2.0 mg Fe/g wet tissue, in reasonable agreement with the value of 1.3–1.5 mg/g wet tissues from the literature [19,30,31]. Upon exposure to gaseous NO, the ISPs were not noticeably degraded but EPR showed the formation of DNIC containing 1–1.5 mg Fe/g of wet tissue. The result shows that significant DNIC was formed in the liver tissue without concomitant loss of ISCs.

The same situation applies to the formation of DNIC from endogenous NO in liver tissue from mice. These animals were injected with bacteria *C. parvum* and the endogenous NO production was stimulated by injection with bacterial lipopolysaccharide [39] (Fig. 6).

Similar results were obtained with liver tissue from mice injected with bacteria *C. sporogenus* treated with nitrite as NO donors [25]. The above body of data attests to a remarkable stability of the ISC in the respiratory chain against disruption induced by NO treatment.

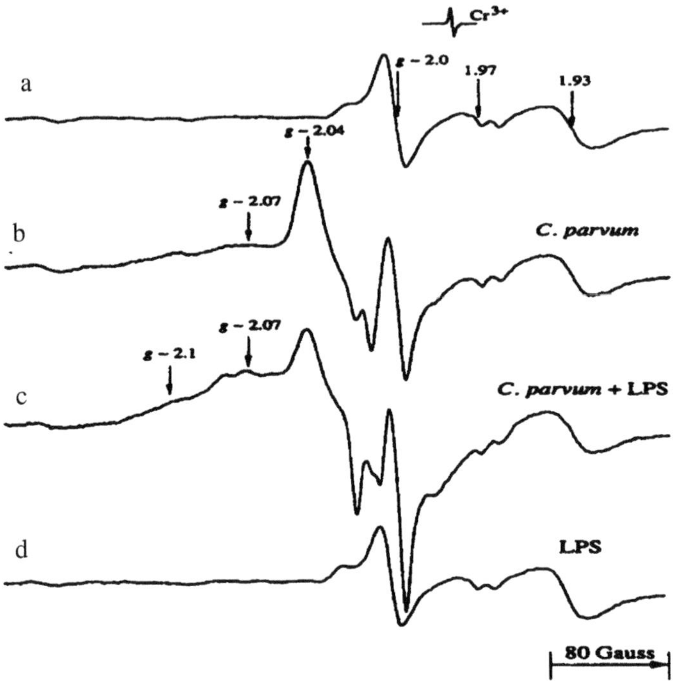

Fig. 6. EPR spectra measured at 77 K of mouse liver. (Curve a) a normal mouse; (curve b) seven days after injection of 1 mg *C. parvum* (the liver was obtained 2 h after saline injection); (curve c) seven days after injection of 1 mg *C. parvum* (the liver was obtained 2 h after 1 μg LPS injection) and (curve d) 2 h after injection of 300 μg LPS. (From Ref. [39].)

However, this stability is not a general rule since certain other ISCs show much higher sensitivity to such type of disruption. For example, *Escherichia coli* contain $[Fe_2S_2]$ SoxR protein, a transcription factor modulating the expression of soxRS gene. In intact *E. coli*, 2-min exposure to gaseous NO led to the formation of the DNICs together with a strong decrease of the EPR signal of reduced SoxR protein. The loss of signal was attributed to the degradation of its active $[Fe_2S_2]$ ICS. The same sensitivity to NO was seen in isolated SoxR protein [15]. Other ISPs like aconitase, ferrochelatase, endonuclease III or *Chromatium vinosum* high-potential iron protein have shown a clear and quantitative correlation between formation of DNIC and loss of ISC induced by exposure to NO [9,12,14,40]. For these cases, the experimental results are compatible with a mechanism as shown in Scheme 1. Details are discussed in Chapter 6. The above body of data can be summarized by noting that various ISCs show widely different susceptibility to degradation induced by NO exposure. Certain ISCs are very sensitive and easily disrupted by the formation of DNIC. For these ISCs, the disruption is plausibly explained by the mechanism of Scheme 1. The ISCs from the respiratory chain are very different in that they are robust against even high levels of NO. For these respiratory ISCs, the reaction shown in Scheme 1 seems irrelevant. Thus, the question arises which mechanism can degrade the ISC in the proteins of the mitochondrial chain?

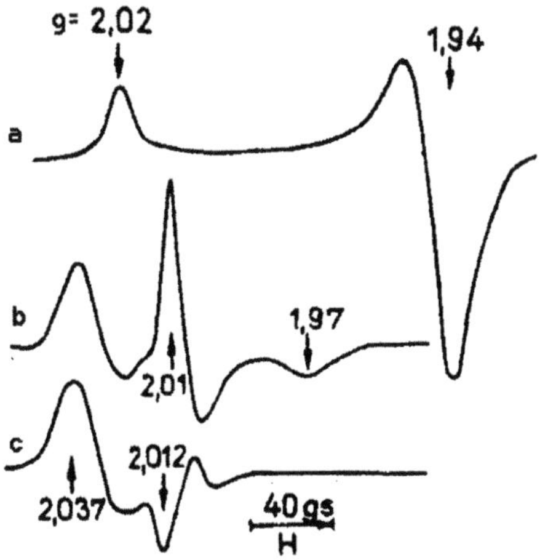

Fig. 7. Changes of EPR spectra of reduced adrenodoxin (Adr) induced by nitrite treatment: (curve a) initial Adr preparation reduced by dithionite, (curve b) Adr treated by freezing–thawing in the presence of 2 mM guanidine-HCl followed by the addition of nitrite + dithionite under anaerobic conditions. (curve c) sample (b) after air exposure. Receiver gain for (a) is two times as high as that for (b) and (c). Recordings were made at 77 K [42].

As a model, we investigated the reaction between NO and [Fe_2S_2] ISC of Adr isolated from bovine adrenal glands. In vertebrates, Adr is a well-characterized [Fe_2S_2] ferredoxin. Adr mediates the electron transfer from NADPH *via* the Adr-reductase flavoenzyme to the cytochrome P_{450} of the respirational chain [41]. Adr has a [Fe_2S_2] ICS as presented in Scheme 1. In the reduced state, the [Fe_2S_2] cluster of Adr contains a ferric and a ferrous iron with antiferromagnetic coupling of the electronic spins to yield a total spin $S = 1/2$. This reduced binuclear cluster is easily observed with EPR and appears usually as an axially symmetric spectrum with main intensity at $g_\perp = 1.938$ and $g_\parallel = 2.022$ [18,19] (Fig. 7, top).

Our first experiments date back to the 1970s, where reduced 0.2 mM Adr [42] demonstrated that the spectral shape and intensity at $g = 1.94$ remained unaffected by repeated cycles of freezing–thawing and by the addition of the NO, producing a mixture of nitrite and dithionite. These experiments showed directly that NO alone did not induce significant loss of ICS in Adr (Fig. 7). Similarly, the combination of nitrite (0.25 mM) and dithionite failed to affect the EPR spectrum of reduced Adr after pre-incubation with protein denaturant guanidine-HCl (0.4 mM) at anaerobic conditions for 3–5 min. However, loss of ISC could be induced by freezing–thawing the solution with reduced Adr and guanidine-HCl prior to the addition of nitrite + dithionite. Under anaerobic conditions, EPR showed the formation of a mixture of two types of paramagnetic DNIC with $\{3d^7\}$ and $\{3d^9\}$ configurations (Fig. 7). These two redox states are easily distinguished from differences in line position and lineshape (cf Chapter 2). The EPR absorption by the ISC of Adr at $g = 1.94$ disappeared completely. Upon exposure to ambient air, all $\{3d^9\}$ DNICs were oxidized to $\{3d^7\}$ and only the 2.03 signal remained (Fig. 7). Double integration of the EPR spectrum showed that the total spin density from both DNIC species accounted for ca 80% of the spin density of the binuclear

1.94 signal. Therefore, the total iron content of the DNIC amounts to ca 40% of the iron originally in the $[Fe_2S_2]$ clusters of Adr.

Thus, the formation of DNIC in these experiments was only initiated after degradation of this center by another mechanism. We propose that Adr is degraded according to Scheme 2. The mechanism involves loss of the bridging inorganic sulfur atoms (S^*) from the iron. Throughout the degradation the iron remains bound to protein thiol groups and the subsequent DNIC appears protein bound as well.

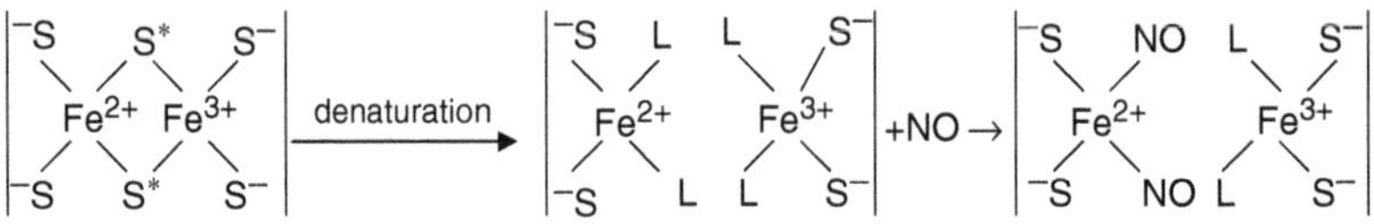

Scheme 2. Proposed mechanism of DNIC formation in denaturated adrenodoxin (Adr) [42].

More recent experiments [21,22] on Adr were consistent with the earlier data [42]. The NO gas used in these latest experiments was purified by the method of low-temperature sublimation in an evacuated system [43] and good care was taken to remove even traces of the nitrogen dioxide radical (this oxidant species rapidly accumulates in NO gas, especially at higher pressures, cf Chapter 2). As shown in Fig. 8 [21], the addition of freshly prepared and purified gaseous NO to 0.2 mM pre-reduced solution of Adr resulted in the appearance of a weak DNIC absorption in the EPR spectrum of the solution. However, the amount of DNIC (2 μM) calculated from the signal was much less than the loss of ca 40 μM ISC as observed at $g = 1.94$.

The exposure to NO also led to the formation of ferrous heme-nitrosyl complexes from heme–protein admixture in Adr preparation as shown by the broad EPR signal in the g-factor range 2.07–1.98 at this treatment (Fig. 8). These results confirm that formation of DNIC and loss of ISC are quite independent processes in Adr.

The degradation of ISC and the formation of DNICs in the presence of NO was sharply accelerated by the addition of 1.8 mM Fe^{2+}-citrate complex to the solutions (Fig. 8, recordings were made at 77 K). The yield of DNICs reached a maximum concentration of 0.3 ± 0.1 mM. This value was already reached at 2–3 min after the addition of iron, and was accompanied with practically complete degradation of the ISC of Adr. Subsequent treatment of the solution with dithionite resulted in a decrease of the intensity of the 2.03 signal and the formation of a new paramagnetic complex with $g_\perp = 2.01$ and $g_\parallel = 1.97$ (Fig. 8). As was mentioned above, this signal was due to reduced form of DNICs with electron configuration $\{3d^9\}$ (cf Chapter 2). The reduction with dithionite did not restore the EPR signal from reduced Adr, thereby showing that the ISC had disintegrated rather than oxidized to diamagnetic state. No effect of Fe^{2+}-citrate complex on the ISC in Adr was observed.

When increasing the measurement temperature from 77 K to room temperature, the shape of the DNIC EPR signal did not change [21] (Fig. 9). This indicated unequivocally that the complexes remained bound to the protein globule: its low mobility was not sufficient to average the g-factor and hyperfine structure (HFS) anisotropy. However, when 5 mM

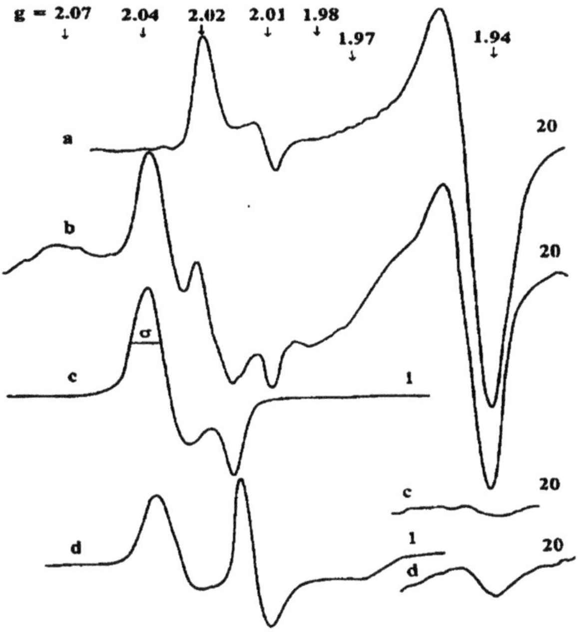

Fig. 8. EPR spectra from a 0.2 mM solution of adrenodoxin (Adr), treated with dithionite (curve a) followed with NO (curve b) or with $^{56}Fe^{2+}$ (1.8 mM) + NO (curve c). (curve d) Treatment of preparation (c) with dithionite. Recordings were made at 77 K. To the right of the spectra, the relative amplifications of the signals are indicated. δ is the width of the signal at half amplitude. Reproduced with permission, from Ref. [21] © the Biochemical Society.

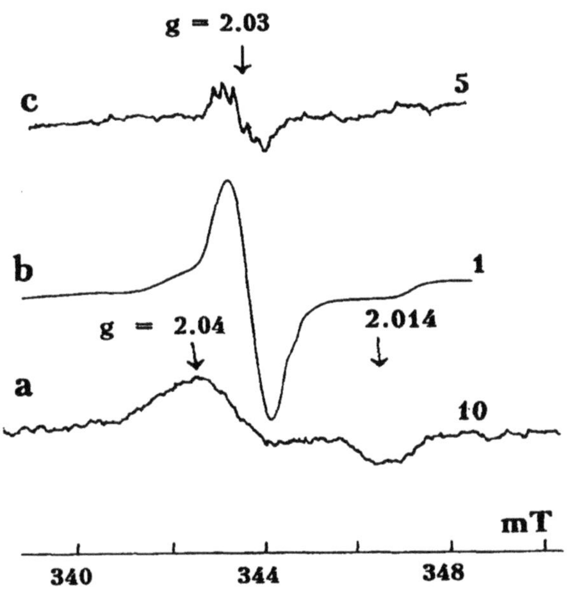

Fig. 9. EPR spectra from 0.2 mM adrenodoxin (Adr) treated with $^{56}Fe^{2+}$ (1.8 mM) + NO (curve a) followed by thiosulfate (curves b,c). Recordings were made at 77 K, microwave power 20 mW and modulation amplitude 0.5 mT (curves a,b) and 0.01 mT (curve c). To the right of the spectra, relative amplifications of the signals are shown. Reproduced with permission, from Ref. [21] © the Biochemical Society.

thiosulfate was added to the solution, a narrow isotropic signal at $g = 2.03$ appeared that was accompanied by a decline in the EPR signal from protein-bound DNIC (Fig. 9).

This was due to the transfer of $Fe^{+}(NO^{+})_2$ moieties from the latter to thiosulfate ions, resulting in the formation of thiosulfate–DNIC of low molecular weight. The mobility of these complexes at ambient temperature was high enough to average the aforementioned anisotropy. Reducing the amplitude of the field modulation from 0.5 to 0.01 mT led to the appearance of a quintet HFS originated from two ^{14}N atoms in NO^{+} ligands (see Chapter 2).

Similar results were obtained in the experiments with Adr in oxidized state [2]. The contact of this protein (0.2 mM) with gaseous NO (at the pressure of 100 mmHg) did not result in the degradation of ISC in Adr. Subsequent reduction with dithionite showed the presence of the same quantity of intact ISC at $g = 1.94$ as in preparations not exposed to NO (Fig. 10). The experimental spectra show some residual broad EPR absorption in the g-factor range 2.07–1.98 from heme-nitrosyl complexes as well as a small EPR signal from DNIC. The intensities of these signals did not exceed 10 and 1 μM, respectively. All indicates that the ISC of Adr is very robust towards NO alone.

The situation changes completely in the presence of exogenous iron: The DNIC formation in this preparation increased the amount of the complexes of 0.2–0.3 mM when 1.8 mM ferrous iron was added to the solution prior to the NO treatment (Fig. 10). Concomitantly, all

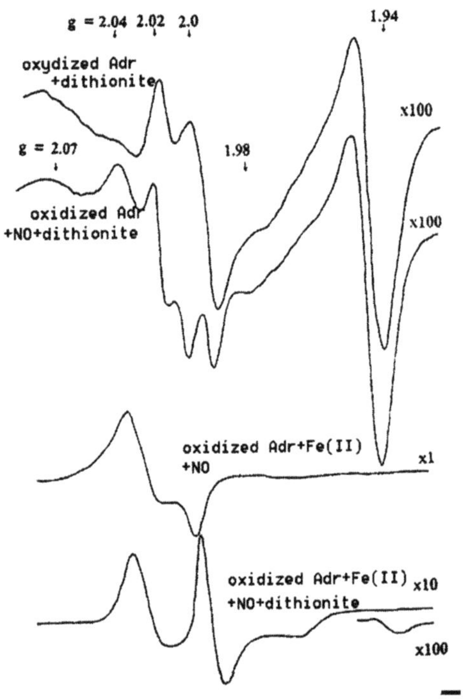

Fig. 10. EPR spectra at 77 K from 0.2 mM oxidized adrenodoxin (Adr) treated with dithionite, NO + dithionite, Fe^{2+} (1.8 mM) + NO, or Fe^{2+} (1.8 mM) + NO + dithionite. Amplification factors are shown on the right side. (From Ref. [22].)

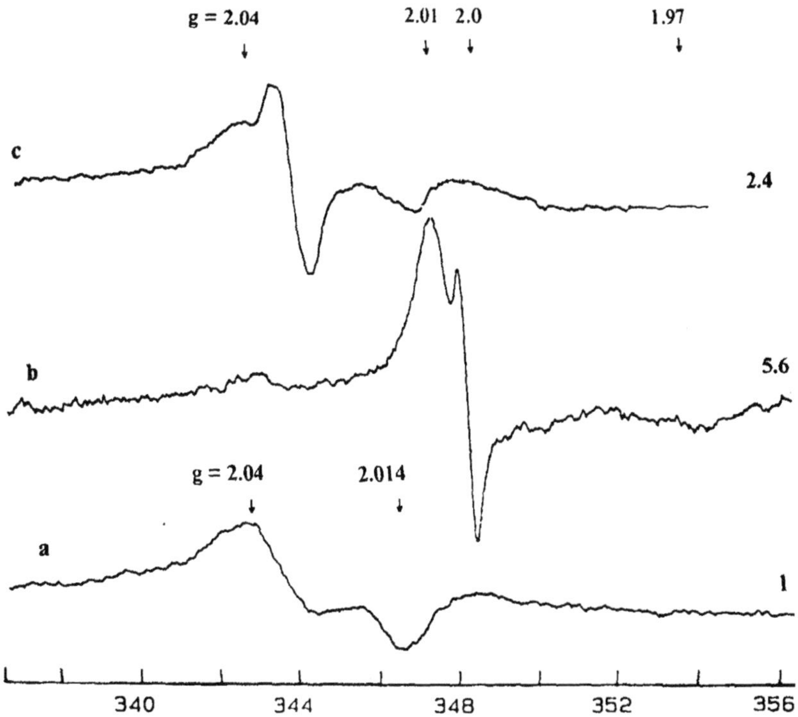

Fig. 11. EPR spectra at room temperature from 0.2 mM oxidized adrenodoxin (Adr) treated with $FeSO_4$. Amplification factors are shown on the right side. (From Ref. [22].)

ISCs in Adr were destroyed. The DNIC was protein bound, because the EPR lineshape remained unchanged when the temperature was raised from 77 K to ambient (Fig. 11, curve a). The treatment of protein-bound DNIC with dithionite resulted in a sharp decrease of the EPR signal of $\{3d^7\}$-DNIC and new EPR absorption with $g_\perp = 2.01$ and $g_\parallel = 1.97$ from $\{3d^9\}$-DNIC (Fig. 11, curve b). The effect of dithionite treatment was reversible: exposure to air restored the signal from $\{3d^7\}$-DNIC (Fig. 11, curve c). However, experiments at room temperature show that the population of $\{3d^7\}$-DNIC was a mixture of protein-bound DNIC and low-weight DNIC. The registration of the narrow symmetric EPR signal at $g = 2.03$ overlapping the anisotropic signal from protein-bound DNIC was due to the formation of low-molecular DNIC with thiosulfate ligands originating from the dithionite.

The strong effect of ferrous iron could be explained by the formation of low-molecular DNICs from the exogenous iron and NO with subsequent action of these DNICs on the $[Fe_2S_2]$ ICS of Adr. Our experiments suggest that the degradation of ICSs proceed *via* a mechanism involving a DNIC of low molecular weight (Scheme 3).

The reaction consumes two low-molecular DNICs and a single ISC and results in the formation of two protein-bound DNICs and the release of endogenous *Fe iron and inorganic

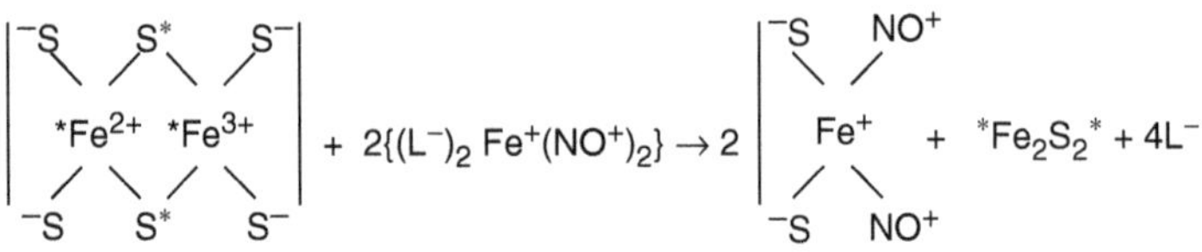

Scheme 3. Degradation of reduced ISC induced by the action of DNIC with low-molecular weight [21]. L^- is a small anionic ligand, for example thiosulfate or cysteine.

sulfur S* atoms. At sufficient levels of L^- and NO, the released iron can again form new low-molecular DNICs, and attack the next ISC. In this way, the DNIC of low molecular weight acts as a catalyst for the denaturation of the ISC in Adr. Fig. 10 suggests that Scheme 3 is applicable to oxidized Adr as well. Possibly, it could apply to tetranuclear $[Fe_4S_4]$ clusters also.

The validity of Scheme 3 was supported by many experimental observations: First, in the absence of NO gas, the addition of 1.8 mM phosphate-DNIC to reduced Adr (0.2 mM) led to full degradation of ISC as observed at $g = 1.94$. EPR demonstrated the complete transfer of the DNIC moiety from low to high molecular weight. Second, the addition of 0.3 mM ^{57}Fe (nuclear spin $I = 1/2$) instead of abundant ^{56}Fe ($I = 0$) to reduced Adr solution in the presence of NO led to a DNIC spectrum with significant isotopic broadening of the $g_\perp$ component due to HFS from ^{57}Fe (for details of this isotopic broadening cf Chapter 2). With 2.1 mT, the width at half amplitude (δ) of the $g_\perp$ component of the ^{57}Fe-DNIC is more than that of ^{56}Fe-DNIC with 1.6 mT (Figs. 8 and 11).

The linewidth proves that the DNIC had predominantly formed from the exogenous iron rather than the endogenous unlabeled iron in the ISC. The isotopic ratio could be shifted by subsequent exposure to gaseous NO: ^{56}Fe-DNIC from the endogenous iron pool began to dominate when the preparation was subsequently treated with NO and dithionite. This was clearly reflected in the sharpening of the DNIC spectrum at $g = 2.03$ (Fig. 12). The change in lineshape proves that the mechanism of DNIC formation and ISC degradation involves a mixing of the endogenous (^{57}Fe) and exogenous (^{56}Fe) iron pools. The increase of the latter in the DNIC composition can point to auto-catalytical mechanism of DNIC formation and ISC degradation.

The relative contributions of exogenous (^{57}Fe) and endogenous (^{56}Fe) iron were investigated by adding an excess quantity of thiosulfate (5 mM) to the preparation resulting in the formation of low-molecular DNIC with thiosulfate (Fig. 9). The kinetics of the process could be monitored with EPR on a timescale of minutes at ambient temperature. The EPR spectra of thiosulfate–DNIC are sensitive to the isotopic labeling with ^{57}Fe. At low amplitude of the field modulation (0.01 mT), the isotropic DNIC-thiosulfate spectrum has resolved quintet HFS with nitrogens of the nitrosyl ligands and additional doublet HFS from ^{57}Fe ($I = 1/2$) (Fig. 13). As expected, the EPR spectra of mixtures of ^{57}Fe and ^{56}Fe isotopes show a complicated HFS pattern (Fig. 13).

A similar complicated HFS pattern was recorded for the solution of pre-reduced Adr (0.2 mM) following the addition of NO and an equimolar mixture of ^{57}Fe- and ^{56}Fe-citrate (0.2 mM), with 5 mM thiosulfate added 5 min later (Fig. 14, curve a). The intensity of the EPR signal increased and the HFS pattern changed when this preparation was additively treated with NO and dithionite (Fig. 14, curve b, right side). The spectrum could be well simulated with a 1.5:1 mixture of ^{56}Fe-DNIC and ^{57}Fe-DNIC (Fig. 14, curve b, left side).

The enrichment of DNIC with ^{56}Fe proves that endogenous iron from ISC began to make contribution to the formation of DNIC-thiosulfate. This conclusion is supported by exposing reduced Adr with NO and exogenous ^{57}Fe (0.3 mM) without ^{56}Fe. It results in the formation of thiosulfate-ligated DNIC with isotopic ratio ^{56}Fe:^{57}Fe = 0.44:1 (Fig. 14, curve c). Subsequent exposure to gaseous NO and dithionite increased the ratio to 0.75:1 (Fig. 14, curve d).

The above data can be summarized as follows: The degradation of the [Fe_2S_2] cluster in Adr and formation of DNIC cannot be achieved by free NO molecules alone. In particular, the mechanism of Scheme 1 is not valid for Adr. *In vitro*, we found that this ferredoxin was very susceptible to degradation by the action of low-molecular DNICs independently from reduced or oxidized state of ISC Adr. This action is rapid and significant already at low DNIC concentrations. Such small quantities of DNIC will form spontaneously in any solution containing small anionic ligands, NO and spurious quantities of free or loosely bound iron. Thiols are known to be particularly effective ligands for the formation of DNIC. We note that all three ingredients are readily available in actual biological systems. Therefore, it seems plausible that the attack of low-molecular DNIC on binuclear ISC be relevant for *in vivo* conditions as well. We propose that the attack of low-molecular DNIC on [Fe_2S_2] of Adr proceed according to Scheme 3. This reaction mechanism forms apo-ISP-bound DNIC.

We have seen above that certain classes of proteins do not show functional correlation between ISC degradation and formation of DNIC. In particular, breakup of the ISC was

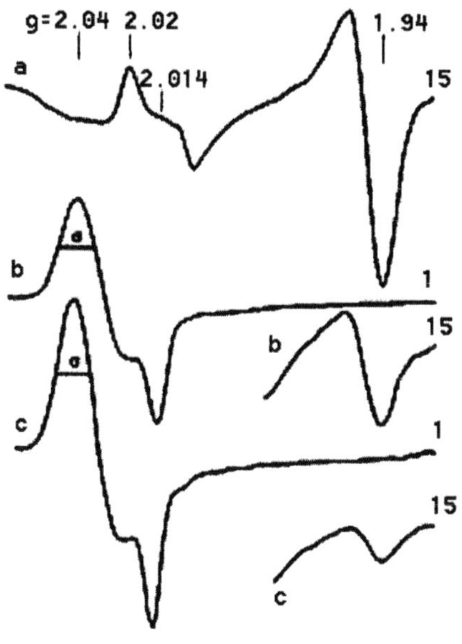

Fig. 12. EPR spectra at 77 K from 0.2 mM reduced adrenodoxin (Adr) (curve a); addition of $^{57}Fe^{2+}$ (0.3 mM) + NO to the preparation (a) (curve b); addition of dithionite to preparation (b) (curve c). Amplification factors are shown on the right. δ is the width of the signal at half amplitude. Reproduced with permission, from Ref. [21] © the Biochemical Society.

not initiated by NO exposure alone. For these NO-resistant proteins, the degradation was achieved by the action of low-molecular DNIC. It is not to be excluded that certain ISCs are resistant to small DNIC as well. For example, geometrical constraints may leave ISC inaccessible to low-molecular DNIC. The latter are small but certainly bulky in comparison with the highly mobile NO radical. Blocked access could potentially protect the active centers of ISPs buried in lipid compartments, for example in the mitochondrial electron transport chain.

It seems tempting to attribute all protein-bound DNIC in cells and tissues as originating from the degradation of ISCs *via* Scheme 3. However, *in vitro* studies have shown that low-molecular DNIC may readily transfer their $Fe(NO^+)_2$ moiety to other thiol groups on the protein, i.e. to thiols not at all involved in the ISC. We are convinced that a significant fraction of protein-bound DNIC is anchored to such non-ISC thiols. It is even conceivable that the majority of protein-bound DNICs be anchored on non-ISC thiols. The presence of various cellular compartments with different polarity and dielectric properties complicates the *in vivo* situation even further. Although the elucidation of these problems needs further investigations, it is already clear that Scheme 1 cannot be taken as a paradigm for the degradation of all types of ISCs, and that the ISC of Adr and the respiratory chain form a different class. *In vitro* experiments with Adr have demonstrated that the presence of small DNIC complexes promotes catalytic breakup of ISC under concomitant formation of DNIC, as well as exchange between the iron pools of ISC an DNIC. The efficiency of this catalytic reaction suggests that it may have physiological relevance for biological systems.

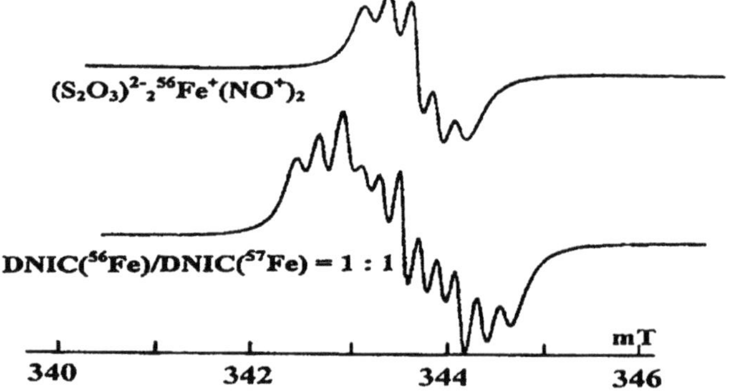

Fig. 13. EPR spectra from the solutions of DNIC with thiosulfate containing ^{57}Fe (upper trace), ^{56}Fe (middle trace) or an equimolar mixture of ^{57}Fe and ^{56}Fe (lower trace). Recordings were made at ambient temperature, microwave power 20 mW and modulation amplitude 0.01 mT. Reproduced with permission, from Ref. [21] © the Biochemical Society.

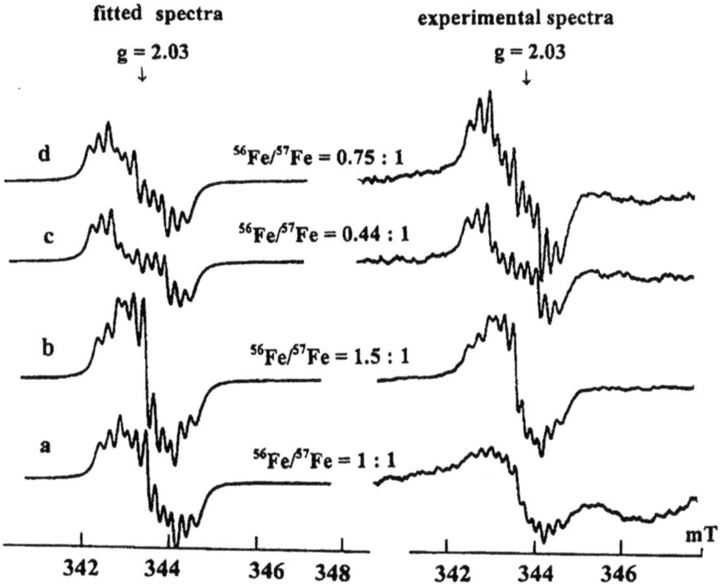

Fig. 14. Right-hand panel: (curve a) EPR spectra from 0.2 mM reduced adrenodoxin (Adr) treated with $^{56}Fe^{2+}$ (0.2 mM) and $^{57}Fe^{2+}$ (0.2 mM) followed by 5 mM thiosulfate. (Curve b) addition of dithionite and additional NO to sample (a). (Curve c) EPR spectrum from 0.2 mM reduced Adr treated with $^{57}Fe^{2+}$ (0.3 mM) + NO followed with subsequent thiosulfate (5 mM) treatment. (Curve d) addition of dithionite and additional NO to sample (c). (Curve e) EPR spectra from 0.2 mM reduced Adr treated with $^{56}Fe^{2+}$ (0.3 mM) + NO followed by subsequent thiosulfate (5 mM) treatment. Recordings were made at ambient temperature, microwave power 20 mW and modulation amplitude 0.01 mT. Reproduced with permission, from Ref. [21] © the Biochemical Society.

ACKNOWLEDGMENT

The work was supported by the Russian Foundation of Basic Researches (Grant 05-04-49383).

REFERENCES

1 Reddy D, Lancaster JR, Conworth DP. Nitrite inhibition of *Clostridium botulinum:* electron spin resonance detection of iron-nitric oxide complexes. Science (Washington) 1983; 21: 769–770.
2 Butler AR, Glidewell C, Hyde AR, Walton JC. Nitrosylation of 2Fe-2S and 4Fe-4S models for iron-sulfur redox proteins. Inorg. Chem. Acta 1985; 106: L7–L8.
3 Drapier J-C, Hibbs JB. Differentiation of murine macrophages to express cytotoxicity for tumor cells results in L-arginine-dependent inhibition of mitochondrial iron-sulfur enzymes in the macrophage affected cells. J. Immunol. 1988; 140: 2829–2841.
4 Drapier J-C, Hibbs JB. Murine cytotoxic activated macrophages inhibits aconitase in tumor cells. J. Clin. Invest. 1986; 78: 790–797.
5 Drapier J-C, Hibbs JB. Murine cytotoxic activated macrophages inhibits aconitase in tumor cells. J. Clin. Invest. 1986; 78: 790–797.
6 Hibbs JB, Taintor RR, Vavrin Z, Rachlin EM. Nitric oxide: a cytotoxic activated macrophage effector molecule. Biochem. Biophys. Res. Comm. 1988; 157: 867–894.

7 Stuehr DJ, Gross SS, Sakuna I, Levi R, Nathan CF. Activated murine macrophages secrete a metabolite of arginine with the bioactivity of endothelium-derived relaxing factor and the chemical reactivity of nitric oxide. J. Exp. Med. 1989; 169: 1011–1020.
8 Stuehr DJ, Nathan CF. Nitric oxide a macrophage product, responsible for cytostatic and respiratory inhibition in tumor target cells. J. Exp. Med. 1989; 169: 1543–1555.
9 Sellers VM, Johnson MK, Daily HA. Function of the [Fe-S] cluster in mammalian ferrochelatase: as possible role as a nitric oxide sensor. Biochemistry 1996; 35: 2699–2704.
10 Lancaster JR, Hibbs JB. EPR demonstration of iron-nitrosyl complex formation by cytotoxic activated macrophages. Proc. Natl. Acad. Sci. USA 1990; 87: 1223–1227.
11 Welter R, Yu L, Yu C-A, The effect of nitric oxide on electron transport complexes. Arch. Biochem. Biophys. 1996; 331: 9–14.
12 Kennedy MC, Antholine WE, Beinert H. An EPR investigation of the products of the reaction of cytosolic and mitochondrial aconitases with nitric oxide. J. Biol. Chem. 1997; 272: 20340–29347.
13 Drapier J-C. Interplay between NO and [Fe-S] clusters: relevance to biological systems. Methods: A Companion to Methods in Enzymology 1997; 11: 319–329.
14 Foster MW, Cowan JA. Chemistry of nitric oxide with protein-bound iron sulfur centers. Insights on physiological reactivity. J. Am. Chem. Soc. 1999; 121: 4093–4100.
15 Ding H, Demple B. Direct nitric oxide signal transduction via nitrosylation of iron-sulfur centers in the SoxR transcription activator. Proc. Natl. Acad. Sci. USA 2000; 97: 5146–5150.
16 McDonald CC, Phillips WD, Mower HF. An electron spin resonance study of some complexes of iron, nitric oxide and anionic ligands. J. Am. Chem. Soc. 1965; 87: 3319–3326.
17 Burbaev DS, Vanin AF, Blumenfeld LA. Electronic and spatial structures of paramagnetic dinitrosyl ferrous complexes. Zhurn. Strukt. Khimii (Rus.) 1971; 2: 252–256.
18 Hall DO, Rao KK, Mullinger R. Biological functions of iron-sulfur proteins. Biochem. Soc. Trans. 1975; 3: 472–479.
19 Beinert H. Iron-sulfur proteins. New insights and unresolved problems. Biochim. Biophys. Acta. Rev. Bioenerg. 1982; 683: 246–277.
20 Gomes CM, Vicente JB, Wasserfallen A, Teixeira M. Spectroscopic studies and characterization of a novel electron-transfer chain from *Escherichia coli* involving a flavorubredoxin and its flavoprotein reductase partner. Biochemistry 2000; 39: 16320–16327.
21 Voevodskaya NV, Serezhenkov VA, Cooper CE, Kubrina LN, Vanin AF. Exogenous ferrous iron is required for the nitric oxide-catalysed destruction of the iron-sulphur center in adrenodoxin. Biochem. J. 2002; 368: 633–639.
22 Voevodskaya NV, Kubrina LN, Serezhenkov VA, Mikoyan VD, Vanin AF. The nitric oxide-mediated degradation of active center in an iron-sulphur protein adrenodoxin. Current Top. Biophys. 1999; 23: 31–37.
23 Vanin AF. Origin of iron forming nitrosyl complexes in animal tissues. Biofizika (Rus.) 1987; 32: 128–131.
24 Vanin AF, Menshikov GB, Moroz IA, Mordvintcev PI, Serezhenkov VA, Burbaev DS. The source of non-heme iron that binds nitric oxide in cultivated macrophages. Biochim. Biophys. Acta 1992; 1135: 275–279.
25 Payne MJ, Woods LFJ, Gibbs P, Cammack R. Electron paramagnetic resonance spectroscopic investigation of the phosphoroclastic of Clostridium sporogens by nitrite. J. Gen. Microbiol. 1990; 136: 2067–2076.
26 Ohnishi T, Ingledew WJ, Shiraishi S. Resolution and functional characterization of two mitochondrial iron-sulphur centers of the "high-potential iron-sulphur protein 1" type. Biochem. J. 12976; 153: 39–48.
27 Burbaev DS, Solozhenkin IP, Zvyagilskaya RA, Blumenfeld LA. EPR study of high potential iron-sulphur centers in mitochondria of yeasts *Endomyces magnusii* and of rat liver. Biofizika (Rus.) 1981; 26: 447–453.
28 Orme-Johnson NR, Hansen RE, Beinert H. Electron paramagnetic resonance-detectable electron acceptors in beef heart mitochondria. Reduced diphosphopyridine nucleotide ubiquinone reductase segment of the electron transfer. J. Biol. Chem. 1974; 249: 1922–1931.

29 Ohnishi T, Hemington JG, La Noue KF, Morris HP, Williamson JK. Electron paramagnetic resonance studies of iron-sulphur centers in mitochondria prepared from three morris hepatomas with different growth rates. Biochem. Biophys. Res. Comm. 1973; 55: 372–381.

30 Albracht SP, Subramanian J. The number of Fe atoms in the iron-sulphur centers of the respiratory chain. Biochim. Biophys. Acta 1977; 462: 36–48.

31 Albracht SPJ, Leenwerik FJ, Van Swal B. The stoichiometry of the iron-sulphur clusters 1a, 1b and 2 of NADH:Q oxireductase as present in beef heart submitochondrial particles. FEBS Lett. 1979; 104: 197–200.

32 Salerno JC, Lin J, Kin TE, Blum H, Ohnishi T. The spatial relationships and structure of the binuclear iron-sulphur clusters in succinate dehydrogenase. J. Biol. Chem. 1979; 245: 4828–4835.

33 DeVries S, Albracht SPJ, Leenwerik FJ. The multiplicity and stoichiometry of the prostetic groups of QH2:cytochrome c oxyreductase as studied by EPR. Biochim. Biophys. Acta 1979; 546: 316–333.

34 Lebedev YS, Muromtsev VI. EPR and stable radical relaxation. Khimia Press, Moscow, Russia, 1972, pp. 42–64.

35 Vanin AF. Nitrosyl non-heme iron complexes in animal tissues and microorganisms. D. Sc. Thesis, Institute of Chemical Physics, Moscow, 1980, p. 66.

36 Nunez MT, Garate MA, Arredondo M, Tapia V, Munoz P. The cellular mechanisms of body iron homeostasis. Biol. Res. 2000; 33: 133–142.

37 Tangeras A, Flatmark T, Bakstrom D. Mitochondrial iron not bound in heme and iron-sulphur centers. Estimation, compartmentation and redox state. Biochim. Biophys. Acta 1980; 589: 162–175.

38 Vakhnina LV, Ruuge EK. The kinetics of the development of the ESR signal at g_{av} 2.03 in mitochondria treated with nitrite. Biofizika (Rus.) 1972; 17: 690–692.

39 Chamulitrat W, Jordan SUJ, Mason RP, Litton AL, Wilson JG, Wood ER, Wolberg G, Molina Y, Vedia L. Targets of nitric oxide in a mouse model of liver inflammation by *Coryebacterium parvum.* Arch. Biochem. Biophys. 1995; 316: 30–37.

40 Roger PA, Eidl L, Klungland I, Ding H. Reversible inactivation of E. coli endonuclease III via modification of its [4Fe-4S] cluster by nitric oxide. DNA Repair 2003; 16: 809–817.

41 Wickramasinghe RH, McIntosh FN. Adrenodoxin, ferredoxin and other iron-sulphur (non-heme) iron proteins. Enzyme 1974; 17: 210–226.

42 Vanin AF, Mardanyan SS, Nalbandyan RM. Nitrosyl non-heme iron complexes in denaturated adrenodoxin. Stud. Biophy. 1973; 38: 13–18.

43 Vanin AF, Malenkova IV, Serezhenkov VA. Iron catalyzes both decomposition and synthesis of S-nitrosothiols: Optical and EPR studies. Nitric Oxide: Biol. & Chem. 1997; 1: 191–203.

Radicals for Life: The Various Forms of Nitric Oxide
E. van Faassen and A.F. Vanin (Eds.)

CHAPTER 6

Products of the reaction of cytosolic and mitochondrial aconitases with nitric oxide

M. Claire Kennedy[1], William E. Antholine[2] and Helmut Beinert[3]

[1]*Department of Chemistry, Gannon University, Erie, PA 16561, USA*
[2]*Biophysics Research Institute, Medical College of Wisconsin, Milwaukee, WI 53226, USA*
[3]*Institute for Enzyme Research, University of Wisconsin—Madison, 1710 University Avenue, Madison, WI 53726-4087, USA*

Aconitases are enzymes that catalyze the stereospecific isomerization of citrate to isocitrate in the tricarboxylic cycle [1]. They are iron–sulfur proteins that contain [4Fe—4S] clusters. In most Fe–S proteins, the cluster is involved in electron transfer without itself undergoing significant structural changes [2,3]. Aconitases, however, are unique in that their Fe–S cluster directly interacts with its substrate [4,5]. Mammalian tissues encode two different aconitases, mitochondrial, m-acon and cytosolic, c-acon. Cytosolic aconitase is a bifunctional protein, which upon the loss of its Fe–S cluster is converted from an enzyme to iron-regulatory protein-1 (IRP1) [4,6,7]. This may occur when the intracellular concentration of iron drops to suboptimal levels. Therefore, the structural integrity of the Fe–S cluster is crucial for understanding the physiological role of aconitases. Interest in aconitases greatly increased after it was recognized that these enzymes play an important role in apoptosis [8] and in addition, that nitric oxide (NO) plays a critical role in the interconversion of c-acon and IRP1 [9]. It is, therefore, not surprising that intensive research efforts were made concerning the action of the catalytic cluster and the mechanisms by which this cluster could be modified or disrupted.

The geometrical structure of the 4Fe cluster in aconitases is that of a distorted cube with alternating Fe and S atoms at its corners. This structure is typical of the cubane type [4Fe—4S] cluster found in a wide range of different proteins [2,3]. In most cubane clusters, the iron atoms are ligated to cysteine thiolates. Aconitases are unusual in that one of the iron atoms, Fe_a, is coordinated to a hydroxyl group in the absence of substrate. During catalysis, the coordination of Fe_a changes from tetrahedral to octahedral when substrate and water are bound [4]. Another important feature of the aconitase cluster is the ease with which Fe_a is lost upon oxidation, resulting in inactivation of the enzyme and formation of a [3Fe—4S] cluster. This arrangement illustrates how the enzymatic activity of aconitase is directly related to the structural geometry of the cluster. The reader interested in more detail on the reaction and the use of *electron paramagnetic resonance* (EPR) and *electron nuclear double resonance* (ENDOR) spectroscopy in the exploration of the mechanisms involved in the functioning of the enzyme may consult the Refs. [2,4,5,10].

There are contradictory reports in the existing literature concerning the reaction of NO with the enzyme aconitase [11–14]. While together at the National Biomedical EPR Center at Milwaukee, the authors decided to seize the opportunity and combine their experience with the enzyme aconitase and EPR, respectively, to determine the sensitivity of aconitase toward NO [14]. As expected, the result was that the active forms of both m-acon and c-acon readily form dinitrosyl-iron complexes (DNICs) containing protein cysteinyl ligands and are thus inactivated. The finding that under some conditions m-acon can also form an analogous dinitrosyl-iron-histidyl complex with NO was unexpected.

In order to pin down the reasons for the contradictory results found in the literature, we will consider in detail the properties of all the reagents and procedures used in the relevant experiments. The most important ingredient is, of course, the protein aconitase. Preparations having "aconitase activity" have been commercially available from Sigma Chemical Company for many years. Out of interest, we acquired a sample of this material that was most likely quite similar to the Sigma aconitase used by other authors. On analysis, we found extremely low enzyme activity in the sample we had received, even after subjecting it to the same activation procedure routinely used with aconitase prepared in our laboratory [15]. Furthermore, the low-temperature EPR spectrum indicated very minor amounts of aconitase present in relation to the high protein content of the commercial sample. In addition, the light absorption spectrum showed features typical of heme compounds, as was also shown in the article by Castro et al. [11], in which no sensitivity of "aconitase" toward NO was found. It is most likely that in such samples NO reacts with the large amounts of heme compounds and other impurities present, such that any aconitase would have been spared.

Another possible source of irreproducibility concerned the mode of addition of NO. We found that adding NO in a solution after gassing under anaerobic conditions is less reproducible than evolving NO from compounds referred to as NONOates and monitoring the NO content with a specific NO electrode. Addition of substrate delayed the decomposition of aconitase to some extent, but did not prevent reaction of the enzyme with NO. We did, however, find that as little as 10 mM citrate had an influence on the rate of decomposition of the NONOate, which was an unexpected complication.

Although the native $[4Fe—4S]^{2+}$ clusters found in active aconitases are in the $S = 0$, EPR-silent state, the oxidative loss of Fe_a from their clusters leads to inactivation with the formation of $[3Fe—4S]^{1+}$ clusters which can be observed by EPR at temperatures below 30 K at $g \approx 2.02$ [16]. In contrast, the EPR signals for the DNIC complexes formed upon addition of NO to the enzyme can be observed at all temperatures. Thus, in order to obtain maximal information from the reaction of the various forms of aconitase and NO, the products of the reaction were monitored by EPR at room (RT) and at low temperatures, i.e. 77 K and < 20 K, which allowed us to decide whether the DNIC signals observed originated from a macromolecular (protein bound) source or from a rapidly tumbling species of low molecular weight. Consideration was also given to the various species of DNIC which might be formed and to their electronic states such as the d^9 state obtained on reduction of the EPR detectable d^7 species as well as the other oxidation states, d^6 or d^8, which are not detected by EPR. Mössbauer spectroscopy would be a suitable method for recognizing these latter species.

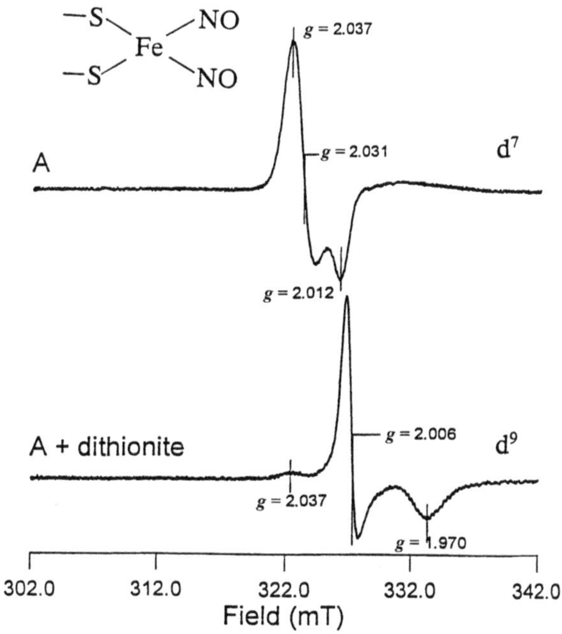

Fig. 1. EPR spectra of the d^7 and d^9 forms of DNI-acon. (Upper curve) EPR spectrum of product formed upon the anaerobic incubation of 120 μM $[4Fe—4S]^{2+}$ active m-acon with 8.3 mM spermine NONOate, pH 6.5, at 23°C for 30 min (no enzyme activity remaining). (Lower curve) EPR spectrum of (a) following the addition of excess dithionite. Conditions of spectroscopy: microwave power and frequency, 0.1 mW and 9.229 GHz; modulation amplitude and frequency, 0.5 mT and 100 kHz; time constant, 0.128 s; scanning time, 20 mT/min; temperature, 15 K.

We shall, in the following, present some illustrations from the experiments that convinced us that both m-acon and c-acon are inactivated by NO through the formation of DNIC complexes. The only EPR signal observed at RT or 77 K upon the bolus injection of a solution of NO at pH 7.5 to active $[4Fe—4S]^{2+}$ m-acon is that of the d^7 DNIC form at $g \approx 2.04$ (Fig. 1). This same signal is observed at pH 6.5 when the NO source is a NONOate solution. These samples, upon reduction with dithionite, yield the d^9 species which upon analysis by EPR gives a spectrum with a signal at $g \approx 2.006$ (Fig. 1). When examined at temperatures below 15 K, the spectrum of these samples is a composite of the $g \approx 2.02$ DNIC and $g \approx 2.04$ $[3Fe—4S]^{1+}$ signals. Quantitation of these signals does not correlate with loss of activity nor does formation of $[3Fe—4S]^{1+}$ aconitase parallel loss of activity (Fig. 2). This may indicate the presence of EPR-silent d^6 or d^8 DNIC species or further disassembly of the Fe—S cluster. Results similar to those obtained for active m-acon were also observed for c-acon. A notable difference was that a higher ratio of NO to enzyme was needed to obtain a similar loss of activity of c-acon and also that a transient EPR signal, tentatively assigned to a thiyl radical was seen on occasion.

The reaction of NO with inactive $[3Fe—4S]^{1+}$ m-acon yields spectra at RT and 77 K that exhibited a signal not found in experiments with the active $[4Fe—4S]^{2+}$ form. This signal appeared early in the reaction and disappeared with time. As shown in Fig. 3, subtraction

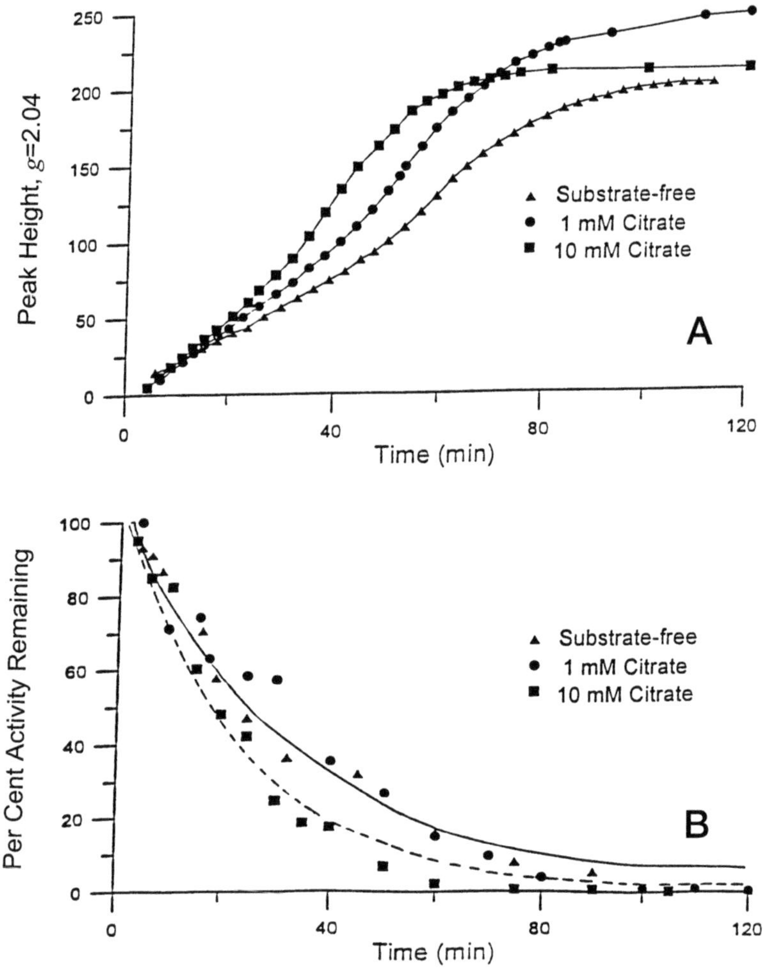

Fig. 2. Time course of inactivation of m-acon and formation of DNI-acon in the presence of spermine NONOate. Spermine NONOate, final concentration 4.5 mM, was added anaerobically to a solution of 200 μM $[4Fe—4S]^{2+}$ m-acon in 0.1 M MES at pH 6.6 and at 23°C. (A) Continuous monitoring of the EPR signal at $g = 2.04$ was made using a flat cell at a microwave power of 100 mW and at a modulation amplitude and frequency of 0.5 mT and 100 kHz. (B) Activity measurements of the same sample as in (A) were performed simultaneously as described in Kennedy et al. [15].

of the $g \approx 2.04$ signal obtained at 75 min at RT from the spectrum of the earlier sample at 15 min yields a new signal with $g'_x = 2.050$, $g'_y = 2.032$ and $g'_z = 2.004$. Similar samples examined at 77 K gave identical results. Characteristics of the EPR spectra suggested that the signal might be due to a histidyl-iron-nitrosyl complex which had been observed by others [17]. The spectrum of a solution containing Fe^{2+}, imidazole and NONOate shown in Fig. 4, curve A, exhibited a mixture of two signals which was converted to a single species upon addition of dithionite to the solution (Fig. 4, curve B). Subtraction of the signal of

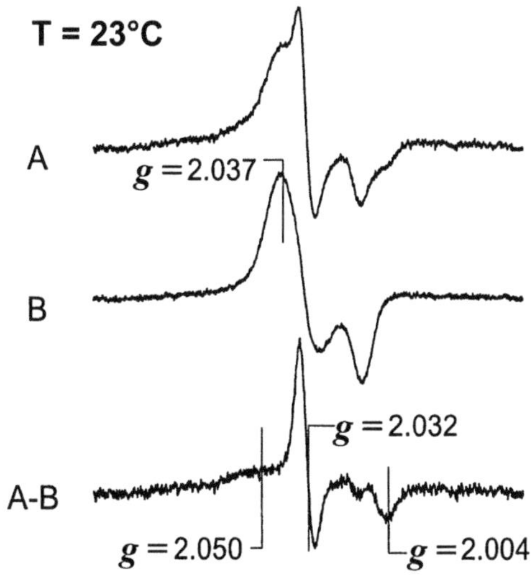

Fig. 3. EPR spectra obtained at 23°C, during the reaction of $[3Fe—4S]^{2+}$ m-acon and spermine NONOate. Spectrum (curve A) is the scan at 15 min and spectrum (curve B) is the scan at 75 min of a mixture of 95 μM $[3Fe—4S]^{1+}$ m-acon with 2.6 mM spermine NONOate. Conditions of spectroscopy as in Fig. 2.

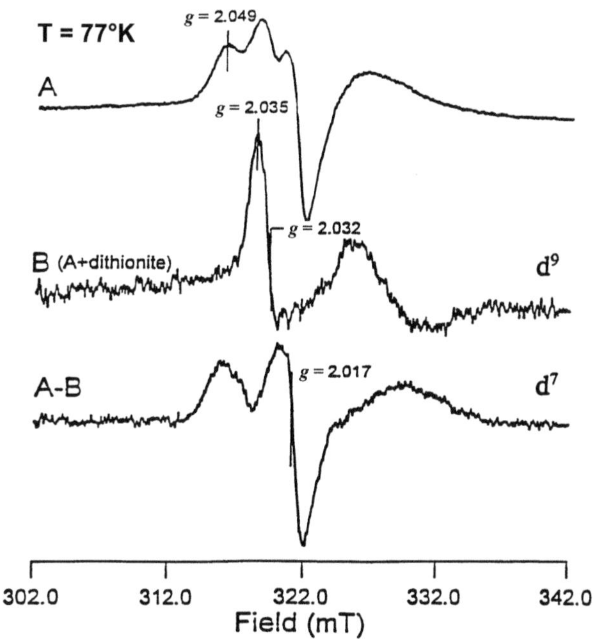

Fig. 4. EPR spectra of the d^7 and d^9 forms of the dinitrosyl-iron-imidazole complex. A spectrum of a solution prepared anaerobically in 0.1 M HEPES, pH 7.5, containing 10 mM imidazole, 0.5 mM ferrous ammonium sulfate and 5.0 mM diethylamine NONOate and incubated 30 min at room temperature. (curve B) Spectrum of sample as in (curve A) to which an excess of dithionite has been added. (Curve A – B) is the difference spectrum of (curve A) and (curve B). The conditions of spectroscopy are as described in the legend of Fig. 1 except that the microwave frequency is 9.090 GHz and microwave power 2 mW.

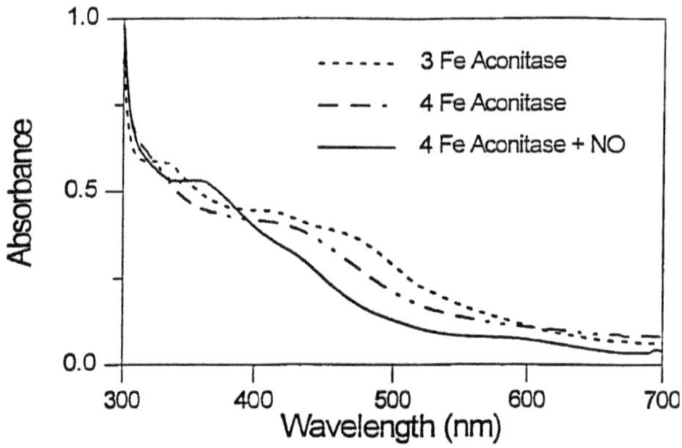

Fig. 5. Optical spectra of [3Fe—4S] and [4Fe—4S] m-acon and 4Fe m-acon inactivated by spermine NONOate. The concentrations of 3Fe and 4Fe m-acon were 30 μM. The spectrum of 4Fe + NO was normalized to the absorbance at 280 nm of the 3Fe and 4Fe forms. All spectra were run against a buffer blank of 0.1 M HEPES, pH 7.5.

the reduced sample (curve B) from the original spectrum (curve A) gives the signal of the d^7 imidazole complex with g values of 2.050 and 2.17. The transient signal observed in the reaction of 3Fe m-acon and NO, Fig. 3, (curve A − B), most closely resembles that of the d^9 state of the imidazole complex, with g values of 2.032 and 2.004. Thus, it was concluded that the new signal formed in this reaction is most likely due to d^7 and d^9 forms of a histidyl-nitrosyl-acon complex. Reaction of NO with 3Fe c-acon did not reveal species other than $g \approx 2.04$ DNI-acon. Changes in the optical spectra of both the 3Fe and 4Fe forms of aconitase upon addition of NO are shown in Fig. 5. The optical spectrum of the product formed is identical for each, which upon analysis by EPR gave evidence for only the $g \approx 2.04$ species present.

The data presented above leave no doubt that the Fe—S clusters of mammalian aconitases are attacked by NO leading to the formation of dinitrosyl-iron complexes.

ACKNOWLEDGMENT

The authors wish to thank Professor Ernst van Faassen for his comments on this manuscript.

REFERENCES

1 Martius C. Über den Abbau der Citronensäure. Hoppe-Seyler's Z. Physiol. Chem. 1937; 247: 104–110.

2 Beinert H, Holm RH, Münck E. Iron-sulfur clusters: nature's modular, multipurpose structures. Science 1997; 277: 653–659.

3 Beinert H, Meyer J, Lill R. Iron-sulfur proteins. In Encyclopedia of Biological Chemistry, Vol. 2 (Lennarz WJ, Lane MD, eds.), Elsevier, Amsterdam, Boston, 2004, pp. 482–489.

4 Beinert H, Kennedy MC, Stout CD. Aconitase as iron-sulfur protein, enzyme, and iron-regulatory protein. Chem. Rev. 1996; 96: 2335–2374.

5 Kennedy MC, Werst M, Telser J, Emptage MH, Beinert H, Hoffman BM. Mode of substrate carboxyl binding to the $[4Fe\text{-}4S]^{+}$ cluster of reduced aconitase as studied by ^{17}O and ^{13}C electron-nuclear double resonance spectroscopy. Proc. Natl. Acad. Sci. USA 1987; 84: 8854–8858.

6 Rouault TA, Stout CD, Kaptain S, Harford JB, Klausner RD. Structural relationship between an iron-regulated RNA-binding protein (IRE-BP) and aconitase: functional implications. Cell 1991; 64: 881–883.

7 Hentze MW, Argos P. Homology between IRE-BP, a regulatory RNA-binding protein, aconitase, and isopropylmalate isomerase. Nucleic Acids Res. 1991; 19: 1739–1740.

8 Tabuchi A, Funaji K, Nakatsubo J, Fukuchi M, Tsuchiya T, Tsuda M. Inactivation of aconitase during the apoptosis of mouse cerebellar granule neurons induced by a deprivation of membrane depolarization. J. Neurosci. Res. 2003; 71: 504–515.

9 Kim S, Ponka P. Role of nitric oxide in cellular iron metabolism. Biometals 2003; 16: 125–135.

10 Hüttermann J, Kappl R. Iron coordination in metalloproteins: structural and electronic aspects. In Electron Paramagnetic Resonance, Vol. 18 (Gilbert BC, Davies MJ, Murphy DM, eds.), Royal Society of Chemistry, Cambridge, UK, 2002, pp. 304–346.

11 Castro L, Rodriguez M, Radi R. Aconitase is readily inactivated by peroxynitrite, but not by its precursor, nitric oxide. J. Biol. Chem. 1994; 269: 29409–29415.

12 Hausladen A, Fridovich I. Superoxide and peroxynitrite inactivate aconitases, but nitric oxide does not. J. Biol. Chem. 1994; 269: 29405–29408.

13 Butler AR, Megson IL. Non-heme iron nitrosyls in biology. Chem. Rev. 2002; 102: 1155–1166.

14 Kennedy MC, Antholine WE, Beinert H. An EPR investigation of the products of the reaction of cytosolic and mitochondrial aconitases with nitric oxide. J. Biol. Chem. 1997; 272: 20340–20347.

15 Kennedy MC, Emptage MH, Dreyer JL, Beinert H. The role of iron in the activation-inactivation of aconitase. J. Biol. Chem. 1983; 258: 11098–11105.

16 Emptage MH, Dreyer JL, Kennedy MC, Beinert H. Optical and EPR characterization of different species of active and inactive aconitase. J. Biol. Chem. 1983; 258: 11106–11111.

17 Lee M, Arosio P, Cozzi A, Chasteen ND. Identification of the EPR-active iron-nitrosyl complexes in mammalian ferritins. Biochemistry 1994; 33: 3679–3687.

CHAPTER 7

Harnessing toxic reactions to signal stress: reactions of nitric oxide with iron–sulfur centers and the informative case of SoxR protein

Bruce Demple[1]*, Huangen Ding[2], Binbin Ren[2] and Tiffany A. Reiter[1]

[1]*Department of Genetics and Complex Diseases, Harvard School of Public Health, 665 Huntington Avenue, Boston, MA 02115, USA*
[2]*Department of Biological Sciences and Molecular Biology, Louisiana State University, Baton Rouge, LA 70803, USA*

INTRODUCTION

As a rule, the loss of iron–sulfur (FeS) clusters is very damaging to the function of FeS proteins. Nitric oxide (NO) radicals provide a prominent pathway for disruption of iron–sulfur clusters, and lead to the formation of dinitrosyl-iron complexes (DNICs). The spectroscopic properties of such DNIC motifs are discussed in Chapter 2, and the formation of protein-bound DNIC is the subject of Chapters 4–6. Usually, the transformation of FeS clusters to DNICs is highly disruptive. However, in the case of the SoxR protein, this transformation converts the protein into a potent activator of gene transcription. SoxR is a homodimer protein containing a pair of [2Fe—2S] clusters, and it becomes activated when cells are exposed to superoxide or NO. Activation of this protein stimulates the transcription of the *soxS* gene as a first step in a cascade to activate regulon promoters. The activity of SoxR is regulated through the state of its [2Fe—2S] clusters. Two distinct regulatory mechanisms have been recognized: First, regulation *via* the redox state since the protein is transcriptionally active only in oxidized state. Second, *via* the reversible assembly–disassembly of the [2Fe—2S] clusters. Interestingly, the reconstruction of the clusters seems to be a tightly controlled process in living cells: Although the DNICs formed in purified SoxR by treatment with pure NO gas are quite stable *in vitro*, the SoxR DNICs formed in intact cells are very rapidly replaced by normal reduced [2Fe—2S] centers, with a consequent deactivation

* Author for Correspondence. E-mail: bdemple@hsph.harvard.edu

of SoxR. This rapid turnover is not exclusive to SoxR. DNICs formed in ferredoxin are also rapidly turned over and replaced by unmodified metal centers. *In vitro*, this turnover can be mimicked by treatment with L-cysteine (which non-enzymatically removes the DNICs) and the IscS cysteine desulfurase, which supplies inorganic sulfide for reconstructing the [2Fe–2S] centers. The active repair of DNICs appears to be a general process that helps counteract the toxicity of NO exposure. The reversibility of the transformation FeS $\leftrightarrow$ DNIC also makes it well suited to a role in signal transduction.

Reactive molecules are generated routinely in biological systems, both actively as cellular products with defined roles, and inadvertently as the hazardous by-products of normal metabolism. Oxygen radicals are endogenously formed in both ways [1]: they are the unavoidable by-products of aerobic metabolism, through autooxidation reactions of electron transport components and other cellular molecules; oxygen radicals are also generated actively in large amounts during the inflammatory response of immune cells such as macrophages and neutrophils. Oxygen radicals actively produced at lower levels are involved in signaling pathways, for example in response to some growth factors [2]. Another physiological example is NO, which at high levels contributes to the cytotoxic action of activated macrophages [3], but at much lower levels is employed for intercellular signaling and perhaps other regulatory purposes [4,5]. NO is also a metabolic intermediate for denitrifying bacteria; disruption of regulation in this process can cause large amounts of the NO intermediate to accumulate, in some circumstances enough to inhibit the growth of the NO-generating bacteria or their neighbors [6].

Radical species cause broad cellular damage. Superoxide ($O_2^{\bullet-}$), the primary product of autooxidation reactions and of the NADPH-consuming oxidases, reacts directly with only a few types of molecules, notably protein FeS centers [7]. Such reactions often inactivate enzymes. Aconitases [8] are a well-studied example, and have been discussed in Chapter 6. Superoxide is rapidly converted to H_2O_2 by superoxide dismutase (SOD); H_2O_2 can then react with reduced transition metals such as iron or copper to generate hydroxyl radical, which can damage all classes of macromolecules [7]. NO also reacts with FeS centers [9,10], as well as with heme [11], in both cases often leading to protein inactivation. Additional NO-dependent damage in cells derives from downstream reaction products of NO with O_2, $O_2^{\bullet-}$ or thiols [12]. The resulting reactive species can damage DNA, proteins, or membranes. Thus, all the major cellular components are at risk for damage from free radicals.

In the face of this threat, complex cellular systems have evolved to offset the toxic threat of oxidative or NO damage [13,14]. SOD and catalase remove $O_2^{\bullet-}$ and H_2O_2, respectively, while small compounds such as glutathione and α-tocopherol neutralize radicals. Protein FeS centers are repaired or replaced [15], and DNA lesions are corrected by complex enzymatic systems [16]. Collectively, these defenses offset the threat of mutations or cell death that could result from the accumulation of free radical damage.

Another level of protection comes from the regulated expression of cellular defense systems [17]. The response to oxidative stress coordinates the regulation of numerous genes and proteins to reduce the formation of radicals, neutralize them more effectively, protect macromolecules from damage, and repair lesions when they do occur [13]. For the systems in which the biochemistry of signal transduction has been clarified, it is often the case that the triggering reaction (e.g. oxidation or nitrosylation of FeS centers) is one that usually inactivates protein function. In the oxidative stress sensors, these damaging reactions have been exploited for purposes of signaling [18].

In this chapter, we will concentrate on a particular type of radical signaling by NO, namely the transformation of the FeS centers in SoxR to protein-bound DNICs. Such transformations have been detected in many different iron–sulfur proteins and usually inhibit the enzymatic activity of the affected protein. SoxR is unusual in this respect as the transformation activates the enzyme. This property is crucial for the role of SoxR as an oxidative stress regulator [19]. In this protein, the reversible formation of DNIC provides an important mechanism for intracellular signaling.

DNICs AND NITRIC OXIDE TOXICITY

The first indication for the NO-mediated modification of FeS proteins *in vivo* came from electron paramagnetic resonance (EPR) spectroscopy studies of *Clostridium botulinum* [20]. Cultured *C. botulinum* cells show a prominent EPR absorption near $g_{av} = 1.94$ from intact FeS proteins. Upon exposure to the combination of sodium nitrite and a reductant such as ascorbate, the EPR signal at $g_{av} = 1.94$ was lost and replaced with a new signal at $g_{av} = 2.04$. The g-factor and lineshape indicated that the FeS clusters were converted to protein-bound DNIC. The conversion was caused by the NO radicals released by the reduction of nitrite. The same EPR signal at $g_{av} = 2.04$ was later observed in lipopolysaccharide-activated macrophages [21–24] and in cultured hepatocytes and macrophages treated with exogenous NO [25]. More recently, Pieper et al. [26] reported that formation of protein-bound DNIC in post-operative day-4 allografts. The formation of DNIC coincided with a decrease of intact FeS clusters, and indicated that FeS proteins are the target of NO cytotoxicity in acute cardiac allograft rejection. More details of this phenomenon can be found in Chapter 19 of this book. Collectively, these studies confirm that pathophysiological levels of NO can transform FeS clusters to protein-bound DNIC in cells and tissues.

In *E. coli* cells alone, thus far over 180 different FeS proteins have been identified [27]. To explore whether all these FeS proteins may be modified by NO in cells, we (H. Ding laboratory) treated *E. coli* cells with NO using the Silastic tubing NO delivery system at a rate of ~50 nM NO per second, as described by Tannenbaum's group [28]. This NO release rate is comparable to that under pathophysiological conditions [28,29]. Cell extracts prepared from the NO-treated *E. coli* cells were then fractionated using gel filtration chromatography (Superdex 200). Each eluted fraction was then analyzed by EPR (Fig. 1A). The amplitude of the EPR signal at $g_{av} = 2.04$ of the protein-bound DNIC and the protein concentration in each fraction were plotted together as a function of the fraction numbers (Fig. 1B). This analysis showed that protein-bound DNICs were present in almost all eluted fractions, indicating that the DNICs were distributed across a broad range of protein sizes in NO-treated *E. coli* cells.

SoxR AS A SENSOR OF OXIDATIVE STRESS OR NITRIC OXIDE

The SoxR protein is the master regulator of a response to oxidative stress triggered by redox-cycling agents such as paraquat, which generate large amounts of $O_2^{\bullet -}$ in cells. SoxR itself

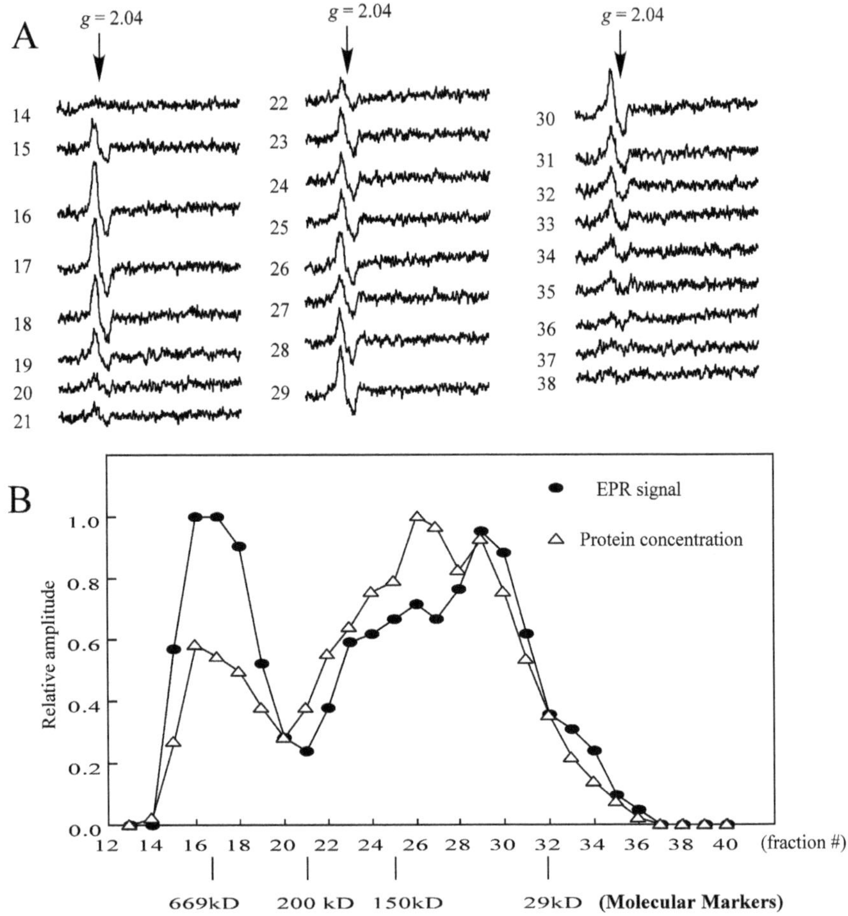

Fig. 1. Diversity of DNIC proteins recovered from NO-treated *E. coli*. A cell extract from bacteria treated with NO was fractionated by gel filtration chromatography. Individual fractions were analyzed by EPR spectroscopy for the $g_{av} = 2.04$ signal characteristic of DNIC [numbered traces shown in (Panel A)]. Panel B shows the 280-nm absorbance and EPR signal amplitude in the gel filtration column fractions.

appears to control only one promoter in *E. coli*, that of the *soxS* gene, which is positioned head-to-head with the *soxR* gene in *E. coli* [30,31] and *Salmonella* [32]. The elevated levels of SoxS protein resulting from SoxR activation stimulate expression of the many genes of the *E. coli soxRS* defense system through specific binding of the target promoters and recruitment of RNA polymerase [33]. The activation of the *soxRS* regulon genes enhances bacterial resistance against redox-cycling agents such as paraquat (but not to simple oxidants such as H_2O_2), as well as to a broad range of antibiotics and to organic solvents [17]. The oxidant resistance phenotype involves many functions, including some with obvious antioxidant roles (e.g. SOD) or functions in cellular repair (e.g. the DNA repair enzyme endonuclease IV). In contrast, the increased antibiotic resistance phenotype was quite unexpected [34]. This SoxR-mediated resistance against antibiotics and organic solvents involved the upregulation of the efflux pumps in the inner membrane [35] and modified the expression of porins in the outer membrane [36]. These effects show that the range of metabolic and biochemical functions affected by the *soxRS* system is very broad, and suggests possible evolutionary origins of the system in response to various environmental exposures [37].

The *soxRS* system was discovered and characterized initially by studies of the response to redox-cycling agents, which generate increased levels of superoxide in the cell [38]. An initial survey of various oxidative and other agents was consistent with a specific response to $O_2^{\bullet-}$ or superoxide-generating agents [39], and even revealed a previously unsuspected potent redox-cycling activity of a DNA-damaging carcinogen [40]. For various reasons, exposure to NO (pure gas, delivered through gas-permeable tubing) caused significant activation of *soxRS* [41]. Notably, this activation in response to NO was independent of oxygen, and was even somewhat more efficient under anaerobic than aerobic conditions [41]. NO donors such as "NONOate" were also able to activate SoxR *in vivo* [42]. Activation of the *soxRS* system by NO improved the chance of survival of phagocytosed *E. coli* in activated murine macrophages [41]. A direct measurement of *soxS* promoter activity in phagocytosed bacteria indicated about 30-fold induction 8 h after their initial uptake by the macrophages. This effect was dependent on a functional *soxR* gene and abrogated by incubation of the bacteria-containing macrophages with the an NO synthase inhibitor [43]. In contrast, *soxRS* did not contribute significantly to the survival of *Salmonella* in activated macrophages [44], although it should be noted in this context that *S. enterica* has evolved to reside in the macrophage phagosome, unlike *E. coli*.

The activation of SoxR by NO was somewhat surprising. Previous studies had established that SoxR could be activated by one-electron oxidation of the protein's [2Fe—2S] centers. This redox activation had been monitored by EPR in intact bacteria, as well as with purified SoxR [45–49]. *In vitro* and *in vivo* studies indicated that the transcriptional activity was dependent on intact [2Fe—2S] centers in the protein [50,51]. Against this background, it seemed unlikely that NO could activate SoxR by formation of the FeS centers into DNICs. However, a detailed study [52] showed that this was indeed the case: an EPR signal characteristic of SoxR-bound DNIC was formed in intact *E. coli* treated with NO. The intensity of this DNIC signal correlated perfectly with the transcriptional activation of SoxR. Furthermore, treatment of purified SoxR yielded a DNIC-containing protein that could be repurified and which exhibited potent transcriptional activity [52]. An interesting observation during these studies was the contrast between the stability of the SoxR DNIC *in vitro*, and its rapid disappearance *in vivo* after termination of the NO exposure [52]. The experiments show that cells actively promote this turnover.

REPAIR OF PROTEIN DNICs

To search for the cellular factors responsible for removing and replacing the protein-bound DNIC, ferredoxin DNICs were prepared from either NO-treated *E. coli* or by treating ferredoxin [2Fe—2S] directly with NO *in vitro* [53]. EPR spectroscopy was used to measure the ferredoxin DNIC [54]. When the ferredoxin DNIC (Fig. 2, trace a) was incubated

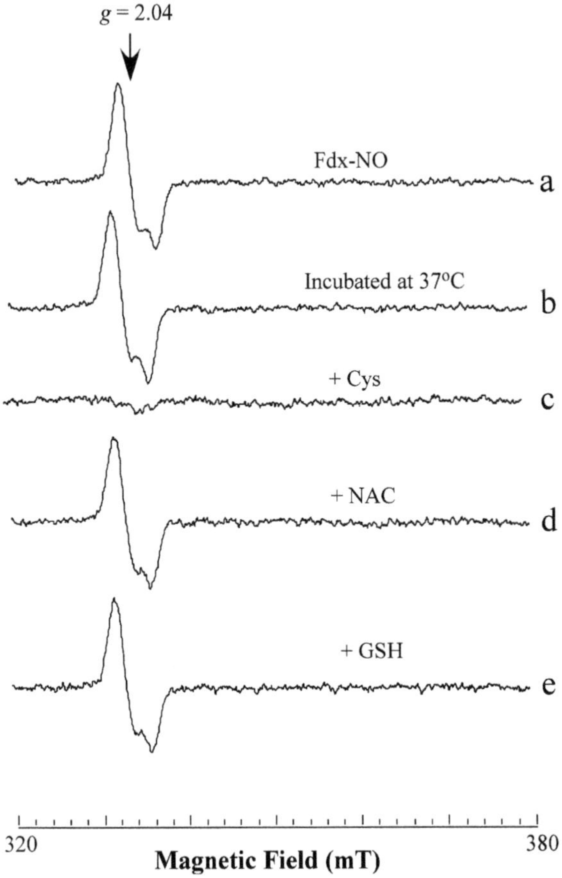

Fig. 2. Effect of biological thiols on protein DNIC. A ferredoxin-DNIC sample (10 μM; trace a) was incubated with no addition (trace b) or with 1 mM L-cysteine (Cys; trace c), 1 mM N-acetyl-L-cysteine (NAC; trace d) or 1 mM glutathione (GSH; trace e) at 37°C for 20 min, followed by EPR spectroscopy. Reproduced from Rogers and Ding [53].

alone, the amplitude of the EPR signal at $g_{av} = 2.04$ of the sample was not affected (Fig. 2, trace b). However, when the ferredoxin DNIC was incubated with L-cysteine, the $g_{av} = 2.04$ EPR signal was completely eliminated (Fig. 2, trace c). Oxygen in solution had little effect, as L-cysteine decomposed the protein-bound DNIC under both aerobic and anaerobic conditions [53].

D-cysteine had the same effect as L-cysteine (data not shown), which indicates that the redox properties of cysteine rather than its stereo configuration are critical for decomposing the ferredoxin DNIC. However, the monothiols N-acetyl-L-cysteine and glutathione had no effect (Fig. 2, traces d,e), which was also true of reduced thioredoxin (not shown). Thus, cysteine seems to have unique activity in eliminating the DNIC EPR signal and perhaps the protein-bound DNICs themselves.

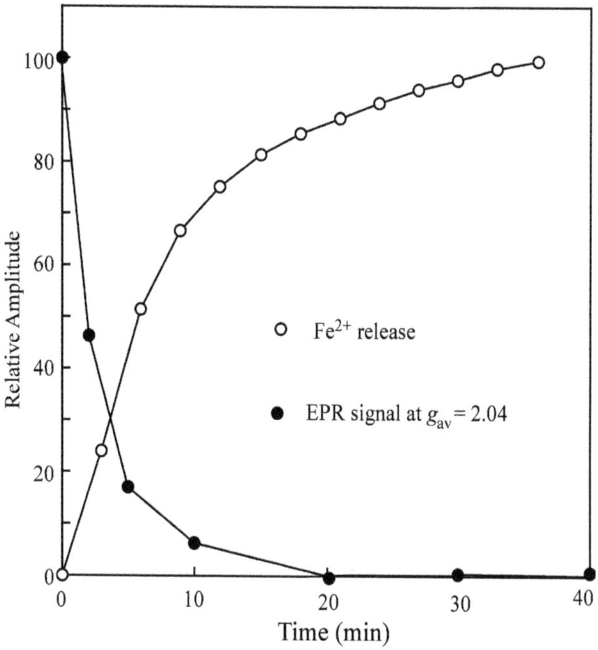

Fig. 3. Correlation of iron release and EPR intensity during incubation of ferredoxin DNIC with L-cysteine at 37°C. Reproduced from Rogers and Ding [53].

L-CYSTEINE RELEASES FERROUS IRON FROM THE PROTEIN-BOUND DNIC

If cysteine decomposes the protein-bound DNIC, it is expected that "free" iron will be released from the complex. To monitor the iron release, the iron indicator α, α-dipyridyl was added to the samples during incubation of the ferredoxin DNIC with cysteine [55]. Very little iron was released from the ferredoxin DNIC during incubation with α, α-dipyridyl alone, but with the inclusion of L-cysteine, most of the iron originally contained in the ferredoxin was released (Fig. 3).

IRON–SULFUR CLUSTERS CAN BE RE-ASSEMBLED TO REPLACE DNIC

To re-assemble the FeS clusters, recombinant cysteine desulfurase (IscS protein) was prepared from *E. coli* [56]. IscS is a member of the recently identified FeS cluster assembly machinery [57]. The enzyme catalyzes desulfurization of L-cysteine and provides sulfide for the synthesis of FeS clusters [58]. When the ferredoxin DNIC (Fig. 4, trace a) was incubated with L-cysteine and IscS, the observed UV–visible absorption spectrum was typical for the ferredoxin [2Fe—2S] clusters, with two absorption peaks at 415 and 459 nm (Fig. 4, trace d). The results clearly show that NO-generated ferredoxin DNIC can be repaired by the addition

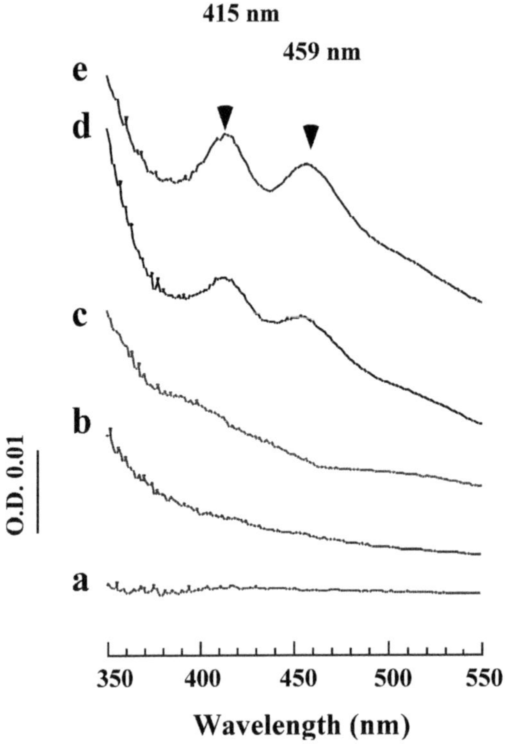

Fig. 4. Repair of ferredoxin DNIC *in vitro*. Formation of the repaired [2Fe—2S] centers was monitored by UV–Visible absorption spectroscopy. NO-treated ferredoxin (trace a) was incubated with L-cysteine alone (trace b), IscS alone (trace c), or both together (trace d); trace e is from undamaged [2Fe—2S] ferredoxin. Reproduced from Yang et al. [56].

of L-cysteine and IscS *in vitro*. Of note, the FeS clusters of the protein are reconstructed without the need for additional iron [56].

A similar process has also been observed for tetranuclear clusters in a protein in which the FeS center appears to have a strictly structural role. Quite unexpectedly, the *E. coli* DNA repair enzyme endonuclease III was found to contain a [4Fe—4S] center [59], providing the first example of this type of metalloprotein among DNA repair enzymes. Homologous proteins have since been identified in mammalian species, including humans [60]. The [4Fe—4S] center in endonuclease III is resistant to oxidation and reduction [59], and structural studies indicate that the metal center instead helps position parts of the protein for substrate recognition [61]. Despite its lack of redox or apparent catalytic activity, the endonuclease III [4Fe—4S] center is susceptible to nitrosylation by NO [15]. FeS nitrosylation inactivates the enzyme, as it does with many other proteins. However, similarly to ferredoxin, intact [4Fe—4S] centers can be restored to the endonuclease III by incubation *in vitro* of the DNIC-containing form with ferrous iron, L-cysteine and IscS, resulting in full recovery of the enzymatic activity [15]. As is the case for SoxR, the endonuclease III DNICs are rapidly removed *in vivo* and replaced by functional FeS centers, without the need for new protein synthesis [15]. Thus, mechanisms to reconstruct FeS clusters from protein DNIC seem to be general.

NEW NITRIC OXIDE SIGNALING PATHWAYS *VIA* FeS CLUSTERS

The studies on the appearance and processing of DNICs upon NO exposure of SoxR and ferredoxin demonstrate that this modification has the key properties necessary to act in a specific stress-signaling role. First, the modification (DNIC) is formed directly by the activating agent (NO). Second, in SoxR the formation of DNIC is activating rather than inhibiting. Third, the modification is rapidly reversible, so that the activity of SoxR and the downstream genetic response can be finely modulated.

A new example of nitrosylation of an iron center in gene control was recently described. The NorR protein governs a defensive response in enterobacteria that induces the synthesis of flavorubredoxin and an associated flavoprotein that constitute an NO reductase activity, which converts toxic NO to the less toxic N_2O [62]. NorR activation appears to be rather specific for NO, in contrast to SoxR (see below). A recent biochemical analysis demonstrated that NorR contains a mononuclear iron center, and that exposure of the protein to NO generates a nitrosylated form [63]. Most notably, the nitrosylation is reversible, which is a key feature of such a regulatory system, as noted earlier. It will be of interest to determine how the nitrosylation is reversed (passively or actively), and whether a similar mechanism is employed in other regulatory scenarios.

A MODEL FOR SoxR ACTIVATION IN RESPONSE TO MULTIPLE SIGNALS

The observed activation of SoxR by diverse signals (oxidation, nitrosylation, iron removal) has prompted efforts to understand the structural basis for signal transduction by this sensor. Although a direct three-dimensional structure for SoxR has not been reported, related protein structures are available to aid with the interpretation of mutational and other analyses [64].

The mechanism by which activated SoxR stimulates transcription of the *soxS* gene is shared by the various members of the eponymous MerR regulatory family. The *soxS* promoter has -10 and -35 elements that are close to consensus sequences, but these elements are separated by a sub-optimal 19-bp spacing [65]. The majority of *E. coli* promoters dependent on σ^{70} RNA polymerase have 17 ± 1 bp spacing. In the *soxS* promoter, shortening the $-10/-35$ spacing to $16-19$ bp gives strong transcription in the absence of SoxR [65], consistent with the spacing as the critical regulatory feature of this promoter. The MerR target promoter has similar properties, and an unusual characteristic is that the inactive and active forms of both MerR [66] and SoxR [48,50] bind their target promoters with approximately equal affinity. SoxR and MerR thus act allosterically, *via* structural transitions in the protein–DNA complex.

Topological experiments with MerR suggested that transcriptional induction is accomplished by localized distortions mediated by the protein [67]. Data consistent with such a mechanism for SoxR emerged from footprinting studies of the changes exerted by SoxR on the *soxS* promoter [51].

The foregoing focuses on changes at the DNA level, but these changes must be driven by structural transitions in the protein in response to signals such as [2Fe—2S] oxidation. It was noted earlier that deletions or insertions affecting the SoxR C-terminus led to a constitutively active protein [68]. Analysis of single amino acid substitutions in constitutively active SoxR proteins also showed a preference for alterations of the SoxR C-terminus [69]. This diversity of activating changes suggested that the C-terminus might act to restrain SoxR transcriptional activity.

More recent mutational analysis revealed a new class of single amino acid changes that disrupt the SoxR signal transduction circuit without preventing DNA binding [42]. Biochemical analysis of the mutant proteins against the backdrop of structural information from a few related proteins [70–72] provided some important insights [73]. This analysis suggested important roles for the center of the subunit interface, formed by helical coils in each of the subunits, and the likely intimate proximity of [2Fe—2S] domain, anchored near the C-terminus, with the DNA-binding domain formed by the N-terminus.

How various signals might result in analogous structural transitions in the SoxR–DNA complex? An additional important clue came from a seemingly discordant result: it was reported that the chelator 1,10-phenanthroline activates SoxR *in vivo*, even under anaerobic conditions [74]. Removal of the SoxR [2Fe—2S] seemed a likely effect of the chelator, and we later confirmed the formation of apo-SoxR in cells treated with an iron chelator that moderately activates SoxR (Ding and Demple, unpublished data). In contrast with these results, various studies showed an absolute requirement of the SoxR [2Fe—2S] centers for transcription *in vitro* [50,51,75,76]. However, SoxR proteins with cysteine-to-alanine substitutions that prevent formation of the [2Fe—2S] centers have modest transcriptional activity *in vivo* that gives up to 10-fold stimulation of *soxS* transcription [76]. Transcriptional stimulation by apo-SoxR *in vitro* was finally achieved when we found that it is strongly dependent on negative supercoiling of the DNA template (E. Hidalgo and Demple, unpublished data).

There are thus at least three active forms of SoxR: the protein with oxidized [2Fe—2S] centers; SoxR containing DNICs; and the apo-protein, dependent on supercoiling. Both apo-SoxR and DNIC-containing SoxR are expected to have disrupted metal-binding domains compared to SoxR with reduced [2Fe—2S] centers. This view suggests that [2Fe—2S] oxidation might also disrupt some aspect of the C-terminus structure. As suggested schematically here (Fig. 5), C-terminal disruption in all three cases would release the SoxR homodimer to take up the active conformation, with the resulting transcriptional activity. Clearly, detailed structural studies will be needed to test this hypothesis.

PERSPECTIVE

There seem to be clear evolutionary advantages of fortuitously adapting toxic reactions for signaling pathways. This may imply that every cytotoxic or genotoxic reaction has such potential. In the case of reactive oxygen and NO, new biological reactions continue to be described. Some of these involve new signaling mechanisms. Others are merely new examples of previously known mechanisms. In the particular case of DNICs, we note their broad occurrence in a wide range of different cells after exposure to NO. The DNICs appear wherever FeS

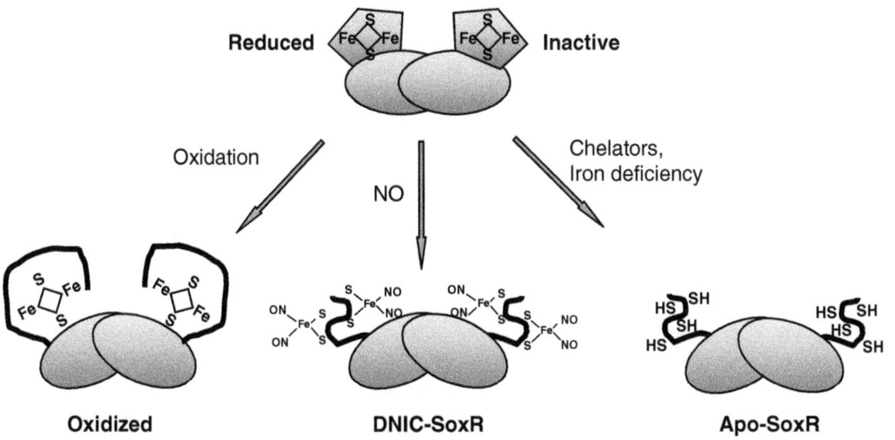

Fig. 5. Scheme for SoxR activation by various stimuli. SoxR with reduced [2Fe—2S] centers binds the *soxS* promoter but does not stimulate transcription. Oxidation of the metal centers converts SoxR to a strongly active form, as does nitrosylation by nitric oxide to form DNIC. Removal of the metal centers by chelation or mutational substitution of the iron-anchoring cysteine residues also activates SoxR, although this form has lower transcriptional activity than oxidized or nitrosylated SoxR. The structural transitions in the C-terminus due to these modifications are symbolized, as is a proposed reorientation of the subunits accompanying activation.

proteins play a role. It implies that the reversible formation of DNIC be involved in many signaling pathways.

ACKNOWLEDGMENTS

Work in the B.D. and H. D. laboratories was supported by grants from the U.S. National Institutes of Health. T.A.R. is a fellow of the Muscular Dystrophy Association. We are grateful to past laboratory members and our colleagues for their contributions to this field and for numerous stimulating discussions.

REFERENCES

1 Finkel T, Holbrook NJ. Oxidants, oxidative stress and the biology of ageing. Nature 2000; 408: 239–47.
2 Ushio-Fukai, M. Localizing NADPH oxidase-derived ROS. Sci STKE. 2006; re8.
3 MacMicking, J, Xie QW, Nathan C. Nitric oxide and macrophage function. Annu. Rev. Immunol. 1997; 15: 323–350.
4 Ignarro, LJ. Nitric oxide: a unique endogenous signaling molecule in vascular biology. Biosci. Rep. 1999; 19: 51–71.
5 Jaffrey, SR, Erdjument-Bromage H, Ferris CD, Tempst P, Snyder SH. Protein S-nitrosylation: a physiological signal for neuronal nitric oxide. Nat. Cell Biol. 2001; 3: 193–197.
6 Choi, PS, Naal Z, Moore C, Casado-Rivera E, Abruna HD, Helmann JD, Shapleigh JP. Assessing the impact of denitrifier-produced nitric oxide on other bacteria. Appl. Environ. Microbiol. 2006; 72: 2200–2205.
7 Fridovich, I. Superoxide radical and superoxide dismutases. Annu. Rev. Biochem. 1995; 64: 97–112.

8 Gardner PR, Fridovich I. Superoxide sensitivity of the Escherichia coli aconitase. J. Biol. Chem. 1991; 266: 19328–19333.
9 Gardner PR, Costantino G, Szabo C, Salzman AL. Nitric oxide sensitivity of the aconitases. J. Biol. Chem. 1997; 272: 25071–25076.
10 Foster MW, Cowan JA. Chemistry of nitric oxide with protein-bound iron sulfur centers. Insights on physiological reactivity. J. Am. Chem. Soc. 1999; 121: 4093–4100.
11 Cooper CE. Nitric oxide and iron proteins. Biochim. Biophys. Acta 1999; 1411: 290–309.
12 Beckman JS, Koppenol WH. Nitric oxide, superoxide, and peroxynitrite: the good, the bad, and ugly. Am. J. Physiol. 1996; 271: C1424–C1437.
13 Storz G, Imlay JA. Oxidative stress. Curr. Opin. Microbiol. 1999; 2: 188–194.
14 Demple B. Signal transduction by nitric oxide in cellular stress responses. Mol. Cell Biochem. 2002; 234–235: 11–18.
15 Rogers PA, Eide L, Klungland A, Ding H. Reversible inactivation of E. coli endonuclease III via modification of its [4Fe—4S] cluster by nitric oxide. DNA Repair (Amst.) 2003; 2: 809–817.
16 Wilson 3rd DM, Sofinowski TM, McNeill DR. Repair mechanisms for oxidative DNA damage. Front Biosci. 2003; 8: d963–d981.
17 Pomposiello PJ, Demple B. Global adjustment of microbial physiology during free radical stress. Adv. Microb. Physiol. 2002; 46: 319–341.
18 Demple B. Radical ideas: genetic responses to oxidative stress. Clin. Exp. Pharmacol. Physiol. 1999; 26: 64–68.
19 Demple B, Ding H, Jorgensen M. Escherichia coli SoxR protein: sensor/transducer of oxidative stress and nitric oxide. Methods Enzymol. 2002; 348: 355–364.
20 Reddy D, Lancaster Jr. JR, Cornforth DP. Nitrite inhibition of Clostridium botulinum: electron spin resonance detection of iron-nitric oxide complexes. Science 1983; 221: 769–770.
21 Lancaster Jr. JR, Hibbs Jr. JB. EPR demonstration of iron-nitrosyl complex formation by cytotoxic activated macrophages. Proc. Natl. Acad. Sci. USA 1990; 87: 1223–1227.
22 Drapier JC, Pellat C, Henry Y. Generation of EPR-detectable nitrosyl-iron complexes in tumor target cells cocultured with activated macrophages. J. Biol. Chem. 1991; 266: 10162–10167.
23 Henry Y, Ducrocq C, Drapier JC, Servent D, Pellat C, Guissani A. Nitric oxide, a biological effector. Electron paramagnetic resonance detection of nitrosyl-iron-protein complexes in whole cells. Eur. Biophys. J. 1991; 20: 1–15.
24 Henry Y, Lepoivre M, Drapier JC, Ducrocq C, Boucher JL, Guissani A. EPR characterization of molecular targets for NO in mammalian cells and organelles. FASEB J. 1993; 7: 1124–1134.
25 Stadler J, Billiar TR, Curran RD, Stuehr DJ, Ochoa JB, Simmons RL. Effect of exogenous and endogenous nitric oxide on mitochondrial respiration of rat hepatocytes. Am. J. Physiol. 1991; 260: C910–C916.
26 Pieper GM, Halligan NL, Hilton G, Konorev EA, Felix CC, Roza AM, Adams MB, Griffith OW. Nonheme iron protein: a potential target of nitric oxide in acute cardiac allograft rejection. Proc. Natl. Acad. Sci. USA 2003; 100: 3125–3130.
27 Johnson MK. Iron-sulfur proteins: new roles for old clusters. Curr. Opin. Chem. Biol. 1998; 2: 173–181.
28 Tamir S, Lewis RS, de Rojas Walker T, Deen WM, Wishnok JS, Tannenbaum SR. The influence of delivery rate on the chemistry and biological effects of nitric oxide. Chem. Res. Toxicol. 1993; 6: 895–899.
29 Li CQ, Trudel LJ, Wogan GN. Nitric oxide-induced genotoxicity, mitochondrial damage, and apoptosis in human lymphoblastoid cells expressing wild-type and mutant p53. Proc. Natl. Acad. Sci. USA 2002; 99: 10364–10369.
30 Amabile-Cuevas CF, Demple B. Molecular characterization of the soxRS genes of Escherichia coli: two genes control a superoxide stress regulon. Nucleic Acids Res. 1991; 19: 4479–4484.
31 Wu J, Weiss B. Two divergently transcribed genes, soxR and soxS, control a superoxide response regulon of Escherichia coli. J. Bacteriol. 1991; 173: 2864–2871.
32 Koutsolioutsou A, Martins EA, White DG, Levy SB, Demple B. A soxRS-constitutive mutation contributing to antibiotic resistance in a clinical isolate of Salmonella enterica (Serovar typhimurium). Antimicrob. Agents Chemother. 2001; 45: 38–43.

33 Li Z, Demple B. SoxS, an activator of superoxide stress genes in Escherichia coli. Purification and interaction with DNA. J. Biol. Chem. 1994; 269: 18371–18377.

34 Demple B. The Nexus of oxidative stress responses and antibiotic resistance mechanisms in Escherichia coli and Salmonella enterica. In Frontiers in Antibiotic Resistance: A Tribute to Stuart (Levy B, White DG, Alekshun MN, McDermott PF, eds.), ASM Press, Washington, DC, 2005, pp. 191–197.

35 White DG, Goldman JD, Demple B, Levy SB. Role of the acrAB locus in organic solvent tolerance mediated by expression of marA, soxS, or robA in Escherichia coli. J. Bacteriol. 1997; 179: 6122–6126.

36 Chou JH, Greenberg JT, Demple B. Posttranscriptional repression of Escherichia coli OmpF protein in response to redox stress: positive control of the micF antisense RNA by the soxRS locus. J. Bacteriol. 1993; 175: 1026–1031.

37 Demple B. Redox signaling and gene control in the Escherichia coli soxRS oxidative stress- - - -a review. Gene 1996; 179: 53–57.

38 Sies H. Oxidative stress: introduction. In Oxidative Stress: Oxidants and Antioxidants (Sies H, ed.), Academic Press, London, 1991, pp. xv–xxii.

39 Nunoshiba T, Hidalgo E, Amabile Cuevas CF, Demple B. Two-stage control of an oxidative stress regulon: the Escherichia coli SoxR protein triggers redox-inducible expression of the soxS regulatory gene. J. Bacteriol. 1992; 174: 6054–6060.

40 Nunoshiba T, Demple B. Potent intracellular oxidative stress exerted by the carcinogen 4-nitroquinoline-N-oxide. Cancer Res. 1993; 53: 3250–3252.

41 Nunoshiba T, deRojas-Walker T, Wishnok JS, Tannenbaum SR, Demple B. Activation by nitric oxide of an oxidative-stress response that defends Escherichia coli against activated macrophages Proc. Natl. Acad. Sci. USA 1993; 90: 9993–9997.

42 Chander M, Raducha-Grace L, Demple B. Transcription-defective soxR mutants of Escherichia coli: isolation and in vivo characterization. J. Bacteriol. 2003; 185: 2441–2450.

43 Nunoshiba T, DeRojas-Walker T, Tannenbaum SR, Demple B. Roles of nitric oxide in inducible resistance of Escherichia coli to activated murine macrophages. Infect. Immun. 1995; 63: 794–798.

44 Fang FC, Vazquez-Torres A, Xu Y. The transcriptional regulator SoxS is required for resistance of Salmonella typhimurium to paraquat but not for virulence in mice. Infect. Immun. 1997; 65: 5371–5375.

45 Ding H, Hidalgo E, Demple B. The redox state of the [2Fe—2S] clusters in SoxR protein regulates its activity as a transcription factor. J. Biol. Chem. 1996; 271: 33173–33175.

46 Hidalgo E, Ding H, Demple B. Redox signal transduction: mutations shifting [2Fe—2S] centers of the SoxR sensor-regulator to the oxidized form. Cell 1997; 88: 121–129.

47 Ding HG, Demple B. In vivo kinetics of a redox-regulated transcriptional switch. Proc. Natl Acad. Sci. USA 1997; 94: 8445–8449.

48 Gaudu P, Weiss B. SoxR, a [2Fe—2S] transcription factor, is active only in its oxidized form. Proc. Natl. Acad. Sci. USA 1996; 93: 10094–10098.

49 Gaudu P, Moon N, Weiss B. Regulation of the soxRS oxidative stress regulon. Reversible oxidation of the Fe-S centers of SoxR in vivo. J. Biol. Chem. 1997; 272: 5082–5086.

50 Hidalgo E, Demple B. An iron-sulfur center essential for transcriptional activation by the redox-sensing SoxR protein. EMBO J. 1994; 13: 138–146.

51 Hidalgo E, Bollinger Jr. JM, Bradley TM, Walsh CT, Demple B. Binuclear [2Fe–2S] clusters in the Escherichia coli SoxR protein and role of the metal centers in transcription. J. Biol. Chem. 1995; 270: 20908–20914.

52 Ding H, Demple B. Direct nitric oxide signal transduction via nitrosylation of iron- sulfur centers in the SoxR transcription activator. Proc. Natl. Acad. Sci. USA 2000; 97: 5146–5150.

53 Rogers PA, Ding H. L-cysteine-mediated destabilization of dinitrosyl iron complexes in proteins. J. Biol. chem. 2001; 276: 30980–30986.

54 Kennedy MC, Antholine WE, Beinert H. An EPR investigation of the products of the reaction of cytosolic and mitochondrial aconitases with nitric oxide. J. Biol. Chem. 1997; 272: 20340–20347.

55 Ding H, Demple B. Glutathione-mediated destabilization in vitro of [2Fe—2S] centers in the SoxR regulatory protein. Proc. Natl. Acad. Sci. USA 1996; 93: 9449–9453.

56 Yang W, Rogers PA, Ding H. Repair of nitric oxide-modified ferredoxin [2Fe—2S] cluster by cysteine desulfurase (IscS). J. Biol. Chem. 2002; 277: 12868–12873.

57 Zheng L, Cash VL, Flint DH, Dean DR. Assembly of iron-sulfur clusters. Identification of an iscSUA-hscBA-fdx gene cluster from Azotobacter vinelandii. J. Biol. Chem. 1998; 273: 13264–13272.

58 Flint DH. Escherichia coli contains a protein that is homologous in function and N-terminal sequence to the protein encoded by the nifS gene of Azotobacter vinelandii and that can participate in the synthesis of the Fe-S cluster of dihydroxy-acid dehydratase. J. Biol. Chem. 1996; 271: 16068–16074.

59 Cunningham RP, Asahara H, Bank JF, Scholes CP, Salerno JC, Surerus K, Munck E, McCracken J, Peisach J, Emptage MH. Endonuclease III is an iron-sulfur protein. Biochemistry 1989; 28: 4450–4455.

60 Aspinwall R, Rothwell DG, Roldan-Arjona T, Anselmino C, Ward CJ, Cheadle JP, Sampson JR, Lindahl T, Harris PC, Hickson ID. Cloning and characterization of a functional human homolog of Escherichia coli endonuclease III. Proc. Natl. Acad. Sci. USA 1997; 94: 109–114.

61 Thayer MM, Ahern H, Xing D, Cunningham RP, Tainer JA. Novel DNA binding motifs in the DNA repair enzyme endonuclease III crystal structure. EMBO J. 1995; 14: 4108–4120.

62 Tucker NP, D'Autreaux B, Spiro S, Dixon R. Mechanism of transcriptional regulation by the Escherichia coli nitric oxide sensor NorR. Biochem. Soc. Trans. 2006; 34: 191–194.

63 D'Autreaux B, Tucker NP, Dixon R, Spiro S. A non-haem iron centre in the transcription factor NorR senses nitric oxide. Nature 2005; 437: 769–772.

64 Newberry KJ, Brennan RG. The structural mechanism for transcription activation by MerR family member multidrug transporter activation, N terminus. J. Biol. Chem. 2004; 279: 20356–20362.

65 Hidalgo E, Demple B. Spacing of promoter elements regulates the basal expression of the soxS gene and converts SoxR from a transcriptional activator into a repressor. EMBO J. 1997; 16: 1056–1065.

66 O'Halloran TV, Frantz B, Shin MK, Ralston DM, Wright JG. The MerR heavy metal receptor mediates positive activation in a topologically novel transcription complex. Cell 1989; 56: 119–129.

67 Ansari AZ, Chael ML, O'Halloran TV. Allosteric underwinding of DNA is a critical step in positive control of transcription by Hg-MerR. Nature 1992; 355: 87–89.

68 Tsaneva IR, Weiss B. soxR, a locus governing a superoxide response regulon in Escherichia coli K-12. J. Bacteriol. 1990; 172: 4197–4205.

69 Nunoshiba T, Demple B. A cluster of constitutive mutations affecting the C-terminus of the redox-sensitive SoxR transcriptional activator. Nucleic Acids Res. 1994; 22: 2958–2962.

70 Godsey MH, Baranova NN, Neyfakh AA, Brennan RG. Crystal structure of MtaN, a global multidrug transporter gene activator. J. Biol. Chem. 2001; 276: 47178–47184.

71 Heldwein EE, Brennan RG. Crystal structure of the transcription activator BmrR bound to DNA and a drug. Nature 2001; 409: 378–382.

72 Changela A, Chen K, Xue Y, Holschen J, Outten CE, O'Halloran TV, Mondragon A. Molecular basis of metal-ion selectivity and zeptomolar sensitivity by CueR. Science 2003; 301: 1383–1387.

73 Chander M, Demple B. Functional analysis of SoxR residues affecting transduction of oxidative stress signals into gene expression. J. Biol. Chem. 2004; 279: 41603–41610.

74 Privalle CT, Kong SE, Fridovich I. Induction of manganese-containing superoxide dismutase in anaerobic Escherichia coli by diamide and 1,10-phenanthroline: sites of transcriptional regulation. Proc. Natl. Acad. Sci. USA 1993; 90: 2310–2314.

75 Hidalgo E, Demple B. Activation of SoxR-dependent transcription in vitro by noncatalytic or NifS-mediated assembly of [2Fe—2S] clusters into apo-SoxR. J. Biol. Chem. 1996; 271: 7269–7272.

76 Bradley TM, Hidalgo E, Leautaud V, Ding H, Demple B. Cysteine-to-alanine replacements in the Escherichia coli SoxR protein and the role of the [2Fe—2S] centers in transcriptional activation. Nucleic Acids Res. 1997; 25: 1469–1475.

CHAPTER 8

Nitric oxide and dinitrosyl iron complexes: roles in plant iron sensing and metabolism

Magdalena Graziano and Lorenzo Lamattina*

Instituto de Investigaciones Biológicas, Facultad de Ciencias Exactas y Naturales, Universidad Nacional de Mar del Plata, CC 1245, (7600) Mar del Plata, Argentina

Abstract

Nitric oxide (NO) is a signaling and physiologically active molecule in plants. However, the molecular mechanism/s involved in transducing the NO signal between cells and tissues is/are still unknown. The formation of low-molecular weight dinitrosyl iron complexes (DNICs) from internal NO sources has been recently demonstrated in plants. In addition, *S*-nitrosoglutathione (GSNO) was shown to be a biologically active compound in plants. Both DNICs and GSNO are candidates for NO storage and/or mobilization between plant tissues and cells. More details on the chemistry of these compounds are given in Chapters 2 and 9 of this book. NO was shown to have a role in plant iron nutrition; therefore, it is proposed that DNICs and GSND may have roles in NO-mediated improvement of iron nutrition in plants growing under iron deficient conditions. Here, we suggest that the formation of DNICs constitutes a key process in plant iron sensing and metabolism. An interconversion between DNICs and GSNO based on the iron and NO status of the plant cell might be the core of a metabolic process leading plant iron homeostasis. Such interconversion has been firmly established *in vitro* and is extensively documented in Chapter 11. The situation is less clear *in vivo*, and this chapter is devoted to reviewing the roles of DNIC and NO in plants.

NITRIC OXIDE FUNCTIONS IN PLANTS

Small, simple and highly toxic nitric oxide (NO) is a molecule with a broad chemistry that involves an array of interrelated redox forms with different chemical reactivities [1]. Some information on the chemistry of NO and other oxidized nitrogen species may be found in Chapter 1. In the 1980s, NO was discovered as an integral part of physiological processes in

* Author for Correspondence. E-mail: Lorenzo Lamattina (lolama@mdp.edu.ar).

animals [2–4]. This discovery led to a major revolution in biomedical research. By the late 1990s, NO was identified as an important signaling molecule in plant (patho)physiological responses [5–7]. NO was subsequently shown to be a crucial player in the regulation of basic plant physiological processes. These include: germination, growth, development, reproduction and responses to abiotic stresses [8–13]. NO is also involved in plant iron homeostasis; it has been shown to mediate iron-dependent ferritin expression in *Arabidopsis thaliana* [14] and to prevent iron deficiency-induced chlorosis in maize, supporting a role for NO in the availability of bioactive iron [15].

NITRIC OXIDE SYNTHESIS IN PLANTS

In mammals, NO is generated primarily by nitric oxide synthase(s) (NOSs). NOSs are a group of evolutionarily conserved cytosolic or membrane-bound isoenzymes that convert L-arginine to L-citrulline and NO [16]. In plants, NOS activity has been detected in several plant species when mammalian NOS inhibitors have been used to inhibit NO production [17]. However, no gene or protein with a sequence similar to mammalian NOS has been found in the sequenced *A. thaliana* genome. Guo et al. (2003) identified an AtNOS1 gene with sequence similarity to a distinct NOS that was characterized in the snail *Helix pomatia*. The role of AtNOS1 in NO synthesis *in vivo* was examined in a homozygous mutant line, in which NO levels were found to be much lower than in wild type plants [18]. However, very recently, a discussion regarding the identity of the AtNOS1 gene was generated [19,20,52] since other laboratory was not able to detect NOS activity in the recombinant protein encoded by plant NOS genes [52]. Authors suggested that the AtNOS1 mutant line display lower levels of NO because AtNOS1 deletion would lead to defects in mitochondrial biogenesis. Therefore, the mechanisms for arginine-dependent NO synthesis in plants is still not clear.

However, plant NO synthesis is not as simple as in animals since also nitrite can be a substrate for NO production. Nitrate reductase (NR) is another enzyme capable of producing NO in plants [21]. This enzyme has a primary activity that converts nitrate to nitrite; however, NR can also convert nitrite to NO, *via* a single electron reduction. It was shown that NO synthesis is stimulated by nitrite [21–23] and this stimulation is attenuated in an *A. thaliana* NR double mutant (*nia1/nia2*) [22]. It was shown that NR plays a significant role in basal NO generation in leaves [23] as well as during several plant physiological processes or environmental stimulus [21,22,24,25]. In addition, NO production from nitrite involves also non-enzymatic sources. It has been demonstrated that a non-enzymatic reduction of nitrite to NO occurs in the apoplast of barley aleurone layers [26]. This NO production requires an acidic pH and was accelerated by reducing agents such as phenolic compounds.

NITRIC OXIDE STORAGE, DELIVERY AND DETOXIFICATION IN PLANTS

NO is extremely labile in physiologic environment where iron and oxygen are present. Lifetime of NO free radical is just 6–10 s [27]. In mammals, *S*-nitrosothiols (RSNO) and dinitrosyl iron complexes (DNICs) were proposed to be the main forms of NO storage and transport [27]. These types of molecules can be both low- or high-molecular-weight (i.e. bound to proteins) compounds. Usually, the protein complexes are more stable as compared to the low-molecular-weight ones [27]. S-nitrosylation of certain proteins gives rise to NO transport–storage forms such as *S*-nitrosoalbumin [28,29] and *S*-nitrosohemoglobin [30]. In plants, *S*-nitrosohemoglobin is endogenously produced, indicating a conserved role of hemoglobin *S*-nitrosothiols in NO metabolism among humans, nematodes and plants [31]. Moreover, hexacoordinate hemoglobin AHb1 from *A. thaliana* metabolizes both NO and GSNO. The nitrate forming reactions involve an Fe^{3+} intermediate, which is efficiently reduced by NADPH. It was proposed that this hemoglobin takes part in NO detoxification, principally during stresses that produce over-accumulation of NO such as hypoxia [31]. However, the rate of S-nitrosohemoglobin decomposition is low at physiological conditions, suggesting that hemoglobin S-nitrosylation could be a means for NO storage. There is no evidence of hemoglobin transport in the plant vascular system and it is highly improbable that hemoglobin function as NO carrier between plant tissues and organs, as it was suggested for mammals [32]. Therefore, the attention is focused to low-molecular-weight NO-containing compounds that could be responsible for NO trafficking in plants.

In the model plant species *A. thaliana*, a GSNO reductase (AtGSNOR1) was characterized. It was demonstrated that AtGSNOR1 controls NO bioactivity, suggesting that GSNO is a molecule present in plants that potentially can act as an NO transporter [33]. It is known that an active interconversion exists between RSNO and iron nitrosyl compounds. RSNO and DNIC are capable of interconverting depending on the concentrations of iron, low-molecular-weight thiols and NO in the cell. The levels of both DNIC and RSNO decrease when thiols content decreases. When NO level decreases while the concentration of thiols and iron remains constant, DNIC and RSNO serve as donors of NO which is transferred to the acceptor cell. In addition, it was reported that iron induces the destabilization of RSNO leading to the formation of DNICs [34–36]. This biological system would need a small amount of a reducing agent. Ascorbate could be a good candidate to maintain Fe in its reduced form Fe^{2+} [37]. Even NO can reduce Fe^{3+} to Fe^{2+} when it is linked to high-molecular-weight compounds.

DNICs IN PLANTS

DNICs have been extensively studied in mammals and measured under various different physiological conditions [38–40]. Detection of DNICs in plants by EPR faces the problem that there are relatively high manganese levels in plant tissues. One of the hyperfine splitting components of EPR signal from manganese complexes mask the 2.03 EPR signal of DNICs [36, Graziano, Simontacchi, Puntarulo and Lamattina, unpublished results]. Therefore, the detection limit for DNICs by EPR in plants is higher than in animals. Formation of DNICs from endogenous-free iron and thiol-containing ligands was first demonstrated in isolated leaves from bean and China rose treated with exogenous gaseous NO [41]. Recently, it was shown that treatment with high concentrations of nitrite, a condition that stimulates NO production, induced the formation of DNICs in isolated China rose leaves [23]. Incubation of the leaves with nitrate, a competitive inhibitor of the NR-dependent NO-synthesis reaction, resulted in decreased signal intensity of DNICs, demonstrating that NR activity is the main source for NO synthesis in those conditions. This landmark work demonstrated that DNICs could be synthesized in plants not only from endogenous thiol-containing ligands and iron pools but also from endogenously-produced NO. The concentration of the complexes accumulated over 1 h was found to be 8 nmol·g^{-1} leaf FW (fresh weight). Thus, the levels of this nitrosylated iron pool are sufficiently high that its formation may function as a metabolic process for maintaining iron bioavailability under conditions of iron starvation [15,23].

PLANT IRON NUTRITION: AN OVERVIEW

Iron fulfills a vital role in virtually all living organisms from bacteria to animals. Iron ability to undertake one electron oxidation–reduction reactions places this molecule in the center of key biochemical processes like oxygen transport, ATP generation and the rapid reaction with free radicals. Iron can vary its redox potential in response to different environmental conditions and place this transition metal as an important intermediate for electron-transfer reactions and thereby with essential properties for electron transport chains in respiration and photosynthesis. The fact that iron deficiency affects more than 30% of the world's population [42] confirms its relevance and places it as a subject for many research projects worldwide.

Iron is a critical element for many central metabolic pathways during the whole plant life cycle. Low iron availability leads to growth arrest and chlorosis while increased amounts of iron leads to increased formation of reactive oxygen species (ROS), loss of selective membrane permeability and generalized tissue damage. Thus, the control of mechanisms that regulate iron homeostasis is crucial for plant survival [43,44]. Under iron-limiting conditions, plants develop iron deficiency symptoms characterized by leaf chlorosis, impaired chloroplast development and decreased expression of photosynthetic proteins [15,45] that could limit plant growth and survival.

In soils and in aerobic environment, iron is mainly in the ferric [Fe^{3+}] form. This iron is not easily accessible for plants because it is highly insoluble at neutral or basic pH. Thereby, plants have evolved mechanisms to facilitate iron acquisition from soil which include the solubilization or chelation of iron into solution. In general, non-graminaceous plants have developed a mechanism named Strategy I based on the reduction of Fe^{3+} to the more soluble form Fe^{2+} and its uptake through specific iron transporters. The proteins involved in iron reduction and transport at the root level were already characterized in several plant species [46]. On the other hand, graminaceous plants use a mechanism termed Strategy II which is based on the chelation of Fe^{3+} through the release of specific Fe^{3+} chelators called siderophores [43,46,47].

Once iron is inside the root cell it is reduced or de-chelated. However, under aerated conditions part of the iron is oxidized and precipitates as hydroxide or phosphate salt, forming an extracellular or apoplastic iron pool [48]. This pool comprises up to 95% of total root iron content in hydroponic culture and can be used under conditions of iron deficiency [44]. Free iron has to be shielded from oxygen to prevent precipitation and generation of oxygen radicals. It has been proposed that organic acids (citrate) or non-proteinogenic amino acids (nicotianamine) are the molecules responsible for chelation and relocation of iron to the aerial part of the plant through the transpiration stream; however, this process is rather unclear. Once in the leaves, Fe^{3+}-citrate is probably a substrate of a ferric-chelate reductase already described in mesophyll cells [49]. The reduced Fe^{2+} form is thought to be internalized by cells. The traffic of iron through phloem is far less documented in spite of the confirmed iron presence in phloem. It was suggested that nicotianamine is the main candidate for iron transport in phloem [50]. Nevertheless, a metal-binding protein was demonstrated to transport iron in *Ricinus communis* [51].

As was highlighted above, much is known about iron uptake mechanisms at the root level, but there is no clear evidence about how plants transport iron between shoot and roots and also between cells and inside the cells. Indeed, iron mobilization inside leaf mesophyll cells is a limiting step for iron acquisition [49]. Moreover, little is known about the molecules and cellular regulatory processes that sense and/or perceive iron status in plants and coordinate the molecular changes that finally execute the adjustment of adaptive responses leading to iron homeostasis. Very poor and fragmented information is also available on the molecules and regulatory mechanisms involved in intracellular iron transport and even in the compartmentalization and delivery of iron within the cells and organelles.

NITRIC OXIDE AND DNICs PARTICIPATION IN PLANT IRON METABOLISM

Recent works have demonstrated that NO is emerging as a new player in plant iron metabolism. It was shown that, in maize plants growing under low-iron conditions, application of the NO donors sodium nitroprusside (SNP) or *S*-nitroso-acetylpenicillamine (SNAP) or even gaseous NO completely prevent the development of iron-deficiency symptoms. Moreover, NO was able to revert the iron deficiency phenotype of the *ys1* and *ys3* maize mutants, impaired in iron acquisition. Interestingly, total iron content in those plants was not increased by NO treatment, suggesting that probably an NO-mediated improved efficiency in iron mobilization within the plant occurs [15]. Thereafter, it was found that iron-deficiency symptoms were also prevented by NO in tomato [Graziano and Lamattina, unpublished results], a plant belonging to the Strategy I responses for iron acquisition. Interestingly, as was previously described in young human patients with anemia [53], plants respond with an induction of NO production when growing under iron-deficient conditions. The source for this NO synthesis might be NR, since NO accumulation was blocked by the NR inhibitor tungstate and not by the NOS-inhibitor L-NAME [Graziano and Lamattina, unpublished results].

In order to explain NO participation in plant iron availability, we have previously proposed that a complex between iron and NO is synthesized in the form of low-molecular-weight DNIC [54]. It was also proposed that there is an interconversion between low-molecular-weight *S*-nitrosothiols (RSNO) and DNIC, which equilibrium depends on the NO and iron state of the plant. During low-iron supply, it has been reported that glutathione (GSH) levels are increased in roots [55] as well as NO production [Graziano and Lamattina, unpublished results]; therefore, RSNO synthesis may be stimulated under iron-deficient conditions in plant roots. Since iron destabilizes RSNO and induces the formation of DNICs, iron uptake from soil to the root cell would favor DNIC synthesis. These low-molecular-weight complexes may be transported to the aerial parts of the plant and also between and inside cells. According to the equilibrium between synthesis and degradation of DNICs, these complexes would decompose inside leaf mesophyll cells releasing iron (Fig. 1). This would be a mechanism for transporting not only iron but also NO between plant organs and cells and even

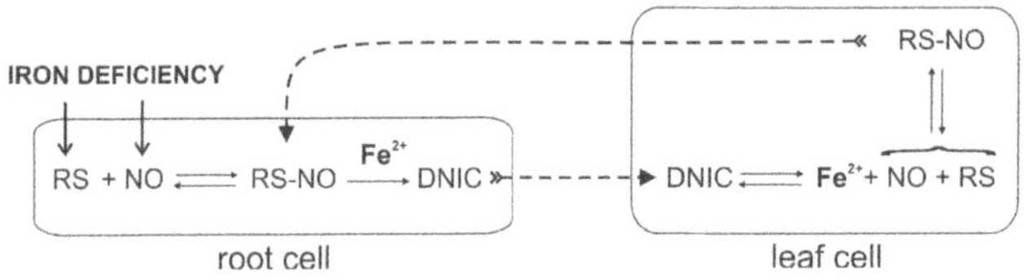

Fig. 1. Simplified model that represents the putative NO and DNIC involvement in plant iron sensing and transport. Iron deficiency stimulates GSH (RS^-) and NO production in root cells, inducing RSNO synthesis. Iron uptake from soil destabilizes RSNO and favors DNIC formation. DNIC is transported to the aerial tissues through the plant vascular system and is decomposed inside leaf mesophyll cells releasing iron. RSNO could be newly synthesized in leaf cells and transported to roots.

within organelles. When NO production is stimulated or NO is exogenously applied, it was shown that DNIC formation is induced [23]. Therefore, the addition of NO donors to plants growing under low-iron supply could stimulate DNIC formation and, consequently, iron mobilization.

PERSPECTIVES

Since the complexity of the network controlling iron bioavailability (uptake, transport, storage and delivery) in the plant cell, any approach to improve the iron content in crops to ameliorate the nutritional quality of foods must take into consideration the multiple steps connecting iron metabolism with the primary cell metabolism. It should be noted that, an uncontrolled iron excess can result in tissue damage by an overproduction of ROS and a consequent yield diminution and/or decrease of the product quality. The finding that both Strategy I and Strategy II plants can grow healthier under iron-deficient conditions if supplemented with exogenous NO could imply the development of molecular tools designed to improve crop yield in soils with iron limiting conditions.

It has been shown that in mammals [56] as well as in plants [57] and in the unicellular green alga *Chlorella sorokiniani* [58], mitochondria reduce nitrite to NO. It was also reported that nitrite is the major source of NO produced by *A. thaliana* when challenged by the pathogen *Pseudomonas syringae* [25]. In parallel, Vanin et al. [23] found that the nitrite/nitrate ratio together with the endogenous level of superoxide control the concentration of NO in plants. Interestingly, it has been found that ascorbate plays an important role by producing more than 90% of the NO yield from peroxynitrite ($ONOO^-$),

$$(NO + O_2^- \rightarrow ONOO^- + \text{ascorbic acid} \rightarrow NO)[59].$$

Here, it is interesting to point out that the 80% of cellular iron and the 80% of cellular ascorbate are thought to be localized in chloroplast where chlorosis, one of the first symptoms of iron deficiency, is developed. The questions are, could Fe be released from DNICs in chloroplast in a process that involves NO and ascorbate? Might iron and ascorbic acid be crucial for chlorophyll biosynthesis through the maintenance of NO homeostasis in chloroplasts?

As a final comment, we would like to emphasize about the fact that, most of the biological processes are a result of the interaction between numerous cellular constituents forming complex networks that are precisely coordinated and synchronized. There, metabolic compensations as well redundancy activities are operating to maintain cellular homeostasis. In this scenario, NO, iron, DNICs, GSNO and ascorbic acid are probably working together in the same act of the play that describes the mechanisms controlling plant iron homeostasis.

ACKNOWLEDGMENTS

This work was supported by Agencia Nacional de Promoción Científica y Tecnológica (ANPCyT, PICTs 1-9767/00 and 1-14457/03 to L.L.), Consejo Nacional de Investigaciones Científicas y Técnicas (CONICET, PIP 0898/98 to L.L.), Fundación Antorchas, and institutional grants from Universidad Nacional de Mar del Plata (UNMdP), Argentina. L.L. is a member of the Permanent Research Staff and M.G. is a Postgraduate Fellow from CONICET, Argentina. L.L. is a fellow from J. S. Guggenheim Foundation.

REFERENCES

1 Stamler JS, Singel DJ, Loscalzo J. Biochemistry of nitric oxide and its redox-activated forms. Science 1992; 258: 1898–1902.
2 Furchgott RF, Zawadski JV. The obligatory role of endothelial cells in the relaxation of arterial smooth muscle by acetylchloline. Nature 1980; 288: 373–376.
3 Ignarro LJ, Buga GM, Wood KS, Byrns RE, Chaudhuri G. Endothelium-derived relaxing factor produced and released from artery and vein is nitric oxide. Proc. Natl. Acad. Sci. USA 1987; 84: 9265–9269.
4 Palmer RM, Ferrige AG, Moncada S. Nitric oxide release accounts for the biological activity of endothelium-derived relaxing factor. Nature 1987; 327: 524–526.
5 Laxalt AM, Beligni MV, Lamattina L. Nitric oxide preserves the level of chlorophyll in potato leaves infected by *Phytophthora infestans*. Eur. J. Plant Pathol. 1997; 73: 643–651.
6 Delledonne M, Xia Y, Dixon RA, Lamb C. Nitric oxide functions as a signal in plant disease resistance. Nature 1998; 394: 585–588.
7 Durner J, Wendehenne D, Klessig DF. Defense gene induction in tobacco by nitric oxide, cyclic GMP, and cyclic ADP-ribose. Proc. Natl. Acad. Sci. USA 1998; 95: 10328–10333.
8 Beligni MV, Lamattina L. Nitric oxide stimulates seed germination and de-etiolation, and inhibits hypocotyl elongation, three light-inducible responses in plants. Planta 2000; 210: 215–221.
9 Garcia-Mata C, Lamattina L. Nitric oxide and abscisic acid cross talk in guard cells. Plant Physiol. 2002; 128: 790–792.
10 Pagnussat GC, Simontacchi M, Puntarulo S, Lamattina L. Nitric oxide is required for root organogenesis. Plant Physiol. 2002; 129: 954–956.
11 Correa-Aragunde N, Graziano M, Lamattina L. Nitric oxide plays a central role in determining lateral root development in tomato. Planta 2004; 218: 900–905.
12 He Y, Tang RH, Hao Y, Stevens RD, Cook CW, Ahn SM, Jing L, Yang Z, Chen L, Guo F, Fiorani F, Jackson RB, Crawford NM, Pei ZM. Nitric oxide represses the Arabidopsis floral transition. Science 2004; 305: 1968–1971.
13 Prado AM, Porterfield DM, Feijo JA. Nitric oxide is involved in growth regulation and re-orientation of pollen tubes. Development 2004; 131: 2707–2714.
14 Murgia I, Delledonne M, Soave C. Nitric oxide mediates iron-induced ferritin accumulation in Arabidopsis. Plant J. 2002; 30: 521–528.
15 Graziano M, Beligni MV, Lamattina L. Nitric oxide improves internal iron availability in plants. Plant Physiol. 2002; 130: 1852–1859.
16 Stuehr DJ. Mammalian nitric oxide synthases. Biochim. Biophys. Acta 1999; 1411: 217–230.

17 del Rio LA, Corpas FJ, Barroso JB. Nitric oxide and nitric oxide synthase activity in plants. Phytochemistry 2004; 65: 783–792.
18 Guo FQ, Okamoto M, Crawford NM. Identification of a plant nitric oxide synthase gene involved in hormonal signaling. Science 2003; 302: 100–103.
19 Crawford NM, Galli M, Tischner R, Heimer YM, Okamoto M, Mack A. Response to Zemojtel et al: Plant nitric oxide synthase: back to square one. Trends Plant Sci. 2006; 11: 526–527.
20 Guo FQ. Response to Zemojtel et al: Plant nitric oxide synthase: AtNOS1 is just the beginning. Trends Plant Sci. 2006; 11: 527–528.
21 Rockel P, Strube F, Rockel A, Wildt J, Kaiser WM. Regulation of nitric oxide (NO) production by plant nitrate reductase *in vivo* and *in vitro*. J. Exp. Bot. 2002; 53: 103–110.
22 Desikan R, Griffiths R, Hancock J, Neill S. A new role for an old enzyme: nitrate reductase-mediated nitric oxide generation is required for abscisic acid-induced stomatal closure in *Arabidopsis thaliana*. Proc. Natl. Acad. Sci. USA 2002; 99: 16314–16318.
23 Vanin AF, Svistunenko DA, Mikoyan VD, Serezhenkov VA, Fryer MJ, Baker NR, Cooper CE. Endogenous superoxide production and the nitrite/nitrate ratio control the concentration of bioavailable free nitric oxide in leaves. J. Biol. Chem. 2004; 279: 24100–24107
24 Hu X, Neill SJ, Tang Z, Cai W. Nitric oxide mediates gravitropic bending in soybean roots. Plant Physiol. 2005; 137: 663–670.
25 Modolo LV, Augusto O, Almeida IM, Magalhaes JR, Salgado I. Nitrite as the major source of nitric oxide production by *Arabidopsis thaliana* in response to *Pseudomonas syringae*. FEBS Lett. 2005; 579: 3814–3820.
26 Bethke PC, Badger MR, Jones RL. Apoplastic synthesis of nitric oxide by plant tissues. Plant Cell 2004; 16: 332–341.
27 Manukhina EB, Smirin BV, Malyshev IY, Stoclet J-C, Muller B, Solodkov AP, Shebeko VI, Vanin AF. Nitric oxide storage in the cardiovascular system. Biol. Bull. 2002; 29: 477–486.
28 Stamler JS, Jaraki O, Osborne J, Simon DI, Keaney J, Vita J, Singel D, Valeri CR, Loscalzo J. Nitric oxide circulates in mammalian plasma primarily as an S-nitroso adduct of serum albumin. Proc. Natl. Acad. Sci. USA 1992; 89: 7674–7677.
29 Ewing JF, Young DV, Janero DR, Garvey DS, Grinnell TA. Nitrosylated bovine serum albumin derivatives as pharmacologically active nitric oxide congeners. J. Pharmacol. Exp. Ther. 1997; 283: 947–954.
30 Jia L, Bonaventura C, Bonaventura J, Stamler JS. S-nitrosohaemoglobin: a dynamic activity of blood involved in vascular control. Nature 1996; 380: 221–226.
31 Perazzolli M, Dominici P, Romero-Puertas MC, Zago E, Zeier J, Sonoda M, Lamb C, Delledonne M. Arabidopsis nonsymbiotic hemoglobin AHb1 modulates nitric oxide bioactivity. Plant Cell 2004; 16: 2785–2794.
32 Pawloski JR, Hess DT, Stamler JS. Export by red blood cells of nitric oxide bioactivity. Nature 2001; 409: 622–626.
33 Feechan A, Kwon E, Yun BW, Wang Y, Pallas JA, Loake GJ. A central role for S-nitrosothiols in plant disease resistance. Proc. Natl. Acad. Sci. USA 2005; 102: 8054–8059.
34 Vanin AF. Dinitrosyl iron complexes and S-nitrosothiols are two possible forms for stabilization and transport of nitric oxide in biological systems. Bio. (Mosc.) 1998; 63: 782–793.
35 Vanin AF, Muller B, Alencar JL, Lobysheva II, Nepveu F, Stoclet JC. Evidence that intrinsic iron but not intrinsic copper determines S-nitrosocysteine decomposition in buffer solution. Nitric Oxide 2002; 7: 194–209.
36 Vanin AF, Papina AA, Serezhenkov VA, Koppenol WH. The mechanisms of S-nitrosothiol decomposition catalyzed by iron. Nitric Oxide 2004; 10: 60–73.
37 Bridges KR, Hoffman KE. The effect of ascorbic acid and the intracellular metabolism of iron and ferritin. J. Biol. Chem. 1986; 26: 14273–14277.
38 Vanin AF, Men'shikov GB, Moroz IA, Mordvintcev PI, Serezhenkov VA, Burbaev D. The source of non-heme iron that binds nitric oxide in cultivated macrophages. Biochim. Biophys. Acta 1992; 1135: 275–279.

39 Mulsch A, Lurie DJ, Seimenis I, Fichtlscherer B, Foster MA. Detection of nitrosyl-iron complexes by proton-electron-double-resonance imaging. Free Radic. Biol. Med. 1999; 27: 636–646.

40 Ueno T, Suzuki Y, Fujii S, Vanin AF, Yoshimura T. *In vivo* distribution and behavior of paramagnetic dinitrosyl dithiolato iron complex in the abdomen of mouse. Free Radic. Res. 1999; 31: 525–534.

41 Aliev DI, Kurbanov IS, Vanin AF. Stud. Biophysica. 1986; 115: 173–180.

42 WHO 2002, htp://www.who.int/nutrition/topics/ida/en

43 Guerinot ML, Yi Y. Iron: nutritious, noxious, and not readily available. Plant Physiol. 1994; 104: 815–820.

44 Hell R, Stephan UW. Iron uptake, trafficking and homeostasis in plants. Planta 2003; 216: 541–551.

45 Nishio JN, Abadía J, Terry N. Chlorophyll-proteins and electron transport during iron nutrition-mediated chloroplast development. Plant Physiol. 1985; 78: 296–299.

46 Curie C, Briat JF. Iron transport and signaling in plants. Annu. Rev. Plant Biol. 2003; 54: 183–206.

47 Römheld V. Different strategies for iron acquisition in higher plants. Physiol. Plant 1987; 70: 231–234.

48 Bienfait HF, van den Briel W, Mesland-Mul NT. Free space iron pools in roots: generation and mobilization. Plant Physiol. 1985; 78: 596–600.

49 Brüggemann W, Maas-Kantel K, Moog PR. Iron uptake by leaf mesophyll cells: the role of the plasma membrane-bound ferric-chelate reductase. Planta 1993; 190: 151–155.

50 Stephan UW, Scholz G. Nicotianamine: Mediator of transport of iron and heavy metals in the phloem? Physiol. Plant 1993; 88: 522–529.

51 Kruger C, Berkowitz O, Stephan UW, Hell R. A metal-binding member of the late embryogenesis abundant protein family transports iron in the phloem of Ricinus communis L. J. Biol. Chem. 2002; 277: 25062–25069.

52 Zemojtel T, Fröhlich A, Palmieri MA, Kolanczyk M, Mikula I, Wyrwicz LS, Wanker EE, Mundlos S, Vingron M, Martasek P, Durner J. Plant nitric oxide synthase: a never-ending story? Trends Plant Sci. 2006; 11: 524–525.

53 Choi JW, Pai SH, Kim SK, Ito M, Park CS, Cha YN. Iron deficiency anemia increases nitric oxide production in healthy adolescents. Ann. Hematol. 2002; 81: 1–6.

54 Graziano M, Lamattina L. Nitric oxide and iron in plants: an emerging and converging story. Trends Plant Sci. 2005; 10: 4–8.

55 Zaharieva TB, Abadia J. Iron deficiency enhances the levels of ascorbate, glutathione, and related enzymes in sugar beet roots. Protoplasma 2003; 221: 269–275.

56 Kozlov AV, Stanick K, Nohl H. Nitrite reductase activity is a novel function of mammalian mitochondria. FEBS Lett. 1999; 454:127–130.

57 Gupta KJ, Stoimenova M, Kaiser WM. In higher plants only root mitochondria, but not leaf mitochondria reduce nitrite to NO, *in vitro* and in situ, J. Exp. Bot. 2005; 56: 2601–2609.

58 Tischer R, Planchet E, Kaiser WM. Mitochondrial electron transport as a source for nitric oxide in the unicellular alga *Chlorella sorokiniana.* FEBS Lett. 2004; 576: 151–155.

59 Barone MC, Darley-Usmar VM, Brookes PS. Reversible inhibition of cytochrome c oxidase by peroxynitrite proceeds through ascorbate-dependent generation of nitric oxide. J. Biol. Chem. 2003; 30: 27520–27524.

PART III

Nitrosospecies and S-*nitrosothiols: Physico-chemical Properties and Biological Activity*

CHAPTER 9

Low-molecular-weight S-nitrosothiols

Ernst van Faassen[1] and Anatoly F. Vanin[2]

[1]*Debye Institute, Section Interface Physics, Ornstein Laboratory, Utrecht University, 3508 TA, Utrecht, The Netherlands*
[2]*Semenov Institute of Chemical Physics, Russian Academy of Sciences, Moscow, 119991, Russian Federation*

S-nitrosothiols (thionitrites, RSNO) appear as endogenous reaction products of NO or NO metabolites with sulfhydryl groups. Prime targets are the sulfhydryl groups on cysteine residues found in many different proteins [1,2] or the thiol moieties on endogenous glutathione (GSH) or cysteine (Cys). The nitrosation of the sulfhydryl moiety leads to *S*-nitrosothiols with high and low molecular weight, respectively. In organisms, the combined pool of *S*-nitrosothiols varies over the compartments. Table 1 gives an overview of various NO stores in rats, and *S*-nitroso moieties in organ tissue are 10–100 nM. Much higher levels up to 250 nM are found in erythrocytes, and plasma levels remain below 10 nM. The overwhelming majority (ca 90%) of all S-nitroso groups are anchored to proteins and belong to the class of high-molecular mass [3,4]. The interest in the *S*-nitrosothiols was motivated by the observation that these compounds form *in vivo* and may elicit physiological responses that are strongly reminiscent of free NO radicals. Well-documented examples are vasodilation and the inhibition of platelet aggregation. The formation and functions of S-nitrosated proteins are discussed in Chapter 10, whereas this chapter will be primarily concerned with the *S*-nitrosothiols of low molecular weight. A number of examples is shown in Fig. 1. Many different forms have been synthesized *in vitro* and tested for use as NO donor [for example *S*-nitroso-*N*-acetyl-pencillamine (SNAP)], but in mammals the most relevant species are *S*-nitrosocysteine (CysNO) and especially *S*-nitrosoglutathione (GSNO). The latter is found in all tissues, with basal concentrations in the range of 1–10 nM, but higher concentrations arise in aortic tissue and erythrocytes (cf Table 1). Given that endogenous GSH concentrations are in the mM range, it appears that the degree of S-nitrosation of GSH is only a percent or less. Therefore, S-nitrosation cannot have an impact on the redox status of tissues by shifting the GSH/GSSG ratio. But the basal GSNO levels appear comparable to those of free NO, so that the pool of *S*-nitrosothiols may be significant for the NO status of the tissue.

The compounds mentioned above are all cysteine based and relevant for mammalian physiology. However, they are definitely not exhaustive for endogenous thiols found in organisms. Other types of cysteine-based thiols like trypanothione (TSH) or mycothiol have been found in certain bacterial strains. A completely independent class of intracellular thiols is based on histidine residues, with very different chemistry. Well-studied examples are ergothioneine

Table 1 Magnitude of various pools of NO-metabolites in Wistar rats (adapted from [9]). The [GSH]/[GSSG] ratio is a marker for the redox state of the tissue. The rightmost column gives NO-metabolite values from human plasma for Ref. [10]

NO-metabolite	Aorta	Brain	Heart	Plasma	Erythrocytes	Human plasma¶
Nitrite (μM)	23 ± 9	1.7 ± 0.3	0.80 ± 0.08	0.29 ± 0.05	0.68 ± 0.06	0.20 ± 0.02
Nitrate (μM)	49 ± 7	6.1 ± 1.1	5.9 ± 1.7	5.7 ± 0.6	10.2 ± 1.2	14.4 ± 1.7
S-nitroso moieties (nM)	96 ± 24	22 ± 6	13 ± 2	1.4 ± 0.5	246 ± 32	7.2 ± 1.1
N-nitroso moieties (nM)	19 ± 11	61 ± 8	14 ± 2	3.5 ± 0.4	95 ± 14	32.3 ± 5.0
Nitrosyl-heme (nM)	Below detection*	160 ± 30	15 ± 1	Below detection*	10.8 ± 1.8	n.d.
GSH (mM)	0.34 ± 0.08	1.28 ± 0.06	0.93 ± 0.12	0.02†	0.8 mM† (whole blood)	n.d.
[GSH]/[GSSG]	6.8	24	14	5°	> 10° (whole blood)	n.d.

* Detection limit 1 nM.
† [11].
¶ [10].
n.d. = not done.

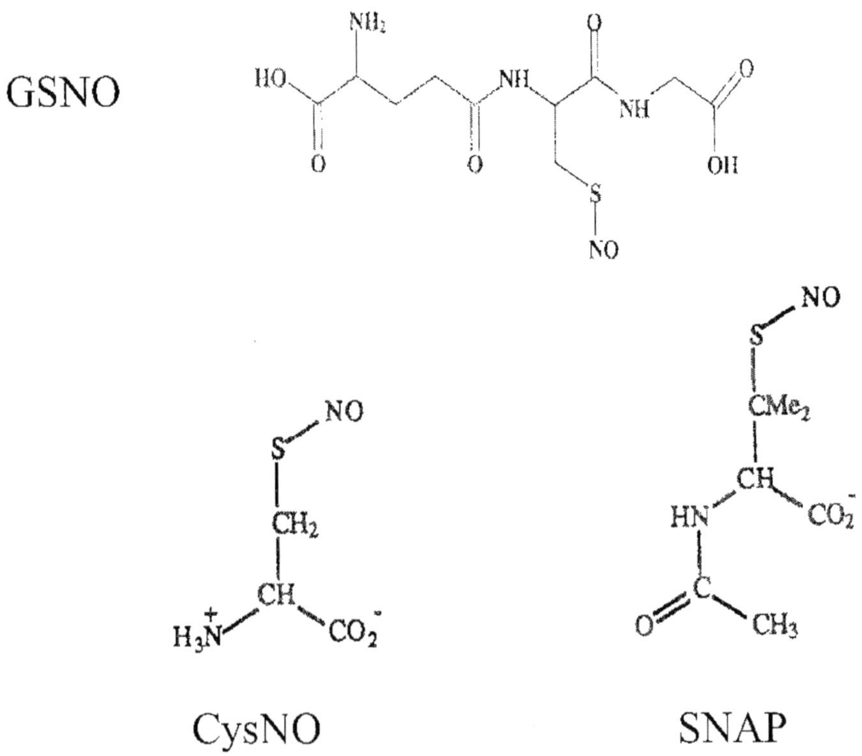

Fig. 1. Molecular structures of CysNO, GSNO and SNAP.

(ESH) and ovothiols [5]. ESH is a low-molecular-weight thiol with established reactivity towards GSNO [6], capacity to scavenge singlet oxygen and hydroxyl radicals, and is found in millimolar concentrations in specific tissues like liver, kidney and erythrocytes [7]. ESH was not reported to form *S*-nitrosothiols, but it may form disulfides with cysteinic thiols. Although the reactivity towards GSNO is intriguing, the biological function of such histidine-based thiols is still unclear. Therefore, such "unconventional" thiols will not be considered further in this chapter.

Finally, it should be mentioned that S-nitrosation is only one of many pathways *via* which nitroso moieties can be incorporated into proteins. Organic molecules and peptides may undergo N-, O-, C- as well as S-nitrosation, depending on the structural motif to which the NO moiety is attached. Mixed nitrosations are of course possible also. An interesting example is albumin which may be N-nitrosated on one of its two tryptophan residues as well as S-nitrosated on the Cys34 residue [8]. Significantly, the *N*-nitrosated tryptophan residue also elicited vasodilatory response from precontracted aortic rings of rabbits [8]. Since the nitrosation was carried out under extreme non-physiological conditions (acidification of nitrite), it remains unclear whether this N-nitrosation of tryptophans has significance for *in vivo* conditions.

In contrast, the physiological significance of S-nitrosation has been proven beyond doubt. Many examples will be given at the end of this chapter. Under biological conditions, S-nitrosation is facile, fast and affects a wide range of proteins *in vitro* and *in vivo*. Judging from the citation numbers in scientific literature, the nitrosation of sulfhydryl groups has highest relevance for physiology so far. Therefore, this chapter will be primarily concerned with S-nitrosocompounds.

SPECTROSCOPIC PROPERTIES

In general, RSNO have red, pink or green color due to strong absorption of visible light. As an example, Fig. 2 shows the UV–VIS spectrum of pink GSNO with two prominent absorption bands from the S–NO bond. The UV peak at 336 nM ($\varepsilon_{336} = 778\ (\text{Mcm})^{-1}$) is attributed to a $\pi \rightarrow \pi^*$ transition, and the secondary band at 545 nM ($\varepsilon_{545} = 34\ (\text{Mcm})^{-1}$) to a $n_N \rightarrow \pi^*$ transition [12,13]. The intensity of these absorption bands is quite sensitive to the type of thiol. In albumin, S-nitrosation of the Cys34 residue leads to a slightly higher extinction of $\varepsilon_{335} = 870\ (\text{Mcm})^{-1}$ [14]. Multiple S-nitrosation of different residues leads to correspondingly higher extinctions. Multiple nitrosated albumin (poly-SNO-albumin) can reach a molar extinction of $\varepsilon_{335} = 3870\ (\text{Mcm})^{-1}$[15]. In the infrared region, the RSNO moiety has additional characteristic absorptions from the N–O stretch vibration (1480–1530 cm^{-1}) and the C–S stretch vibration (600–730 cm^{-1}).

SYNTHESIS AND DETECTION *IN VITRO* AND *IN VIVO*

The easiest synthetic pathway for *S*-nitrosothiols is *via* the equilibrium with nitrous acid, obtained by acidification of a solution of nitrite [16,17].

$$\text{RSH} + \text{HNO}_2 \rightleftharpoons \text{RSNO} + \text{H}_2\text{O} \qquad (1)$$

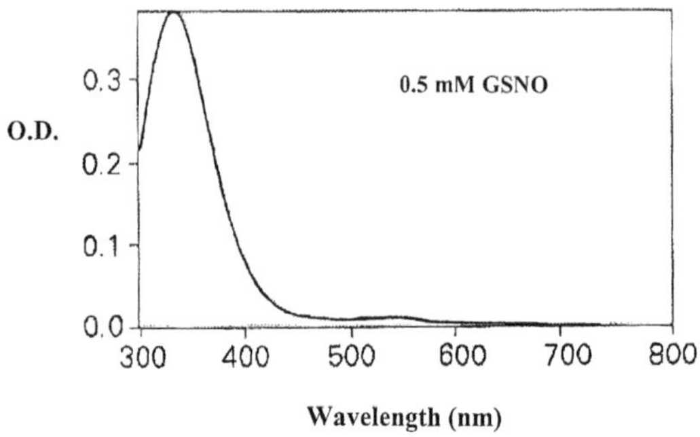

Fig. 2. The UV–VIS spectrum of 0.5 mM GSNO in buffer (pH = 7.4). O.D. refers to optical density.

A typical stock of CySNO is obtained by adding equimolar amounts (ca 50 mM) of CySH and sodium nitrite to 0.1 M HCl. The reaction is rather fast and requires less than a minute to complete. The kinetics of CysNO formation at various pH can be seen in Fig. 3. The reaction products always appear as a mixture of CysNO and disulfides, with the relative balance being strongly dependent on pH (cf Table 2). At physiological pH the disulfide is dominant, but at pH $\sim$ 1 the CysNO yield is nearly 100%. Near pH = 1, the conversion of Cys to CysNO is completed within a minute at 37°C, at room temperature within a few minutes.

As *S*-nitrosothiols are prone to photolysis and thermolysis, the reaction should proceed in a dark and cool location. The formation of CysNO and GSNO is indicated by its intense

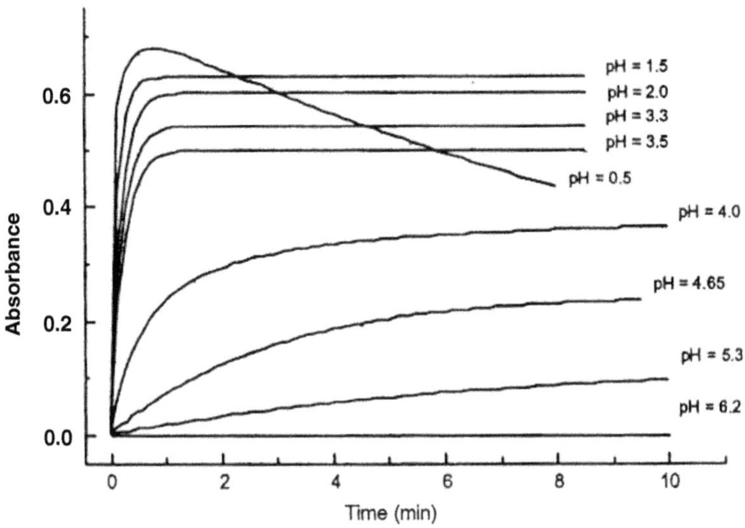

Fig. 3. Kinetics of CysNO formation from acid reduction of nitrite as measured by the optical absorbance at 543 nM in buffered solutions. The starting concentrations were 40 mM CysH and 40 mM nitrite. The temperature was 37°C. (From Ref. [18].)

Table 2 Yields of CysNO and cystine obtained in a reaction mixture of 0.04 M CysH with 0.04 M $NaNO_2$ in aqueous buffers at 37°C. The yields show that all cysteine is incorporated in either *S*-nitrosothiol or the cystine disulfide. Data adapted from Ref. [18]

pH	0.5	1.5	2.0	2.5	3.5	4.65
CysNO%	100	92.1	90.2	85.8	75.1	35.1
Cystine%	0	7.9	9.8	14.2	24.9	64.9

pink color. The formation of small gas bubbles with time indicates the escape of a certain quantity of NO gas, and the final yield should be verified spectrophotometrically just prior to use. Stock solutions are best freshly prepared and should not be stored in liquid form for more than a few hours. Storage up to a week is possible in frozen state, but longer term storage is possible only for refrigerated crystalline GSNO or CysNO (pink powders).

It should be noted that the equimolar mixture of thiol and nitrite always leaves some residual deprotonated thiol in the solution. This holds for synthesis of any kind of RSNO. The presence of a certain quantity of RS^- in assays with RSNO is highly significant because of its capacity to reduce spurious copper ions in the solution

$$Cu^{2+} + RS^- \rightarrow Cu^+ + RS^\bullet \tag{2}$$

This reaction leads to thiyl radicals and a small quantity of monovalent copper Cu^+. The latter is an effective catalyst for RSNO decomposition *(vide infra)* and the pathway of spurious copper often introduces artefacts. The problem can be avoided by inhibiting the redoxactivity of the copper ions by chelators like EDTA or neocuproine.

Although the reaction Eq. (1) has been known for a long time, its molecular mechanism is still not fully understood. Recent reports [18] showed that the major nitrosating species at low pH < 3.5 is nitrosonium NO^+ and the reaction is fast. At higher pH the release of nitrosonium is blocked, and N_2O_3 acts as the main nitrosating species in the reaction

$$GSH + N_2O_3 \rightarrow GSNO + NO_2^- + H^+$$

This nitrosation by N_2O_3 is very fast with a second order rate of $k = 6.6 \times 10^7\ (Ms)^{-1}$ [19]. The rate of formation of N_2O_3 is much lower (cf Chapter 1) and acts as the rate limiting step.

Interestingly, the NO radical itself does not show significant reactivity towards the sulfhydryl group in anaerobic conditions [20–22]. Phrased otherwise, NO is not a *S*-nitrosating compound by itself in the absence of oxygen. Instead, the reaction of NO with thiols leads to formation of disulfides RS–SR [23–25]

$$RSH + NO \rightarrow RS-N-OH$$

$$2RS-N-OH \rightarrow RS-SR + HONNOH \rightarrow RS-SR + N_2O + H_2O$$

$$RS-N-OH + NO \rightarrow RSOH + N_2O$$

$$RSOH + RSH \rightarrow RS-SR + H_2O \tag{3}$$

It should be noted that this sequence of reactions is pH dependent since the thiolate anion shows far higher reactivity than the protonated thiol. For $[NO] \ll [GSH]$, the sequence (3) was reported [25] to have an apparent second order reaction rate of $0.080 \pm 0.008\ (\mathrm{Ms})^{-1}$ (37°C, pH 7.4). Table 1 shows that physiological [GSH] $\sim$ 1 mM. At this concentration, reaction (3) would give NO radicals a lifetime of ca 10^4 s. Therefore, it is unlikely that the anoxic reaction (3) cause significant loss of NO *in vivo*.

In this context it is important to note that acid dissociation constant pK_a of thiols shows a large variation, depending on the functionalization of the cysteine residues. The thiolate moiety of free cysteine has $pK_a = 8.3$. In glutathione, the constant for the cysteine residue has been raised to $pK_a = 8.7$. The constant is drastically lowered for the cysteine residues of many proteins, like protein tyrosine phosphatase (PTP1) with $pK_a \sim 5.4$ or protein disulfide isomerase with $pK_a \sim 3.5$. This large variation in dissociation constant implies a large variation in the balance between deprotonated/protonated forms of the sulfhydryl group, with concomitant influence on reaction rates and yields of S-nitrosation.

So how are *S*-nitrosothiols formed *in vivo*? A modest release of RSNO from catalytic action of ceruloplasmin and NOS itself has been reported [26], but potent enzymatic pathways for S-nitrosation seem not to have been identified so far [2]. It suggests that non-enzymatic pathways dominate *in vivo*. The transformation of RSH to RSNO requires the appearance of S-nitrosating intermediates like nitrosonium NO^+, NO_2 or N_2O_3. The reaction pathways for the formation of these species from free NO were discussed in Chapter 1. Although the reactions are well studied *in vitro*, there remains considerable uncertainty as to the dominant mechanism for S-nitrosation of sulfhydryls *in vivo* [22,27]. *Via* what pathway do S-nitrosocompounds appear in live tissue? This question cannot be addressed without noting that *S*-nitrosothiols forms only one class of nitrosocompounds that is closely linked to other pools *via* slow chemical equilibria.

In-vitro experiments indicate that aerobic mixtures of NO/GSH or GSNO/GSH release a certain quantity of highly reactive oxygen species (ROS). The release of free ROS was manifested from significant strand breaking in DNA and protection against strand breaking by scavengers like catalase and superoxide dismutase (SOD) [28]. The reaction mechanisms still remain controversial. A mechanism for S-nitrosation involving the reduction of O_2 to superoxide O_2^- was proposed and confirmed *in vitro* [29]:

$$\begin{aligned} &\mathrm{RSH} + \mathrm{NO} \rightarrow \mathrm{RS{-}N{-}OH} \\ &\mathrm{RS{-}N{-}OH} + \mathrm{O_2} \rightarrow \mathrm{RS{-}NO} + \mathrm{O_2^-} + \mathrm{H^+} \qquad (4) \\ &\mathrm{O_2^-} + \mathrm{NO} \rightarrow \mathrm{ONOO^-} \end{aligned}$$

This mechanism does not involve trace metal ions. However, such trace metal ions are known to accelerate both formation and decomposition of *S*-nitrosothiols *in vitro* [30] and *in vivo*, possibly by catalyzing the formation of nitrosating species like NO^+, NO_2 or N_2O_3 from NO in the presence of oxygen. This makes it unlikely that a mechanism like Eq. (4) be the dominant pathway for formations of *S*-nitrosothiols *in vivo*. Certain enzymes like ceruloplasmin and NOS itself were seen to catalyze the formation of *S*-nitrosothiols [26] but yields remain low. Alternatively, dinitrosyl iron complexes (DNICs) are known to participate in a complex reaction equilibrium between thiols, NO and *S*-nitrosothiols. This equilibrium is discussed in detail in Chapter 11 of this book. *In-vitro* experiments have confirmed that the presence of iron induce formation of RSNO from NO *via* intermediate DNIC even in anoxic conditions [31]. Intriguingly, the formation of small quantities of DNIC from endogenous iron has been detected in many biological systems after exposure to exogenous NO or after stimulus of endogenous NO production. Many examples were reported from cell cultures to plant and animal tissues. However, decisive proof of the physiological relevance of this DNIC pathway is still lacking.

From the above it is clear that thiols participate in many reaction pathways with endogenous reactants like oxygen or the various nitrogen oxides. The relative importance of these pathways is determined by the corresponding reaction rates. A useful selection of reaction rates for glutathione has been collected in Ref. [32].

Detection and quantification of *S*-nitrosothiols poses a considerable challenge in biological systems. Special techniques for S-nitrosated proteins have been recently reviewed in Volume 396 of *Methods in Enzymology*. A widely used technique is homolytic cleavage of the S–NO bond by copper/iodide, and detection of the gaseous NO *via* chemiluminescence with an ozone analyzer. Alternatively, the NO may be detected electrochemically with an NO electrode, or by spin-trapping with electron paramagnetic resonance (cf Chapter 18). Selectivity for high or low molecular mass may be obtained *via* filtering with ultrafiltration membranes [4] or high performance liquid chromatography (HPLC). Photolysis-chemiluminescence spectroscopy [33] is a sensitive method to analyze mixtures of *S*-nitrosothiols. It consists of an instrumental cascade comprising a HPLC pump, photolysis chamber and a chemiluminescence spectrometer or ozone analyzer. The method works by homolytic cleavage of the S–NO bond by irradiation in the photolysis chamber, followed by detection and quantification of the escaping gaseous NO with the ozone analyzer.

Alternative methods are based on fluorescence detection of decomposition products of the *S*-nitrosothiol. In the Saville reaction [34], mercuric chloride is used to catalyze the release of free NO and subsequent trapping with 2,7-dichlorofluorescein (DCF) or 2,3-diaminonaphthalene (DAN). The fluorescence detection is highly sensitive, but the specificity of the detection is compromised by artefacts and impostors for true NO [35].

To conclude this section, it should be remarked that the reaction of S-nitrosation in a complex biological system might proceed significantly different from that in a homogenous aqueous phase like a buffered solution. The coexistence of hydrophobic and hydrophilic compartments in cells, tissues and even individual proteins may modify the reactions as known in water. The main reason is the tendency of neutral species like NO, O_2 and N_2O_3 to accumulate in hydrophobic compartments like the membrane fraction [26,36] or hydrophobic pockets of proteins. This effect could promote the S-nitrosation of the β_{93} cysteine of hemoglobin which is located in a hydrophobic region of the protein [32]. In addition, it has been recognized that the presence of nearby amino acids affects the rate of S-nitrosylation by modifying the polarity and effective acidity in proximity of the cysteine residue [2]. It has been shown that the presence of a lipid compartment accelerates the oxidation of NO by O_2 by orders of magnitude [36].

STABILITY OF GSNO *IN VITRO*

Stability of RSNO is compromised by thermal decomposition, photolysis and the catalytic decomposition by trace metal ions like iron and copper. Although the various RSNO share the common C–S–N=O motif, they show a big variation in intrinsic stability when in solution.

The reason for this variation is not well understood. Given this instability, aqueous solutions of RSNO tend to release a certain quantity of gaseous NO and accumulate typical decomposition products like thiols, disulfides and reaction products of thiyl radicals. It was recently discovered that, in presence of oxygen, a certain quantity of disulfide *S*-oxides (GS(O)SG) and disulfide *S*-dioxides (GS(O_2)SG) [37,38] is formed as well. These metabolites of GSH are effective agents for S-glutathiolation of sulfhydryl groups. They can inactivate glyceraldehyde 3-phosphate and alcohol dehydrogenases, and release zinc from metallothionin and zinc finger proteins. Such mixtures of decomposition products result from a combination of pathways operating simultaneously. In practical applications usually more than just a single pathway contributes significantly. The basic decomposition reactions will be discussed below.

In solution, RSNO undergo slow thermal decomposition into a disulfide and free NO according to

$$2RSNO \rightleftarrows RS-SR + 2NO \tag{5}$$

The rate of thermal decomposition depends on the concentration and type of thiol [39,13]. At starting concentrations of 50 mM, the thermal decomposition of GSNO and CysNO exceeds 10% after ca 6 and 1 h, respectively. At low concentrations in absence of UV or blue light, the stability of RSNO is quite high. In presence of the metal chelator DTPA, even CysNO achieves a half life of 11 h [40]. Therefore, Eq. (5) is not significant unless catalyzed by trace metal ions (see below).

Illumination with UV or blue light causes homolytic cleavage of the S–N bond. The probability for photolysis is highest at wavelengths near the absorption maximum of RSNO at 336 nM. The quantum yield for photolysis is quite high and exceeds that of photodissociation of the nitrosyl ligand from heme. Fig. 4. shows the photolysis of GSNO by successive sets

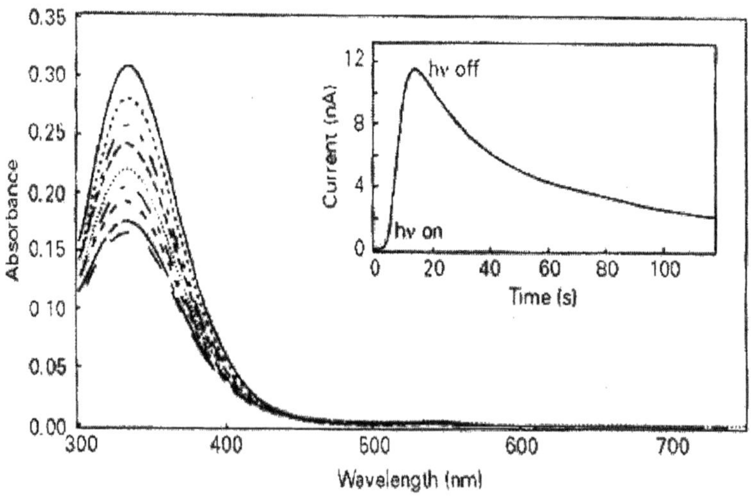

Fig. 4. Photolysis of GSNO by pulsed UV irradiation at 355 nM at room temperature in oxygenated PBS buffer. The successive curves differ by 10 pulses of 50 mJ/pulse. The inset shows the simultaneous photolytic release of free NO radicals as measured with an NO electrode. (From Ref. [41].)

of 10 laserpulses (355 nM, 50 mJ/pulse). The photolytic decomposition of S–NO bonds has raised interest in the application of S-nitrosated complexes for photodynamic therapy of cancers [41]. It should be mentioned that *S*-nitrosothiols are not the only endogenous species showing such photolytic release of NO: Alongside endogenous *S*-nitrosothiols, also nitrite was shown [42] to contribute significantly to the light-induced release of NO from vascular tissues of rats.

Reduced trace metal ions like Fe^{2+} and Cu^{+} are efficient catalysts for the decomposition of RSNO [43,44]. Fig. 5 shows the effect of copper and iron on the kinetics of CysNO. The decomposition of GSNO is slower, but qualitatively similar.

The reaction mechanism [46] is given by the following sequence

$$\begin{aligned} &RSNO + Cu^{+} \rightarrow RS^{-} + NO + Cu^{2+} \\ &Cu^{2+} + RS^{-} \rightarrow Cu^{+} + RS^{\bullet} \\ &2RS^{\bullet} \rightarrow RS-SR \end{aligned} \tag{6}$$

The reaction releases disulfides and free NO. Fe^{2+} also can catalyze a similar sequence.

The above reactions might leave the erroneous impression that iron and copper act on *S*-nitrosothiols in similar ways. In reality the iron-catalyzed decomposition of RSNO is complicated by a competing reaction mechanism involving iron complexes carrying two nitrosyl ligands. In principle, copper also may form (di)nitrosyl species like EPR silent

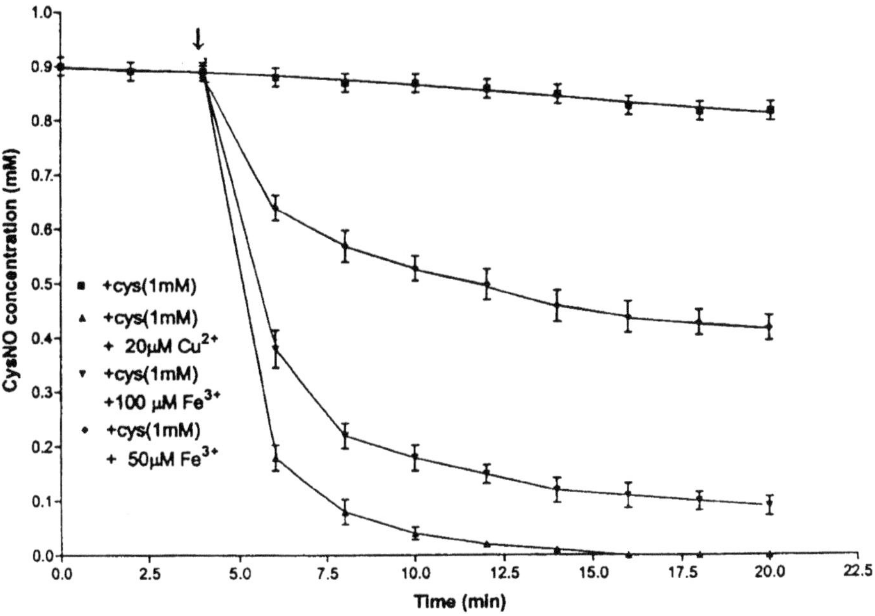

Fig. 5. Effect of copper and iron on the decomposition of 0.9 mM CysNO in HEPES buffer (150 mM, pH = 7.4) as monitored by the optical absorption at 340 nM. The decomposition was studied in the presence of 1 mM free cysteine. (From Ref. [45].)

$Cu^{+}(NO)_2$ or paramagnetic $Cu^{+}(NO)$ or $Cu^{2+}(NO)_2$. However, these species have so far only been reported as adsorbants in dry anoxic porous solids [47] but not in aqueous solutions. Therefore, in solutions, iron alone can interact with *S*-nitrosothiols to form such DNICs. In the presence of iron, a pool of DNIC will form at the expense of *S*-nitrosothiols [43,45]. These reactions were well studied *in vitro* and are described in Chapter 11.

Transnitrosation to other types of thiols may have significant effect on the lifetime of *S*-nitrosothiols. The basic transnitrosation reaction involves the transfer of a nitrosonium NO^{+} moiety

$$R_1-S-NO+{}^{-}S-R_2 \leftrightarrow R_1-S^{-}+NO-S-R_2 \quad (7)$$

Here $R_1{-}S^{-}$, $R_2{-}S^{-}$ can be various deprotonated thiols, from small cysteine anions to sulfhydryl groups on macromolecular proteins. The rates for this reversible transfer vary considerably with the type of thiol, pH and temperature. Transnitrosation rates for cysteine and glutathione were reported [14] at around 80 $(Ms)^{-1}$ (37°C, pH 7.4). Lower rates of 3–9 $(Ms)^{-1}$ apply for transnitrosation from *S*-nitrosoalbumin to cysteine and glutathione. The transfer rates increase significantly when pH is raised and a larger fraction of thiols is deprotonated. It was shown that 50 μM CysNO and *S*-nitrosoalbumin establish their equilibrium by transnitrosation on a timescale of several minutes *in vitro* (cf Fig. 6) [48]. Transnitrosation from GSNO is slower due to a combination of smaller transfer rate and orders of magnitude lower concentration of GSNO (Table 1). Therefore, the characteristic timescale for equilibration with GSNO should be many minutes. In human plasma, albumin was confirmed to be the dominant target for transnitrosation from CysNO and GSNO [49], with metal chelators having only insignificant effect on the extent and rate of S-nitrosation of the albumin target. In whole blood, the reaction balance is shifted considerably because Hb may be S-nitrosated or nitrosylated at the heme. In addition, oxyHb is an efficient scavenger

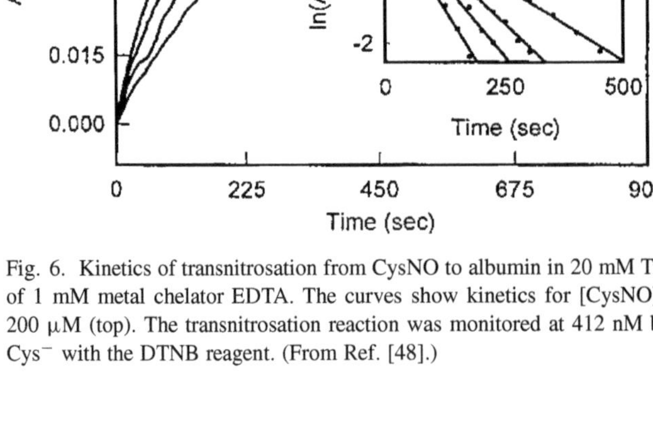

Fig. 6. Kinetics of transnitrosation from CysNO to albumin in 20 mM Tris buffer (pH 7.4, 23°C) in presence of 1 mM metal chelator EDTA. The curves show kinetics for [CysNO] = 50 μM (bottom), 100, 150 and 200 μM (top). The transnitrosation reaction was monitored at 412 nM by photometric detection of released Cys^{-} with the DTNB reagent. (From Ref. [48].)

of free NO (but not for *S*-nitrosothiols, see below). In whole blood, S-transnitrosation from GSNO to *S*-nitrosohemoglobin was reported [49] to remain below a few percent.

The previous *in-vitro* results should not leave the impression that S-transnitrosation be an exclusively passive process with timescales largely determined by the chemical characteristics of the exchanging thiols. In biological systems, the crossing of cellular membranes will act as a significant barrier for S-nitrosation and affect the rate of exchange between the intra- and extracellular compartments. This effect has been shown *in vivo*: Protein disulfide isomerase (PDI) is a protein in cellular membranes that normally catalyzes thiol-disulfide exchange. The activity of the PDI protein was found to enhance the intracellular pool of *S*-nitrosothiols in cultured human erythroleukemia cells [50]. The mechanism of membrane crossing of *S*-nitrosothiols will be discussed later in this chapter.

The process of transnitrosation from low to high molecular weight has been demonstrated *in vivo* in rabbits [51]. A collection of transnitrosation rates between various thiols has been collected in Ref. [14]. Analogous results were obtained with transnitrosation from *S*-nitrosoglutathionyl-sepharose beads [52] to free thiols in solution. The transnitrosation to cysteine and glutathione was rapid and accelerated by an order of magnitude when pH was raised from 5 to 9. In contrast, transnitrosation from the beads towards bovine serum albumin was negligible.

It should be noted that reaction (7) proceeds in the presence of a competing pathway which leads to the formation of disulfide bonds and the release of some free NO. It has recently been shown [53] that the sulfhydryl groups of certain proteins can also be modified by nucleophilic attack of the protein thiolate on the sulfur of GSNO rather than on the NO^+ moiety. This reaction pathway amounts to S-glutathiolation of the sulfhydryl moiety. Several proteins underwent a combination of S-nitrosation and S-glutathiolation when exposed to GSNO. In contrast, bovine serum albumin, actin and alcohol dehydrogenase were only S-nitrosated by fresh GSNO. S-glutathiolation of intracellular proteins has been demonstrated *in vivo*: upon incubation with exogenous CysNO, NIH-3T3 fibroblasts underwent combined S-nitrosation and S-glutathiolation of the cysteine residues of H-ras protein [54]. High capacity of S-glutathiolation was attributed [37] to glutathione disulfide *S*-oxide (GS(O)SG) which appears as one of the decomposition products of GSNO itself.

The lifetime of hours for GSNO may be shorted to minutes by addition of other thiols, in particular SNAP and cysteine [55–57]. The effect may be inhibited by the thiol-blocking compound *N*-ethylmaleimide [58]. Two different mechanisms have been identified for this phenomenon: transnitrosation and formation of disulfides. In the first case, the GSNO is depleted by the transnitrosation to the shorter lived CysNO. In the same spirit, transnitrosation to more stable species may stabilize the pool of *S*-nitrosothiols up to a certain degree. The second mechanism is mediated by the reductive nature of the thiol anions and involves the formation of a disulfide bridge under release of a nitroxyl anion [24,59]:

$$\mathrm{RSNO} + \mathrm{GS}^- \rightarrow \mathrm{RS{-}SG} + \mathrm{NO}^- \qquad (8)$$

If RS represents glutathione, the reaction releases GS–SG disulfide at a rather slow rate with a second order rate constant of $k = 8.3 \times 10^{-3}$ $(\mathrm{Ms})^{-1}$ [24]. Given that tissues contain GSH below mM concentrations (Table 1), this decay channel of GSNO appears insignificant *in vivo*. If on the other hand RS represents a sulfhydryl group on a protein, the reaction (8) amounts to S-glutathiolation of the protein. It has been confirmed that the coincubation with a mixture of GSNO and glutathione causes S-glutathiolation of the sulhydryl groups on a wide range of different proteins. [53,60]. More detailed studies [37] have shown that intermediate oxides like GS(O)SG or GS(O_2)SG are better S-glutathiolating agents than GSNO itself (see below). Therefore, the process of S-glutathiolation may in fact be more complex than suggested by Eq. (8). S-glutathiolation maybe reversed by dithiothreitol [60].

Release of NO radicals from RSNO can be caused by reductants, exogenous as well as endogenous. Ascorbate was found to have a pronounced effect on the lifetime of RSNO [39]. Without chelation of spurious copper, small quantities of ascorbate reduced Cu^{2+} to Cu^{+} and initialized the reaction Eq. (6) with formation of disulfides and the release of free NO. After chelation of spurious copper, the true reduction of RSNO became apparent with the release of free NO and thiols instead of disulfides. The effective rate constant accounts for the reduction pathway plus a thermal decomposition rate k_T

$$k_e = k[Asc] + k_T \tag{9}$$

At pH = 7.4, the value of k was 0.25, 0.015 and 0.032 $(Ms)^{-1}$ for CysNO, GSNO and SNAP, respectively. Careful observations showed that the ascorbate monoanion HA^- and the dianion A^{2-} have different rates of reduction. Therefore, the rate k was highly dependent on pH and increased over four orders of magnitude when the pH was changed from 3.6 to 11 [39].

It might seem plausible to expect that the stability of nitrosothiols be inversely related to its potency as vascular effector. More concretely, one might expect that the more rapidly decaying nitrosothiols elicit stronger physiological responses for relative short times. When tested for the responses of vasorelaxation and platelet aggregation, no such correlation was found [61]. This result confirms that *S*-nitrosothiols are potent physiological effectors in their own respect. In particular, they can act without having to release free NO.

STABILITY OF RSNO *IN VIVO*

In biological systems, additional pathways exist for removal of RSNO. The consumption of S-nitrosogroups (S-denitrosation) by endogenous enzymes has not been often reported in the literature. Mammalian physiology does not seem to require a dedicated system to remove excess quantitites of RSNO. A modest capacity for S-denitrosation has been recently reported for the protein disulfide isomerase enzyme [62] and for anaerobic xanthine oxidase [63]. Cu–Zn superoxide dismutase was capable to denitrosate low concentrations of GSNO *in vitro*, but the significance of this finding is questionable since the reaction was inhibited by *in-vivo* levels of GSH (Table 1). Thioredoxin reductase (TR) [64] and glutathione peroxidase [65] will denitrosate GSNO to GSH under release of free NO. Finally, the γ-glutamyl transferase enzyme was implicated in the stereoselective effect of L-CysNO on posthypoxic ventilation of mice [66]. The preceding examples do not consume large quantities of LMW nitrosothiols and are not expected to have significance for GSNO levels *in vivo*.

In contrast, *S*-nitrosoglutathione reductase (GSNOR) was found to have an impact on GSNO metabolism *in vivo*. In the literature, this enzyme is also referred to as alcohol dehydrogenase-3 (ADH3) or as GSH-dependent formaldehyde dehydrogenase. It can consume large quantities of GSNO [4,67–70]. GSNOR activity is found in many different tissues and found highest in liver [71]. The physiological significance of the GSNOR pathway was impressively demonstrated by experiments with GSNOR knockout mutant mice [4], which showed greatly enhanced levels of RSNO and became hypotensive under anesthesia. Human asthmatics show the combination of enhanced GSNOR levels and depressed GSNO in the bronchial fluids. It suggests that chemically incorporated NO stores are depleted in asthmatic patients. Since the remainder of NO is exhaled, the unbalance between NO and S-nitrosocompounds may be the reason for the observed increase of exhaled NO in asthma [70,72]. Recent studies on the decomposition of low-molecular-weight *S*-nitrosothiols in tissue homogenates of rats have provided more evidence that enzymatic pathways contribute significantly to the process [73]. The above publications leave no doubt that significant enzymatic pathways for S-denitrosation of LMW nitrosothiols operate in mammals. Their role in the regulation of endogenous pool of *S*-nitrosothiols remains the subject of intense research for the moment being.

In addition to this enzymatic pathway, GSNO may be depleted by superoxide according to the third order reaction [74]

$$\frac{d}{dt}[\mathrm{GSNO}] = -k[\mathrm{GSNO}]^2 g[\mathrm{O}_2^-] \tag{10}$$

The third order reaction rate *in vitro* was reported as $k = 6 \times 10^8\ \mathrm{M}^{-2}\ \mathrm{s}^{-1}$. With GSNO concentrations in tissue below ca 50 nM (cf Table 1) and superoxide levels below micromolar range, the loss of GSNO *via* this pathway is insignificant in comparison to loss by catalytic action of spurious metal ions. This observation supports the hypothesis that the transient formation of RSNO may help to increase the effective lifetime of NO in biological systems by protection against the reaction with superoxide.

The hydroxyl radical $OH^{\bullet}$ is a very reactive oxygen species that can be generated in solutions when hydrogen peroxide undergoes a Fenton reaction with spurious ferrous iron. Using pulse radiolysis for generation of the hydroxyl radicals at pH = 7, the main reaction products were found to be disulfides at nitrite. The reaction was second order with diffusion limited reaction rates as usually found for hydroxyl. The reaction rates for CysNO, GSNO and ACysNO were 2.27, 1.46 and 1.94 $\times$ 10^{10} $(\mathrm{Ms})^{-1}$, respectively [75].

Finally, the reaction with oxy-heme proteins, in particular oxy-hemoglobin (oxyHb) and oxy-myoglobin (oxyMb) should be considered. Oxy-hemoglobin is known as the major sink for free NO radicals in the vascular system. The reaction products are methemoglobin and nitrate

$$\mathrm{oxyHb} + \mathrm{NO} \rightarrow \mathrm{metHb} + \mathrm{NO}_3^-$$

$$\mathrm{oxyMb} + \mathrm{NO} \rightarrow \mathrm{metMb} + \mathrm{NO}_3^- \tag{11}$$

The reaction is very fast with second order reaction rates of $8.9 \times 10^7\ (\mathrm{Ms})^{-1}$ for oxyHb and $4.4 \times 10^7\ (\mathrm{Vs})^{-1}$ for oxyMb (pH = 7, 20°C) [76]. Therefore, one might expect a significant oxidation of *S*-nitrosothiols by oxyHb as well. However, *in-vitro* experiments show that the NO moiety of *S*-nitrosothiols is well protected against oxidation to nitrate by oxy-heme. Exposure of a small quantity of CysNO to oxyHb leads to S-nitrosation of the Cysβ93 residue of the protein but not to the formation of nitrate [77,78]. In fact, the rates and extent of transnitrosation from CysNO were very similar for oxyHb and metHb.

This transnitrosation reaction to oxyHb was completely inhibited by the addition of the metal chelators neocuproine and DTPA [78]. Therefore, the transnitrosation of oxyHb seems to require the presence of catalytic traces of copper, either in free form or in the form of Cu–Zn dismutase.

Clearly, reaction (11) is highly significant for free NO radicals but does not apply to *S*-nitrosothiols. Phrased otherwise, a small quantity of low-molecular-weight *S*-nitrosothiols can survive for a very long time in the presence of oxy-heme. This property is a very significant distinction with free NO. It explains why significant quantities of *S*-nitrosothiols can coexist with oxygenated erythrocytes and blood (Table 1), in clear contrast to free NO. Oxidation to nitrate becomes feasible only if GSNO is supplied in excess over Cysβ93 residues [78]. This situation is very unlikely to happen *in vivo*.

Under anaerobic conditions, GSNO can be reduced by deoxyhemoglobin in a slow and irreversible reaction [79]. In presence of excess deoxyHb, the half life of GSNO is about an hour with concomitant release of equimolar quantities of GSH and nitrosylated ferric hemoglobin. The reaction rate was dependent on the conformer state (R or T) of the Hb tetramer. Interestingly, transnitrosation from GSNO to the Cysβ93 residue of Hb did not occur in this assay. As with oxyHb, the reaction pathway with deoxyHb seems not relevant for the lifetime of GSNO *in vivo*.

Recapitulating these various reactions, we recall that *in vivo* RSNO concentrations remain below ca 100 nM (Table 1). At such low levels, the thermolytic and photolytic pathways are negligible in comparison with the effect of trace metal ions, superoxide and the chemical equilibria with other stores of NO. Significantly, *S*-nitrosothiols are essentially stable against oxidation by oxyHb, whereas free NO is rapidly oxidized to nitrate.

TRANSPORT OF LMW *S*-NITROSOTHIOLS

Long-distance transport of *S*-nitrosothiols is possible *via* the blood flow along the vascular tree, but the kinetics of transport on a cellular scale in tissues is still being investigated. The transport of *S*-nitrosothiols across cell membranes was particularly controversial [50,80,81]. Experimental evidence suggests that molecules like GSNO or GSH cannot cross cellular membranes, neither by passive diffusion nor by an active transport mechanism. The carboxyl groups of glutathione are only weakly acid, so that GSNO and GSH should be a dynamic mixture of neutral, anionic and dianionic forms. But even the neutrals are not capable of crossing into the intracellular compartment. Studies [82] on NIH/3T3 fibroblasts showed that exposure to extracellular GSNO or SNAP failed to increase intracellular S-nitrosation. Low levels of extracellular CysNO (<0.1 mM) enhanced intracellular cysteine levels but failed to raise S-nitrosation. Significant S-nitrosation was only observed at high levels (1 mM) of extracellular CysNO. S-nitrosation of the intracellular compartment took place on a timescale of about 10 min [82]. NMR studies [83] confirmed that no significant exchange of GSH takes place over the membrane of human erythrocytes. Moreover, the extracellular glutathione affected neither the intracellular GSH/GSSG ratio nor the glucose metabolism of the erythrocytes. This indicates that extracellular GSH did not provoke intracellular release of thiols from cleavage of disulfide bonds in the interior compartment. Taken together, this study showed that intracellular GSH cannot be raised by suppletion of extracellular GSH.

The cellular transport of small *S*-nitrosothiols was extensively studied by Hogg et al. [84,85]. They showed that extracellular GSNO and SNAP did not significantly increase the pool of intracellular *S*-nitrosothiols in RAW 264.7 macrophages. However, the coincubation of GSNO with L-cysteine or its disulfide (L-cystine) greatly enhanced the intracellular S-nitrosation. The *S*-nitroso moieties were predominantly located on the cysteine residues of proteins. Significant intracellular S-nitrosation was also observed after incubation with only extracellular L-CysNO. The data suggested that L-CysNO was transported intact across the membrane, and initiated subsequent transnitrosation towards the intracellular proteins. The effect of L-cysteine on extracellular GSNO was plausibly explained by the following sequence of events: First, transnitrosation from GSNO to L-CysNO. Second, active transport of L-CysNO across the membrane into the interior compartment. Third, transnitrosation from L-CysNO to the sulfhydryl groups of the proteins. The transport of L-CysNO was stereospecific for the L-isomer and could be blocked by inhibitors of the cellular transport system of amino acids. The uptake of L-CysNO was specifically attributed to the amino acid transporter system L (L-AT) [81,85].

The amino acid transporter system of the cellular membrane is not the only factor determining entry of *S*-nitrosothiols into the interior compartment and the timescale of intracellular transnitrosation. The cell membrane itself is rich in thiol-containing proteins, and each one of these is a potential carrier for cellular entry of NO. Protein disulfide isomerase (PDI) is a membrane enzyme that normally catalyzes thiol–disulfide exchange reaction. This protein was found to raise the level of *S*-nitrosothiols inside human erythroleukemia cells [50]. Additional pathways for the entry of *S*-nitrosothiols will undoubtedly be recognized in the future.

BIOLOGICAL ACTIONS OF LMW NITROSOTHIOLS

A large body of data attests to the effect of low-molecular-weight RSNOs on various physiological and biochemical processes. We can roughly classify these effects in three categories. In the first, RSNO operate as donors of free NO and the NO subsequently acts as the true effector, just as would NO released by NO donors or by NOS enzymes. In the second, RSNO acts as a transnitrosator of other sulfhydryls as found on proteins, and the effects result from the S-nitrosation of the group. Effects in this category can also be evoked by exposure to other nitrosating agents like N_2O_3. In the third category, RSNO itself acts as an effector in its own right, for example as activator of guanylate cyclase enzymes (GC). We also include S-glutathiolation into this third category.

First category effects

The first category contains the majority of effects mediated by the guanylate cyclase-dependent pathway in tissues and cellular assays [86,87]. The RSNO can be of low molecular weight like GSNO or high molecular weight like *S*-nitrosohemoglobin. It is now widely accepted that the release of free NO from cleavage of the S–NO bond in cells and tissues is dominated by catalytic action of copper [88]. Recent studies [86] report that GSNO and

S-nitrosated oxy-hemoglobin is vasodilating without release of free NO (cf Fig. 7). The sensitivity of aortic tissue to GSNO is very high, and of the same order of magnitude as the sensitivity to the physiological stimulant acetyl choline (Ach) (Fig. 8). The vasodilation of aortic rings is considerably reduced when superoxide levels in the organ bath are artificially raised by coincubation with pyrogallol [89], a generator of superoxide radicals. Seen as NO donors various RSNOs can produce both beneficial and adverse effects on cells and tissues. This state of affairs reminds us of the ongoing controversy surrounding the benefits/disadvantages of upregulation of NOS enzymes. Many detailed studies *in vitro* and *in vivo* suggest that raising levels of free NO be beneficial in vascular ischemia, but harmful in brain ischemia. Accordingly, many therapeutic treatments of ischemia involve NO donors for vasculature, and NOS inhibitors for brain. Experiments confirm many similarities between the effect on ischemia of LMW *S*-nitrosothiols and true NO donors. The similarities include relaxation of blood vessels or bronchial smooth muscle, inhibition platelet aggregation, etc. Just as with true NO donors, high doses of RSNO can be harmful and activate the apoptotic chain [90–92]. The mechanism is still being investigated. A significant contribution is attributed to the oxidation of excess NO to the highly noxious peroxinitrite. The latter induces so-called nitrosative stress [93].

In many cases, moderate doses of *S*-nitrosothiols are seen to provide protection against oxidative stress: Just like true NO donors, GSNO stimulated the recycling of ascorbate by human erythrocytes [94]. The mechanism was not fully understood but did not significantly depend on the presence of free transition metal ions in solution.

It is interesting to note that the iron status of cells is an important parameter in the response of cultured cells to exogenous RSNO. NO is implicated in the activation of the apoptotic chain. Details of this mechanism are discussed in Chapter 12. Interestingly, the apoptotic action of NO depends on the presence of iron. Cellular apoptosis could also be abrogated in a range of cultured human tumor cell lines by the addition of exogenous iron to the medium [91].

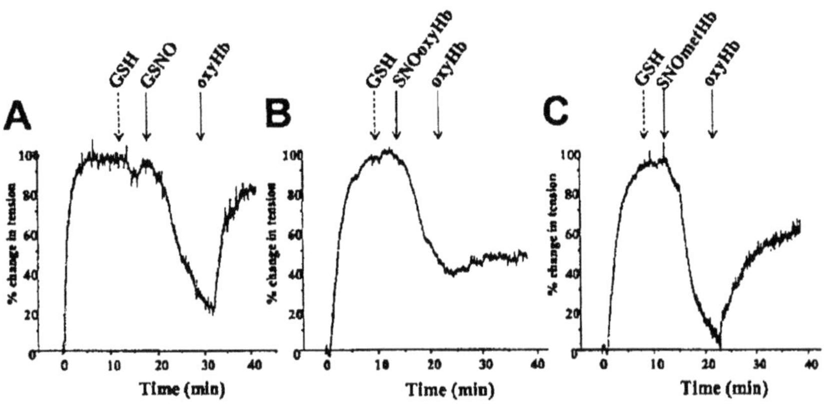

Fig. 7. Vasorelaxation by GSNO and S-nitrososated hemoglobin in precontracted aortic rings from rats. The relaxation was stimulated by 100 μM GSH, followed by either 50 nM GSNO (A); SNOoxyHb (B); SNOmetHb (C). At a later timepoint 10 μM oxyHb was added to scavenge free NO. Experiments in Krebs buffer at 37°C at high oxygen tension (95%O_2 + 5%CO_2). (From Ref. [86].)

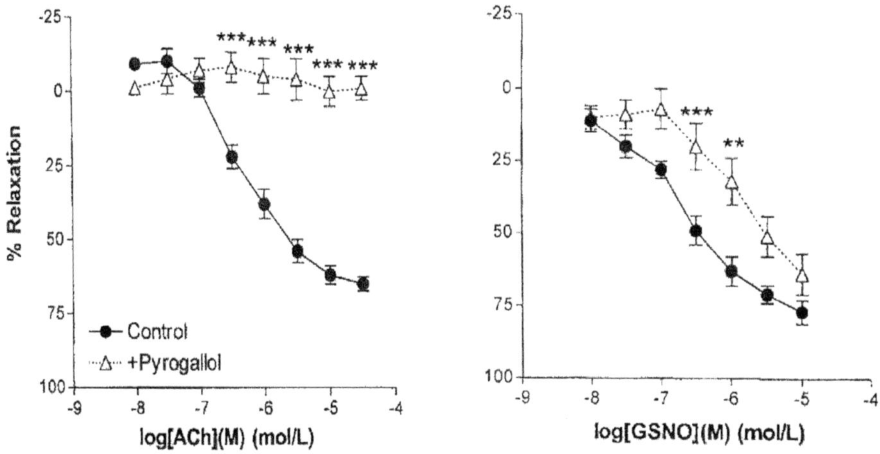

Fig. 8. Dose response curves for relaxation of precontracted rat aortic rings by acetylcholine (left panel) and GSNO (right panel). The presence of 0.3 mM pyrogallol (open triangles) cancels the stimulus from Ach, and reduces the sensitivity to GSNO by an order of magnitude. Experiments in Krebs buffer at 37°C at high oxygen tension (95%O_2 + 5%CO_2). (From Ref. [89].)

RAW264.7 macrophages contain a comparatively low amount of non-heme iron. These macrophages proved quite susceptible to the pro-apoptotic action of RSNOs. Preincubation with exogenous iron reinforced their resistance to apoptosis and EPR spectroscopy revealed the formation of intracellular DNIC with thiol-containing ligands [90]. These DNIC were evidently assembled from RSNO and non-heme iron. In contrast, hepatocytes contain fairly large amounts of endogenous iron and show much higher resistance to the pro-apoptotic effect of NO [90]. The cellular reaction pathways for DNIC are considered in detail in Chapter 11. Intracellular formation of DNIC was also observed after addition of ferrous iron to cultured leukemic cells incubated with *S*-nitrosocysteine (Vanin, A.F., non-published data). The incubation of cells with Cys-DNIC or GS-DNIC did not affect the cell viability.

The mechanism of apoptotic protection was investigated in cellular assays [90]. It was shown that apoptosis was suppressed by inhibition of caspase 3 by S-nitrosation of this protein. The S-nitrosation could be reversed by the addition of dithiothreitol. It demonstrated that extracellular DNIC enhanced S-nitrosation of the intracellular proteins. Details of the mechanism remain unclear since transport of intact DNIC across cellular membranes remains unproven at this moment.

We just noted that both RSNO and DNIC can act as NO donors in their own respect. Usually, activation of a physiological response requires higher doses of GSNO than DNIC [90,95]. As donors, they are clearly distinguished by their response to alterations in the pool of free iron and copper in the solutions. Addition of a metal chelator like bathophenanthroline disulfonate (BPDS) will remove the iron from DNIC and induce instantaneous release of the nitrosyl ligands into the solution. The effect on GSNO is exactly opposite in that its lifetime is greatly extended by chelation of the free metal ions. But the pools of RSNO and

DNIC are always coupled in a dynamic equilibrium by the presence of free iron ([96] and Chapter 11). It should be noted that tissues always contain a certain quantity of loosely bound iron (experimental evidence for this iron pool in tissues is reviewed in Chapter 2).

Second category effects

Besides acting as NO donors, RSNOs can modulate physiological and biochemical processes by transnitrosation. Certain effects were found to be stereoselective and are specific for the L-isomer of *S*-nitrosocysteine. Stereoselectivity was observed for neuronal stimulation in brain of conscious rats [97] and in transnitrosation of the intracellular compartment [84,85]. Mounting evidence suggests that these effects are manifestations of a cascade of events involving the exchange of S-nitrosation between exogenous or endogenous low-molecular RSNOs and sulfhydryl moieties on intracellular proteins. [84,98]. This chemical modification is reversible and, in principle, potentially useful for regulation of the enzymatic activity inside cells.

Effects of S-nitrosation on enzymatic activity have now been confirmed for dozens of enzymes *in vitro* as well as *in vivo*. The involvement of S-nitrosation is usually established beyond doubt by being reversible under treatment with the denitrosating agent dithiothreitol. S-nitrosation can inhibit glutathione reductase [99], seven members of the caspase family [100], including the inducer of apoptosis, caspase 3 [100,101], creatine kinase [102], glutathione-*S*-transferase [103], adenosyltransferase [104], transcription factor Yin Yang [105], catepsin K [106], glyceraldehyde-3-phosphate dehydrogenase [107], alcohol dehydrogenase [108], c-Jun *N*-terminal kinase 2 [109], HIF-1α protein [110], protein tyrosine phosphatase [111], inactivate aconitase [112], activate the ryanodine receptor for calcium release [113], inhibit adenyl cyclase enzyme in rat thymocytes [81,114], etc. Also S-nitrosation of crucial thiol groups by exogenous or endogenous low-molecular RSNOs can inhibit the ligand-binding ability of glucocorticoid receptor [115], disrupt the mitochondrial electron-transfer chain [116], induce cytostasis or cytotoxicity [117], inhibit NF-κB binding to DNA [118,119], inhibit papain [111] and ornithine decarboxilase [120], inhibit Complex I from the mitochondrial respiratory chain [121], regulate the redox state and anti-apoptotic function of thioredoxin [122]. This list is exemplary only and by no means exhaustive. As a rule, these inhibitions are reversible, and enzymatic activity may be restored by S-denitrosation, for example by adding excess glutathione or dithiothreitol. It should also be noted that enzymes may interact with nitrosospecies in more than just one way. A well-known example is Hb which undergoes nitrosylation of the heme and also S-nitrosation of its Cysβ93 residue. Another is Complex I of the respirational chain which undergoes nitrosylation of the catalytic Cu-B/heme-a3 site as well as multiple S-nitrosation of cysteine residues [94].

Before concluding the effects in the second category, we recall that DNIC plays a very important role in the process of protein S-nitrosation by LMW RSNOs. As was mentioned earlier, the probability of transnitrosation from low-molecular RSNOs to protein thiol groups is low and takes many minutes to establish equilibrium (Fig. 6). Spurious iron and copper help significantly, but the formation of DNIC really accelerates transnitrosation. It was shown by Boese et al. [123] that Cys-DNIC is much more effective for S-nitrosation of serum albumin than CysNO itself. It is conceivable that loosely bound iron and DNIC rather than spurious copper determine the rate and extent of S-nitrosation of the proteins in cultured cells and tissues.

Third category effects

Here we list a number of physiological responses *via* mechanisms other than S-nitrosation. More precisely, processes for which S-nitrosation does not play a role or in which S-nitrosation has not been recognized yet. *S*-nitrosothiols provide neuroprotection [124], protect against amyloid β-peptide neurotoxicity [125], induce bacteriostatic effect of α_1-protease inhibitor [126], interrupt the replication of the coronavirus [127], inactivate HIV-1 protease [128], inhibit dystrophin proteolysis by Coxsackieviral protease 2A [129], promote expression of the 5-lipoxygenase (5-LO) in several human bronchial cells types [130] and stimulate noradrenalin release in rat brain [81]. Nitrosprusside and SNAP activate a complex extracellular signal-regulated kinase (ERK) pathway *via* an as yet unknown mechanism [131]. *S*-nitrosothiols are often found to influence the life cycle of viruses, for example human immunodeficiency [132], Herpes Simplex [2] and Epstein-Barr [133]. In many cases the mechanism has not been fully understood, and it is speculated that S-nitrosation of thiols might have significant effect on virion maturation [2,127]. Upon incubation with *S*-nitrosothiols, the cysteine residues of proteins often show a certain degree of S-glutathiolation. The S-glutathiolation pathway is often operating simultaneously with S-nitrosation. The process of S-glutathiolation requires the presence of oxygen and seems to be dominated by intermediate oxides like GS(O)SG or $GS(O_2)SG$ [37]. S-glutathiolation of intracellular proteins has been demonstrated *in vivo*: upon incubation with exogenous CysNO, NIH-3T3 fibroblasts underwent combined S-nitrosation and S-glutathiolation of the cysteine residues of H-ras protein [54]. S-glutathiolation may be reversed by dithiothreitol which is an effective reductor for disulfide bonds [60]. Again, the above list is not exhaustive.

This growing body of evidence attests to the signaling role of various RSNOs. These agents are clearly involved in the control of enzyme activity and in various intracellular messenger pathways.

THERAPEUTIC USES OF LMW NITROSOTHIOLS

Hitherto, therapeutic application of LMW nitrosothiols was limited because they elicit a rapid and strong vasodilatory response which exceeds the margins of safety for humans. In short, they are too effective as vasodilators. Alternative donors like organic nitrates or nitroglycerine are considered safer for clinical applications as they provide better control over the vasodilation. These alternative NO donors are described in detail in Chapter 17 of this book. The risk of excessive hypotension makes that clinical studies of the effect of GSNO in humans are very few. A clinical study into the effects of infusion of GSNO into the bloodstream of pregnant women with severe preeclampsia [134] has shown promising improvement in maternal blood pressure (Fig. 9) and uterine arterial resistance (Table 3).

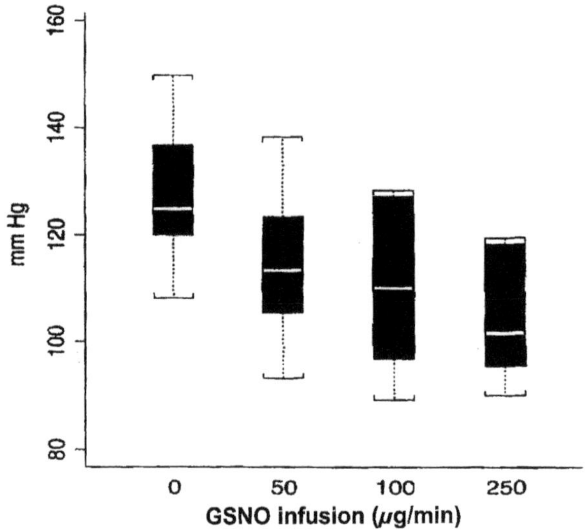

Fig. 9. Drop in arterial blood pressure of preeclamptic pregnancies upon infusion of increasing doses of GSNO. (From Ref. [134].)

RSNO has been implicated in a number of respiratory diseases. Inflammatory state as in pneumonia gives higher RSNO levels in the lungs. Asthmatic patients exhale high levels of NO but have depressed RSNO levels in the airways. The modulation of RSNO status for clinical therapy is only now being considered [66,70,135].

Injection of modest doses of GSNO has also been found to promote wound healing in rats. In this case, repeated intraperitoneal injection of GSNO was found to accelerate healing of skin lesions by promoting deposition of new collagen in the affected areas [136] (Fig. 10).

External application of RSNO avoids the risk of exceedingly strong vasodilation and life threatening loss of blood pressure. Very promising results have been recently reported for external application of hydrogels containing GSNO to promote skin repair [137].

Table 3 Effect of venal infusion of GSNO on blood circulation parameters of pregnant women with severe preeclampsia. PI and RI are pulsatility and resistance index, respectively. P-selectin expression is a marker for activation of blood platelets and has a value 1.1 ± 0.2% in healthy pregnancies. Adapted from Ref. [134]

Parameter	Before infusion	During infusion
Mean arterial pressure (mm Hg)	125 ± 5	104 ± 4
Pulse rate (beats/min)	74 ± 6	90 ± 4
Mean uterine artery RI	0.76 ± 0.03	0.70 ± 0.03
Umbilical artery PI	1.9 ± 0.2	1.6 ± 0.2
Fetal thoracic aorta PI	2.5 ± 0.2	2.3 ± 0.2
P-selectin expression (%)	3.0 ± 0.5	1.2 ± 0.2

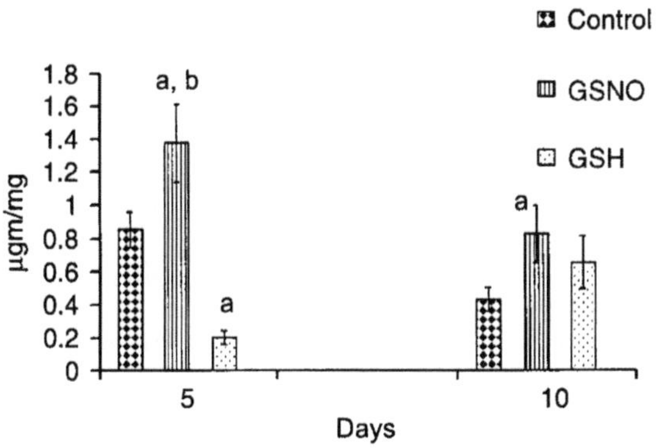

Fig. 10. Hydroxyproline content of scar tissue at two time points after dorsal incision in rats. Hydroxyproline is the main component of collagen. After incision GSNO and GSH were injected intraperitoneally at 24 h intervals with dose of 0.3 mg/kg. The treatment with GSNO significantly enhances the collagen deposition in the affected tissue. (From Ref. [137].)

The beneficial action was attributed to a combination of factors. First, the sterilizing action of NO radicals released from the GSNO. Second, the vasodilation and ensuing improvement of blood supply to the affected tissues.

The older literature on potential therapeutic applications of *S*-nitrosothiols was reviewed in Refs. [138–140].

Summarizing the above results, it appears surprising that such a powerful vasodilator has found relatively few practical therapeutical applications. As mentioned before, this state of affairs is primarily caused by the practical problems regarding the need to keep complete control over the dosage to avoid life threatening loss of systemic blood pressure. *S*-nitrosothiols, whether they be of low or high molecular weight, clearly elicit strong physiological responses. Some of these responses are acute (e.g. vasodilation) and some induced on a longer timescale (like apoptosis). But strong that they are and since the *S*-nitrosothiols are endogenously formed, these compounds deserve constant attention and consideration by researchers in the field of NO.

REFERENCES

1 Foster M, Stamler J. New insights into protein S-nitrosylation. J. Biol. Chem. 2004; 279: 25891–25897.

2 Miersch S, Mutus B. Protein S-nitrosation: Biochemistry and characterization of protein thiol-NO interactions as cellular signals. Clin. Biochem. 2005; 38: 777–791.

3 Stamler J, Lamas S, Fang F. Nitrosylation: the prototypic redox-based signalling mechanism. Cell 2001; 106: 675–683.

4 Liu L, Yan Y, Zeng M, Zhang J, Hanes M, Ahearn G, MCMahon T, Dickfeld T, Marshall H, Que L, Stamler JS. Essential roles of S-nitrosothiols in vascular homeostasis and endotoxic shock. Cell 2004; 116: 617–628.

5 Hand C, Taylor N, Honek J. Ab initio studies of the properties of intracellular thiols ergothioneine and ovothiol. Bioorg. Med. Chem. Lett. 2005; 15: 1357–1360.

6 Misiti F, Castagnola M, Zuppi C, Giardina B, Messana I. Role of ergothioneine on S-nitrosoglutathion catabolism. Biochem. J. 2001; 356: 799–804.

7 Kaneko I, Takeuchi Y, Yamaoka Y, Tanaka Y, Fukuda T, Fukumori Y, Mayumi T, Hama T. Quantitative determination of ergothioneine in plasma and tissues by TLC densitometry. Chem. Pharm. Bull. 1980; 28: 3093–3097.
8 Zhang Y-Y, Xu A-M, Nomen M, Walsh M, Keany J, Loscalzo J. Nitrosation of tryptophan residue(s) in serum albumin and model dipeptides. J. Biol. Chem. 1996; 271: 14271–14279.
9 Bryan N, Rassaf T, Maloney R, Rodriguez C, Saijo F, Rodriguez J, Feelish M. Cellular targets and mechanisms of nitros(yl)ation: an insight into their nature and kinetics *in vivo*. Proc. Natl. Acad Sci. USA 2004; 101: 4308–4313.
10 Rassaf T, Bryan N, Kelm M. Concomitant presence of N-nitroso and S-nitroso proteins in human plasma. Free Rad. Biol. Med. 2002; 33: 1590–1596.
11 Halliwell B, Gutteridge J. Free radicals in biology and medicine, Clarendon Press, Oxford, 2nd ed., 1989.
12 Sexton D, Muruganandam A, McKennedy D, Mutus B. Visible light photochemical release of nitric oxide from S-nitrosoglutathion. Photochem. Photobiol. 1994; 59: 463.
13 de Oliveira M, Shishido S, Seabra A, Morgon N. Thermal stability of primary S-nitrosothiols: roles of autocatalysis and structural effects on the rate of nitric oxide release. J. Phys. Chem. 2002; A106: 8963–8970.
14 Meyer DJ, Kramer H, Ozer N, Coles B, Keterrer B. Kinetics and equilibria of S-nitrosothiol-thiol exchange between glutathione, cysteine, penicillamines and serum albumin. FEBS Lett. 1994; 345: 177–180.
15 Scorza G, Pietraforte D, Minetti M. Role of ascorbate and protein thiols in the release of nitric oxide S-nitrosoalbumin and S-nitrosoglutathion in human plasma. Free Rad. Biol. Med. 1997; 22: 633–642.
16 Hart T. Some observations concerning the S-nitroso and S-phenylsulphonyl derivatives of L-cysteine and glutathione. Tetrahedron Lett. 1985; 26: 2013–2016.
17 Beloso P, Williams D. Reversibility of S-nitrosothiol formation. Chem. Commun. 1997; 1: 89–91.
18 Grossi L, Montevecchi PC. S-nitrosocysteine and cystine from reaction of cysteine with nitrous acid. A kinetic investigation. J. Org. Chem. 2002; 67: 8625–8630.
19 Keshive M, Singh S, Wishnok J, Tannenbaum S, Deen W. Kinetics of S-nitrosation of thiols in nitric oxide solutions. Chem. Res. Toxicol. 1996; 9: 988–993.
20 Wink D, Nims R, Darbyshire J et al. Reaction kinetics for nitrosation of cysteine and glutathione in aerobic nitric oxide solutions at neutral pH—-insights into the fate and physiological effects of intermediates generated in the NO/O2 reaction. Chem. Res. Toxicol. 1994; 7: 519–525.
21 Demaster E, Quast B, Redfern B, Nagasawa H. Reaction of nitric oxide with the free sulfhydryl group of human serum albumin yields sulfenic acid and nitrous oxide. Biochemistry 1995; 34: 11494–11499.
22 Kharitonov V, Sundquist A, Sharma V. Kinetics of nitrosation of thiols by nitric oxide in the presence of oxygen. J. Biol. Chem. 1995; 270: 28158–28164.
23 Pryor W, Church D, Govindan C, Crank G. Oxidation of this thiols by nitric oxide and nitrogen dioxide synthetic utility and toxicological implications J. Org. Chem. 1982; 47: 156–159.
24 Hogg N, Singh R, Kalyanaraman B. The role of glutathione in the transport and catabolism of nitric oxide. FEBS Lett. 1996; 382: 223–228.
25 Folkes L, Wardman P. Kinetics of the reaction between nitric oxide and glutathione: implications for thiol depletion in cells. Free Rad. Biol. Med. 2004; 37: 549–556.
26 Carver J, Doctor A, Zaman K, Gaston B. S-nitrosothiol formation. Meth. Enzymol. 2005; 396: 95–105.
27 Goldstein S, Czapski G. Mechanism of the nitrosation of thiols and amines by oxygenated NO solutions: the nature of the nitrosating intermediates. J. Am. Chem. Soc. 1996; 118: 3419–3425.
28 Kikugawa K, Oikawa N, Miyazawa A, Shinda K, Kato T. Interaction of nitric oxide with glutathione or cysteine generates reactive oxygen species causing DNI single strand breaks. Biol. Pharm. Bull. 2005; 28: 990–1003.
29 Gow AJ, Buerk DG, Ischiropoulos H. A novel reaction mechanism for the formation of S-nitrosothiol in vivo. J. Biol. Chem. 1997; 272: 2841–2845.
30 Stubauer G, Guiffre A, Sarti P. Mechanism of S-nitrosothiol formation and degradation by copper ions. J. Biol. Chem. 1999; 274: 28128–28133.

31 Vanin A, Malenkova I, Serezhenkov V. Iron catalyzes both decomposition and synthesis of S-nitrosothiols: Optical and electron paramagnetic resonance studies. Nitric Oxide 1997; 1: 191–203.
32 Herold S, Rock G. Mechanistic studies of S-nitrosothiol formation by NO/O2 and by NO/methemoglobin. Arch. Biochem. Biophys. 2005; 436: 386–396.
33 Alpert C, Ramdev N, George D, Loscalzo J. Detection of S-nitrosothiols and other nitric oxide derivatives by photolysis-chemiluminescence spectroscopy. Anal. Biochem. 1997; 245: 1–7.
34 Saville B. A scheme for the colorimetric determination of microgram amounts of thiols. Analyst 1958; 83: 670–672.
35 von Bohlen und Halbach O. Nitric oxide imaging in living neuronal tissues using fluorescent probes. Nitric Oxide 2003; 9: 217–228.
36 Liu X, Miller M, Joshi M, Thomas D, Lancaster J. Accelerated reaction of nitric oxide with O_2 within the hydrophobic interior of biological membranes. Proc. Natl. Acad. Sci. USA 1998; 95: 2175–2179.
37 Tao L, English A. Protein S-glutathionylation triggered by decomposed S-nitrosoglutathion. Biochemistry 2004; 43: 4028–4038.
38 Jacob C, Lancaster J, Giles G. Reactive sulphur species in oxidative signal transduction. Biochem. Soc. Trans. 2004; 32: 1015–1017.
39 Holmes A, Williams DL. Reaction of ascorbic acid with S-nitrosothiols: clear evidence for two distinct reaction pathways. J. Chem. Soc. Perkin Trans. 2000; 2: 1639–1644.
40 Tullett J, Hodson H, Gescher A, Shuker D, Moncada S. Stability of S-nitrosothiols related to glutathione. Endothelium 1995; 3: s66.
41 Rotta J, Lunardi C, Tedesco A. Nitric oxide release from the S-nitrosothiol zinc phthalocyanine complex by flash photolysis. Brazil. J. Med. Biol. Res. 2003; 36: 587–594.
42 Rodriguez J, Maloney R, Rassaf T, Bryan N, Feelisch M. Chemical nature of nitric oxide storage forms in rat vascular tissue. Proc. Natl. Acad. Sci. USA 2003; 100: 336–341.
43 Vanin A, Papina A, Serezhenkov V, Koppenol W. The mechanisms of S-nitrosothiol decomposition catalyzed by iron. Nitric Oxide 2004; 10: 60–73.
44 Dicks A, Swift H, Williams D, Butler A, Al-Sadoni H, Cox B. Identification of Cu^+ as the effective reagent in nitric oxide formation from S-nitrosothiols. J. Chem. Soc. Perkin Trans. 1996; 2: 481–487.
45 Vanin A, Muller B, Alencar J, Lobysheva I, Nepveu F, Stoclet J-C. Evidence that intrinsic iron but not intrinsic copper determines S-nitrosocysteine decomposition in buffer solution. Nitric Oxide 2002; 7: 194–209.
46 McAninly J, Williams D, Askew S, Butler A, Russel C. Metal ion catalysis in nitrosothiol decomposition. J. Chem. Soc. Chem. Commun. 1993; 1758–1759.
47 Prestipino C, Berlier G et al. An in situ temperature dependent IR, EPR and high resolution XANES study on the NO/Cu^+-ZSM-5 interaction. Chem. Phys. Lett. 2002; 363: 389–396.
48 Zhang H, Means G. S-nitrosation of serum albumin: spectrophotometric determination of its nitrosation by simple S-nitrosothiols. Anal. Biochem. 1996; 237: 141–144.
49 Jourdheuil D, Hallen K, Feelisch M, Grisham M. Dynamic state of S-nitrosothiols in human plasma and whole blood. Free Rad. Biol. Med. 2000; 28: 409–417.
50 Zai A, Rudd M, Scribner A, Loscalzo J. Cell-surface protein disulfide isomerase catalyzes transnitrosation and regulates intracellular transfer of nitric oxide. J. Clin. Invest. 1999; 103: 393–399.
51 Scharfstein J, Keany J, Slivka A, Welch G, Vita J, Stamler J. In vivo transfer of nitric oxide between a plasma protein-bound rservoir and low molecular weight thiols. J. Clin. Invest. 1994; 94: 1432–1439.
52 Liu Z, Rudd M, Freedman J, Loscalzo J. S-transnitrosation reactions are involved in the metabolic fate and biologic actions of nitric oxide. J. Pharm. Exp. Ther. 1998; 284: 526–534.
53 Giustarini D, Milzani A, Aldini G, Carini M, Rossi R, Dalle-Donne I. S-nitrosation versus S-glutathionylation of protein sulfhydryl groups by S-nitrosoglutathione. Antiox. Redox Signal. 2005; 7: 930–939.
54 Mallis R, Buss J, Thomas J. Oxidative modification of H-ras: S-thiolation and S-nitrosylation of reactive cysteines. Biochem. J. 2001; 355: 145–153.
55 Singh R, Hogg N, Joseph J, Kalyanaraman B. Photosensitized decomposition of S-nitrosothiols and 2-methyl-2-nitrosopropane. Possible use for site-directed nitrite oxide production. FEBS lett. 1995; 360: 47–51.

56 Askew S, Butler A, Flitney F, Kemp G, Megson I. Chemical mechanisms underlying the vasodilator and platelet anti-aggregating properties of S-nitroso-N-acetyl-DL-pencillamine and S-nitrosoglutathione. Bioorg. Med. Chem. 1995; 3: 1–9.
57 Hu T-M, Chou T-C. The kinetics of thiol-mediated decomposition of S-nitrosothiols. AAPS J. 2006; 8, article 57.
58 Marley R, Patel R, Orie N, Ceaser E, Darley V, Moore K. Formation of nanomolar concentrations of S-nitrosoalbumin in human plasma by nitric oxide. Free Rad. Biol. Med. 2001; 31: 688–696.
59 Hogg N. Biological chemistry and clinical potential of S-nitrosothiols. Free Rad. Biol. Chem. 2000; 28: 1478–1486.
60 Sadidi M, Geddes T, Kuhn D. S-thiolation of tyrosine hydroxylase by reactive nitrogen species in the presence of cysteine of glutathione. Antiox. Redox Signal. 2005; 7: 863–869.
61 Tullett J, Rees D, Shuker D, Gescher A. Lack of correlation between the observed stability and pharmacological properties of S-nitroso derivatives of glutathione and cysteine-related peptides. Biochem. Pharmacol. 2001; 62: 1239–1247.
62 Sliskovic I, Raturi A, Mutus B. Characterization of the S-denitrosation activity of protein-disulfide dismutase. J. Biol. Chem. 2005; 280: 8733–8741.
63 Trujillo M, Alvarez M, Peluffo G, Freeman B, Radi R. Xanthine oxidase mediated decomposition of S-nitrosothiols. J. Biol. Chem. 1998; 273: 7828–7834.
64 Nikitovic D, Holmgren A. S-nitrosoglutathione is cleaved by the thioredoxin system with liberation of glutathione and redox regulating nitric oxide. J. Biol. Chem. 1996; 272: 19180–19185.
65 Freedman J, Frei B, Welch G, Loscalzo J. Glutathione peroxidase potentiates the inhibition of platelet function by S-nitrosothiols. J. Clin. Invest. 1995; 96: 394–400.
66 Lipton A, Johnson M, MacDonald T, Lieberman M, Gozal D, Gaston B. S-nitrosothiols signal the ventilatory response to hypoxia. Nature 2001; 413: 171–174.
67 Xiong Y, Karupiah G, Hogan S, Fosters P, Ramsay A. Inhibition of allergic airway inflammation in mice lacking nitric oxide synthase 2. J. Immunol. 1999; 162: 445–452.
68 Liu L, Hausladen A, Zeng M, Que L, Heitman J, Stamler J. A metabolic enzyme for S-nitrosothiol conserved from bacteria to humans. Nature 2001; 410: 490–494.
69 Hedberg J, Griffiths W, Nilsson S, Hoog J-O. Reduction of S-nitrosoglutathion by human alcohol dehydrogenase-3 is an irreversible reaction as analysed by electrospray mass spectrometry. Eur. J. Biochem. 2003; 270: 1249–1256.
70 Que L, Liu L, Yan Y, Whitehead G, Gavett S, Schwartz D, Stamler J. Protection from experimental asthma by an endogenous bronchodilator. Science 2005; 308: 1618–1621.
71 Gow A, Chen Q, Hess D, Day B, Ischiropoulos H, Stamler J. Basal and stimulated protein-S-nitrosylation in multiple cell types and tissues. J. Biol. Chem. 2002; 277: 9637–9640.
72 Gerard C. Asthmatics breathe easier when it's SNO-ing. Science 2006; 308: 1560–1561.
73 Mani A, Ebrahimkhani M, Ippolito S, Ollosson R, Moore K. Metalloprotein-dependent decomposition of S-nitrosothiols: studies on the stabilization and measurement of S-nitrosothiols in tissues. Free Rad. Biol. Med. 2006; 40: 1654–1663.
74 Jourdheuil D, Mai C, Laroux F, Wink D, Grisham M. The reaction of S-nitrosoglutathione with superoxide. Biochem. Biophys. Res. Commun. 1998; 244: 525–530.
75 Manoj V, Aravindakumar C. Reaction of hydroxyl radical with S-nitrosothiols: determination of rate constants and product analysis. Org. Biol. Chem. 2003; 1: 1171–1175.
76 Herold S, Exner M, Nauser T. Kinetic and mechanistic studies of the NO mediated oxidation of oxymyoglobin and oxyhemoglobin. Biochemistry 2001; 40: 3385–3395.
77 Palmerini C, Palombari R, Arienti G. Electrochemical assay of human haemoglobin S-nitrosylation by nitrosocysteine. Amino Acids 2003; 25: 59–62.
78 Patel R, Hogg N, Spenser N, Kalyanaraman B, Matalon S, Darley-Usmar V. Biochemical characterization of human S-nitrosohemoglobin. Effects of oxygen binding and transnitrosation. J. Biol. Chem. 1999; 274: 15487–15492.
79 Spencer N, Zeng H, Patel R, Hogg N. Reaction of S-nitrosoglutathione with the heme group of deoxyhemoglobin. J. Biol. Chem. 2000; 47: 36562–36567.

80 Ramachandran N, Root P, Jiang X, Hogg P, Mutus B. Mechanism of transfer of NO from extracellular S-nitrosothiols into the cytosol by cell-surface protein disulfide isomerase. Proc. Natl. Acad. Sci. USA 2001; 98: 9539–9544.

81 Nemoto T, Shimma N, Horie S, Saito T, Okuma Y, Nomura Y, Murayama T. Involvement of the system L amino acid transporter on uptake of S-nitroso-L-cysteine, an endogenous S-nitrosothiol, in PC12 cells. Eur. J. Pharmacol. 2003; 458: 17–24.

82 Mallis R, Thomas J. Effect of S-nitrosothiols on cellular glutathione and reactive protein sulfhydryls. Arch. Biochem. Biophys. 2000; 383: 60–69.

83 Kennett E, Bubb W, Bansal P, Alewood P, Kuchel P. NMR studies of exchange between intra- and extracellular glutathione in human erythrocytes. Redox Rep. 2005; 10: 83–90.

84 Zhang Y, Hogg N. The mechanism of membrane S-nitrosothiol transport. Proc. Natl. Acad. Sci. USA 2004; 101: 7891–7896.

85 Zhang Y, Hogg N. S-nitrosothiols: cellular formation and transport. Free Rad. Biol. Med. 2005; 38: 831–838.

86 Crawford J, White C, Patel R. Vasoactivity of S-nitrosohemoglobin: role of oxygen, heme and NO oxidation states. Blood 2003; 101: 4408–4415.

87 Mathews MR, Kerr SW. Biological activity of S-nitrosothiols, J. Pharmacol. Exp. Ther. 1993; 267: 1529–1526.

88 Al-Sa'Doni HH, Megson IL, Bisland S, Butler AR, Flitney FW. Neocuproine, a selective Cu(I) chelator and the relaxation of rat vascular muscle by S-nitrosothiols. Br. J. Pharmacol. 1997; 121: 1047–1050.

89 Hanspal I, Magid K, Webb D, Megson I. The effect of oxidative stress on endothelium-dependent and nitric oxide donor-induced relaxation: implications for nitrate tolerance. Nitric Oxide 2002; 6: 263–270.

90 Kim Y-M, Chung H-T, Simmons RL, Billiar TR. Cellular non-heme iron content is a determinant of nitric oxide-mediated apoptosis, necrosis, and caspase inhibition. J. Biol. Chem. 2000; 275: 10954–10961.

91 Feger F, Ferry-Dumazet H, Matsuda MM, Bordenave J, Dupouy M, Nussler AK, Arock M, Devevey L, Nafziger J, Guilosson JJ, Reiffers J, Mossalayi MD. Role of iron in tumor cel protection from pro-apoptotic effect of nitric oxide. Cancer Res. 1993; 61: 5289–5294.

92 Lin D, Ma W, Duan S, Zhang Y, Du L. Real time imaging of viable apoptotic switch in GSNO-induced mouse thymocyte apoptosis. Apoptosis 2006; 11: 1289–1298.

93 Hausladen A, Privalle CT, Keng T, DeAngelo J, Stamler JS. Nitrosative stress: Activation of the transcription factor OxyR. Cell 1996; 86: 719–729.

94 Spagnuolo M, Carlucci A, Cigliano L, Abrescia P. Nitric oxide stimulates the erythrocyte for ascorbate recycling. Nitric Oxide 2006; 14: 272–227.

95 Boldyrev AA, Bulygina ER, Kramarenko GG, Vanin AF. Effect of nitrosocompounds on Na/K-ATPase. Biochim. Biophys. Acta 1997; 1321: 243–251.

96 Severina I, Bussygina O, Pyatakova N, Malenkova I, Vanin A. Activation of soluble guanylate cyclase by NO donors-S-nitrosothiols and dinitrosyl-iron complexes with thiol-containing ligands. Nitric oxide: Biol. Chem. 2003; 8: 155–163.

97 Davisson RL, Travis MD, Bates JN, Johnson AK, Lewis SJ. Stereoselective actions of S-nitrosocysteine in the central nervous system of conscious rats. Am. J. Physiol. 1997; 272, H2361–H2368. Ibid 273: H1493–H1501.

98 Stamler JS. Redox signalling: Nitrosylation and related target intetractions of nitric oxide. Cell 1994; 78: 931–936.

99 Butzer U, Weidenbach H, Gansauge S, Gansauge F, Berger HG, Andreas AK. Increased oxidative stress in the RAW 264.7 macrophage cell line is partially mediated viavia S-nitrosothiol-induced inhibition of glutathione reductase. FEBS Lett. 1999; 445: 274–278. Biochem. Pharmacol. 37: 3199–3201.

100 Li J, Billiar TR, Talanian RV, Kim YK. Nitric oxide inhibits seven members of the caspase family via S-nitrosation. Biochem. Biophys. Res. Comm. 1997; 240: 419–424.

101 Mohr S, Zech B, Lapetina EG, Brune B. Inhibition of caspase-3 by S-nitrosation and oxidation caused by nitric oxide. Biochem. Biophys. Res. Comm. 1997; 238: 387–391.

102 Wolosker H, Panizutti R, Engelender S. Inhibition of creatine kinase by S-nitrosoglutathione. FEBS Lett. 1996; 392: 274–276.

103 Ji Y, Toader V, Bennet B. Regulation of microsomal and cytosolic glutathione-S-transferase activities by S-nitrosylation. Biochem. Pharmacol. 2002; 63: 1397–1406.
104 Ruiz F, Corrales FJ, Miqueo C, Mato JM. Nitric oxide inactivates rat hepatic methionine adenosyltransferase in vivo by S-nitrosylation. Hepatology 1998; 28: 1051–1057.
105 Hongo F, Garban H, Huerta-Yepez S, Vega M, Jazirehi AR, Mizutani Y, Miki T, Bonavida B. Inhibition of the transcription factor Yin Yang activity by S-nitrosation. Biochem. Biophys. Res. Comm. 2005; 336: 692–701.
106 Percival MD, Ouellet M, Campagnolo C, Claveau D, Li C. Inhibition of catepsin K by nitric oxide donors: evidence for the formation of mixed disulfide and sulfenic acid. Biochemistry 1999; 38: 13574–13583.
107 Mohr S, Stamler JS, Brune B. Posttranslational modification of glyceraldehydes-3-phosphate dehydrogenase by S-nitrosylation and subsequent NADH attachement. J.Biol. Chem. 1996; 271: 4209–4214.
108 Gergel D, Cederbaum A. Inhibition of catalytic activity of alcohol degydrogenase by nitric oxide is associated with S-nitrosylation and the release of zinc. Biochemistry 1996; 35: 16186–16194.
109 So HS, Park RK, Kim MS, Lee SR, Jung BH, Chung SY, Jun CD, Chung HT. Nitric oxide inhibits c-Jun N-terminal kinase 2 (JNK2) via S-nitrosylation. Biochem. Biophys. Res. Comm. 1998; 247: 809–813.
110 Sumbaev V, Budde A, Zhou J, Brune B. HIF-1α protein as a target for S-nitrosation. FEBS Lett. 2003; 535: 106–112.
111 Xian M, Chen X, Liu Z, Wang K, Wang PG. Inhibition of papain by S-nitrosothiols. J. Biol. Chem. 2000; 275: 20467–20473.
112 Bouton C, Raveau M, Drapier J-C. Modulation of iron regulatory protein functions. Further insights into the role of nitrogen- and oxygen-derived reactive species. J. Biol. Chem. 1996; 271: 2300–2306.
113 Xu L, Eu JP, Meissner G, Stamler JS. Activation of cardiac calcium release channel (Ryanodine receptor) by poly-S-nitrosylation. Science 1998; 279: 234–236.
114 Miyakoshi M, Yamada T, Katayama H, Murayama T, Nomura Y. Regulation of cyclic GMP and cyclic AMP production by S-nitrosocysteine in rat thymocytes. Eur. J. Pharmacol. 1998; 359: 235–241.
115 Galigniana MD, Piwien-Pilipuk G, Assreuy J. Inhibition of glycocorticoid receptor binding by nitric oxide. Mol. Pharmacol. 1999; 55: 317–323.
116 Clementi E, Brown GC, Feelish M, Moncada S. Persistent inhibition of cell respiration by nitric oxide: Crucial role of S-nitrosylation of mitochondrial complex I and protective action of glutathione. Proc. Natl. Acad. Sci. USA 1998; 95: 7631–7636.
117 deGroote MA, Granger D, Xu Y, Cambell G, Pronce R, Fang FC. Genetic and redox determinants of nitric oxide toxicity in a *salmonella typhimurium* model. Proc. Natl. Acad. Sci. USA 1995; 92: 6399–6403.
118 Matthews JR, Botting CH, Panico M, Morris HR, Hay RT. Inhibition of NF-kB DNA binding by nitric oxide. Nucleic Acids Res. 1996; 24: 2236–2242.
119 Shoshodia S, Aggarwal B. Nuclear factor κB: a friend or a foe in cancer? Biochem. Pharmacol. 2004; 68: 1071–1080.
120 Bauer PM, Fukuto JM, Buga GM, Pegg AE, Ignarro LJ. Nitric oxide inhibits ornithine decarboxilase by S-nitrosylation. Biochem. Biophys. Res. Comm. 1999; 262: 355–358.
121 Burwell L, Nadtochiy S, Tompkins A, Young S, Brookes P. Direct evidence for S-nitrosation of mitochondrial complex I. Biochem. J. 2006; 394: 627–634.
122 Haendeler J, Hoffmann J, Tischler V, Berk B, Zeiher A, Dimmeler S. Redox regulatory and anti-apoptotic functions of thioredoxin depend on S-nitrosylation at cysteine 69. Nat. Cell Biol. 2002; 4: 743–749.
123 Boese M, Mordvintcev PI, Vanin AF, Busse R, Mulsch A. S-nitrosation of serum albumin by dinitrosyl-iron complex. J. Biol. Chem. 1995; 270: 29244–29249.
124 Rauhala P, Andoh T, Chiueh CC. Neuroprotective properties of nitric oxide and S-nitrosoglutathione. Toxicol. Appl. Pharmacol. 2005; 207: S91–S95.
125 Ju T-C, Chen S-D, Liu C-C, Yang D-I. Protective effects of S-nitrosoglutathione against amyloid b-peptide neurotoxicity. Free Rad. Biol. Med. 2005; 38: 938–949.
126 Miyamoto Y, Akaike T, Maeda H. S-nitrosylated human α1-protease inhibitor. Biochim. Biophys. Acta 2000; 1477: 90–97.

127 Akerström S, Mousavi-Jazi M, Klingström J, Leijon M, Lundkvist A, Mirazimi A. Nitric oxide inhibits the replication cycle of severe acute respiratory syndrome coronavirus. J. Virology 2005; 79: 1966–1969.
128 Persichini T, Colasanti M, Lauro G, Ascenti P. Cysteine nitrosylation inactivates the HIV-1 protease. Biochem. Biophys. Res. Comm. 1998; 250: 575–576.
129 Badorff C, Fichtlscherer B, Rhoads RE, Zeiher AM, Muelsch A, Dimmeler S, Knowlton KU. Nitric oxide uninhibits dystrophin proteolysis by Coxsackieviral protease 2A through S-nitrosylation. A protective mechanism against enteroviral cardiomyopathy. Circulation 2000; 102: 2276–2281.
130 Zaman K, Hanigan M, Smith A, Vaughan J, McDonald T, Jones D, Hunt J, Gaston B. Endogenous S-nitrosoglutathione modifies 5-lipoxygenase expression in airway epithelial cells. Am. J. Respir. Cell Mol. Biol. 2006; 34: 387–393.
131 Lander H, Jacovina A, Davis R, Tauras J. Differential activation of mitogen activated protein kinases by nitric oxide related species. J. Biol. Chem. 1996; 271: 19705–19709.
132 Mannick J, Stamler J, Teng E et al. Nitric oxide modulates HIV-1 replication. J. Acquir. Immune Defic. Syndr. 1999; 22: 1–9.
133 Colasanti M, Persechini T, Venturini G, Ascenzi P. S-nitrosylation of viral proteins: Molecular bases for antiviral effect of nitric oxide. IUBMB Life 1999; 48: 25–31.
134 Lees C, Langford E, Brown A et al. The effect of S-nitrosoglutathion on platelet activation, hypertension and uterine and fetal Doppler in severe preeclampsia. Obstet. Gynecol. 1996; 88: 14–19.
135 Gerard C. Asthmatics breathe easier when it's SNO-ing. Science 2005; 308: 1560–1561.
136 Achuth H, Moochala S, Mahendran R, Tan W. Nitrosoglutathion triggers collagen deposition in cutaneous wound repair. Wound Repair and Regeneration 2005; 13: 383–389.
137 Seabra A, Fitzpatrick A, Paul J, de Oliveira M, Weller R. Topically applied S-nitroso-containing hydrogels as experimental and pharmacological nitric oxide donors in human skin. Br. J. Derm. 2004; 151: 977–983.
138 Al-Sa'Doni H, Ferro A. S-nitrosothiols: a class of nitric oxide donor drugs. Clin. Sci. 2000; 98: 507–520.
139 Richardson G, Benjamin N. Potential therapeutic uses for S-nitrosothiols. Clin. Sci. 2002; 102: 99–105.
140 Foster M, McMahon T, Stamler J. S-nitrosylation in health and disease. Trends in Mol. Med. 2003; 9: 160–168.

Radicals for Life: The Various Forms of Nitric Oxide
E. van Faassen and A.F. Vanin (Eds.)

CHAPTER 10

S-nitrosated proteins: formation, metabolism, and function

Yi Yang and Joseph Loscalzo*

From the Cardiovascular Division, Department of Medicine, Brigham and Women's Hospital, Harvard Medical School, 75 Francis Street Boston, MA 02115

INTRODUCTION

S-nitrosothiols, also known as thionitrites, are simple organic esters of nitrite and sulfhydryls. First synthesized in 1909 [1] as nitrite esters of alcohols, these compounds have been known to have nitrovasodilator properties for over 20 years. With the identification of nitric oxide as endothelium-derived relaxing factor [2,3], the possibility that *S*-nitrosothiols form endogenously was proposed. Ignarro and colleagues suggested that *S*-nitrosothiols were the pharmacological intermediates in the thiol-dependent activation of guanylyl cyclase by nitrosoguanidine and sodium nitroprusside [4,5]. We first showed that thiols potentiate the action of endothelial nitric oxide [6,7] and, subsequently, that *S*-nitrosothiols, including GSNO and protein *S*-nitrosothiols, form *in vivo* [8,9]. Altered blood level of *S*-nitrosothiols have been associated with cardiovascular disease [10–13], and dysregulation of protein S-nitrosation is associated with various pathobiological states [9,14,15]. S-nitrosation of protein cysteinyl residues is reversible and rapidly gaining recognition as a major form of post-translational modification of the proteome.

BIOLOGICAL CHEMISTRY

Synthesis

In general, *S*-nitrosothiols can be readily synthesized by the reaction of a thiol with acidified nitrous acid, which generates the nitrosonium cation, NO^+, or with an alkyl nitrite.

* Author for Correspondence. E-mail: jloscalzo@partners.org

Both aryl and alkyl thionitrites can also be synthesized quantitatively by the reaction of the parent thiol with dinitrogen tetroxide [16]. More details on the reaction chemistry of S-nitrosation can be found in Chapters 1 and 9 of this book.

Transport

S-nitrosothiols may both enter cells by passive diffusion and be metabolized at the cell surface to facilitate entry into the cell by trans-S-nitrosation, likely in the membrane micro-environment. The stereoselective effects of *S*-nitrosothiols [17,18] suggest that their bioactivity is mediated by interaction with cell-surface receptors. We have shown that cell-surface protein disulfide isomerase (PDI) catalyzes the trans-S-nitrosative decomposition of extracellular low-molecular-weight *S*-nitrosothiols [19,20], which appear to undergo denitrosation with subsequent reaction with molecular oxygen in the cell membrane to form the nitrosating species, nitrous anhydride (N_2O_3), subsequently leading to S-nitrosation of intracellular thiols [20].

Recent reports show that the cell membrane anion transporter can import *S*-nitrosocysteine (CysNO), but not *S*-nitrosoglutathione (GSNO), into the cytosol. Cysteine facilitates the ability of *S*-nitrosothiols like GSNO or *S*-nitroso-*N*-acetyl-penicillamine to increase intracellular *S*-nitrosothiols [21]. Intracellular *S*-nitrosothiol formation depends on trans-S-nitrosation between GSNO and L-cysteine to form L-CysNO, with uptake of L-CysNO *via* the amino acid transport system L, a process inhibited by L-leucine but not D-leucine [22–24].

Biological formation

The cellular metabolism of *S*-nitrosothiols is complex and has not yet been fully elucidated. In general, the levels of *S*-nitrosothiols formed *in vivo* are determined by the NO flux, i.e., activity of NOS and availability of substrate; rate of formation of the *S*-nitrosating species; and competition of different thiols and other reactive motifs for reaction with the nitrosating species. We found that *S*-nitrosated proteins can form within endothelial cells from an exogenous *S*-nitrosothiol donor or from endogenous production of NO by endothelial NO synthase activity [25]. In comparison to CysNO, NO is a relatively poor nitrosating agent in cultured cells [22,26,27]. It is important to point out that the reaction of nitric oxide with thiols under anaerobic conditions does not yield an *S*-nitrosothiol, but disulfide and nitroxyl anion. Nitrosating intermediates are formed in the course of aerobic oxidation of NO. It is commonly assumed that nitrosation of thiols by NO in the presence of oxygen involves electrophilic substitution by N_2O_3. In contrast to NO auto-oxidation, the reaction between NO and superoxide is first order in both reactants, occurs at near diffusion limits [28–30], and greatly enhances protein S-nitrosation [31]. Nitrosation is maximal when the molar ratio of superoxide to NO is near-equivalent [32].

By confocal microscopy and *in situ* *S*-nitrosothiol fluorescence labeling, we found that *S*-nitrosoproteins exist mainly in the mitochondria, which is the primary superoxide source in resting cells [25]. PseudoRho$_0$ cells, which are essentially devoid of mitochondria, demonstrate much less *S*-nitrosoprotein formation than do cells with a normal complement of mitochondria. Cellular protein *S*-nitrosothiol formation is also diminished when cells are treated with electron transport inhibitors to decrease superoxide generation. These studies suggest that superoxide anion reacts with endogenous NO or exogenous NO to form the nitrosating species peroxynitrite or N_2O_3. In addition, physiological concentrations of CO_2 modestly enhance levels of NO_2 formation *via* formation of nitrosoperoxocarbonate from peroxynitrite [32], thereby affecting nitrosation potential.

Other possible mechanisms for *S*-nitrosothiols formation include the role of metal ions [33] and dinitrosyl iron complexes [34]. Protein thiols may donate one electron to transition metals like copper and become thiyl radicals in the process, which then react with nitric oxide to form *S*-nitrosothiols [34–37]. The S-nitrosating agent NO^+ may also form in the presence of transition metals [38]. Several proteins containing transition metals also catalyze the formation of *S*-nitrosothiols. Romeo suggested that superoxide dismutase targets NO from GSNO to Cysβ93 of oxyhemoglobin [39], and ceruloplasmin showed potent RSNO-forming activity in cell culture [40]. *S*-nitrosothiols may also be formed from nitrite anion through heme protein-mediated nitrite reductase activity [41–43]. Such enzymatic nitrite reductase pathways are discussed in Chapters 14 and 15 of this volume. Xanthine oxidase also catalyzes anaerobic transformation of organic nitrates to nitric oxide and *S*-nitrosothiols [44].

S-nitrosothiol formation is also controlled by the competition between GSH and different proteins. Decreased intracellular GSH concentration markedly enhances protein *S*-nitrosothiol formation [25], while overexpression of metallothionein, which also exists in a very reduced apo-form in the cell [45], protects proteins from S-nitrosation [46].

Metabolism of *S*-nitrosothiols

Cleavage of the S—N bond in *S*-nitrosothiols can occur homolytically or heterolytically, thereby yielding NO, NO^+, or NO^-, depending on redox conditions. Homolytic cleavage of the S—NO bond photochemically was first shown in 1966 [47]. This photolytic cleavage is likely responsible, at least in part, for the action spectrum of vascular tissue, and supports the importance of *S*-nitrosothiol formation *in vivo*. Furchgott and colleagues first showed that light can induce relaxation of blood vessels [48], which is a consequence of release of nitric oxide from a photolabile store [49] and requires adequate glutathione stores [50–52]. Homolytic cleavage of the S—NO bond by thermal decomposition has also been reported [53]. Recent data, however, suggest that this mechanism of decomposition is not relevant under physiological conditions [54]. In plasma, NO release from S–NO–albumin and S–NO–glutathione may be regulated by heterolytic NO^+ transfer and reductive activation to NO, rather than by homolytic decomposition of labile *S*-nitrosothiols [55].

GSNO is the model compound used for most studies of *S*-nitrosothiol metabolism. GSNO is fairly stable under physiological conditions in plasma and buffer solutions, while other low-molecular-weight *S*-nitrosothiols have tissue half-lives ranging from seconds to minutes, much shorter than protein *S*-nitrosothiols [25]. More information on the relevant reaction pathways and kinetic constants is given in Chapter 9. The ability of the tissue homogenate supernatants to facilitate decomposition of GSNO remains unchanged after removal of proteins and other high-molecular-weight compounds [56], suggesting that the decomposition rate of GSNO is determined by low-molecular-weight compounds; other studies suggest that enzymatic catalysis is involved in GSNO metabolism [14,19,20,57–59].

Superoxide anion [60] and transition metal ions [33,38,61–63] were also found to accelerate greatly *S*-nitrosothiol decomposition *in vitro*; interestingly, these factors also facilitate its formation. Reduced metal ions (e.g., Cu^{+}) decompose *S*-nitrosothiols more rapidly than oxidized metal ions (e.g., Cu^{2+}), indicating that reducing agents such as glutathione can stimulate decomposition of *S*-nitrosothiol by chemical reduction of contaminating transition metal ions [64,65]. CuZn-SOD, but not Mn-SOD, catalyzes the decomposition of GSNO and the formation of NO in the presence of GSH at concentrations present in extracellular fluids [66].

Ascorbate exists in cellular systems at relatively high concentration, and is a very important cellular antioxidant. Decomposition rates of *S*-nitrosothiols in the presence of ascorbate [67] increase drastically with increasing pH, signifying that the most highly ionized form of ascorbate is the more reactive species [68]. The reaction is also accelerated in the presence of trace redox-active copper [69]. Ascorbate has been shown to block S-nitrosation of proteins [25,31,70–72], and in plasma and tissue, ascorbate is one of the critical determinants of the overall *S*-nitrosothiol pool [55,73]. The vasodilator activity of GSNO is significantly enhanced by ascorbic acid [69]. Ascorbate has also been shown to be necessary for transformation of organic nitrates to nitric oxide by xanthine oxidase, where it releases NO from *S*-nitrosothiol intermediates [44].

Despite these findings, however, there are only a few studies on the cellular metabolism of these compounds. Cells consume GSNO at a much greater rate than the spontaneous GSNO decomposition rate, which depends on protein thiols [74,75]. Gordge suggested that this active decomposition of GSNO is mediated by "GSNO lyase" [59]. One candidate enzyme for this activity is γ-glutamyltranspeptidase, which accelerates the decomposition of GSNO by hydrolyzing the glutamyl moiety to yield *S*-nitrosocysteinylglycine (CG-SNO). The latter is susceptible to transition metal ion-dependent decomposition and release of NO [75]. A glutathione-dependent formaldehyde dehydrogenase has recently been identified as a major protein responsible for GSNO-metabolism [58], *viz.*, the *S*-nitrosoglutathione reductase activity [76], in eukaryotes. This protein is conserved across phyla [77], and was renamed GSNO reductase (GSNOR). GSNOR has been shown to protect yeast cells against nitrosative stress both *in vitro* [58] and *in vivo* [57]. In a GSNOR-deficient mouse, *S*-nitrosothiols were markedly increased, and the liver, immune system, and cardiovascular system manifested S-nitrosative stress [14] under these conditions. The effect of GSH, the substrate for GSNOR, in the metabolism of *S*-nitrosothiols is still debatable. Xu and colleagues suggested that GSH shortens GSNO half-life [69], while other studies show that buthionine sulfoximine treatment, which decreases intracellular GSH levels, does not affect GSNO decay [59,74].

It is not yet known whether there is any enzyme that can metabolize a protein *S*-nitrosothiol directly. Thioredoxin or thioredoxin reductase may be one candidate, which has been shown to restore S-nitrosation-induced inhibition of protein kinase C activity in pulmonary endothelial cells [78].

Physiological distribution of *S*-nitrosothiols

Protein *S*-nitrosothiols and low-molecular-weight *S*-nitrosothiol compounds like GSNO comprise two pools of intracellular *S*-nitrosothiols. Protein *S*-nitrosothiols tend to be more stable than low-molecular-weight *S*-nitrosothiols. GSNOR-deficient yeast or mice have higher protein *S*-nitrosothiol content, suggesting that GSNO exists in equilibrium with SNO-proteins [14,58]. The exchange of *S*-nitrosothiols between proteins and GSH occurs through trans-S-nitrosation in which a thiolate anion nucleophilically attacks the nitrogen atom of an *S*-nitrosothiol, resulting in the transfer of the nitroso group to the nucleophilic thiol, providing the exchange of *S*-nitrosothiols between proteins and GSH, the most abundant intracellular free thiol.

Depending on the localization and cell type, *S*-nitrosothiols have different distributions between these two pools. GSNO can accrue in extracellular fluids, e.g., airway lining fluid [9], while albumin represents a principal *S*-nitrosoprotein in that same compartment [8,79,80]. Mallis and Thomas showed that over half of the cellular glutathione pool can be S-nitrosated in immortalized fibroblasts [21], while in other cell types, 95% of *S*-nitrosothiols are *S*-nitrosoproteins [27] or cannot be detected readily [58,81]. This difference could be explained by the fact that GSNO is more rapidly metabolized in some cells than others.

By confocal microscopy and *in situ* *S*-nitrosothiol fluorescence labeling, we found that *S*-nitrosoproteins exist mainly in the mitochondria and peri-mitochondrial compartment [25], which is a consequence of localization of the nitrosative reagent to the mitochondria and of the greater oxidative potential of that compartment compared with the cytosol. Mannick and colleagues observed that S-nitrosation of caspases is higher in mitochondria than in cytosol [82], supporting our observations. Another report suggests that S-nitrosation also occurs in the nucleus [83], although this observation has not yet been confirmed. Plasma membrane-bound eNOS is more easily S-nitrosated than that in cytosolic fractions, and VEGF-stimulated translocation of eNOS from the plasma membrane to the cytosol induces rapid denitrosation of *S*-nitrosothiols [84,85].

The pattern of *S*-nitrosothiol distribution in tissue has also been studied. During glyceryl trinitrate (GTN) biotransformation *in vivo*, the erythrocyte appears to be the predominant site of GTN-induced S-nitrosation. In plasma, nitrosation mainly reflects RSNO formation, whereas in the aorta, only 30% of the nitrosated pool consists of RSNOs, the remaining 70% being RNNOs [86].

DETECTION

More than ten years after our initial report, there appears to be no consensus about the true physiological levels of *S*-nitrosothiols in human plasma, which range from 10 μM [8,87] to 10 nM [88,89]. The differences have their origin in the diverse methodological approaches used to detect them [90]. Challenges about the physiological importance of *S*-nitrosothiols [91,92] arise from the difficulties and technical limitations of detecting them [93]. In particular, *S*-nitrosothiols formed under physiological conditions by either endogenous NO production or exogenous NO donors has remained an analytical morass, largely owing to the variable stability of the protein —*S*—N bond. *S*-nitrosothiols can form from thiols in the presence of nitrite at pH less than 7.4. In the presence of ascorbate, glutathione, and transition metals in biological fluids or buffer, *S*-nitrosothiols decompose rapidly and trans-S-nitrosation between protein *S*-nitrosothiols and low-molecular-weight thiols can occur. Endogenous *S*-nitrosothiols in biological fluid and tissue have not been detected directly in intact form, but only after conversion to stable derivatives [93]. An ideal assay for *S*-nitrosothiols would be rapid, specific, and effective *in situ*, avoiding transfer or loss of the *S*-nitrosothiol signal in the process.

Photolysis-chemiluminescence

Photolytic cleavage of *S*-nitrosothiols generates nitric oxide, which then can be detected by the chemiluminescence of its reaction product with ozone or directly electrochemically by a nitric oxide-selective electrode. Alternatively, NO can be released from *S*-nitrosothiols in the presence of copper and cysteine [94,95] or acidic triiodide [96,97]. As low as 10^{-13} M can be detected by chemiluminescence methods, compared to 10^{-9} M electrochemically. Owing to the sensitivity, photolysis-chemiluminescence has been a dominant method for studying *S*-nitrosothiols for many years. Unfortunately, the photolysis-chemiluminescence technique used in these studies is too complex to facilitate widespread application. In addition, the method has limited specificity owing to interference by nitrite, iron-nitrosyl complexes, and *O*-nitroso and *N*-nitroso compounds [23,73,96]. Another disadvantage of the technique is that it measures only the overall concentration of *S*-nitrosothiols without a preceeding separation step.

Saville–Griess method

In the presence of mercury chloride, nitrosonium ion can be released from *S*-nitrosothiols, which reacts with sulfanilamide to form diazonium ion [98]. The resulting diazonium then reacts with aminonaphthalene to form a colored azo-complex. Alternatively, 2,3-diaminoaphthalene is used to form a highly fluorescent product with diazonium to measure indirectly NO derived from *S*-nitrosothiols with a resulting high degree of sensitivity [98].

Diaminofluorescein gels

The Saville and chemiluminescence methods described above cannot distinguish between *S*-nitrosothiol signals that derive from proteins and those that derive from low-molecular-weight species. King and colleagues developed a new gel-based system to detect S-nitrosated proteins [99]. In this method, proteins are separated by non-reducing SDS-PAGE, after which NO is photolytically released from S-nitrosated proteins by UV light. The released NO is detected using the NO-sensitive fluorescent probes 4,5-diaminofluorescein and 3-amino, 4-aminomethyl-2′7′-difluorescein. This assay, however, is not sufficiently sensitive to detect endogenous *S*-nitrosoproteins.

Mass spectrometry

Occasional reports have appeared in which mass spectrometry is directly used to detect *S*-nitrosothiols by measuring the mass changes after S-nitrosation, beginning in 1999 [100].

Zech and colleagues, for example, detected poly-S-nitrosation of caspase 3 with approximately two molecules of NO bound per enzyme by mass spectrometry [101].

Immunohistochemistry

A polyclonal antibody raised against an S-nitrosated glutaraldehyde conjugate of BSA and cysteine has been used in immunoblotting and immunochemistry [102]; however, the lack of specificity of the antibody limits its application.

Chemical labeling

Jaffrey and colleagues recently published a sensitive chemical method in which thiols are first covalently blocked, after which *S*-nitrosothiols are gently reduced to thiols with ascorbate that react with *N*-[6-(biotinamido)hexyl]-3-(2-pyridyldithio)-propionamide, a biotinylating reagent specific for sulfydryl groups [103]. As straightforward as this method appears, its success depends critically on the extent of blockade of protein thiols in the first step. Inefficiency in this reaction could lead to false-positive labeling of proteins, and, thus, inappropriate identification of protein targets for S-nitrosation. We found that increasing the concentration of the thiol blocking (alkylating) reagent effectively decreased false-positive background detection in endothelial cells. Another potential problem for this method is the low conversion rate of *S*-nitrosothiols to free thiols by ascorbate. We found that ascorbate is a very inefficient reducing agent [25,104]; therefore, very high concentrations of ascorbate should be used to facilitate this reduction. To minimize any background labeling, we use Texas Red-2-sulfonamidoethyl methanethiosulfonate (MTSEA Texas red) fluorescent label, which has the same reaction chemistry as the thiol blocker, methyl methanethiosulfonate (MMTS). By optimizing this methodology, we have acquired the first image of intracellular protein *S*-nitrosothiols [25], and showed that S-nitrosated proteins can form within endothelial cells from an exogenous *S*-nitrosothiol donor or from endogenous production of NO by endothelial NO synthase. Both MMTS and MTSEA Texas Red are electrochemically neutral, thereby improving the selectivity and efficiency of labeling the *S*-nitrosothiol functionalities.

PROTEOMICS

A few studies have been published using proteomic approaches that involve chemical labeling of *S*-nitrosated proteins, affinity isolation, one or two-dimensional gel electrophoresis, and mass spectrometry. These studies have consistently shown that a limited number of *S*-nitrosoproteins exist in cells treated with NO-donors, suggesting a selectivity of S-nitrosation [25,103–105]. In these reports, peroxiredoxin [25,105], actin [25,103,105], and GAPDH [25,103,105] were identified as *S*-nitrosoproteins. The function of peroxiredoxin relies on an active cysteine that is critical for the decomposition of peroxide [106].

Both GAPDH and actin are known to contain reactive cysteinyl residues [107]. Under physiological conditions, S-nitrosation of one or a few cysteinyl residues changes the function of OxyR [108], L-type calcium channels [109], NSF ATPase [110], and caspase [111]. These findings suggest that there is specificity to S-nitrosation of the cysteinyl functionality that depends on its pK. Typically ~8.4 in aqueous environments, the pK of cysteinyl sulfhydryl functionality can become anomalously low within the constraints of the tertiary structure of a protein and proximity to charged side-chains, especially histidine and aspartate, leading to thiolate anion formation. For example, the pK of serum albumin's single cysteinyl group (cys34) is ~4.5 [112], rendering it susceptible to electrophilic attack by the partially positively charged nitric oxide moiety of a low-molecular-weight *S*-nitrosothiol such as *S*-nitrosoglutathione [79,80,113]. These S-nitrosation-specific thiols may also change by protein allostery [114,115], controlled by binding of metal ions or other cofactors.

Rhee and colleagues studied the *S*-nitroso-proteome in *Mycobacterium tuberculosis* [116], which is a target for antimicrobial endogenous reactive nitrogen species. Twenty-nine *S*-nitroso proteins were identified from *Mycobacterium tuberculosis* that infected bone marrow-derived primary mouse macrophages, and all are enzymes important for invasion and replication of the bacterium.

FUNCTION OF *S*-NITROSOTHIOLS

The role of *S*-nitrosothiols in cell signaling was initially viewed principally as important for targeting activation of guanylyl cylase *via* donated NO, first demonstrated in 1980 [4]. In plasma, serum albumin serves as a carrier and buffer function for the transport of nitric oxide [8,79,80]. GSNO, which is a naturally occurring *S*-nitrosothiol [9], has been implicated in responses ranging from airway dilation [9] and central breathing stimulation [117] to regulation of the cystic fibrosis transmembrane regulator [118] and host defense [57]. None of the commercially available *S*-nitrosothiol compounds has as yet been used therapeutically in animals or humans, however, mainly owing to their unpredictable rate of decomposition [119].

During recent years, S-nitrosation has become recognized as one of the major temporal and spatially regulated post-translational protein modifications in cell signaling. The total list of *S*-nitrosoproteins that have been identified to date is extensive (see Ref. [120] for review). Most of these proteins have been identified in targeted studies involving *in vitro* S-nitrosation of purified candidate proteins or by analysis of NO release from immunoprecipitated proteins. The latter method may allow for detection of *S*-nitroso-adducts in proteins that are S-nitrosated at low levels, compared to the proteomic methods; however, it requires pre-selection of candidate proteins. Importantly, the existence of S-nitrosation in a protein does not necessarily indicate associated functional modulation, and one cannot exclude *post-hoc* (i.e., following cell lysis) formation of *S*-nitrosoproteins identified by conventional methodologies. To elucidate the regulatory function of *S*-nitrosothiols from that of disulfides or sulfenic acid, one can utilize sensitivity to ascorbate and dithiothreitol [121]. As *S*-nitrosothiols undergo photolytic cleavage, light treatment is another method to validate participation of *S*-nitrosothiols in cell signaling [122].

Energy metabolism

Prolonged exposure to NO or GSNO results in a gradual and persistent inhibition of complex I and mitochondrial respiration, which is reversible by high intensity light and reduced glutathione, indicating involvement by an S-nitrosation mechanism [122–124]. This implication is further confirmed directly by analysis of isolated complex I from hearts subjected to ischemic preconditioning [125]. The cytochrome-c oxidase, complex IV, is well known to be reversibly inhibited by NO, which is competitive with oxygen [126], and a recent study shows that it is also S-nitrosated and, as a result, inhibited [127].

One group has proposed that S-nitrosation of the β chain of hemoglobin occurs *in vivo* and serves to modulate allosterically oxygen release from the heme group [128]. Recent data, however, refute this hypothesis, showing that *S*-nitroso-hemoglobin is unstable in the reductive environment of erythrocytic cytosol and lacks allosteric function [129].

Protein degradation and apoptosis

Ubiquitination and proteasomal degradation represent the major protein degradation pathway in the cell. Two classes of ubiquitin enzymes, E2 ubiquitin-conjugating enzyme UbcH7 [25] and the E3 ubiquitin ligase protein parkin [15,130] or HDM3 [131], were found to be S-nitrosated. S-nitrosation of the target protein may also change its sensitivity to ubiquitination and proteasomal degradation [132]. By contrast, S-nitrosation of iron regulatory protein 2 promotes its ubiquitination and proteasomal degradation [133]. Interestingly, a recent study shows that S-nitrosation of GAPDH augments its binding and stability, as well as promotes nuclear translocation of the ubiquitin ligase Siah1, facilitating its degradation of nuclear proteins [134]. Recently, nitric oxide has been shown to increase HIF-1 and p53 stability by preventing their proteasomal degradation [135–138].

Beginning in 1997, several studies showed that one mechanism by which NO exerts its antiapoptotic effect is by S-nitrosation of the reactive cysteine of caspases [139–141]. It has also been shown that most of the caspases can be S-nitrosated, with the consequent inhibition of their enzymatic activity [82]. The inhibition of S-nitrosated caspases could be reversed by denitrosation, which is induced by two activators of apoptosis, Fas ligand and TNFα [142,143]. In another study, thioredoxin was shown to catalyze S-nitrosation of the caspase 3 active site cysteine [144]. S-nitrosation may also inhibit the caspase-recruitment domain (CARD) interactions between Apaf-1 and procaspase, preventing apoptosome formation and the sequential caspase cascade [145].

Transcription factors

S-nitrosation of critical cysteines essential for DNA binding, transformation, and transcriptional transactivation usually inhibit functions of transcription factors, e.g., c-Myb [146], AP-1 [147], and heterogeneous nuclear ribonucleoprotein [148]. Many transcription factors contain cysteine-coordinated zinc-finger domains, which are prone to S-nitrosation. The zinc-finger domain of yeast transcription activator LAC9 [149] and the estrogen receptor [150] are disrupted by nitric oxide. When S-nitrosated, zinc release and DNA binding ability of the transcription factors are inhibited. Other zinc-finger type transcription factors, glucocorticoid receptor [151], and Yin Yang 1 [152] are also inhibited by S-nitrosation. The zinc–sulfur moieties of many other proteins, including NOS, alcohol dehydrogenase, and metallothionein, can also be S-nitrosated and bound zinc released [84,85,153–156]. SoxR has iron–sulfur centers, which are also disrupted by nitric oxide; however, protein-bound dinitrosyl-iron-dithiol adducts form instead of *S*-nitrosothiols [157] as a result.

The effect of S-nitrosation on NF-κβ activation is more complicated. NF-κβ complexes with I-κβ protein and remains in an inactive state until I-κβ is phosphorylated by I-κβ kinase, ubiquitinated, and degraded by the 26S proteasome. Many studies show that S-nitrosation of NF-κβ with endogenous NO inhibits DNA binding, promoter activity, and gene transcription [158–163]. By contrast, S-nitrosation of IKKβ [164] leads to phosphorylation of I-κβ, stabilizes the complex, and inhibits NF-κβ [165,166]. Additionally, S-nitrosation and consequent activation of p21ras has been shown to be involved in NF-κβ activation [167].

HIF1α function is controlled by O_2-regulated hydroxylation, rapid ubiquitination, and proteolysis by the proteasome. The HIF1α protein accumulates under normoxic conditions upon NOS induction or treatment with NO donors [135,136], which could be explained by the fact that the activity of prolyl hydroxylase activity is inhibited by GSNO [168]. Additionally, exposure to GSNO or induction of iNOS results in the S-nitrosation of Cys800 of HIF1α, enhancing its binding to its transcriptional coactivator CREB and increasing its transcriptional activity [31,169].

Importantly, S-nitrosation of transcription factors also control the expression of different forms of NOS [159,170–174] and other proteins involved in S-nitrosothiol metabolism. In *E. coli*, OxyR is S-nitrosated and activated when cells are exposed to *S*-nitrosothiols [108,175], leading to nitrosative stress adaptation. Deletion of OxyR renders cells hypersensitive to *S*-nitrosothiols.

Membrane ion channels

Many studies show that ion channel activities are directly modulated by NO *via* an unknown, cyclic GMP-independent molecular mechanism. The neuronal NMDA receptor (NMDAR) is critical for development, learning, and memory in the central nervous system by mediating postsynaptic calcium influx and activation of neuronal NOS. As a negative feedback loop, S-nitrosation of a critical cysteine residue (Cys399) under physiological conditions inhibits its function [176,177], and attenuates overactivation of the NMDAR with consequent neurotoxicity [178,179]. In cell systems expressing NMDARs with mutant NR2A subunits in which this single cysteine is replaced by an alanine, the effect of endogenous NO is abrogated [177]. Nitric oxide also blocks the fast, slow, and persistent Na^+ channels in C-type DRG neurons through S-nitrosation [180].

There are also many ion channels, including cyclic nucleotide-gated (CNG) channels [103,181], ATP-sensitive potassium channels [182], and Ca^{2+}-activated K^+ (BKCa) channels [183], that are activated by S-nitrosation. Ryanodine receptors (RyR) mediate Ca^{2+} release from intracellular stores and are essential for muscle contraction. RyR is found to bind eNOS [184], and activation of RyR by S-nitrosation [185] occurs by different mechanisms. NO derived from endogenous eNOS nitrosates only Cys3635 [186], which is also sensitive to an oxygen-induced allosteric effect [114,187], while GSNO results in S-nitrosation or S-glutathiolation of four other thiols [187,188]. An additional set of thiols may be oxidized in response to cellular oxidant stress [189,190]. The distinction among these sets of reactive thiols indicates the complexity of the regulation of RyR activity.

Kinases and phosphatase

S-nitrosation of kinases usually suppresses their activity. S-nitrosation of a single thiol inhibits JNK-mediated phosphorylation and activation of Jun [191,192], and IKKβ-mediated TNFα-induced IKK activity [164]. The tyrosine kinase activity of EGFR and protein kinase C [78] is also inhibited by S-nitrosation [193]. One study showed that S-nitrosation of a single cysteinyl residue inhibits binding of ASK1 to its principal downstream substrates, MKK3 and MKK6, and, thus, inhibits apoptosis. Other kinases, MKK3, MKK6, or p38, are not S-nitrosated [194]. Another study showed that S-nitrosation of thioredoxin induces dissociation and activation of apoptosis signal-regulating kinase-1 (ASK1) [195]. Yet another example of changes in protein function by S-nitrosation is c-Src kinase, which is activated by S-nitrosation [196].

Protein tyrosine phosphatases (PTPase) contain reactive cysteines in their active sites, which are prone to undergo oxidation. S-nitrosation of the cysteine by exposure of the intact cell to an exogenous nitric oxide donor or induction of iNOS inhibits PTPase activity and results in increased levels of phosphotyrosyl-containing proteins [196–199].

CONCLUSIONS

S-nitrosothiols as a chemical class have been known for more than one hundred years, but their significance in biology has only been recognized over the past two decades. In the context of nitric oxide biology, these nitrosated thiol derivatives serve two functions: a buffer pool of bioactive nitric oxide which is more resistant to oxidative inactivation than nitric oxide itself, and, more importantly, a unique mechanism of cell signaling through reversible post-translational modification of protein thiol targets. The development of the biotin switch method has allowed for an indirect, highly specific method for detection of S-nitrosated proteins, which can now be visualized by microscopy or identified through proteomic methods using mass spectrometry. S-nitrosation is rapidly gaining recognition as a major form of protein post-translational modification, and dysregulation of this biochemical process is associated with various diseases. The total list of *S*-nitrosoproteins that have been identified to date is extensive, including receptors, enzymes, and transcription factors; however, authenticity and the functional consequences of *S*-nitrosothiols in these proteins should be rigorously examined. The structural and biochemical determinants of cysteinyl functionalities that render them susceptible to S-nitrosation *vs.* thiolation remain to be defined, and enzyme catalysts that govern protein *S*-nitrosothiol formation, cellular transport, and catabolism remain to be identified. Clearly, this field offers extraordinary potential for understanding the complex biological chemistry and molecular biology of nitric oxide, low-molecular-weight thiols, and the thiol proteome.

ACKNOWLEDGMENTS

This work was supported by NIH Grants HL58976, HL55993, HL61828, HV28178, and HL81587. The authors wish to thank Stephanie Tribuna for expert technical assistance.

REFERENCES

1 Tasker HS, Jones HO. Action of mercaptans on acid chlorides. I. Oxalychloride; the mono- and dithioxalates. J. Chem. Soc. 1909; 95: 1904–1909.
2 Ignarro LJ, Byrns RE, Buga GM, Wood KS. Endothelium-derived relaxing factor from pulmonary artery and vein possesses pharmacologic and chemical properties identical to those of nitric oxide radical. Circ. Res. 1987; 61: 866–879.
3 Furchgott RF, Zawadzki JV. The obligatory role of endothelial cells in the relaxation of arterial smooth muscle by acetylcholine. Nature 1980; 288: 373–376.
4 Ignarro LJ, Edwards JC, Gruetter DY, Barry BK, Gruetter CA. Possible involvement of S-nitrosothiols in the activation of guanylate cyclase by nitroso compounds. FEBS Lett. 1980; 110: 275–278.
5 Ignarro LJ, Gruetter CA. Requirement of thiols for activation of coronary arterial guanylate cyclase by glyceryl trinitrate and sodium nitrite: possible involvement of S-nitrosothiols. Biochim. Biophys. Acta 1980; 631: 221–231.
6 Stamler J, Mendelsohn ME, Amarante P Smick D, Andon N, Davies PF, Cooke JP, Loscalzo J. N-acetylcysteine potentiates platelet inhibition by endothelium-derived relaxing factor. Circ. Res. 1989; 65: 789–795.
7 Cooke JP, Stamler J, Andon N, Davies PF, McKinley G, Loscalzo J. Flow stimulates endothelial cells to release a nitrovasodilator that is potentiated by reduced thiol. Am. J. Physiol. 1990; 259: H804–H812.
8 Stamler JS, Jaraki O, Osborne J, Simon DI, Keaney J, Vita J, Singel D, Valeri CR, Loscalzo J. Nitric oxide circulates in mammalian plasma primarily as an S-nitroso adduct of serum albumin. Proc. Natl. Acad. Sci. USA 1992; 89: 7674–7677.
9 Gaston B, Reilly J, Drazen JM, Fackler J, Ramdev P, Arnelle D, Mullins ME, Sugarbaker DJ, Chee C, Singel DJ, et al. Endogenous nitrogen oxides and bronchodilator S-nitrosothiols in human airways. Proc. Natl. Acad. Sci. USA 1993; 90: 10957–10961.
10 Datta B, Tufnell-Barrett T, Bleasdale RA, Jones CJ, Beeton I, Paul V, Frenneaux M, James P. Red blood cell nitric oxide as an endocrine vasoregulator: a potential role in congestive heart failure. Circulation 2004; 109: 1339–1342.
11 Massy ZA, Borderie D, Nguyen-Khoa T, Drueke TB, Ekindjian OG, Lacour B. Increased plasma S-nitrosothiol levels in chronic haemodialysis patients. Nephrol. Dial. Transplant. 2003; 18: 153–157.
12 Massy ZA, Fumeron C, Borderie D, Tuppin P, Nguyen-Khoa T, Benoit MO, Jacquot C, Buisson C, Drueke TB, Ekindjian OG, Lacour B, Iliou MC. Increased pasma S-nitrosothiol concentrations predict cardiovascular outcomes among patients with end-stage renal disease: a prospective study. J. Am. Soc. Nephrol. 2004; 15: 470–476.
13 James PE, Lang D, Tufnell-Barret T, Milsom AB, Frenneaux MP. Vasorelaxation by red blood cells and impairment in diabetes: reduced nitric oxide and oxygen delivery by glycated hemoglobin. Circ. Res. 2004; 94: 976–983.
14 Liu L, Yan Y, Zeng M, Zhang J, Hanes MA, Ahearn G, McMahon TJ, Dickfeld T, Marshall HE, Que LG, Stamler JS. Essential roles of S-nitrosothiols in vascular homeostasis and endotoxic shock. Cell 2004; 116: 617–628.
15 Chung KK, Thomas B, Li X, Pletnikova O, Troncoso JC, Marsh L, Dawson VL, Dawson TM. S-nitrosylation of parkin regulates ubiquitination and compromises parkin's protective function. Science 2004; 304: 1328–1331.
16 Oae S, Kim YH, Fukushima D, Shinhama K. New syntheses of thionitrites and their chemical reactivities. J. Chem. Soc. Perkin. Trans. 1978; 1: 913–917.
17 Travis MD, Davisson RL, Bates JN, Lewis SJ. Hemodynamic effects of L- and D-S-nitroso-beta,beta-dimethylcysteine in rats. Am. J. Physiol. 1997; 273: H1493–1501.

18 Davisson RL, Travis MD, Bates JN, Lewis SJ. Hemodynamic effects of L- and D-S-nitrosocysteine in the rat. Stereoselective S-nitrosothiol recognition sites. Circ. Res. 1996; 79: 256–262.

19 Zai A, Rudd MA, Scribner AW, Loscalzo J. Cell-surface protein disulfide isomerase catalyzes transnitrosation and regulates intracellular transfer of nitric oxide. J. Clin. Invest. 1999; 103: 393–399.

20 Ramachandran N, Root P, Jiang XM, Hogg PJ, Mutus B. Mechanism of transfer of NO from extracellular S-nitrosothiols into the cytosol by cell-surface protein disulfide isomerase. Proc. Natl. Acad. Sci. USA 2001; 98: 9539–9544.

21 Mallis RJ, Thomas JA. Effect of S-nitrosothiols on cellular glutathione and reactive protein sulfhydryls. Arch. Biochem. Biophys. 2000; 383: 60–69.

22 Zhang Y, Hogg N. The mechanism of transmembrane S-nitrosothiol transport. Proc. Natl. Acad. Sci. USA 2004; 101: 7891–7896.

23 Zhang Y, Hogg N. S-Nitrosothiols: cellular formation and transport. Free Radic. Biol. Med. 2005; 38: 831–838.

24 Satoh S, Kimura T, Toda M, Maekawa M, Ono S, Narita H, Miyazaki H, Murayama T, Nomura Y. Involvement of L-type-like amino acid transporters in S-nitrosocysteine-stimulated noradrenaline release in the rat hippocampus. J. Neurochem. 1997; 69: 2197–2205.

25 Yang Y, Loscalzo J. S-nitrosoprotein formation and localization in endothelial cells. Proc. Natl. Acad. Sci. USA 2005; 102: 117–122.

26 Ramachandran A, Ceaser E, Darley-Usmar VM. Chronic exposure to nitric oxide alters the free iron pool in endothelial cells: role of mitochondrial respiratory complexes and heat shock proteins. Proc. Natl. Acad. Sci. USA 2004; 101: 384–389.

27 Zhang Y, Hogg N. Formation and stability of S-nitrosothiols in RAW 264.7 cells. Am. J. Physiol. Lung Cell Mol. Physiol. 2004; 287: L467–L474.

28 Huie RE, Padmaja S. The reaction of NO with superoxide. Free Radic. Res. Commun. 1993; 18: 195–199.

29 Kissner R, Nauser T, Bugnon P, Lye PG, Koppenol WH. Formation and properties of peroxynitrite as studied by laser flash photolysis, high-pressure stopped-flow technique, and pulse radiolysis. Chem. Res. Toxicol. 1997; 10: 1285–1292.

30 Koppenol WH, Moreno JJ, Pryor WA, Ischiropoulos H, Beckman JS. Peroxynitrite, a cloaked oxidant formed by nitric oxide and superoxide. Chem. Res. Toxicol. 1992; 5: 834–842.

31 Sumbayev VV, Budde A, Zhou J, Brune B. HIF-1 alpha protein as a target for S-nitrosation. FEBS Lett. 2003; 535: 106–112.

32 Espey MG, Thomas DD, Miranda KM, Wink DA. Focusing of nitric oxide mediated nitrosation and oxidative nitrosylation as a consequence of reaction with superoxide. Proc. Natl. Acad. Sci. USA 2002; 99: 11127–11132.

33 Vanin AF, Malenkova IV, Serezhenkov VA. Iron catalyzes both decomposition and synthesis of S-nitrosothiols: optical and electron paramagnetic resonance studies. Nitric Oxide 1997; 1: 191–203.

34 Boese M, Mordvintcev PI, Vanin AF, Busse R, Mulsch A. S-nitrosation of serum albumin by dinitrosyl-iron complex. J. Biol. Chem. 1995; 270: 29244–29249.

35 Kharitonov VG, Sundquist AR, Sharma VS. Kinetics of nitrosation of thiols by nitric oxide in the presence of oxygen. J. Biol. Chem. 1995; 270: 28158–28164.

36 Tao L, English AM. Mechanism of S-nitrosation of recombinant human brain calbindin D28K. Biochemistry 2003; 42: 3326–3334.

37 Stamler JS, Singel DJ, Loscalzo J. Biochemistry of nitric oxide and its redox-activated forms. Science 1992; 258: 1898–1902.

38 Stubauer G, Giuffre A, Sarti P. Mechanism of S-nitrosothiol formation and degradation mediated by copper ions. J. Biol. Chem. 1999; 274: 28128–28133.

39 Romeo AA, Capobianco JA, English AM. Superoxide dismutase targets NO from GSNO to Cysbeta93 of oxyhemoglobin in concentrated but not dilute solutions of the protein. J. Am. Chem. Soc. 2003; 125: 14370–14378.

40 Inoue K, Akaike T, Miyamoto Y, Okamoto T, Sawa T, Otagiri M, Suzuki S, Yoshimura T, Maeda H. Nitrosothiol formation catalyzed by ceruloplasmin. Implication for cytoprotective mechanism in vivo. J. Biol. Chem. 1999; 274: 27069–27075.

41 Cosby K, Partovi KS, Crawford JH, Patel RP, Reiter CD, Martyr S, Yang BK, Waclawiw MA, Zalos G, Xu X, Huang KT, Shields H, Kim-Shapiro DB, Schechter AN, Cannon RO, 3rd, Gladwin MT. Nitrite reduction to nitric oxide by deoxyhemoglobin vasodilates the human circulation. Nat. Med. 2003; 9: 1498–1505.

42 Reutov VP, Sorokina EG. NO-synthase and nitrite-reductase components of nitric oxide cycle. Biochemistry (Mosc.) 1998; 63: 874–884.

43 Luchsinger BP, Rich EN, Gow AJ, Williams EM, Stamler JS, Singel DJ. Routes to S-nitroso-hemoglobin formation with heme redox and preferential reactivity in the beta subunits. Proc. Natl. Acad. Sci. USA 2003; 100: 461–466.

44 Li H, Cui H, Liu X, Zweier JL. Xanthine oxidase catalyzes anaerobic transformation of organic nitrates to nitric oxide and nitrosothiols: characterization of this mechanism and the link between organic nitrate and guanylyl cyclase activation. J. Biol. Chem. 2005; 280: 16594–16600.

45 Yang Y, Maret W, Vallee BL. Differential fluorescence labeling of cysteinyl clusters uncovers high tissue levels of thionein. Proc. Natl. Acad. Sci. USA 2001; 98: 5556–5559.

46 St Croix CM, Wasserloos KJ, Dineley KE, Reynolds IJ, Levitan ES, Pitt BR. Nitric oxide-induced changes in intracellular zinc homeostasis are mediated by metallothionein/thionein. Am. J. Physiol. Lung Cell Mol. Physiol. 2002; 282: L185–L192.

47 Barrett J, Fitzgibbons LJ, Glauser J, Still RH, Young PNW. Photochemistry of the S-nitroso derivatives of hexane-1-thiol and hexane-1,6-dithiol. Nature 1966; 211: 848.

48 Furchgott RF, Ehrreich SJ, Greenblatt E. The photoactivated relaxation of smooth muscle of rabbit aorta. J. Gen. Physiol. 1961; 44: 499–519.

49 Matsunaga K, Furchgott RF. Interactions of light and sodium nitrite in producing relaxation of rabbit aorta. J. Pharmacol. Exp. Ther. 1989; 248: 687–695.

50 Megson IL, Holmes SA, Magid KS, Pritchard RJ, Flitney FW. Selective modifiers of glutathione biosynthesis and 'repriming' of vascular smooth muscle photorelaxation. Br. J. Pharmacol. 2000; 130: 1575–1580.

51 Ogulener N, Ergun Y. Neocuproine inhibits the decomposition of endogenous S-nitrosothiol by ultraviolet irradiation in the mouse gastric fundus. Eur. J. Pharmacol. 2004; 485: 269–274.

52 Ogulener N, Ergun Y. A putative role for S-nitrosoglutathione as the source of nitric oxide in photorelaxation of the mouse gastric fundus. Eur. J. Pharmacol. 2002; 450: 267–275.

53 Grossi L, Montevecchi PC, Strazzari S. Decomposition of S-nitrosothiols: unimolecular versus autocatalytic mechanism. J. Am. Chem. Soc. 2001; 123: 4853–4854.

54 Bartberger MD, Mannion JD, Powell SC, Stamler JS, Houk KN, Toone EJ. S-N dissociation energies of S-nitrosothiols: on the origins of nitrosothiol decomposition rates. J. Am. Chem. Soc. 2001; 123: 8868–8869.

55 Scorza G, Pietraforte D, Minetti M. Role of ascorbate and protein thiols in the release of nitric oxide from S-nitroso-albumin and S-nitroso-glutathione in human plasma. Free Radic. Biol. Med. 1997; 22: 633–642.

56 Kashiba-Iwatsuki M, Kitoh K, Kasahara E, Yu H, Nisikawa M, Matsuo M, Inoue M. Ascorbic acid and reducing agents regulate the fates and functions of S-nitrosothiols. J. Biochem. (Tokyo). 1997; 122: 1208–1214.

57 de Jesus-Berrios M, Liu L, Nussbaum JC, Cox GM, Stamler JS, Heitman J. Enzymes that counteract nitrosative stress promote fungal virulence. Curr. Biol. 2003; 13: 1963–1968.

58 Liu L, Hausladen A, Zeng M, Que L, Heitman J, Stamler JS. A metabolic enzyme for S-nitrosothiol conserved from bacteria to humans. Nature 2001; 410: 490–494.

59 Gordge MP, Addis P, Noronha-Dutra AA, Hothersall JS. Cell-mediated biotransformation of S-nitrosoglutathione. Biochem. Pharmacol. 1998; 55: 657–665.

60 Aleryani S, Milo E, Rose Y, Kostka P. Superoxide-mediated decomposition of biological S-nitrosothiols. J. Biol. Chem. 1998; 273: 6041–6045.

61 Vanin AF, Papina AA, Serezhenkov VA, Koppenol WH. The mechanisms of S-nitrosothiol decomposition catalyzed by iron. Nitric Oxide 2004; 10: 60–73.

62 Vanin AF, Muller B, Alencar JL, Lobysheva, II, Nepveu F, Stoclet JC. Evidence that intrinsic iron but not intrinsic copper determines S-nitrosocysteine decomposition in buffer solution. Nitric Oxide 2002; 7: 194–209.

63 Sorenson E, Skiles EH, Xu B, Aleryani S, Kostka P. Role of redox-active iron ions in the decomposition of S-nitrosocysteine in subcellular fractions of porcine aorta. Eur. J. Biochem. 2000; 267: 4593–4599.
64 Singh RJ, Hogg N, Joseph J, Kalyanaraman B. Mechanism of nitric oxide release from S-nitrosothiols. J. Biol. Chem. 1996; 271: 18596–18603.
65 Gorren AC, Schrammel A, Schmidt K, Mayer B. Decomposition of S-nitrosoglutathione in the presence of copper ions and glutathione. Arch. Biochem. Biophys. 1996; 330: 219–228.
66 Jourd'heuil D, Laroux FS, Miles AM, Wink DA, Grisham MB. Effect of superoxide dismutase on the stability of S-nitrosothiols. Arch. Biochem. Biophys. 1999; 361: 323–330.
67 Dasgupta TP, Smith JN. Reactions of S-nitrosothiols with L-ascorbic acid in aqueous solution. Methods Enzymol. 2002; 359: 219–229.
68 Aquart DV, Dasgupta TP. Dynamics of interaction of vitamin C with some potent nitrovasodilators, S-nitroso-N-acetyl-d,l-penicillamine (SNAP) and S-nitrosocaptopril (SNOCap), in aqueous solution. Biophys. Chem. 2004; 107: 117–131.
69 Xu A, Vita JA, Keaney Jr. JF, Ascorbic acid and glutathione modulate the biological activity of S-nitrosoglutathione. Hypertension 2000; 36: 291–295.
70 Jaffrey SR, Snyder SH. The biotin switch method for the detection of S-nitrosylated proteins. Sci. STKE. 2001; 2001: PL1.
71 Kim SF, Huri DA, Snyder SH. Inducible nitric oxide synthase binds, S-nitrosylates, and activates cyclooxygenase-2. Science 2005; 310: 1966–1970.
72 Borutaite V, Brown GC. Nitric oxide induces apoptosis via hydrogen peroxide, but necrosis via energy and thiol depletion. Free Radic. Biol. Med. 2003; 35: 1457–1468.
73 Bryan NS, Rassaf T, Maloney RE, Rodriguez CM, Saijo F, Rodriguez JR, Feelisch M. Cellular targets and mechanisms of nitros(yl)ation: an insight into their nature and kinetics in vivo. Proc. Natl. Acad. Sci. USA 2004; 101: 4308–4313.
74 Zeng H, Spencer NY, Hogg N. Metabolism of S-nitrosoglutathione by endothelial cells. Am. J. Physiol. Heart Circ. Physiol. 2001; 281: H432–H439.
75 Hogg N, Singh RJ, Konorev E, Joseph J, Kalyanaraman B. S-Nitrosoglutathione as a substrate for gamma-glutamyl transpeptidase. Biochem. J. 1997; 323 (Pt 2): 477–481.
76 Jensen DE, Belka GK, Du Bois GC. S-Nitrosoglutathione is a substrate for rat alcohol dehydrogenase class III isoenzyme. Biochem. J. 1998; 331 (Pt 2): 659–668.
77 Uotila L, Koivusalo M. Gluthatione: chemical, biochemical and medical aspects. John Wiley & Sons, New York, 1989.
78 Kahlos K, Zhang J, Block ER, Patel JM. Thioredoxin restores nitric oxide-induced inhibition of protein kinase C activity in lung endothelial cells. Mol. Cell Biochem. 2003; 254: 47–54.
79 Lane P, Hao G, Gross SS. S-nitrosylation is emerging as a specific and fundamental posttranslational protein modification: head-to-head comparison with O-phosphorylation. Sci. STKE. 2001; 2001: RE1.
80 Stamler JS, Simon DI, Osborne JA, Mullins ME, Jaraki O, Michel T, Singel DJ, Loscalzo J. S-nitrosylation of proteins with nitric oxide: synthesis and characterization of biologically active compounds. Proc. Natl. Acad. Sci. USA 1992; 89: 444–448.
81 Eu JP, Liu L, Zeng M, Stamler JS. An apoptotic model for nitrosative stress. Biochemistry 2000; 39: 1040–1047.
82 Mannick JB, Schonhoff C, Papeta N, Ghafourifar P, Szibor M, Fang K, Gaston B. S-Nitrosylation of mitochondrial caspases. J. Cell Biol. 2001; 154: 1111–1116.
83 Ckless K, Reynaert NL, Taatjes DJ, Lounsbury KM, van der Vliet A, Janssen-Heininger Y. In situ detection and visualization of S-nitrosylated proteins following chemical derivatization: identification of Ran GTPase as a target for S-nitrosylation. Nitric Oxide 2004; 11: 216–227.
84 Erwin PA, Lin AJ, Golan DE, Michel T. Receptor-regulated dynamic S-nitrosylation of endothelial nitric-oxide synthase in vascular endothelial cells. J. Biol. Chem. 2005; 280: 19888–19894.
85 Erwin PA, Mitchell DA, Sartoretto J, Marletta MA, Michel T. Subcellular targeting and differential S-nitrosylation of endothelial nitric-oxide synthase. J. Biol. Chem. 2006; 281: 151–157.
86 Janero DR, Bryan NS, Saijo F, Dhawan V, Schwalb DJ, Warren MC, Feelisch M. Differential nitros(yl)ation of blood and tissue constituents during glyceryl trinitrate biotransformation in vivo. Proc. Natl. Acad. Sci. USA 2004; 101: 16958–16963.

87 Tyurin VA, Liu SX, Tyurina YY, Sussman NB, Hubel CA, Roberts JM, Taylor RN, Kagan VE. Elevated levels of S-nitrosoalbumin in preeclampsia plasma. Circ. Res. 2001; 88: 1210–1215.
88 Rassaf T, Bryan NS, Kelm M, Feelisch M. Concomitant presence of N-nitroso and S-nitroso proteins in human plasma. Free Radic. Biol. Med. 2002; 33: 1590–1596.
89 Marley R, Patel RP, Orie N, Ceaser E, Darley-Usmar V, Moore K. Formation of nanomolar concentrations of S-nitroso-albumin in human plasma by nitric oxide. Free Radic. Biol. Med. 2001; 31: 688–696.
90 Rassaf T, Feelisch M, Kelm M. Circulating NO pool: assessment of nitrite and nitroso species in blood and tissues. Free Radic. Biol. Med. 2004; 36: 413–422.
91 Rossi R, Giustarini D, Milzani A, Colombo R, Dalle-Donne I, Di Simplicio P. Physiological levels of S-nitrosothiols in human plasma. Circ. Res. 2001; 89: E47.
92 Gladwin MT, Shelhamer JH, Schechter AN, Pease-Fye ME, Waclawiw MA, Panza JA, Ognibene FP, Cannon RO, 3rd. Role of circulating nitrite and S-nitrosohemoglobin in the regulation of regional blood flow in humans. Proc. Natl. Acad. Sci. USA 2000; 97: 11482–11487.
93 Tsikas D. Measurement of physiological S-nitrosothiols: a problem child and a challenge. Nitric Oxide 2003; 9: 53–55.
94 Doctor A, Platt R, Sheram ML, Eischeid A, McMahon T, Maxey T, Doherty J, Axelrod M, Kline J, Gurka M, Gow A, Gaston B. Hemoglobin conformation couples erythrocyte S-nitrosothiol content to O_2 gradients. Proc. Natl. Acad. Sci. USA 2005; 102: 5709–5714.
95 Marley R, Feelisch M, Holt S, Moore K. A chemiluminescense-based assay for S-nitrosoalbumin and other plasma S-nitrosothiols. Free Radic. Res. 2000; 32: 1–9.
96 Yang BK, Vivas EX, Reiter CD, Gladwin MT. Methodologies for the sensitive and specific measurement of S-nitrosothiols, iron-nitrosyls, and nitrite in biological samples. Free Radic. Res. 2003; 37: 1–10.
97 Samouilov A, Zweier JL. Development of chemiluminescence-based methods for specific quantitation of nitrosylated thiols. Anal. Biochem. 1998; 258: 322–330.
98 Kostka P, Park JK. Fluorometric detection of S-nitrosothiols. Methods Enzymol. 1999; 301: 227–235.
99 King M, Gildemeister O, Gaston B, Mannick JB. Assessment of S-nitrosothiols on diaminofluorescein gels. Anal. Biochem. 2005; 346: 69–76.
100 Mirza UA, Chait BT, Lander HM. Monitoring reactions of nitric oxide with peptides and proteins by electrospray ionization-mass spectrometry. J. Biol. Chem. 1995; 270: 17185–17188.
101 Zech B, Wilm M, van Eldik R, Brune B. Mass spectrometric analysis of nitric oxide-modified caspase-3. J. Biol. Chem. 1999; 274: 20931–20936.
102 Gow AJ, Chen Q, Hess DT, Day BJ, Ischiropoulos H, Stamler JS. Basal and stimulated protein S-nitrosylation in multiple cell types and tissues. J. Biol. Chem. 2002; 277: 9637–9640.
103 Jaffrey SR, Erdjument-Bromage H, Ferris CD, Tempst P, Snyder SH. Protein S-nitrosylation: a physiological signal for neuronal nitric oxide. Nat. Cell Biol. 2001; 3: 193–197.
104 Zhang Y, Keszler A, Broniowska KA, Hogg N. Characterization and application of the biotin-switch assay for the identification of S-nitrosated proteins. Free Radic. Biol. Med. 2005; 38: 874–881.
105 Martinez-Ruiz A, Lamas S. Detection and proteomic identification of S-nitrosylated proteins in endothelial cells. Arch. Biochem. Biophys. 2004; 423: 192–199.
106 Wood ZA, Schroder E, Robin Harris J, Poole LB. Structure, mechanism and regulation of peroxiredoxins. Trends Biochem. Sci. 2003; 28: 32–40.
107 Loscalzo J, Reed GH. Spectroscopic studies of actin-metal-nucleotide complexes. Biochemistry 1976; 15: 5407–5413.
108 Kim SO, Merchant K, Nudelman R, Beyer Jr. WF, Keng T, DeAngelo J, Hausladen A, Stamler JS. OxyR: a molecular code for redox-related signaling. Cell 2002; 109: 383–396.
109 Campbell DL, Stamler JS, Strauss HC. Redox modulation of L-type calcium channels in ferret ventricular myocytes. Dual mechanism regulation by nitric oxide and S-nitrosothiols. J. Gen. Physiol. 1996; 108: 277–293.
110 Matsushita K, Morrell CN, Cambien B, Yang SX, Yamakuchi M, Bao C, Hara MR, Quick RA, Cao W, O'Rourke B, Lowenstein JM, Pevsner J, Wagner DD, Lowenstein CJ. Nitric oxide regulates exocytosis by S-nitrosylation of N-ethylmaleimide-sensitive factor. Cell 2003; 115: 139–150.

111 Kim YM, Talanian RV, Billiar TR. Nitric oxide inhibits apoptosis by preventing increases in caspase-3-like activity via two distinct mechanisms. J. Biol. Chem. 1997; 272: 31138–31148.

112 Lewis SD, Misra DC, Shafer JA. Determination of interactive thiol ionizations in bovine serum albumin, glutathione, and other thiols by potentiometric difference titration. Biochemistry 1980; 19: 6129–6137.

113 Keaney Jr. JF, Simon DI, Stamler JS, Jaraki O, Scharfstein J, Vita JA, Loscalzo J. NO forms an adduct with serum albumin that has endothelium-derived relaxing factor-like properties. J. Clin. Invest. 1993; 91: 1582–1589.

114 Eu JP, Sun J, Xu L, Stamler JS, Meissner G. The skeletal muscle calcium release channel: coupled O_2 sensor and NO signaling functions. Cell 2000; 102: 499–509.

115 Lai TS, Hausladen A, Slaughter TF, Eu JP, Stamler JS, Greenberg CS. Calcium regulates S-nitrosylation, denitrosylation, and activity of tissue transglutaminase. Biochemistry 2001; 40: 4904–4910.

116 Rhee KY, Erdjument-Bromage H, Tempst P, Nathan CF. S-nitroso proteome of *Mycobacterium tuberculosis*: Enzymes of intermediary metabolism and antioxidant defense. Proc. Natl. Acad. Sci. USA 2005; 102: 467–472.

117 Lipton AJ, Johnson MA, Macdonald T, Lieberman MW, Gozal D, Gaston B. S-nitrosothiols signal the ventilatory response to hypoxia. Nature 2001; 413: 171–174.

118 Zaman K, McPherson M, Vaughan J, Hunt J, Mendes F, Gaston B, Palmer LA. S-nitrosoglutathione increases cystic fibrosis transmembrane regulator maturation. Biochem. Biophys. Res. Commun. 2001; 284: 65–70.

119 Richardson G, Benjamin N. Potential therapeutic uses for S-nitrosothiols. Clin. Sci. (Lond.) 2002; 102: 99–105.

120 Hess DT, Matsumoto A, Kim SO, Marshall HE, Stamler JS. Protein S-nitrosylation: purview and parameters. Nat. Rev. Mol. Cell Biol. 2005; 6: 150–166.

121 Xian M, Chen X, Liu Z, Wang K, Wang PG. Inhibition of papain by S-nitrosothiols. Formation of mixed disulfides. J. Biol. Chem. 2000; 275: 20467–20473.

122 Borutaite V, Budriunaite A, Brown GC. Reversal of nitric oxide-, peroxynitrite- and S-nitrosothiol-induced inhibition of mitochondrial respiration or complex I activity by light and thiols. Biochim. Biophys. Acta 2000; 1459: 405–412.

123 Clementi E, Brown GC, Feelisch M, Moncada S. Persistent inhibition of cell respiration by nitric oxide: crucial role of S-nitrosylation of mitochondrial complex I and protective action of glutathione. Proc. Natl. Acad. Sci. USA 1998; 95: 7631–7636.

124 Beltran B, Orsi A, Clementi E, Moncada S. Oxidative stress and S-nitrosylation of proteins in cells. Br. J. Pharmacol. 2000; 129: 953–960.

125 Burwell LS, Nadtochiy SM, Tompkins AJ, Young SM, Brookes PS. Direct evidence for S-nitrosation of mitochondrial complex I. Biochem. J. 2005.

126 Cleeter MW, Cooper JM, Darley-Usmar VM, Moncada S, Schapira AH. Reversible inhibition of cytochrome c oxidase, the terminal enzyme of the mitochondrial respiratory chain, by nitric oxide. Implications for neurodegenerative diseases. FEBS Lett. 1994; 345: 50–54.

127 Zhang J, Jin B, Li L, Block ER, Patel JM. Nitric oxide-induced persistent inhibition and nitrosylation of active site cysteine residues of mitochondrial cytochrome-c oxidase in lung endothelial cells. Am. J. Physiol. Cell Physiol. 2005; 288: C840–C849.

128 Jia L, Bonaventura C, Bonaventura J, Stamler JS. S-nitrosohaemoglobin: a dynamic activity of blood involved in vascular control. Nature 1996; 380: 221–226.

129 Gladwin MT, Wang X, Reiter CD, Yang BK, Vivas EX, Bonaventura C, Schechter AN. S-Nitrosohemoglobin is unstable in the reductive erythrocyte environment and lacks O_2/NO-linked allosteric function. J. Biol. Chem. 2002; 277: 27818–27828.

130 Yao D, Gu Z, Nakamura T, Shi ZQ, Ma Y, Gaston B, Palmer LA, Rockenstein EM, Zhang Z, Masliah E, Uehara T, Lipton SA. Nitrosative stress linked to sporadic Parkinson's disease: S-nitrosylation of parkin regulates its E3 ubiquitin ligase activity. Proc. Natl. Acad. Sci. USA 2004; 101: 10810–10814.

131 Schonhoff CM, Daou MC, Jones SN, Schiffer CA, Ross AH. Nitric oxide-mediated inhibition of Hdm2-p53 binding. Biochemistry 2002; 41: 13570–13574.

132 Chanvorachote P, Nimmannit U, Wang L, Stehlik C, Lu B, Azad N, Rojanasakul Y. Nitric oxide negatively regulates Fas CD95-induced apoptosis through inhibition of ubiquitin-proteasome-mediated degradation of FLICE inhibitory protein. J. Biol. Chem. 2005; 280: 42044–42050.
133 Kim S, Wing SS, Ponka P. S-nitrosylation of IRP2 regulates its stability via the ubiquitin-proteasome pathway. Mol. Cell Biol. 2004; 24: 330–337.
134 Hara MR, Agrawal N, Kim SF, Cascio MB, Fujimuro M, Ozeki Y, Takahashi M, Cheah JH, Tankou SK, Hester LD, Ferris CD, Hayward SD, Snyder SH, Sawa A. S-nitrosylated GAPDH initiates apoptotic cell death by nuclear translocation following Siah1 binding. Nat. Cell Biol. 2005; 7: 665–674.
135 Palmer LA, Gaston B, Johns RA. Normoxic stabilization of hypoxia-inducible factor-1 expression and activity: redox-dependent effect of nitrogen oxides. Mol. Pharmacol. 2000; 58: 1197–1203.
136 Sandau KB, Fandrey J, Brune B. Accumulation of HIF-1alpha under the influence of nitric oxide. Blood 2001; 97: 1009–1015.
137 Calmels S, Hainaut P, Ohshima H. Nitric oxide induces conformational and functional modifications of wild-type p53 tumor suppressor protein. Cancer Res. 1997; 57: 3365–3369.
138 Brune B, von Knethen A, Sandau KB. Transcription factors p53 and HIF-1alpha as targets of nitric oxide. Cell Signal 2001; 13: 525–533.
139 Dimmeler S, Haendeler J, Nehls M, Zeiher AM. Suppression of apoptosis by nitric oxide via inhibition of interleukin-1beta-converting enzyme (ICE)-like and cysteine protease protein (CPP)-32-like proteases. J. Exp. Med. 1997; 185: 601–607.
140 Melino G, Bernassola F, Knight RA, Corasaniti MT, Nistico G, Finazzi-Agro A. S-nitrosylation regulates apoptosis. Nature 1997; 388: 432–433.
141 Tenneti L, D'Emilia DM, Lipton SA. Suppression of neuronal apoptosis by S-nitrosylation of caspases. Neurosci. Lett. 1997; 236: 139–142.
142 Mannick JB, Hausladen A, Liu L, Hess DT, Zeng M, Miao QX, Kane LS, Gow AJ, Stamler JS. Fas-induced caspase denitrosylation. Science 1999; 284: 651–654.
143 Kim JE, Tannenbaum SR. S-Nitrosation regulates the activation of endogenous procaspase-9 in HT-29 human colon carcinoma cells. J. Biol. Chem. 2004; 279: 9758–9764.
144 Mitchell DA, Marletta MA. Thioredoxin catalyzes the S-nitrosation of the caspase-3 active site cysteine. Nat. Chem. Biol. 2005; 1: 154–158.
145 Zech B, Kohl R, von Knethen A, Brune B. Nitric oxide donors inhibit formation of the Apaf-1/caspase-9 apoptosome and activation of caspases. Biochem. J. 2003; 371: 1055–1064.
146 Brendeford EM, Andersson KB, Gabrielsen OS. Nitric oxide (NO) disrupts specific DNA binding of the transcription factor c-Myb in vitro. FEBS Lett. 1998; 425: 52–56.
147 Tabuchi A, Sano K, Oh E, Tsuchiya T, Tsuda M. Modulation of AP-1 activity by nitric oxide (NO) in vitro: NO-mediated modulation of AP-1. FEBS Lett. 1994; 351: 123–127.
148 Gao C, Guo H, Wei J, Mi Z, Wai P, Kuo PC. S-nitrosylation of heterogeneous nuclear ribonucleoprotein A/B regulates osteopontin transcription in endotoxin-stimulated murine macrophages. J. Biol. Chem. 2004; 279: 11236–11243.
149 Kroncke KD, Fehsel K, Schmidt T, Zenke FT, Dasting I, Wesener JR, Bettermann H, Breunig KD, Kolb-Bachofen V. Nitric oxide destroys zinc-sulfur clusters inducing zinc release from metallothionein and inhibition of the zinc finger-type yeast transcription activator LAC9. Biochem. Biophys. Res. Commun. 1994; 200: 1105–1110.
150 Garban HJ, Marquez-Garban DC, Pietras RJ, Ignarro LJ. Rapid nitric oxide-mediated S-nitrosylation of estrogen receptor: regulation of estrogen-dependent gene transcription. Proc. Natl. Acad. Sci. USA 2005; 102: 2632–2636.
151 Galigniana MD, Piwien-Pilipuk G, Assreuy J. Inhibition of glucocorticoid receptor binding by nitric oxide. Mol. Pharmacol. 1999; 55: 317–323.
152 Hongo F, Garban H, Huerta-Yepez S, Vega M, Jazirehi AR, Mizutani Y, Miki T, Bonavida B. Inhibition of the transcription factor Yin Yang 1 activity by S-nitrosation. Biochem. Biophys. Res. Commun. 2005; 336: 692–701.
153 Mitchell DA, Erwin PA, Michel T, Marletta MA. S-Nitrosation and regulation of inducible nitric oxide synthase. Biochemistry 2005; 44: 4636–4647.

154 Pearce LL, Gandley RE, Han W, Wasserloos K, Stitt M, Kanai AJ, McLaughlin MK, Pitt BR, Levitan ES. Role of metallothionein in nitric oxide signaling as revealed by a green fluorescent fusion protein. Proc. Natl. Acad. Sci. USA 2000; 97: 477–482.
155 Gergel D, Cederbaum AI. Inhibition of the catalytic activity of alcohol dehydrogenase by nitric oxide is associated with S nitrosylation and the release of zinc. Biochemistry 1996; 35: 16186–16194.
156 Ravi K, Brennan LA, Levic S, Ross PA, Black SM. S-nitrosylation of endothelial nitric oxide synthase is associated with monomerization and decreased enzyme activity. Proc. Natl. Acad. Sci. USA 2004; 101: 2619–2624.
157 Ding H, Demple B. Direct nitric oxide signal transduction via nitrosylation of iron-sulfur centers in the SoxR transcription activator. Proc. Natl. Acad. Sci. USA 2000; 97: 5146–5150.
158 DelaTorre A, Schroeder RA, Punzalan C, Kuo PC. Endotoxin-mediated S-nitrosylation of p50 alters NF-kappa B-dependent gene transcription in ANA-1 murine macrophages. J. Immunol. 1999; 162: 4101–4108.
159 Park SK, Lin HL, Murphy S. Nitric oxide regulates nitric oxide synthase-2 gene expression by inhibiting NF-kappaB binding to DNA. Biochem. J. 1997; 322 (Pt 2): 609–613.
160 DelaTorre A, Schroeder RA, Kuo PC. Alteration of NF-kappa B p50 DNA binding kinetics by S-nitrosylation. Biochem. Biophys. Res. Commun. 1997; 238: 703–706.
161 Matthews JR, Botting CH, Panico M, Morris HR, Hay RT. Inhibition of NF-kappaB DNA binding by nitric oxide. Nucleic Acids Res. 1996; 24: 2236–2242.
162 Marshall HE, Stamler JS. Inhibition of NF-kappa B by S-nitrosylation. Biochemistry 2001; 40: 1688–1693.
163 Kim IY, Stadtman TC. Inhibition of NF-kappaB DNA binding and nitric oxide induction in human T cells and lung adenocarcinoma cells by selenite treatment. Proc. Natl. Acad. Sci. USA 1997; 94: 12904–12907.
164 Reynaert NL, Ckless K, Korn SH, Vos N, Guala AS, Wouters EF, van der Vliet A, Janssen-Heininger YM. Nitric oxide represses inhibitory kappaB kinase through S-nitrosylation. Proc. Natl. Acad. Sci. USA 2004; 101: 8945–8950.
165 Peng HB, Libby P, Liao JK. Induction and stabilization of I kappa B alpha by nitric oxide mediates inhibition of NF-kappa B. J. Biol. Chem. 1995; 270: 14214–14219.
166 Marshall HE, Stamler JS. Nitrosative stress-induced apoptosis through inhibition of NF-kappa B. J. Biol. Chem. 2002; 277: 34223–34228.
167 Lander HM, Ogiste JS, Pearce SF, Levi R, Novogrodsky A. Nitric oxide-stimulated guanine nucleotide exchange on p21ras. J. Biol. Chem. 1995; 270: 7017–7020.
168 Metzen E, Zhou J, Jelkmann W, Fandrey J, Brune B. Nitric oxide impairs normoxic degradation of HIF-1alpha by inhibition of prolyl hydroxylases. Mol. Biol. Cell 2003; 14: 3470–3481.
169 Yasinska IM, Sumbayev VV. S-nitrosation of Cys-800 of HIF-1alpha protein activates its interaction with p300 and stimulates its transcriptional activity. FEBS Lett. 2003; 549: 105–109.
170 Melillo G, Musso T, Sica A, Taylor LS, Cox GW, Varesio L. A hypoxia-responsive element mediates a novel pathway of activation of the inducible nitric oxide synthase promoter. J. Exp. Med. 1995; 182: 1683–1693.
171 Palmer LA, Semenza GL, Stoler MH, Johns RA. Hypoxia induces type II NOS gene expression in pulmonary artery endothelial cells via HIF-1. Am. J. Physiol. 1998; 274: L212–L219.
172 Ambs S, Hussain SP, Harris CC. Interactive effects of nitric oxide and the p53 tumor suppressor gene in carcinogenesis and tumor progression. Faseb J. 1997; 11: 443–448.
173 Davis ME, Grumbach IM, Fukai T, Cutchins A, Harrison DG. Shear stress regulates endothelial nitric-oxide synthase promoter activity through nuclear factor kappaB binding. J. Biol. Chem. 2004; 279: 163–168.
174 Lowenstein CJ, Alley EW, Raval P, Snowman AM, Snyder SH, Russell SW, Murphy WJ. Macrophage nitric oxide synthase gene: two upstream regions mediate induction by interferon gamma and lipopolysaccharide. Proc. Natl. Acad. Sci. USA 1993; 90: 9730–9734.
175 Hausladen A, Privalle CT, Keng T, DeAngelo J, Stamler JS. Nitrosative stress: activation of the transcription factor OxyR. Cell 1996; 86: 719–729.

176 Lipton SA, Choi YB, Pan ZH, Lei SZ, Chen HS, Sucher NJ, Loscalzo J, Singel DJ, Stamler JS. A redox-based mechanism for the neuroprotective and neurodestructive effects of nitric oxide and related nitroso-compounds. Nature 1993; 364: 626–632.

177 Choi YB, Tenneti L, Le DA, Ortiz J, Bai G, Chen HS, Lipton SA. Molecular basis of NMDA receptor-coupled ion channel modulation by S-nitrosylation. Nat. Neurosci. 2000; 3: 15–21.

178 Kim WK. S-nitrosation ameliorates homocysteine-induced neurotoxicity and calcium responses in primary culture of rat cortical neurons. Neurosci. Lett. 1999; 265: 99–102.

179 Kim WK, Choi YB, Rayudu PV, Das P, Asaad W, Arnelle DR, Stamler JS, Lipton SA. Attenuation of NMDA receptor activity and neurotoxicity by nitroxyl anion, NO. Neuron 1999; 24: 461–469.

180 Renganathan M, Cummins TR, Waxman SG. Nitric oxide blocks fast, slow, and persistent Na^+ channels in C-type DRG neurons by S-nitrosylation. J. Neurophysiol. 2002; 87: 761–775.

181 Broillet MC. A single intracellular cysteine residue is responsible for the activation of the olfactory cyclic nucleotide-gated channel by NO. J. Biol. Chem. 2000; 275: 15135–15141.

182 Lin YF, Raab-Graham K, Jan YN, Jan LY. NO stimulation of ATP-sensitive potassium channels: Involvement of Ras/mitogen-activated protein kinase pathway and contribution to neuroprotection. Proc. Natl. Acad. Sci. USA 2004; 101: 7799–7804.

183 Lang RJ, Harvey JR, Mulholland EL. Sodium (2-sulfonatoethyl) methanethiosulfonate prevents S-nitroso-L-cysteine activation of $Ca2^+$-activated K^+ (BKCa) channels in myocytes of the guinea-pig taenia caeca. Br. J. Pharmacol. 2003; 139: 1153–1163.

184 Martinez-Moreno M, Alvarez-Barrientos A, Roncal F, Albar JP, Gavilanes F, Lamas S, Rodriguez-Crespo I. Direct interaction between the reductase domain of endothelial nitric oxide synthase and the ryanodine receptor. FEBS Lett. 2005; 579: 3159–3163.

185 Xu L, Eu JP, Meissner G, Stamler JS. Activation of the cardiac calcium release channel (ryanodine receptor) by poly-S-nitrosylation. Science 1998; 279: 234–237.

186 Sun J, Xin C, Eu JP, Stamler JS, Meissner G. Cysteine-3635 is responsible for skeletal muscle ryanodine receptor modulation by NO. Proc. Natl. Acad. Sci. USA 2001; 98: 11158–11162.

187 Sun J, Xu L, Eu JP, Stamler JS, Meissner G. Nitric oxide, NOC-12, and S-nitrosoglutathione modulate the skeletal muscle calcium release channel/ryanodine receptor by different mechanisms. An allosteric function for O_2 in S-nitrosylation of the channel. J. Biol. Chem. 2003; 278: 8184–8189.

188 Aracena P, Sanchez G, Donoso P, Hamilton SL, Hidalgo C. S-glutathionylation decreases $Mg2^+$ inhibition and S-nitrosylation enhances $Ca2^+$ activation of RyR1 channels. J. Biol. Chem. 2003; 278: 42927–42935.

189 Sun J, Xu L, Eu JP, Stamler JS, Meissner G. Classes of thiols that influence the activity of the skeletal muscle calcium release channel. J. Biol. Chem. 2001; 276: 15625–15630.

190 Zable AC, Favero TG, Abramson JJ. Glutathione modulates ryanodine receptor from skeletal muscle sarcoplasmic reticulum. Evidence for redox regulation of the $Ca2^+$ release mechanism. J. Biol. Chem. 1997; 272: 7069–7077.

191 So HS, Park RK, Kim MS, Lee SR, Jung BH, Chung SY, Jun CD, Chung HT. Nitric oxide inhibits c-Jun N-terminal kinase 2 (JNK2) via S-nitrosylation. Biochem. Biophys. Res. Commun. 1998; 247: 809–813.

192 Park HS, Huh SH, Kim MS, Lee SH, Choi EJ. Nitric oxide negatively regulates c-Jun N-terminal kinase/stress-activated protein kinase by means of S-nitrosylation. Proc. Natl. Acad. Sci. USA 2000; 97: 14382–14387.

193 Estrada C, Gomez C, Martin-Nieto J, De Frutos T, Jimenez A, Villalobo A. Nitric oxide reversibly inhibits the epidermal growth factor receptor tyrosine kinase. Biochem. J. 1997; 326 (Pt 2): 369–376.

194 Park HS, Yu JW, Cho JH, Kim MS, Huh SH, Ryoo K, Choi EJ. Inhibition of apoptosis signal-regulating kinase 1 by nitric oxide through a thiol redox mechanism. J. Biol. Chem. 2004; 279: 7584–7590.

195 Sumbayev VV. S-nitrosylation of thioredoxin mediates activation of apoptosis signal-regulating kinase 1. Arch. Biochem. Biophys. 2003; 415: 133–136.

196 Akhand AA, Pu M, Senga T, Kato M, Suzuki H, Miyata T, Hamaguchi M, Nakashima I. Nitric oxide controls src kinase activity through a sulfhydryl group modification-mediated Tyr-527-independent and Tyr-416-linked mechanism. J. Biol. Chem. 1999; 274: 25821–25826.

197 Caselli A, Chiarugi P, Camici G, Manao G, Ramponi G. In vivo inactivation of phosphotyrosine protein phosphatases by nitric oxide. FEBS Lett. 1995; 374: 249–252.
198 Callsen D, Sandau KB, Brune B. Nitric oxide and superoxide inhibit platelet-derived growth factor receptor phosphotyrosine phosphatases. Free Radic. Biol. Med. 1999; 26: 1544–1553.
199 Xian M, Wang K, Chen X, Hou Y, McGill A, Zhou B, Zhang ZY, Cheng JP, Wang PG. Inhibition of protein tyrosine phosphatases by low-molecular-weight S-nitrosothiols and S-nitrosylated human serum albumin. Biochem. Biophys. Res. Commun. 2000; 268: 310–314.

Radicals for Life: The Various Forms of Nitric Oxide
E. van Faassen and A.F. Vanin (Eds.)

CHAPTER 11

Chemical equilibria between S-nitrosothiols and dinitrosyl iron complexes with thiol-containing ligands

Anatoly F. Vanin[1] and Ernst van Faassen[2]

[1] *Semenov Institute of Chemical Physics, Russian Academy of Sciences, Moscow, 119991, Russian Federation*
[2] *Debye Institute, Section Interface Physics, Ornstein Laboratory, Utrecht University, 3508 TA, Utrecht, The Netherlands*

INTRODUCTION

S-nitrosothiols are ubiquitous in biological systems from cells to whole animals [1–6]. On basis of molecular weight, they can be roughly classified as either low molecular weight (below 1 kd), or attached to a large molecule (above 1 kd) like a protein. Resting levels of *S*-thiolated proteins in cells are rather low and remain below 1% of the total protein pool (see Chapter 9). Far higher levels of transient S-thiolation occurs in cells close to stimulated immune cells, or in cells exposed to toxic levels of oxidants. The *S*-nitrosothiols are endogenously formed by nitrosation of low-molecular-weight thiols like glutathione (GSH), or by nitrosation of the cysteinyl side chains of proteins. It is interesting to note that the NO radical itself is a poor nitrosating agent in biological systems. Exposure of cultured cells to NONOate donors does not significantly increase the intracellular pool of *S*-nitrosothiols [7,8], although the neutral NO radical easily crosses the cell membrane to reach the interior compartment. However, the presence of small quantities of transition metal ions has often been found to enhance nitrosation of the endogenous thiols after exposure to NO [3,5,9–12]. This chapter will be primarily concerned with such metal-catalyzed pathways of S-nitrosation of thiol groups. When reading this chapter, it should be kept in mind that additional pathways for nitrosation exist: certain metabolites of NO like nitrogen dioxide (NO_2), nitrosonium (NO^+) or dinitrogen trioxide (N_2O_3) have strong nitrosative capacity [7] and could contribute to the endogenous formation of *S*-nitrosothiols *in vivo*. More details on the chemistry of these compounds are given in Chapter 1.

S-nitrosothiols participate in a range of kinetic equilibria involving transport, synthesis, decomposition and transformation. Small *S*-nitrosothiols may enter viable cultured cells *via* a stereoselective mechanism, as discussed in Chapter 9. Transnitrosation from small

S-nitrosothiols to protein thiols or *vice versa* provides a mechanism of exchange between the various intracellular thiol pools. Finally, it has been observed that both the decomposition and formation of *S*-nitrosothiols is strongly affected by the presence of transition metal ions, in particular iron [9,10,13–19] and copper [1,3,11,13,14,20–23]. However, it should not be taken for granted that the differences between copper and iron may be understood from redox activity alone. Iron is the only compound that forms dinitrosyl iron complexes (DNIC) when exposed to free nitric oxide and thiols. These DNIC are implied in the complex chemistry of nitrosothiols and, in fact, are found to participate in reaction equilibria between the nitrosothiols and their constituents. This chapter will be primarily concerned with the role of DNIC in the reaction equilibria of the *S*-nitrosothiols. The kinetics of these reactions will be studied using *in-vitro* experiments, and their relevance for *in-vivo* situations will be discussed at the end.

THE REDOX REACTIONS OF COPPER

Except for conditions of strong oxidative stress or high iron levels, the endogenous copper pool in cells and tissues is considered now as the determinant for the formation and decomposition of RS-NO in biosystems [1,3,13,20]. The redox couple Cu^{2+}/Cu^{+} has a rather small redox potential $E_o = +0.15$ V. Therefore, in principle, copper can catalyze the decomposition of RS-NO by shuttling between its two redox states according to the reaction as in Scheme 1.

$$RS^{-}\text{-}NO^{+} + Cu^{+} \longrightarrow RS^{-} + NO + Cu^{2+}$$
$$Cu^{2+} + RS^{-} \longrightarrow Cu^{+} + RS^{\bullet}$$
$$2\,RS^{\bullet} \longrightarrow RS\text{ - }SR$$

net reaction: $$RS^{-}\text{-}NO^{+} \xrightarrow[Cu^{+/2+}]{} NO + \tfrac{1}{2}(RS\text{-}SR)$$

Scheme 1. The reductive mechanism of RS-NO decomposition catalyzed by monovalent copper [13].

One-electron reduction of RS-NO by the monovalent Cu^{+} ion destabilizes the RS-NO molecule and initiates its decomposition by release of the NO moiety. Subsequent oxidation of the RS^{-} anion by Cu^{2+} ion results in the release of a thiyl radical, and regeneration of copper to its original monovalent charge state. The thiyl radicals are not very reactive to other species, but easily combine by forming a covalent disulfide bond. The formation of disulfides like cysteine makes the decomposition of RS-NO irreversible. The reductive decomposition does not involve oxygen and proceeds in oxygenated as well as anoxic solutions.

In principle, the decomposition of RS-NO may be initiated by divalent copper ions. This reaction requires the presence of an oxidizer like oxygen and could proceed according to Scheme 2:

$$RS^{-}\text{-}NO^{+} + Cu^{2+} \longrightarrow RS^{\bullet} + NO^{+} + Cu^{+}$$
$$Cu^{+} + O_2 \longrightarrow Cu^{2+} + O_2^{-}$$

net reaction: $$RS^{-}\text{-}NO^{+} + O_2 \xrightarrow[Cu^{2+/+}]{} NO^{+} + \tfrac{1}{2}(RS\text{-}SR) + O_2^{-}$$

Scheme 2. The oxidative mechanism of RS-NO decomposition catalyzed by divalent copper.

This decomposition pathway is also irreversible due to the formation of disulfides and the subsequent hydrolysis of nitrosonium NO^+ to nitrite NO_2^-.

Evidently, the oxidative mechanism of RS-NO decomposition can be dominant at high levels of oxygen in the solution. However, under normoxic physiological conditions as encountered in tissues and biological samples, the reductive mechanism seems dominant. Experiments have shown that GS-NO decomposition involves the release of NO in neutral radical state rather than nitrosonium, and that the decomposition may be inhibited by scavenging Cu^+ ions with neocuproine. Both observations provide strong experimental evidence for the dominance of reductive decomposition according to Scheme 1.

The synthesis of RS-NO catalyzed by Cu^{2+} ions can proceed *via* two pathways shown in Schemes 3 and 4:

$$RS^- + Cu^{2+} \longrightarrow RS^\bullet + Cu^+$$
$$NO + RS^\bullet \longrightarrow RS\text{-}NO$$
$$Cu^+ + O_2 \longrightarrow Cu^{2+} + O_2^-$$
$$\text{net reaction:} \quad RS^- + O_2 + NO \xrightarrow[Cu^{2+/+}]{} RS\text{-}NO + O_2^-$$

Scheme 3. The first pathway for RS-NO synthesis catalyzed by copper [11].

Second possible mechanism of RS-NO synthesis induced by Cu^{2+} ions is shown in Scheme 4:

$$NO + Cu^{2+} \longrightarrow NO^+ + Cu^+$$
$$NO^+ + RS^- \longrightarrow RS\text{-}NO$$
$$Cu^+ + O_2 \longrightarrow Cu^{2+} + O_2^-$$
$$\text{net reaction:} \quad RS^- + NO + O_2 \xrightarrow[Cu]{} RS\text{-}NO + O_2^-$$

Scheme 4. The second pathway for RS-NO synthesis catalyzed by copper [5,11].

Thus, the oxidative mechanisms are characteristic of RS-NO synthesis for both pathways. However, two arguments suggest that the oxidative mechanisms are not significant under physiological conditions. First, thiyl radicals as formed in Scheme 3 will react not only with free NO, but with each other (disulfide formation) as well as with other compounds. Second, in the presence of water, free nitrosonium ions as released in Scheme 4 rapidly hydrolyzed to nitrite. Therefore, nitrosonium could S-nitrosate thiol moieties only at high thiol concentrations or if strong binding with copper ions protects it against the hydrolysis.

DECOMPOSITION AND SYNTHESIS OF *S*-NITROSOTHIOLS BY IRON

In principle, the same oxidative pathways of RS-NO synthesis and decomposition similar to shown in Schemes 2, 3 and 4 could also be catalyzed by iron. Just like Cu^{2+} ions, ferric iron could initiate the reactions. However, high redox potential of Fe^{3+}/Fe^{2+} pair ($E_o = +0.77$ V) makes improbable that the reductive mechanism of RS-NO decomposition as in Scheme 1 actually works for iron as well. This redoxpotential is the value for aqueous solution and quoted in many textbooks. However, the redoxcouple of iron is notoriously sensitive to the ligand field around the iron as caused by ligands like thiols, solvent molecules like H_2O or OH^-, or small solutes like phosphate or Cl^-. The wide variation in redoxpotentials found in iron enzymes is a good illustration of this sensitivity. In addition, even when thermodynamically allowed by the redoxpotential, the actual transfer of electrons for certain redox reactions might be inhibited by geometrical constraints imposed by the ligands. Therefore, the presence of iron in an aqueous solution should be regarded as a superposition of iron pools spanning a considerable range of redoxpotentials.

Actual experiments demonstrate that ferrous iron catalyzes both decomposition and synthesis of RS-NO [10,15–19]. In both cases, ferrous iron does not change its redox state. It acts only as a complex-forming metal capable of binding NO, RS^- or RS-NO ligands. The d-electrons of the iron provide electronic coupling between the reactants necessary for the S-nitrosation/denitrosation reactions to proceed. Invariably, the combination of iron, thiols and free NO leads to the formation of paramagnetic dinitrosyl iron complexes (DNIC) with thiol-containing ligands [thiol-DNIC with formula $\{(RS^-)_2Fe^+(NO^+)_2\}^+$ were considered in Chapter 2]. The pathways of DNIC formation were investigated mainly by our groups [10,19,24], other investigators have not devoted extensive attention to these reactions. Since endogenous NO levels usually remain far below the micromolar range, the formation of a significant pool of a dinitrosyl species was expected to be highly improbable. However, our investigations have shown that such dinitrosyl complexes play an important role in the iron-catalyzed reactions between the pools of thiols and *S*-nitrosothiols. Up to our knowledge, there are only a few investigations devoting to the role of ferrous iron in the processes of RS-NO decomposition or synthesis [9,14–16]. The main objective of this chapter is to review the evidence that not only copper but also iron, particularly ferrous iron, determine the status of *S*-nitrosothiols in cells and tissues.

THE EXPERIMENTS WITH NEOCUPROINE SHOW THAT DECOMPOSITION AND SYNTHESIS OF RS-NO ARE CATALYZED BY COPPER AS WELL AS IRON

Experiments have shown that neocuproine (2,9-dimethyl phenanthroline) blocks the decomposition of RS-NO in cells and tissues. Neocuproine which is often considered to be a selective chelating agent for Cu^+ and its effect is often interpreted as proof that the presence of monovalent Cu^+ dominates the decomposition of nitrosothiols [20,22]. However, as discussed in Chapter 2, neocuproine interacts with the pool of loosely bound iron as well. EPR spectroscopy has demonstrated that the combination of nitric oxide, ferrous iron and neocuproine results in the formation of dinitrosyl iron complexes (DNIC) with neocuproine ligands. In fact, the presence of NO initiates the binding of neocuproine to ferrous iron. Depending on the pH, two species of DNIC could be distinguished with EPR [25]. The EPR spectra of these species are presented in Chapter 2. It was proved that neocuproine can compete with cysteine for $Fe^+(NO^+)_2$ moiety in DNIC: The addition of 12.5 mM neocuproine

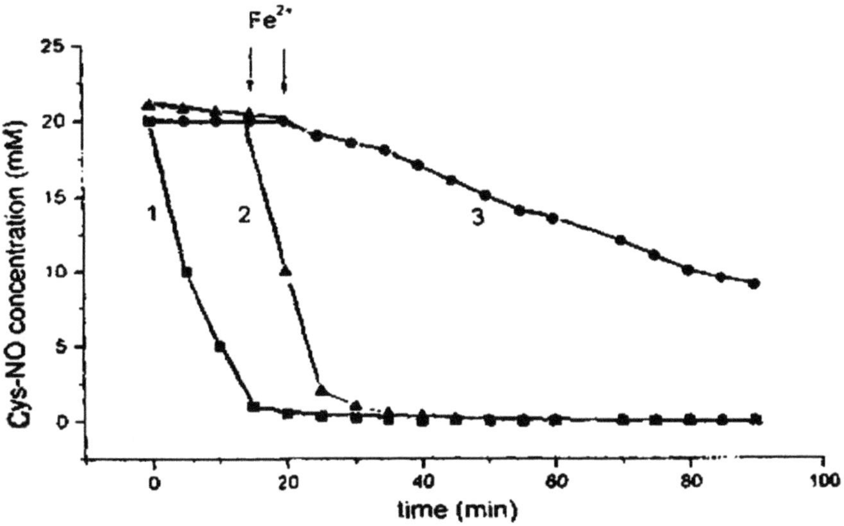

Fig. 1. Kinetics of the decomposition of 2 mM Cys-NO at ambient temperature as monitored by the optical absorption at 548 nm. (Curve 1) Without any additives, (Curve 2) With 2 mM cysteine and (Curve 3) With 2 mM neocuproine. The arrows indicate time points at which 0.5 mM Fe^{2+} was added. (From Ref. [25].)

to the 1.8 mM solution of DNIC with cysteine containing 7.2 mM cysteine at pH = 7.4 led to practically full transformation of Cys-DNIC into neocuproine-DNIC [25].

Interestingly, neocuproine is capable of protecting Cys-NO against the catalytic decomposition by ferrous iron (Fig. 1) [25]: The experiments were performed on 20 mM solutions of Cys-NO in 10 mM HEPES buffer, pH 7.4. The addition of 2 mM cysteine to the solution prevented rapid (for 10 min) decomposition of Cys-NO (Fig. 1, curves 1,2).

The decomposition was initiated at this condition by bolus addition of 0.5 mM ferrous iron and could be followed by optical absorption. The decomposition was quite fast and completed after ca 5 min (Fig. 1, curve 2). The decomposition could be inhibited if 2 mM neocuproine was added prior to ferrous iron (Fig. 1, curve 3). In presence of Cys-NO and neocuproine, the addition of ferrous iron induces formation of significant quantities of DNIC. Aliquots frozen immediately after the addition of iron showed intense EPR absorption by neocuproine-DNIC, observable at $g = 2.02$ (Fig. 2, curve b). The nature of this complex was unambiguously identified by the broadening of the EPR spectrum when ferrous ^{57}Fe iron was added to the solution. This broadening is characteristic for the hyperfine structure (HFS) from this iron isotope (Fig. 2, curve a).

These EPR experiments have revealed two important properties of neocuproine: First and foremost, that neocuproine is capable of chelating ferrous Fe^{2+} as well as Cu^{+}. Second, that the presence of free NO or nitrosothiols results in the formation of neocuproine-DNIC. The paramagnetism of the neocuproine-DNIC shows that the complex has an electronic configuration where the unpaired electrons of the nitrosyl ligands are largely transferred towards the iron. In Enemark–Feltham notation, the electronic state of the complex is written

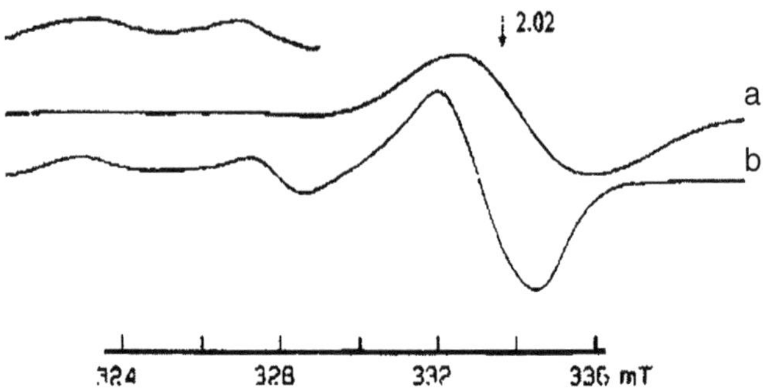

Fig. 2. The shape of the EPR spectra from ^{57}Fe-(or ^{56}Fe)–DNIC neocuproine (A,B). Doublet spectra above spectra A is low-field part of the latter recorded at higher amplification. The spectra were recorded at 77 K. (From Ref. [25].)

as $\{3d\}^7$ and the iron center acquires the character of Fe^+. In the absence of NO, the Fe^{2+} ions are not chelated by neocuproine because of rapid formation of water insoluble iron hydroxide complexes [25].

Bathocuproine disulfonate (BCS) is a highly soluble cuproine derivative. Vasodilatory studies of Cys-NO have shown that exogenous copper and iron inhibit the relaxation of precontracted endothelium-denuded rat aorta rings by Cys-NO, and that the inhibitory action of both metal ions is cancelled by BCS [17,18]. The addition of this chelator enhanced the duration of the vessel relaxation induced by Cys-NO in the presence of ascorbate, copper or iron (Fig. 3, curves f,g) [17].

Bathophenanthroline disulfonate (BPDS) is a potent chelator of transition metal ions and optical absorption experiments confirmed that it binds both iron and copper. Nevertheless, BPDS imparted significant protection of Cys-NO against iron only. It showed that chelation by BPDS inhibits the pathway for iron-catalyzed decomposition of Cys-NO thereby enhancing the duration of vessel relaxation (Fig. 3, curve d). In contrast, the copper-catalyzed pathway proceeded practically unimpaired (Fig. 3, curve e) [17].

The stability of Cys-NO was also studied *in vitro* in solutions containing ascorbate, copper or iron. *In vitro*, BCS protected Cys-NO effectively against catalytic decomposition by both metals but strong iron chelator bathophenanthroline disulfonate (BPDS) protected against ferrous iron only (Fig. 4) [17]. The observations were attributed to full saturation of the coordination sphere in the Fe^{2+}–BPDS complexes (i.e. the binding of three BPDS ligands to each iron) and preventing the iron from participating in redox reactions with Cys-NO [17,18]. In contrast, the monovalent copper ions bind only two BPDS ligands, and leaving the copper atom accessible to small molecules in the solution [26]. The experimental observations suggest that Cys-NO can penetrate into $Cu(BPDS)_2$ complexes and be decomposed *via* reduction by Cu^+ (Fig. 4) [17].

The catalytic decomposition of Cys-NO by copper–BPDS in the presence of BPDS and ascorbate was noticeable down to very low copper concentrations of about 2 μM (Fig. 4) [17,18]. This makes it questionable to attribute the protection of Cys-NO by cuproine derivatives to their sequestration of intrinsic Cu^+. For example, Fig. 5 (curve b) [17] shows that BPDS gives significant protection of Cys-NO in vasodilation experiments in

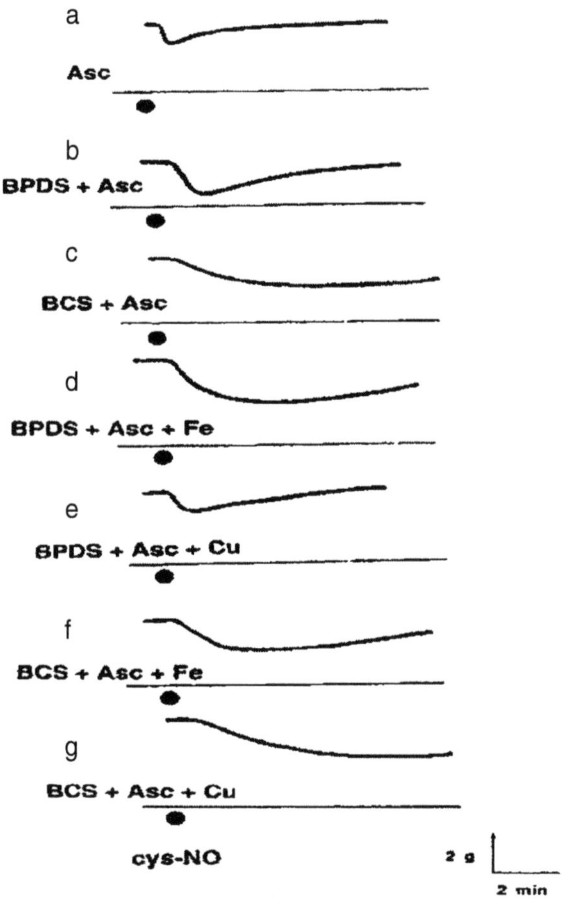

Fig. 3. Representative traces of the relaxant effect of Cys-NO (30 nM) in endothelium-denuded rat aortic rings precontracted with noradrenaline (0.1 μM). (Curve a) with 0.5 mM ascorbate only, (curve b) With 0.1 mM bathophenanthroline disulfonate (BPDS) and 0.5 mM ascorbate, (curve c) With 0.1 mM bathocuproine sulfonate (BCS) and 0.5 mM ascorbate, (curve d) BPDS + ascorbate + 250 nM Fe^{2+}, (curve e) BPDS + ascorbate + 250 nM Cu^{2+}, (curve f) BCS + ascorbate + 250 nM Fe^{2+} and, (curve g) BCS + ascorbate + 250 nM Cu^{2+}. (From Ref. [17].)

Krebs buffers. It means that intrinsic copper did not form the main pathway for the decomposition of Cys-NO in these experiments. We estimate the intrinsic copper content to be lower than 2 μM. Instead, the protection of Cys-NO was attributed to the sequestration and inactivation of intrinsic iron in the solutions. Phrased otherwise, intrinsic ferrous iron rather than intrinsic copper was seen to dominate the decomposition of Cys-NO.

In experiments with nitrosothiols, the presence of spurious quantities of reduced iron and copper is often found to be significant. Sheu et al. [27] investigated the iron and copper levels of sample solutions containing cysteine, glutathione (1–2 mM) or phosphate buffer (100 mM) and reported values of [Fe] ~ 1.8 μM and [Cu] ~ 0.05 μM. Therefore, spurious

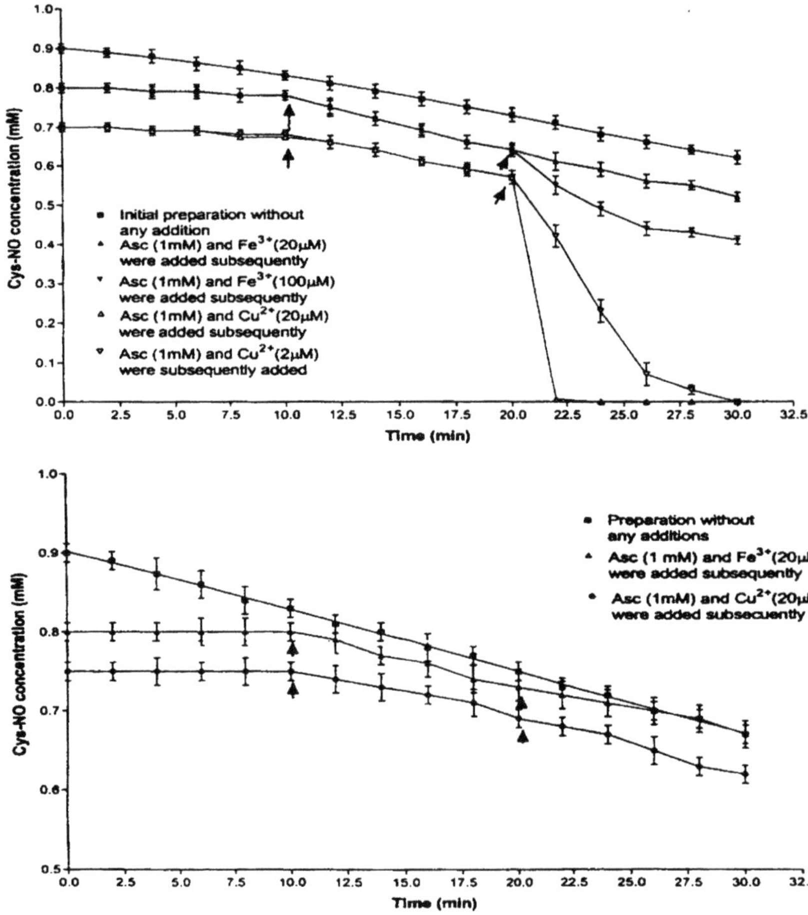

Fig. 4. Influence of subsequent additions (shown by arrows) of ascorbate, copper and iron on the stability of Cys-NO in the presence of 1 mM bathophenanthroline disulfonate (top) or bathocuproine sulfonate (bottom). Results are expressed as mean ± SE of three experiments [17].

iron comfortably exceeded copper (Fig. 4). Under these conditions, only the addition of BPDS would bind the intrinsic iron, extend the lifetime of Cys-NO and thereby make the vessel relaxation last longer.

The addition of BCS also prolongs the vasodilatory action of Cys-NO significantly (Fig. 5, curve c) [17], but the mechanism is very different. As remarked before, the combination of iron, Cys-NO and BCS leads to the formation of BCS–DNIC. This BCS–DNIC is quite stable and long-lived and was shown to induce a long-lasting vasodilation in aortic rings from which the endothelium had been removed. The vasorelaxation observed with BCS–DNIC was sustained significantly longer than that induced by DNIC with phosphate ligands (Fig. 6, curves a,b) [17].

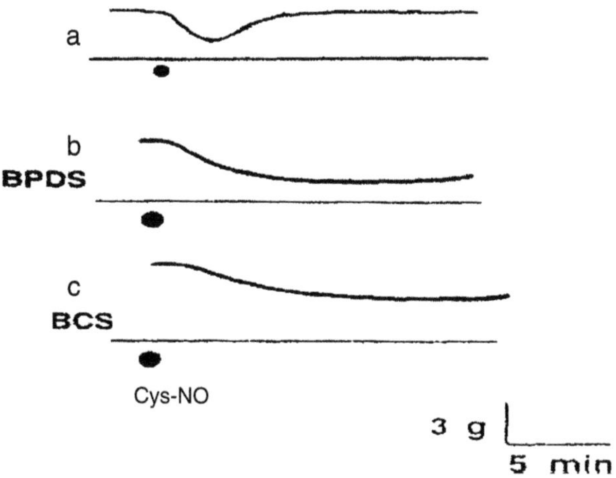

Fig. 5. Representative traces of the relaxant effect of Cys-NO (30 nM) in endothelium-denuded rat aortic rings precontracted with noradrenaline (0.1 μM). (Curve a) Effect of Cys-NO alone, (Curve b) Effect of Cys-NO in the presence of 0.1 mM BPDS and (Curve c) Effect of Cys-NO in the presence of 0.1 mM BCS. (From Ref. [17].)

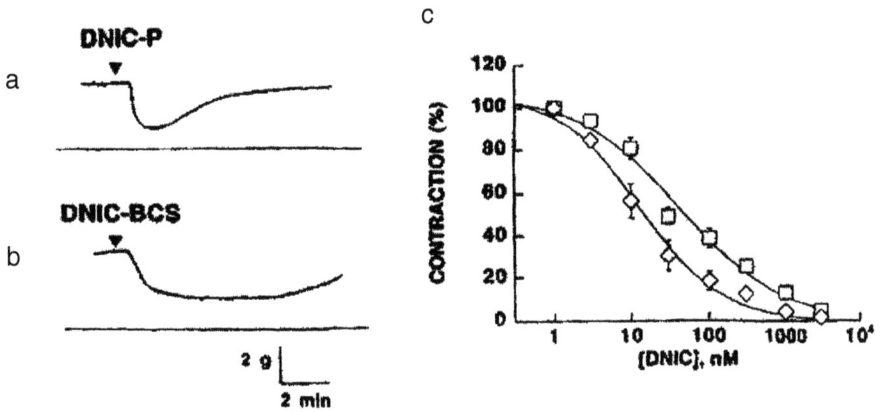

Fig. 6. Representative traces of the vasorelaxant effect of DNIC complexes in endothelium-denuded aortic rings from rats. The rings had been precontracted with 0.1 μM noradrenaline. (Curve a) 100 nM DNIC with phosphate ligands, (Curve b) 100 nM DNIC with BCS ligands and (Curve c) Dose response curve for maximal relaxation by DNIC with phosphate (diamonds) or BCS ligands (squares). (From Ref. [17].)

THE MECHANISM OF CATALYTIC DECOMPOSITION OF RS-NO BY FERROUS IRON

The pathway for copper-catalyzed decomposition of RS-NO was shown in Scheme 1. The pathway for catalytic decomposition by ferrous iron must be considerably more complex. This is clear from the appearance of paramagnetic dinitrosyl iron complexes (DNIC) in the

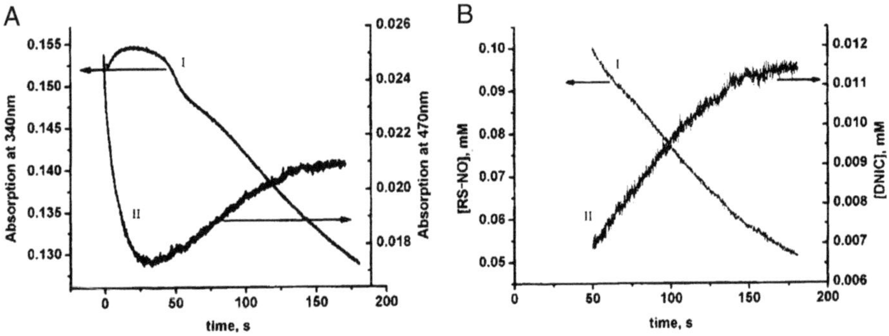

Fig. 7. (A) Kinetics of optical absorption at 340 nm (curve I) and 470 nm (curve II) in a solution of 40 μM Fe^{2+}, 125 μM S-nitrosocysteine and 1 mM cysteine in 100 mM HEPES (100 mM, pH = 7.4). (B) The same kinetics expressed in concentrations of Cys-NO (curve I) and Cys-DNIC (curve II). (From Ref. [19].)

mixtures of ferrous iron and nitrosothiols [17–19]. The amount of oxygen is always found to be a determining factor in the reaction chemistry. A typical experiment [19] involves the addition of 40 μM ferrous iron to an anaerobic solution of 125 μM Cys-NO + 1 mM cysteine in strong HEPES buffer (150 mM, pH 7.4). The optical absorption demonstrated rapid decomposition of Cys-NO within ca 2 min, with simultaneous formation of Cys-DNIC (Fig. 7). The decomposition of Cys-NO was followed by optical absorption at 340 nm ($\varepsilon_{340} = 778$ $\mathrm{Mcm^{-1}}$) and the formation of DNIC was followed at 470 nm ($\varepsilon_{470} = 990$ $\mathrm{Mcm^{-1}}$, cf Chapter 2). The optical absorption at 340 nm appears as a mixture of several species during the initial stages in time scale ca 50 ms (Fig. 7A). This first rapid step was attributed to a transient quantity of ferrous mononitrosyl complexes (Cys-MNIC, formula $\{(Cys^-)_4Fe^+NO^+\}^{2-}$ [28,29]), that subsequently transformed into Cys-DNIC.

As mentioned in Chapter 2, the stability of Cys-MNIC complexes depends on the ratio between NO and ferrous iron. With substoichiometric quantities of NO and Fe, the Cys-MNIC is fairly stable and long-lived. At ambient temperature in liquid solution, the complex gives an isotropic EPR signal at $g = 2.04$ with resolved triplet HFS from the NO ligand [19]. Precisely this EPR lineshape was observed in aliquots drawn 10 min after adding Cys-NO (0.4 mM final) to a solution of 0.5 mM Fe^{2+} and 10 mM cysteine in HEPES buffer (100 mM, pH 7.4) (Fig. 8A). The intensity of this EPR signal decreased during the time that was accompanied with the increase of the EPR signal from Cys-DNIC (Fig. 8B).

When Cys-NO was present in sufficient excess with respect to ferrous iron, formation of Cys-DNIC was nearly quantitative (Fig. 9A). Quantitative incorporation of all iron into Cys-DNIC was also observed when Cys-NO was replaced with 0.2 mM NO-proline (Fig. 9B). This NO donor releases two free NO molecules within seconds.

Addition of deoxy-Hb, an effective scavenger of free NO, did not influence the formation of DNIC from Cys-NO in presence of ferrous iron and cysteine but completely inhibited the formation of DNIC from NO-proline (Fig. 10D,B, respectively).

This proves that Cys-NO promotes DNIC formation by direct transfer of the NO moiety to the iron. This transfer proceeds by direct contact and not *via* release of free NO as an

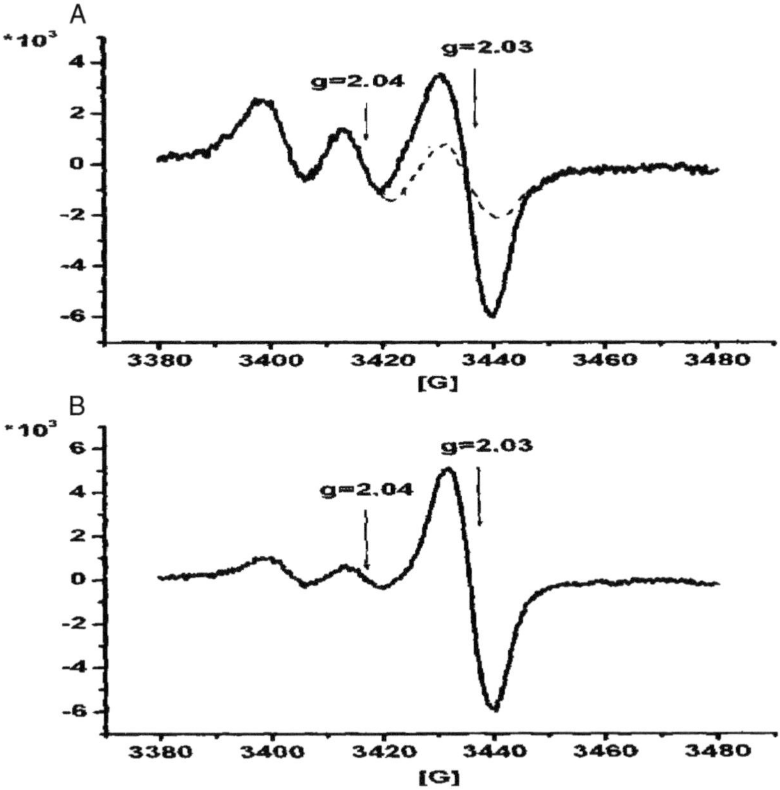

Fig. 8. EPR spectra at room temperature of aliquots of 500 μM Fe^{2+}, 10 mM cysteine, and 400 μM Cys-NO in HEPES (100 mM, pH 7.4). Panel A: 10 min after mixing. Panel B: 20 min after mixing. The triplet signal at $g = 2.04$ is ferrous Cys-MNIC. The singlet at $g = 2.03$ is Cys-DNIC. (From Ref. [19].)

intermediate step. Once Cys-DNIC had been formed, its lifetime was significantly shortened to a few minutes by the presence of deoxy-Hb (Fig. 10C,D). We attribute the decay of Cys-DNIC to scavenging of NO ligands from the DNIC by deoxy-Hb. This interpretation is compatible with the concept of a kinetic equilibrium between DNIC and its constituents as given in Scheme 5. This equilibrium implies the presence of a small quantity of free NO molecules in the solution:

The kinetics of Fig. 10D show that the destruction of Cys-DNIC by deoxy-Hb does not start immediately, but start after an "induction period" of about 3 min. We attribute this "induction period" to the presence of excess Cys-NO. The latter provides a pool of NO to regenerate Cys-DNIC, thereby maintaining Cys-DNIC at a stationary level until the Cys-NO is exhausted.

It is remarkable that the presence of a NO scavenger like deoxy-Hb does not influence the formation of DNIC from Cys-NO and ferrous iron. In particular, the yields of Cys-DNIC yields are unaffected (Fig. 10D). It proves that the mechanism of formation does not require free NO radicals, and does not involve the equilibrium reaction of Scheme 5. The reaction

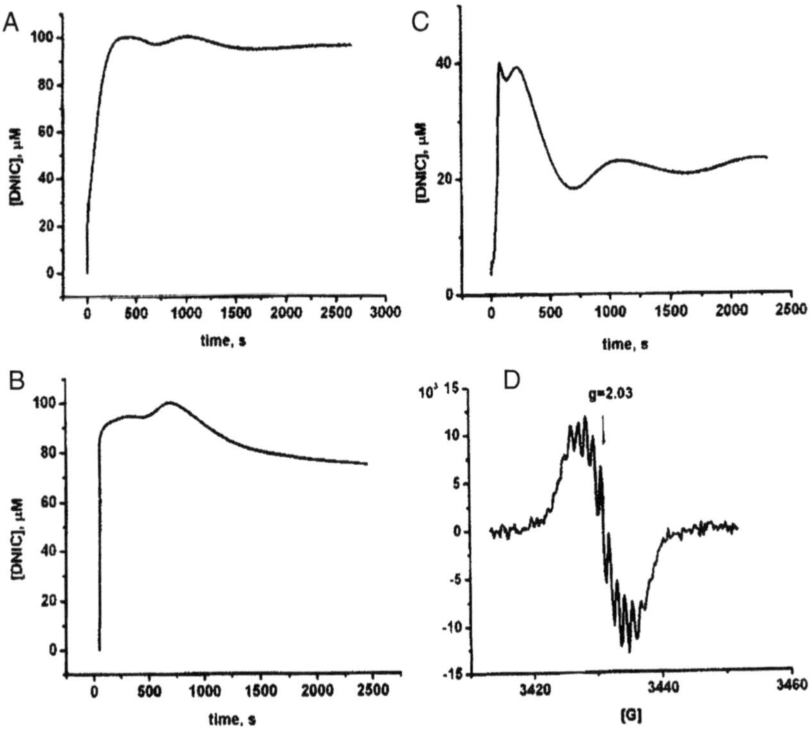

Fig. 9. Kinetics of the formation of Cys-DNIC in HEPES buffer (100 mM, pH = 7.4) as detected by EPR. (A) Mixture of 20 mM cysteine, 400 μM Cys-NO, 100 μM Fe^{2+} and 500 μM sodium citrate. (B) Mixture of 20 mM cysteine, 200 μM NO-proline, 100 μM Fe^{2+} and 500 μM sodium citrate. (C) Mixture of 10 mM cysteine, 400 μM Cys-NO, 50 μM Fe^{2+} and 250 μM sodium citrate and (D) Complete EPR spectrum of the solution (A). (From Ref. [19].)

kinetics was studied with EPR and optical stopped-flow techniques. The analysis showed that formation of Cys-DNIC was first order in ferrous iron and second order in Cys-NO

$$\frac{d}{dt}[\text{DNIC}] = k[\text{Fe}][\text{Cys-NO}]^2$$

The calculation of the third-order rate of formation gave $k = (1.0 \pm 0.2) \times 10^5\ \text{M}^{-2}\text{s}^{-1}$ (estimated by EPR method) or $(2.0 \pm 0.1) \times 10^5\ \text{M}^{-2}\text{s}^{-1}$ (estimated by optical method). The possible reason for the small discrepancy was considered in [19]. It should be noted that DNIC is formed from Cys-NO at a fourfold faster rate than from free NO molecules [19]. This is significant because free NO has a far higher diffusive mobility than Cys-NO due to its smaller size.

Once formed, the Cys-DNIC is not rigorously stable, but has a finite lifetime that depends on the levels of its reaction partners in the chemical equilibria. The excess concentration of free thiol ligands is a particularly important parameter: At lower concentration of free cysteine, the lifetime of paramagnetic Cys-DNIC forming from Cys-NO and iron notably diminished

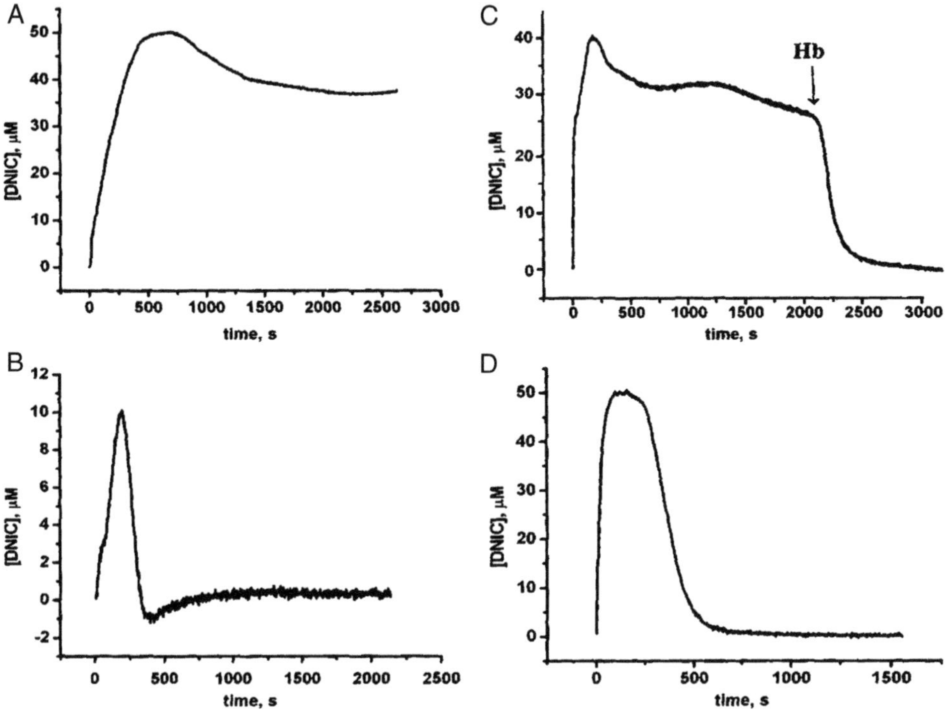

Fig. 10. Kinetics of Cys-DNIC as detected with EPR in HEPES (100 mM, pH 7.4). (A) With 20 mM cysteine, 200 μM NO-proline, 50 μM Fe^{2+} and 250 μM sodium citrate. (B) With 20 mM cysteine, 200 μM NO-proline, 100 μM Hb, 50 μM Fe^{2+} and 250 μM sodium citrate. (C) With 20 mM cysteine, 400 μM Cys-NO, 50 μM Fe^{2+} and 250 μM sodium citrate. At the arrow, 100 μM Hb is added to scavenge NO and (D) With 20 mM cysteine, 400 μM Cys-NO, 100 μM Hb, 50 μM Fe^{2+} and 250 μM sodium citrate. (From Ref. [19].)

$$\{(RS^-)_2Fe^+(NO^+)_2\}^+ \leftrightarrow Fe^{2+} + NO + \underbrace{(RS^- + NO^+)}_{\downarrow\ \text{RS-NO}} + RS^-$$

Scheme 5. The kinetic equilibrium between thiol-DNIC and its constituents [19].

(Fig. 11A,B). However, subsequent addition of 20 mM cysteine completely recovered the EPR signal from Cys-DNIC (Fig. 11C).

The excess free cysteine was consumed during the regeneration of Cys-DNIC from Cys-NO. As described in Chapter 2, the drop in the free cysteine must shift the equilibrium between monomeric and dimeric Cys-DNIC towards the dimeric diamagnetic form. Phrased otherwise, the equilibrium of Scheme 6 is shifted towards left-hand side.

For glutathione, the same reaction equilibria apply as for the cysteine discussed above, albeit with slower reaction rates and lower yields. The yields of GS-DNIC from S-nitrosoglutathione (GS-NO) and ferrous iron in the presence of glutathione were an order

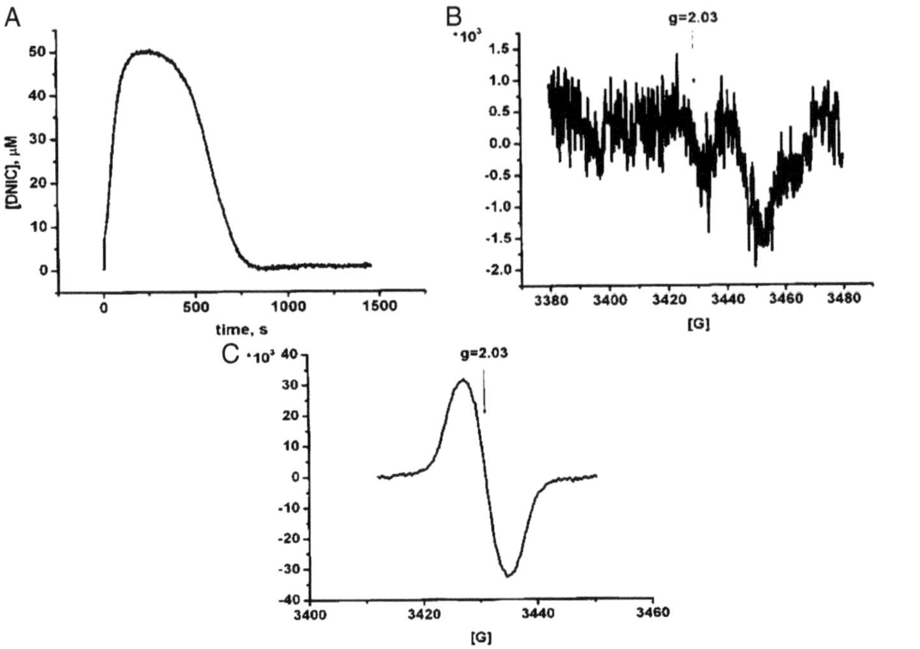

Fig. 11. Formation of Cys-DNIC as detected with EPR in HEPES (100 mM, pH 7.4) at room temperature. (A) Kinetics of Cys-DNIC formation in the mixture of 1 mM cysteine, 400 μM Cys-NO, 50 μM Fe^{2+} and 250 μM sodium citrate. (B) The EPR spectrum of the solution at the end of reaction at 1500 s and (C) EPR spectrum of sample (B) after addition of 20 mM cysteine. (From Ref. [19].)

$$2\ \{(RS^-)_2Fe^+(NO^+)_2\}^+ \underset{+RS^-}{\overset{-RS^-}{\rightleftharpoons}} \{(RS^-)_2Fe^+(NO^+)_2\}_2^{2+}$$

Scheme 6. The equilibrium between monomeric and dimeric forms of thiol-DNIC is determined by the quantity of excess free thiol.

of magnitude lower: EPR showed that only a small fraction of GS-DNIC was of paramagnetic monomeric form, whereas the majority was dimeric diamagnetic DNIC. This finding is in line with Chapter 2, where the monomeric/dimeric ratio at neutral pH was found to depend on the Fe^{2+}/thiol ratio, and on the thiol used. For our conditions, cysteine preferably forms monomeric DNIC, whereas other thiols like glutathione preferably form dimeric DNIC.

The experimental data could be explained by the following mechanism for the formation of Cys-DNIC from Cys-NO and iron (Scheme 7):
First rapid step is the formation of intermediate paramagnetic Cys-MNIC:

$$(RS^-)_nFe^{2+} + \text{NO-RS}^\bullet \Longleftrightarrow (RS^-)_nFe^{2+}....\text{NO-RS}^\bullet$$
$$\Longleftrightarrow (RS^-)_4Fe^+\text{-NO}^+ + RS^\bullet + (RS^-)_{n-4}$$

Second step is the formation of DNIC by transfer of a nitrosonium from the nitrosothiol to the iron:

$$(RS^-)_4Fe^+\text{-}NO^+ + NO^+\text{-}RS^- \Longleftrightarrow (RS^-)_4Fe^+\text{-}NO^+(NO^+\text{-}RS^-)$$

$$\longrightarrow (RS^-)_2Fe^+(NO^+)_2 + RS^-$$

Taken together, the net reaction can be presented as:

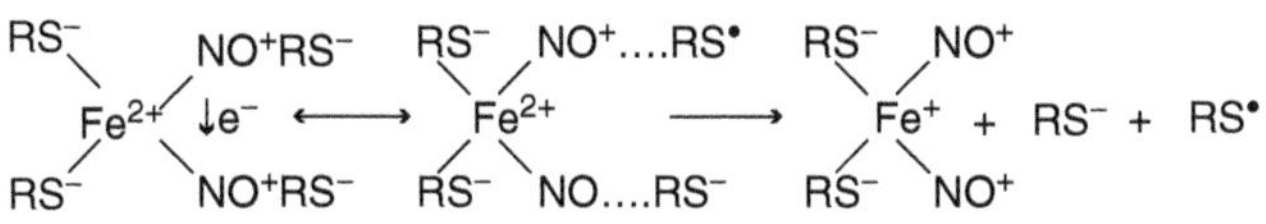

Scheme 7. The formation of DNIC with cysteine from Cys-NO and ferrous iron.

Taken into account the quasi-equilibrium between DNIC and its constituents (Scheme 5), the net process can be presented as follows:

$$RS\text{-}NO \xrightarrow[Fe^{2+}]{} NO + \tfrac{1}{2}(RS\text{-}SR)$$

In this form, the result resembles the reductive degradation of RS-NO catalyzed by copper ions (Scheme 1). However, copper acts by true redox cycling, whereas iron acts by cycling through a DNIC intermediate.

It is significant that the upgrade from MNIC to DNIC is achieved by transfer of a nitrosonium moiety from an *S*-nitrosothiol rather than by a free NO. Therefore, the formation of DNIC does not require successive capture of two free NO molecules, which would be difficult to achieve at the nanomolar NO concentrations under normal physiological conditions. Instead, the second nitrosyl moiety is taken as a nitrosonium from an *S*-nitrosothiol, which is found at far higher concentrations in the low micromolar range (cf Chapters 9 and 10).

For cysteine ligands, all steps in the process of Cys-DNIC formation could be studied *in vitro*. We propose that the same reaction equilibria also apply to mixtures of iron with other thiols and their *S*-nitrosothiols.

A less stable DNIC with phosphate appeared when the solutions of Fe^{2+}-citrate in 100 mM phosphate buffer (pH 7.4) were mixed with Cys-NO (Fig. 12A). The EPR signal of the complex recorded at ambient temperature had a singlet shape at $g = 2.032$ with septet hyperfine structure from nitrogen atoms of two NO ligands and phosphorus atoms of two phospate ligands (Fig. 12D). The rapid formation of DNIC with phosphate was followed by its rapid decay.

Similar kinetics was observed if NO-proline was added instead of Cys-NO (Fig. 12C). Low stability of DNIC with phosphate could be due to hydrolysis of NO^+ groups in the complex as shown in Scheme 8:

Thiol-DNIC has far higher stability and longer lifetime in aqueous solution. We attribute this to electrostatic binding of thiol anions with NO^+ moieties and formation of coordinated RS-NOs. However, due to electron migration from Fe^+ to one NO^+, one RS-NO ligand can be decomposed resulting in the chemical equilibrium (Scheme 9) similar to that shown in Scheme 5:

The mechanism of RS-NO decomposition catalyzed by iron and presented in Scheme 7 can be true for the proteins containing thiol ligands. Evidently due to high affinity of protein-bound

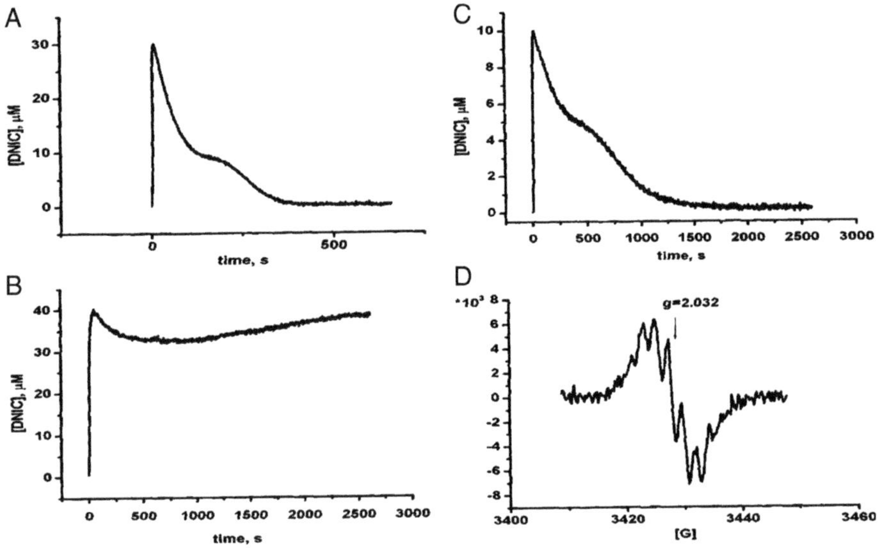

Fig. 12. Formation of the DNIC-phosphate complex as detected by EPR at room temperature in HEPES. (A) Kinetics of the DNIC-phosphate complex in a solution with 100 mM phosphate, 400 μM Cys-NO, 50 μM Fe^{2+} and 250 μM sodium citrate. (B) Idem with 20 mM cysteine. (C) 100 mM phosphate, 400 μM NO-proline, 50 μM Fe^{2+} and 250 μM sodium citrate and (D) The EPR signal of the solution (C). (From Ref. [19].)

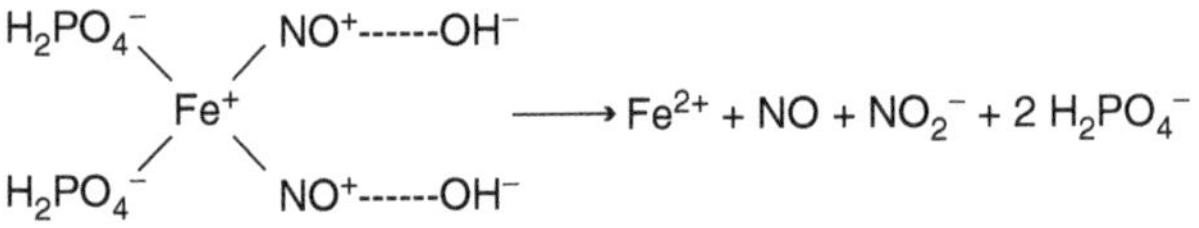

Scheme 8. The proposed mechanism of the decomposition of DNIC with phosphate [18,19].

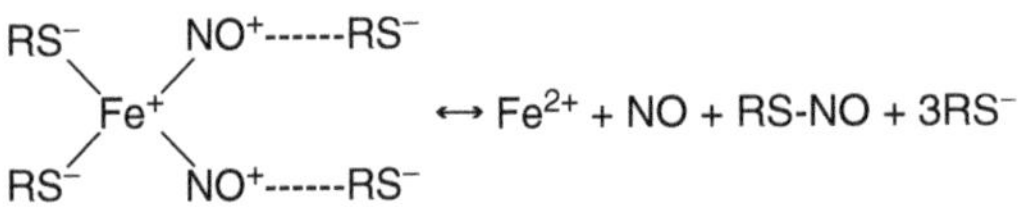

Scheme 9. The structure of DNIC with thiol-containing ligands in equilibrium with its constituents.

thiol ligands to $Fe^+(NO^+)_2$ groups [30], forming DNICs could be bound mainly with protein globula. The studies showed that really the addition of Cys-NO alone to the animal tissue preparations resulted in the formation of protein-bound DNIC [31]. However, if excess cysteine or other low-molecular thiols are added to the protein-bound RS-NO, the EPR spectra showed that low-molecular DNIC became predominant. The transformation to the low-molecular form was evident from the motional narrowing of the EPR signal with $g = 2.03$ when the spectra were recorded at room temperature [30].

The above reactions proceeded in deoxygenated solutions. The admission of oxygen significantly changes the kinetics of the reactions [32]. Figs. 13 and 14 document the decomposition

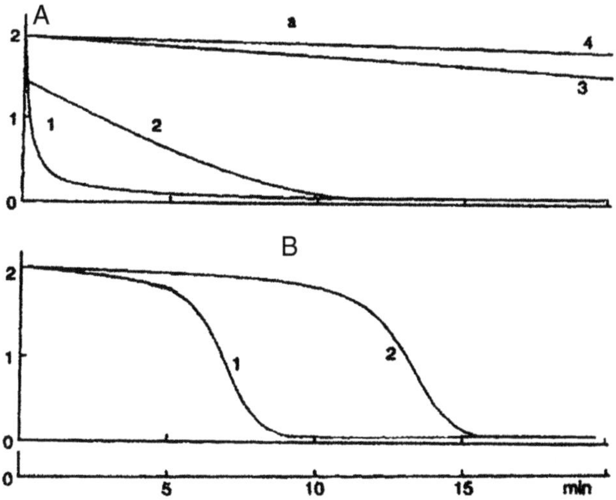

Fig. 13. Effect of desferal and cysteine on the decomposition of Cys-NO as monitored by the optical absorption at 340 nm. The kinetics was studied at room temperature in HEPES open to ambient air (15 mM, pH 7.4). Panel A: (1) decomposition of 2 mM Cys-NO without additives; (2) decomposition of 1.5 mM Cys-NO in presence of 0.5 mM cysteine; (3) 2 mM Cys-NO in presence of 1 mM desferal; (4) 2 mM Cys-NO in presence of 1 mM cysteine. Panel B: (1) 2 mM Cys-NO in presence of 0.1 mM cysteine; (2) 2 mM Cys-NO in presence of 0.25 mM cysteine. (From Ref. [32].)

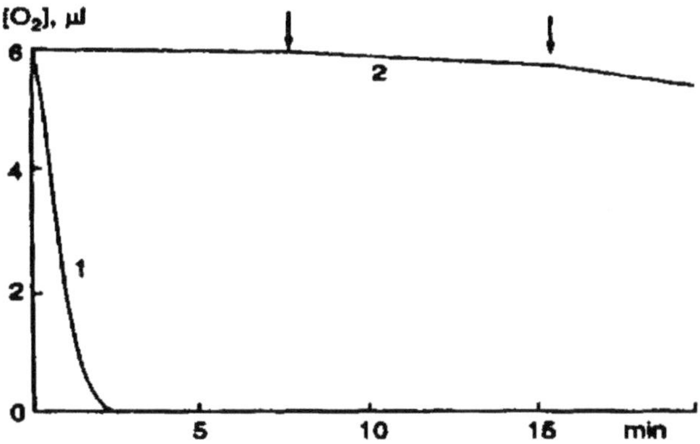

Fig. 14. Kinetics of O_2 consumption by 1 ml solutions of Cys-NO in HEPES (15 mM, pH 7.4). (Curve 1) 2 mM Cys-NO gives rapid consumption of oxygen and (curve 2) 5 mM cysteine does not consume oxygen until Cys-NO is added at 8 min ([Cys-NO] = 2 mM final) or 15 min (4 mM final). (From Ref. [32].)

of 2 mM Cys-NO in HEPES buffer (15 mM, pH 7.4) open to ambient air and at room temperature.

At these conditions, the oxygen concentration is ca 0.25 mM, and the decomposition of Cys-NO was completed within several minutes (Fig. 13, Panel A curves 1,2). The decomposition was inhibited by 1 mM iron chelator desferal or by 2 mM cysteine (Fig. 13, Panel A curves 3,4). Cysteine was found to inhibit only at concentrations significantly higher than the oxygen concentration of 0.25 mM. At lower doses, Cys-NO was stabilized only for a short period of time after which rapid decomposition took place (Fig. 13, Panel B curves 1,2).

In the absence of excess free cysteine, the decomposition of 2 mM Cys-NO caused the loss of oxygen from the solution (Fig. 14, curve 1). This consumption of oxygen was halted if the decomposition of Cys-NO was inhibited by either desferal or excess cysteine (Fig. 14, curve 2). Cysteine by itself did not detectably affect the oxygen concentration in HEPES buffer. In the presence of Cys-NO and excess cysteine ligand, a small quantity of ca 1 μM Cys-DNIC was detected by EPR on frozen aliquots of the buffer solution. Since no iron had been added, the DNIC was attributed to the pool of spurious iron in the solutions and reaction vessels. The chelation experiment with desferal shows that this pool of spurious iron was responsible for the observed decomposition of Cys-NO.

The observations are interpreted by the following hypothesis [32]: In oxygenated solutions, the spurious iron is largely ferric. In the absence of free cysteine, the decomposition of Cys-NO is initiated by oxidation by ferric iron as shown in Scheme 10. This scheme is similar to Scheme 2 for the oxidative mechanism of RS-NO decomposition catalyzed by copper. The oxygen rapidly restores the iron to ferric state by oxidation

$$Fe^{3+} + RS\text{-}NO \longrightarrow Fe^{2+} + RS^{\bullet} + NO^{+}$$

$$Fe^{2+} + O_2 \longrightarrow Fe^{3+} + 2O_2^{-}$$

$$\text{net reaction: } O_2 + RS\text{-}NO \xrightarrow[Fe]{} \tfrac{1}{2}(RS\text{-}SR) + NO^{+} + O_2^{-}$$

Scheme 10. The oxidative mechanism of RS-NO decomposition catalyzed by iron [32].

The reaction is made irreversible by combination of the thiyl radicals to disulfide and rapid hydrolysis of NO^+ to nitrite.

THE MECHANISM OF RS-NO SYNTHESIS CATALYZED BY IRON

In principle, one could expect that ferric iron catalyzes the synthesis of RS-NO from thiols and NO through the oxidative pathways similar to those shown for Cu^{2+} ions in Schemes 3 and 4. However, cells and tissues contain mM levels of glutathione. With [GSH]/[GS-SG] $\sim$ 100 firmly on the reduced side, one would expect that endogenous iron be kept dominantly ferrous state. In buffered solutions, the redox state of iron is easily determined but the situation in biological systems is not clear, and the existence of an endogenous pool of ferrous iron in cells and tissues is still a matter of debate. EPR spectroscopy of frozen tissues usually shows a significant quantity of ferric iron complexes in high-spin state (Fe^{3+} with $S = 5/2$ is observable with EPR at the characteristic position $g \sim 4.3$. cf Chapters 2 and 5). But there is consensus that ferrous irons provide at least a large or dominant fraction of the loosely bound iron pool. In recent years, EPR spectroscopy has provided strong experimental support for the catalytic action of ferrous iron in the synthesis of endogenous RS-NO, but the reaction mechanism is very different from that of monovalent Cu^{+} in that EPR always shows the formation of some quantities of dinitrosyl–iron complexes. Chapter 2 discussed the properties of DNICs with thiol-containing ligands and noted the possibility that the iron atom mediate the disproportionation of the nitrosyl ligands into nitrosonium cation and nitroxyl anion, respectively. Nitrosonium has known capability to S-nitrosate thiols. We propose that the DNIC contribute significantly to the catalytic S-nitrosation of thiols. The disproportion is an important intermediate step in this catalytic cycle and was demonstrated in DNIC with thiol ligands as well as with ligands of the non-thiol class (Chapter 2).

We demonstrated [33,10] that the formation of Cys-NO and GS-NO from thiols and free NO is catalytically accelerated by the presence of ferrous iron. These experiments were carried out in anaerobic aqueous solutions. The first investigation [33] found that exposure of 1–3 mM cysteine or glutathione in HEPES buffer (15 mM, pH 7.0) caused formation of Cys-NO or GS-NO. The *S*-nitrosothiols were detected by their optical absorption at 340 nm. The yield of RS-NO calculated on a thiol basis reached 30–40%. The formation of RS-NO was enhanced 1.5–2 times by the addition of ferrous salt (to 20 μM) and completely suppressed by the selective iron chelator *o*-phenanthroline (250 μM). These results prove that ferrous iron, whether from spurious contaminations or exogenously added, contributes to the process.

The effects of pH and oxygen were studied in a subsequent investigation [10]. The yields of Cys-NO or GS-NO are higher at low pH (Fig. 15, curve a). The experiments were performed with 50 mM solutions of cysteine or glutathione in 2 ml deoxygenated HEPES buffer (15 mM) in an evacuated Thunberg vial with 100 ml volume. With an NO gas pressure of 125 mm Hg, the quantity of gaseous NO in the head space of the solution was 1.00 ± 0.05 mmol. No exogenous iron was added. Under exclusion of oxygen, only 4–6% of the thiol was converted to nitrosothiol. The conversion increased to 25% when 8 μmol of O_2 was added together with 1 mmol NO in acid and neutral thiol solutions (Fig. 15, curve b). The pH remained stable, showing that quantities of NO_2, if formed at all, remained far below the buffering capacity of the HEPES. The conversion ratio was larger if the initial thiol concentration was decreased. Table 1 quotes the conversion ratios for glutathione in HEPES.

In solutions with higher thiol concentration the conversion was incomplete, but could be enhanced by repeated treatment of these solutions with the $NO + O_2$ mixture. The conversion has a clear maximum as a function of the NO dose. At a dose of 8 μmol O_2, the RS-NO yield was highest at NO pressures of 60–150 mm Hg (0.5–1.2 mmol NO in total). At still higher pressures the RS-NO yield decreased, finally approaching zero at a NO pressure of 600–700 mm Hg.

The addition of selective Fe^{2+} chelator, *o*-phenanthroline (250 μM) prior to addition of NO or O_2 completely inhibited the formation of RS-NO. The effect is illustrated in Fig. 16. It shows the effect of *o*-phenanthroline on the optical absorption of a solution with 50 mM glutathione. The formation of GS-NO is completely inhibited by the presence of *o*-phenanthroline. It shows that NO itself is not a good S-nitrosating agent by itself. Rather, the catalytic action of ferrous iron is needed to achieve the S-nitrosation of glutathione. Further evidence is the enhancement of RS-NO yields by the addition of extra Fe^{2+} to thiol solutions. GS-NO yields were enhanced by a factor 1.5 if 20–200 μM Fe^{2+} was added to the solutions prior to the exposure to the mixture of NO and O_2. These remarks apply to pH < 7 (Fig. 15).

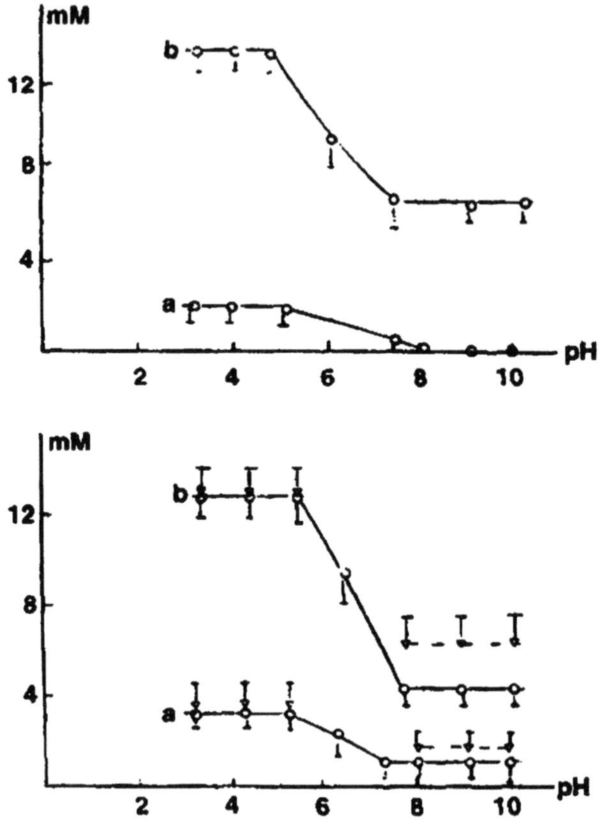

Fig. 15. The pH dependence of the yields of Cys-NO (top) and GS-NO (bottom) in 2 ml HEPES after exposure to nitric oxide. The thiol concentration was 50 mM. (Curve a) Low yields after exposure to 1 mmol NO in absence of oxygen and (Curve b) Much higher yields after exposure to a mixture of 1 mmol NO + 8 μmol O_2. The triangles in the lower panel show the yields if 50 μmol $FeSO_4$ was added prior to the treatment with NO or NO + O_2, respectively. (From Ref. [10].)

Table 1 Ratios of conversion of glutathione to GS-NO by spurious iron in 2 ml HEPES (15 mM, pH = 8.0) by gaseous NO in presence of oxygen. The NO and O_2 doses are 0.65 mmol and 8 μmol, respectively. The conversion ratios were determined with optical absorption spectroscopy. The reaction proceeded at room temperature for 10 min. The conversion could be inhibited by 250 μM *o*-phenanthroline

Starting [GSH] (mM)	1	5	10	50
Conversion of GSH (%)	100	80	60	10

Addition of extra 20–200 μM Cu^{2+} did not significantly affect the GS-NO yields. More details of the S-nitrosation of thiols are given in Chapter 9 of this book.

Evidently, if ferrous iron catalyzes the formation of RS-NO, it is unavoidable that a certain quantity of DNIC is formed from spurious iron and free NO molecules. When liganded to iron, the electronic coupling between the nitrosyl ligands induces a certain degree of dismutation,

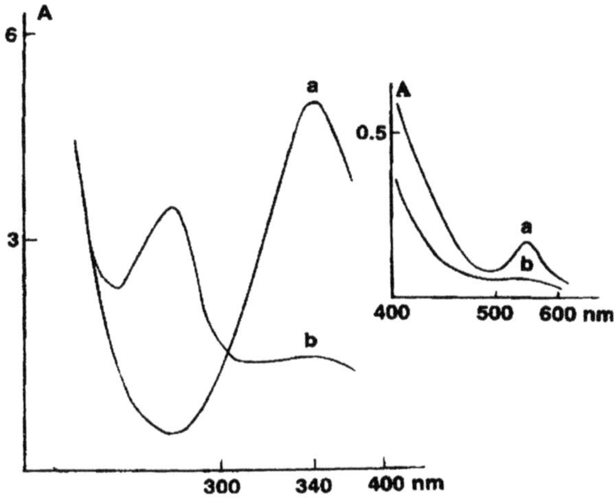

Fig. 16. UV/Vis spectra from 50 mM glutathione in HEPES (15 mM, pH 7.4). 2 ml of the solution was exposed for 5 min to a gas mixture of NO (1 mmol) and air (8 μmol O_2 administered with air). (Curve a) In absence of phenanthroline and (Curve b) In presence of 250 μM *o*-phenanthroline.

and imparts the character of nitrosonium to one of the nitrosyl ligands in the DNICs (the non-equivalence of the two nitrosyl ligands in DNIC was discussed in Chapter 2). Nitrosonium itself is known as a powerful nitrosating agent [7], and may react readily with cysteine to form Cys-NO. It is important to note the stoichiometry of the reactions. In [31], the formation of Cys-NO from Cys-DNIC was shown to be reversible as well as quantitative. It was induced by rapid acidification of the Cys-DNIC solution from pH 7 to 1. Initially, the solution had green color characteristic of Cys-DNIC. This color changed rapidly to pink and a new optical absorption band at 340 nm demonstrated the formation of free Cys-NO. EPR spectroscopy confirmed that the DNIC signal at $g = 2.03$ had vanished. Within the experimental accuracy of ca 10%, the yield of Cys-NO was equimolar to that of the original Cys-DNIC. The reverse transformation to Cys-DNIC was induced by raising the pH from 1 to 7. The pink color reverted to green and EPR showed that some 50% of the original Cys-DNIC was recovered. We attribute the recovery of Cys-DNIC to the reaction of two Cys-NO molecules with ferrous iron as illustrated above by Scheme 7. The formation of equimolar Cys-NO was attributed to S-nitrosation of one free cysteine molecule by the nitrosonium ligand in Cys-DNIC as shown in Scheme 9.

S-nitrosation of GSH can also be induced by exposing DNIC to citrate, a good chelator for iron [32]. The reaction was demonstrated in the solution of 0.4 mM dimeric Cys-DNIC. As described in Chapter 2, dimeric Cys-DNIC is obtained by bubbling purified NO gas through a solution of ferrous iron and cysteine (Fe: Cys = 1:2). The vial was subsequently evacuated to remove dissolved free NO from the solution. The formation of dimeric DNIC was confirmed by the optical absorbance at 310 and 360 nm (cf Chapter 2). At this stage, nearly all NO is sequestered in the form of dimeric DNIC.

In absence of free NO, the dimeric DNIC will slowly decompose according to the equilibrium in Scheme 5, releasing 2NO and $2NO^+$ moieties from one dimer. Strong quantitative support for Scheme 5 came from experiments where the decomposition was accelerated by the addition of an iron chelator, in this case citrate. Subsequently, the complexes were incubated with 1.6 M sodium citrate for 90 min. one mM of glutathione was also added to scavenge nitrosonium (NO^+) ions released from the DNIC. This scavenging reaction has the end product GS-NO. The optical absorbance at 310 and 360 nm confirmed that citrate completely decomposed the dimeric Cys-DNIC. A new band at 340 nm showed the formation of the Fe^{3+}–citrate complex ($\varepsilon_{340} = 1800\ M^{-1}cm^{-1}$), and the optical absorption also confirmed the formation of ca 0.80 ± 0.05 mM *S*-nitrosothiols. This concentration confirms that at this stage, all available NO^+ has been incorporated into the pool of *S*-nitrosothiols, probably as a mixture of Cys-NO and GS-NO. As before, the NO ligands are believed to be lost as NO gas into the headspace of the reaction vessel.

After subsequent addition of 20 mM cysteine and 0.5 mM Fe^{2+}, the solution took the green color of monomeric DNIC. EPR spectroscopy on a frozen aliquot confirmed the formation of ca 0.40 ± 0.04 mM of monomeric DNIC. Clearly, all available NO in the pool of *S*-nitrosothiols had been incorporated into the monomeric DNIC. The formation of the DNIC was evidently due to the reaction of cysteine and iron ions with the GS-NO and Cys-NO formed during the decomposition of the initial DNIC 1:2.

The preceding experiments document the reversible S-nitrosation of low-molecular-weight thiols like GSH or cysteine. However, the same reaction pathways also apply to thiol groups in proteins like bovine serum albumin (BSA) or horse hemoglobin (Hb) [10]. Exposure of 1 mM protein-bound DNICs to a combination of NO + O_2 causes loss of DNIC and S-nitrosation of the protein [10]. The complexes were obtained by addition of 1 mM DNIC

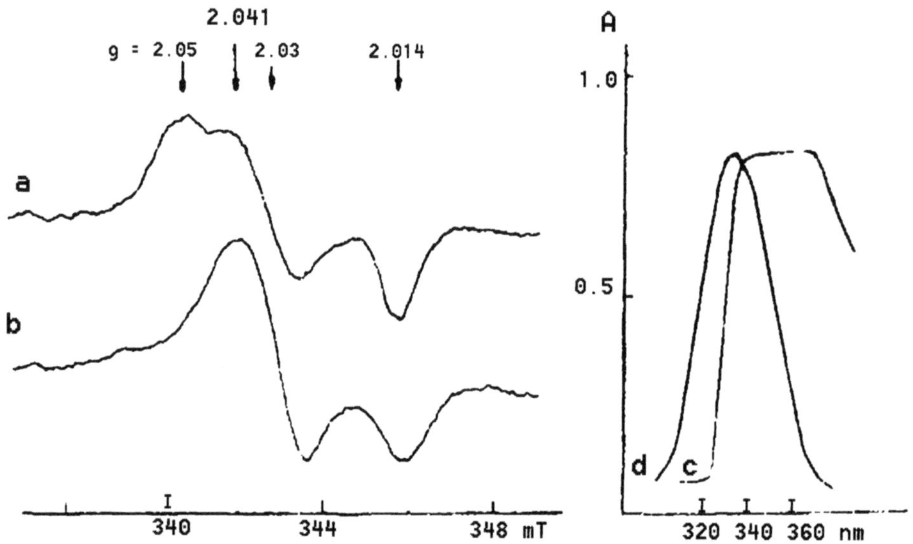

Fig. 17. The left panel shows the EPR spectra from 1 mM solutions or bovine serum albumin (a) or horse hemoglobin (b) after addition of 1.0 mM DNIC with phosphate ligands. The spectra were recorded at ambient temperature. The right panel shows the optical absorptions of BSA (d) and horse Hb (d) after exposure f 5 min to a gas mixture of NO + air. The optical spectra were recorded with the reference cell filled with 1 mM albumin or hemoglobin solutions pretreatred with the same gas mixture but avoiding the preliminary contact of protein with DNIC. (From Ref. [10].)

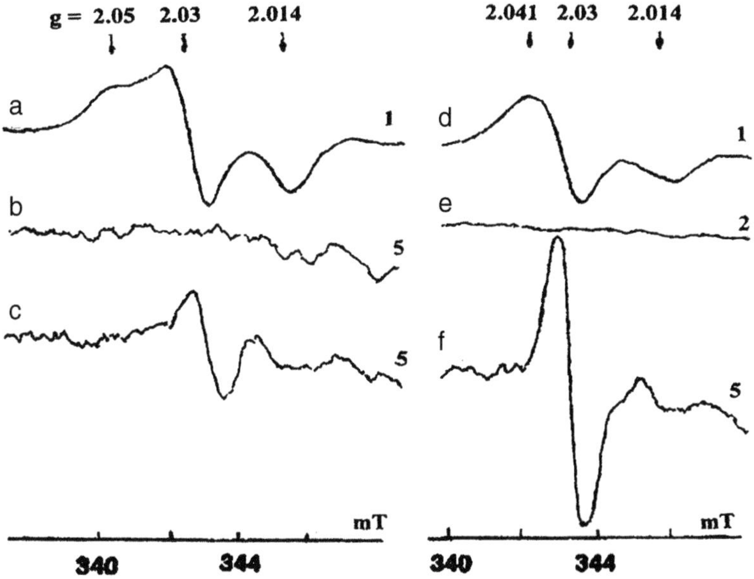

Fig. 18. EPR spectra from 1 mM solutions of bovine serum albumin (left column) and horse hemoglobin (right column). (Samples a,d) After addition of 1 mM phosphate-DNIC. (Samples b,e) Samples a and d after exposure to NO + air gas mixture (as described in the legend to Fig. 17) and (Samples c,f) Samples b and e after addition of 50 mM cysteine +1 mM $FeSO_4$. The EPR spectra were taken at ambient temperature. Multiplication factors are shown on the right side. (From Ref. [10].)

with phosphate to the solutions of 1 mM BSA (Hb) at neutral pH. The disappearance of DNICs was associated with a broadband at 340–360 nm in the optical absorption spectra of the BSA or Hb solutions (Fig. 17).

Given the above *in-vitro* reaction equilibria, we attribute disintegration of DNIC to a destabilizing influence of NO + O_2 on the complex leading to the shift of the quasi-equilibrium between DNIC and its constituents to the right as shown in Schemes 5 and 9.

The transformation of protein-bound DNIC into RS-NO was reversible. The band of optical absorption at 340–360 nm disappeared with the addition of 50 mM cysteine and 1 mM Fe^{2+}. Furthermore, in the EPR spectrum of the solutions recorded at ambient temperature, a narrow signal appeared at $g = 2.03$ (Fig. 18), which was characteristic of Cys-DNIC.

It is interesting that incubation of proteins with DNIC does not cause significant S-nitrosation of the proteins. We conclude that DNIC alone is a poor nitrosating agent, just like NO itself. The S-nitrosation of proteins is started only after addition of extra NO and O_2 [10].

THE EQUILIBRIA BETWEEN THIOL-DNIC AND RS-NO

The considered data demonstrate that the ferrous iron ions are capable of catalyzing both the degradation and the synthesis of RS-NO. The advent of DNIC with thiol-containing ligands precedes the formation of RS-NO in the reaction of NO with ferrous irons in the presence of thiols and accompanies the degradation of RS-NO catalyzed by the same ions. Scheme 11 summarizes the system of coupled reactions:

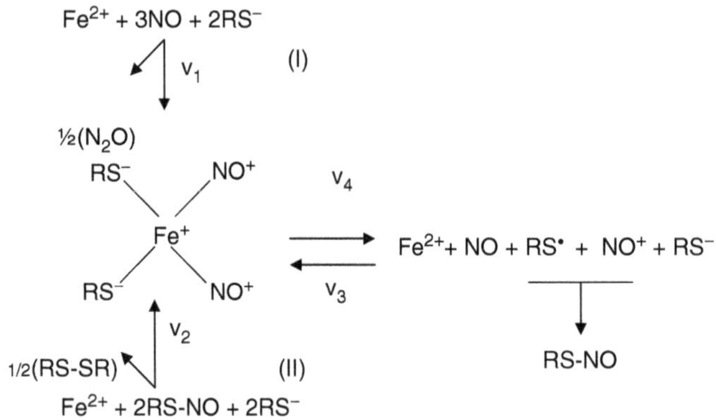

Scheme 11. System of coupled reactions involving mutual transformations of DNIC and RS-NO.

Note that thiol-DNIC is formed as an intermediate in both synthesis (v_1) as well as decomposition (v_2) of RS-NO. Both processes are irreversible under normal physiological conditions due to formation of N_2O or disulfides, respectively (see Chapter 2). The equilibrium $v_3 - v_4$ between DNIC and its constituents is considered as a reversible reaction.

The balance between free thiols and free NO molecules is an important factor in this chemical system. The coupling between the NO and thiol pool is largely determined by the presence of free iron, and involves the formation of intermediate DNIC. As a result, the reaction system reaches a steady-state equilibrium. However, two irreversible processes destroy the stationary state of the reaction system after all. Such irreversible steps include formation of disulfide and dinitrogen monoxide (N_2O), respectively. It means that finally the system ensures reduction of NO to N_2O under consumption of thiols, i.e. the thiols function as a "fuel" of the system. Evidently, oxygen addition to the system could influence it by decreasing the amount of thiols due to their oxidation.

Numerical simulation [19] of the coupled reaction equations described by Scheme 11 showed that reaction conditions can be chosen such that the DNIC concentration starts to oscillate with time. Such oscillations are well-known from other chemical systems and are caused by the nonlinear concentration dependence of higher order reaction rates. Fig. 19 shows the result of such a simulation.

Oscillatory time dependence of the DNIC concentration was experimentally observed with real-time EPR spectroscopy of 50 μl aliquots of reaction volume (Fig. 19). These experiments involved the anaerobic decomposition of Cys-NO by ferrous iron in HEPES-buffered solutions at room temperature. We expect that similar oscillatory reaction kinetics will be observed *in vitro* for RS-NO synthesis in anaerobic mixtures of ferrous iron, free NO radicals and other thiols. The onset, frequency and amplitude of such oscillations depends sensitively on the choice of initial concentrations. It would be very interesting to investigate whether such conditions are within the physiological range as found in cells and tissues.

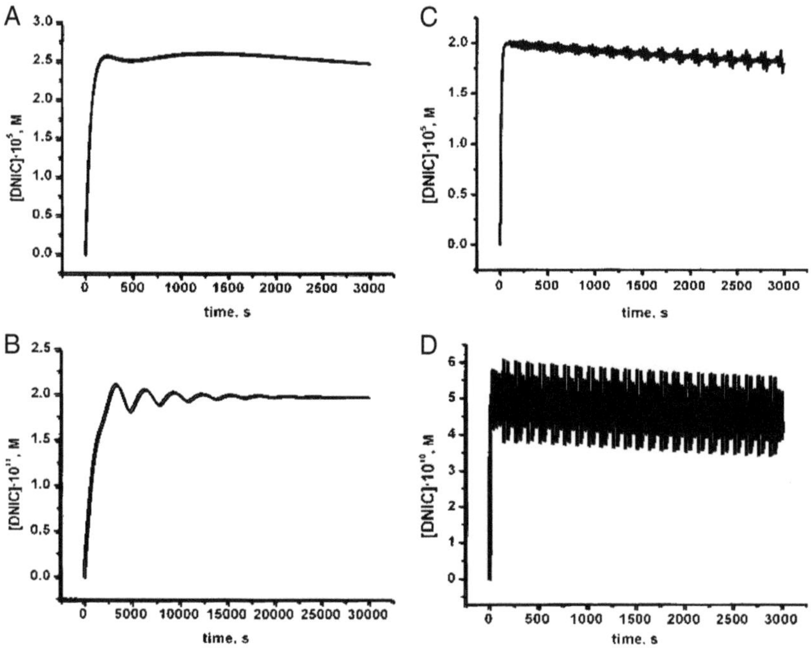

Fig. 19. Simulated kinetics curves of DNIC formation in mixtures of Fe^{2+}, *S*-nitrosothiol and free thiol at various initial concentrations and rate constants. For certain starting concentrations of the reagents, the simulations predict oscillatory time dependence of the DNIC concentration. (From Ref. [19].)

NITROSOTHIOLS ARE NITRIC OXIDE DONORS THROUGH THEIR TRANSFORMATION INTO DNICs

It is well-known that RS-NO can act as NO donors in the presence of spurious metal ions. The phenomenon is usually attributed to redox activity of the metal ions. As discussed above, this viewpoint is valid for copper, but does not apply to iron. Experimental data, in particular EPR, showed that RS-NO in the presence of ferrous iron forms a quantity of DNIC rather than release NO. Instead, free NO is released upon the disintegration of DNIC according to the equilibrium of Scheme 5. The preceding *in-vitro* data showed that DNIC is a crucial transient agent in the iron-catalyzed release of NO from RS-NO. We hypothize that the same situation applies to biological systems. Phrased otherwise, under the action of spurious iron, the true NO donor might not be RS-NO itself, but the RS-DNIC intermediate that is formed from RS-NO under the action of ferrous iron. Experimental support for this hypothesis will be given in this section.

The activation of soluble guanylate cyclase (sGC) is a well-known test for NO donors. We studied the activation of sGC from human platelets by GS-NO or dimeric GS-DNIC [34]. The dependence of sGC activation from the amount of these compounds (dose-response dependence) is shown in Fig. 20. The highest enzymatic activity was achieved at donor concentrations around 10^{-6} M for both donors. When concentrations of both NO donors

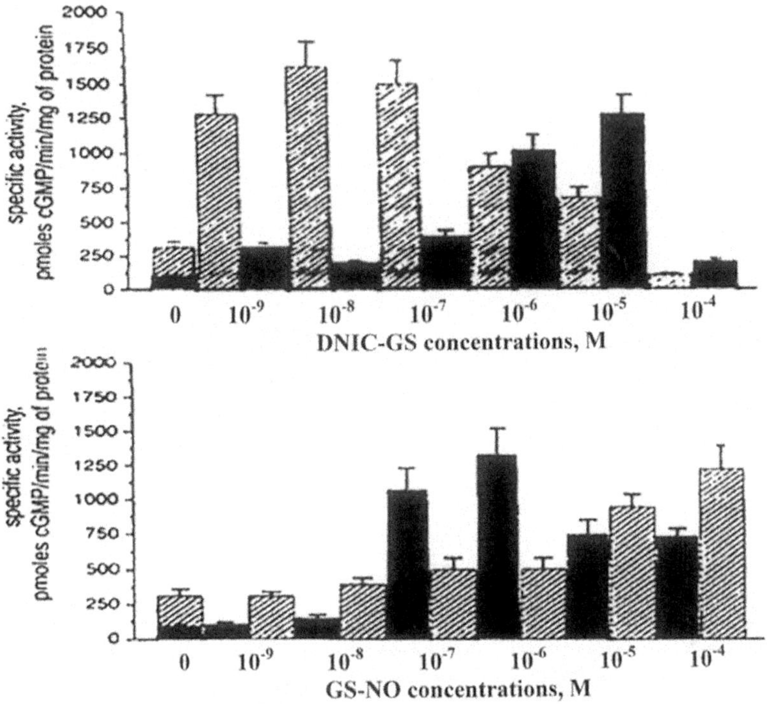

Fig. 20. Activation of soluble guanylate cyclase activation by GS-DNIC or S-nitrosoglutathione. The activation is studied in absence (black columns) or in presence of bathophenanthroline disulfonate (shaded columns). (From Ref. [34].)

were further increased, the specific sGC activity is even reduced, supposedly due to enzyme redox deactivation by NO and oxygen [35]. However, the dose-response dependence for both NO donors changed sharply when enzyme preparations were treated with GS-DNIC or GS-NO, 3–4 min after supplementation of the potent iron chelator bathophenatrolin disulfonate (BPDS, 10^{-4} M). At this condition, the optimal enzyme activation was achieved at concentration of GS-DNIC = 10^{-8} M (Fig. 20).

This value was around two orders of magnitude smaller than that observed in enzyme preparation without BPDS. For GS-NO, BPDS addition had an opposite effect: optimal enzymatic activity was achieved at concentration of GS-NO = 10^{-4} M (Fig. 20), that was two orders of magnitude higher than that found in the absence of BPDS. These changes can be explained by the following way. The iron chelator BPDS extracted iron from the DNIC complex, thereby destroying latter and releasing the NO ligands. This rapid forced NO release made enzyme activation by much lower DNIC concentration. As to BPDS effect on GS-NO activity, it can be rationalized by the capacity of ferrous iron from admixture to transform GS-NO to DNIC. BPDS could suppress the intrinsic iron-mediated destruction of GS-NO. The chelation of iron clearly prevented the formation of DNIC and increased the activation threshold of sGC by GS-NO by massive two orders of magnitude. Clearly GS-DNIC is a much better activator of sGC than GS-NO. We also note that the presence of some free iron greatly enhances the potency of GS-NO to activate sGC.

In principle, intrinsic copper could also contribute to NO release from RS-NO according to Scheme 1. However, the data reported in [17,18] demonstrated that BPDS did not inhibit the activation of sGC by RS-NO. Taken together, these data indicate that GS-NO activates sGC *via* a sequence of successive steps. The first step involves the formation of intermediate GS-DNIC according to Scheme 7. In the second step, GS-DNIC acts as the true activator of the enzyme.

Other examples can be given where GS-NO, RS-NOs and low-molecular DNIC induce similar "NO-like" responses. It is striking to note that DNIC always appears to be far more effective, with responses triggered at DNIC concentrations of orders of magnitude lower than with GS-NO.

sGC is not the only enzyme to have such sensitivity to DNIC. A similar situation was noted for the enzymatic inhibition of brain and kidney Na/K-ATPase when comparing GS-NO with Cys-DNIC and GS-DNIC [36]. Inhibition of this enzyme activity by both types of nitroso compounds was accompanied by a decrease in amount of free thiol groups on the protein and changed its sensitivity to the substrate concentration. Incubation with 100 μM GS-NO suppressed activity of Na/K-ATPase by 12%, whereas 50 μM GS-DNIC or Cys-DNIC inhibited by 45 or 70%, respectively. The low effectiveness of GS-NO shows that it does not get transformed into DNIC to any significant measure, presumably because the level of spurious iron was too small (1–2 μM). This changed drastically when exogenous iron was added: The inhibitory activity of the GS-NO preparations sharply increased when 50 μM ferrous-citrate was added to the solution. The effect cannot be attributed to some iron-catalyzed release of free NO, as the latter is a poor S-nitrosator. We attribute the effect to the formation of a significant quantity of DNIC. The formation of DNIC in these solutions was confirmed with EPR [36].

The key role of non-heme iron in RS-NO function as a NO donor was demonstrated in the paper with the characteristic title: "Cellular non-heme iron content is a determinant of nitric oxide-mediated apoptosis, necrosis, and caspase inhibition" [37]. This paper compares the apoptotic response of RAW264.7 macrophage cell line with low endogenuos non-heme iron, and hepatocytes with high endogenous non-heme iron. The apoptotic stimulus is given with NO donor *S*-nitroso-acetyl-DL-penicillamine (SNAP). The hepatocytes have much higher resistance to the apoptotic stimulus by SNAP. Preincubation of the RAW264.7 cells with exogenous iron reinforced their resistance to apoptosis by SNAP. The resistance could be raised to the hepatocyte level by reinforcing iron. Concomitant with apoptotic resistance, high iron status led to the formation of DNIC in both cell lines. The authors attribute the formation of DNIC to the scavenging of free NO released by SNAP. This interpretation sits uneasily with our conclusion that DNIC formation is dominated by the direct reaction of iron with nitrosothiols. Therefore, we propose that the decomposition of SNAP degradation depends on the level of non-heme iron: In macrophages with low non-heme iron, SNAP could release free NO under catalytic activity of copper and the free NO triggered the apoptotic chain. The apoptotic chain was not triggered in hepatocytes and iron loaded macrophages. It suggests that the NO levels on these cells remain low throughout the experiment. In these cells, a prominent quantity of DNIC could be formed by the equilibrium reactions between non-heme iron and the nitrosothiol (SNAP or GS-NO), without the involvement of free NO.

Other studies of apoptotic processes in cultured cells have provided further support for the significance of the reaction equilibria between *S*-nitrosothiols, iron and DNIC. NO, GS-NO or SNAP-mediated apoptosis are well-established models in various tumor cultured cells. It has been often reported that the activation of the apoptotic chain by NO donors could be inhibited by coadministration of iron [38,39]. Closely related, exogenous iron also reversed NO- or SNAP-mediated inhibition of cell growth and caspase activation. The authors have interpreted the effects of iron without considering the possibility that DNIC might be involved. However, it is highly likely that significant quantities of DNIC were formed in these cell cultures. After all, DNIC formation is routinely observed with EPR in cultured cells after exposure to RS-NO or other NO donors [31,37,40–42].

Nitric oxide is known to stimulate the expression of the soxRS- or SOS regulons in *Escherichia coli*. Alternatively, the expression may be stimulated by exposure to NO donors or GS-NO [41,43]. The main functions of these regulons are to protect bacteria against oxidative stress or to provide DNA repair, respectively. EPR spectroscopy confirmed that the exposure to NO or GS-NO induced the formation of intracellular protein-bound DNIC. In all cases, prior addition of strong iron chelator, *o*-phenanthroline to cell suspension blocked completely both the formation of DNIC in cells and the expression of the soxRS- and SOS-regulons. Interestingly, regulon expressions were also activated if the bacteria were exposed to an extracellular dose of Cys-DNIC or GS-DNIC. The activity was also inhibited by *o*-phenanthroline. It confirms the crucial role of iron in expression of these genes as induced by DNIC, GS-NO or free NO also. Given the reaction equilibria of Scheme 11, it seems conceivable that many physiological effects of NO or GS-NO are potentiated by the formation of endogenous DNIC from intracellular spurious iron.

Thus, the data presented above emphasize that NO, RS-NO and DNIC cannot be considered separately from each other. Instead, the pools of these compounds are coupled *via* the pool of spurious iron. The reaction equilibria were investigated *in-vitro* studies, but a multitude of experiments in cultured cells, tissue extracts and live animals suggests that the same equilibria be significant *in vivo* as well. Moreover, it is not to be excluded that RS-NO function *via* a mechanism involving the transformation of RS-NOs into DNICs as a NO and NO^+ donors. However, the interest to the DNICs is low up to now. Most of the investigators consider the RS-NOs as a most important endogenous nitroso compounds which determine the NO function in cells and tissues. Moreover, it is suggested that the RS-NOs can ensure the basis for designing of the medicines using NO biological activity. Endowing the DNICs by this role seems to be disputable. There is an opinion that iron releasing from DNICs can initiate the formation of hydroxyl radicals *via* the Fenton reaction. Really, that is a serious argument against using the DNICs as a new type of the medicines. So, the intensive investigations are needed to test the proposed adverse action of DNICs on the organisms. A sole contra-argument against this feature of DNICs is as follows. The investigations demonstrated that free iron incorporation into DNICs significantly attenuates Fenton's activity of iron ions [42,44].

ACKNOWLEDGMENT

The work was supported by the Russian Foundation of Basic Researches (Grant 05-04-49383).

REFERENCES

1 Butler AR, Rhodes P. Chemistry, analysis, and biological roles of *S*-nitrosothiols. Anal. Biochem. 1997; 249: 1–9.

2 Gaston B. Nitric oxide and thiol groups. Biochim. Biophys. Acta 1999; 1411: 323–333.

3 Al-Sàdoni H, Ferro A. *S*-nitrosothiols: a class of nitric oxide-donor drugs. Clinical Sci. 2000; 98: 507–520.

4 Liu L, Hausladen A, Zeng M, Que L, Heltman J, Stamler JS. A metabolic enzyme for S-nitrosothiol conserved from bacteria to humans. Nature 2001; 410: 490–494.

5 Foster MW, McMahon TJ, Stamler JS. S-nitrosylation in health and disease. Trends in Mol. Med. 2003; 9: 160–168.

6 Martinez-Ruiz A, Lamads S. S-nitrosylation: a potential paradigm in signal transduction. Cardovasc. Res. 2004; 62: 43–52.

7 Williams DLH, Nitrosation reactions and the chemistry of nitric oxide. Elsevier Amsterdam, 2004.

8 Dahm CC, Moore K, Murphy MP. Persistent S-nitrosation of complex I and other mitochondrial membrane proteins by *S*-nitrosothiols but not nitric oxide or peroxynitrite. J. Biol. Chem. 2006; 281: 10056–10065.

9 Kharitonov VG, Sandquist AR, Sharma VS. Kinetics of nitrosation of thiols by nitric oxide in the presence of oxygen. J. Biol. Chem. 1995; 270: 28158–28164.

10 Vanin AF, Malenkova IV, Serezhenkov VA. Iron catalyzes both decomposition and synthesis of *S*-nitrosothiols: optical and electron paramagnetic resonance studies. Nitric Oxide 1997; 1: 191–203.

11 Stubauer G, Guiffre A, Sarti P. Mechanism of S-nitrosothiol formation and degradation mediated by copper ions. J. Biol. Chem. 1999; 274: 28128–28133.

12 Romeo AA, Filosa A, Capobianco JA, English AM. Metal chelastors inhibits S-nitrosation of Cysβ93 in oxyhemoglobin. J. Am. Chem. Soc. 2001; 123: 1782–1783.

13 McAninly J, Williams DLH, Ascew SC, Butler AR, Russel C. Metal ion catalyzes of nitrosothiol (RSNO) decomposition. J. Chem. Soc. Chem. Commun. 1993; 93: 1758–1759.

14 Bannenberg G, Xue J, Engman L, Cotgreave I, Moldeus, Ryrfeldt A. Characterization of bronchodilator effects and fate of *S*-nitrosothiols in the isolated perfused and ventilated guinea pig lung. J. Pharmacol. Exp. Therap. 1995; 272: 1238–1245.

15 Sorenson E, Skiles EH, Xu B, Aleryani S, Kostka P. Role of redox-active iron ions in the decomposition of S-nitrosocysteine in subcellular fractions of porcine aorta. Eur. J. Biochem. 2000; 267: 4593–4599.

16 Constanzo S, Menage S, Purello R, Bonomo RP, Fontecave M. Re-examination of the formation of dinitrosyl-iron complexes during reaction of *S*-nitrosothiols with Fe(II). Inorg. Chim. Acta 2001; 318: 1–7.

17 Vanin AF, Muller B, Alencar JL, Lobysheva II, Nepveu F, Stoclet J-C. Evidence that intrinsic iron but not intrinsic copper determines S-nitrosocysteine decomposition in buffer solution. Nitric Oxide: Biol. Chem. 2002; 7: 194–209.

18 Vanin AF, Muller B, Alencar JL, Lobysheva II, Nepveu F, Stoclet J-C. Influence of transition metals on stability of various *S*-nitrosothiols. Curr. Top. Biophys. 2002; 26: 101–113.

19 Vanin AF, Papina AA, Serezhenkov VA, Koppenol WH. The mechanisms of S-nitrosothiol decomposition catalyzed by iron. Nitric Oxide: Biol. Chem. 2004; 10: 60–73.

20 Gorge MP, Meyer DJ, Hothersall J, Neild GH, Payne NN, Norohna-Dutra A. Copper chelator-induced reduction of biological activity of *S*-nitrosothiols. Br. J. Pharmacol. 1995; 114: 1038–1049.

21 Williams DLH. The mechanism of nitric oxide formation from *S*-nitrosothiols (thionitrite). Chem. Comm. 1996; 1: 1085–1091.

22 Dicks AP, Swift HR, Williams DLH, Butler AR, Al-Sàdoni HH, Cox BG. Identification of Cu^{+} as the effective reagent in nitric oxide formation from *S*-nitrosothiols (RSNO). J. Chem. Soc. Perkin Trans. 1996; 2: 481–487.

23 Gorren AF, Schrammel A, Schmidt K, Mayer B. Decomposition of S-nitrosoglutathione in the presence of copper and glutathione. Arch. Biochem. Biophys. 1996; 330: 219–228.

24 Vanin AF, Stukan RA, Manukhina EB. Physical properties of dinitrosyl iron complexes with thiol-containing ligands in relation with their vasodilatory activity. Biochim. Biophys. Acta 1995; 1295: 5–12.

25 Vanin AF, Serezhenkov VA, Malenkova IV. Nitric oxide initiates iron binding to neocuproine. Nitric Oxide: Biol. Chem. 2001; 5: 166–175.

26 Bush PM, Whitehead JP, Pink CC, Gramm EC, Eglin JN, Watton SP, Pence LE. Electronic and structural variations among copper (II) complexes with substituted phenantrolines. Inorg. Chem. 2001; 40: 1871–1877.

27 Sheu F-W, Zhu W, Fung PCW. Direct observation of trapping and release of nitric oxide by glutathione and cysteine with electron paramagnetic resonance spectroscopy. Biophys. J. 2000; 78: 1216–1226.

28 Burbaev SS, Vanin AF. On modeling of non-heme iron complexes from biological objects. Doklady Akademii Nauk SSSR (Rus.) 1970; 190: 1348–1350.

29 Vanin AF. On the stability of the dinitrosyl-iron-cysteine complex, a candidate for the endothelium-derived telexing factor. Biochemistry (Moscow) 1995; 60: 225–229.

30 Vanin AF, Serezhenkov VA, Mikoyan VD, Genkin MV. The 2.03 signal as an indicator of dinitrosyl-iron complexes with thiol-containing ligands. Nitric Oxide: Biol. Chem. 1998; 2: 224–234.

31 Vanin AF, Malenkova IV, Mordvintcev PI, Mülsch A. Dinitrosyl iron complexes with thiol-containing ligands and their reversible conversion into nitrosothiols. Biol. Chem. (Rus.) 1993; 58: 1094–1103.

32 Vanin AF. Roles of iron ions and cysteine in formation and decomposition of S-nitrosocysteine and S-nitrosoglutathione. Biochemistry (Moscow) 1995; 60: 441–447.

33 Vanin AF, Malenkova IV. Iron is a catalyst of cysteine and glutathione S-nitrosation on contact with nitric oxide in aqueous solutions at neutral pH. Biochemistry (Moscow) 1996; 61: 374–379.

34 Severina IS, Bussygina OG, Pyatakova NV, Malenkova IV, Vanin AF. Activation of soluble guanylate cyclase by NO donors - *S*-nitrosothiols, and dinitrosyl-iron complexes with thiol-containing ligands. Nitric Oxide: Biol. Chem. 2003; 8: 155–163.

35 Dierks EA, Burstyn JN, The deactivation of soluble guanylate cyclase by redox-agents. Arch. Biochem. Biophys. 1998; 351: 1–7.

36 Boldyrev AA, Bulygina ER, Kramarenko GG, Vanin AF. Effect of nirosocompounds on Na/K-ATPase. Biochim. Biophys. Acta 1997; 1321: 243–251.

37 Kim YM, Chung HT, Simmons RL, Billiar TR. Cellular non-heme iron is a determinant of nitric oxide-mediated apoptosis, necrosis and caspase inhibition. J. Biol. Chem. 2000; 275: 10954–10961.

38 Feger F, Ferry-Dumazet H, Mammani-Matsuda M, Bordenave J, Dupouy M, Nussler AK, Arock M, Devevey L, Nafziger J, Guillosson JJ, Reiffers J, Mossalayi MD. Role of iron in tumor cell protection from pro-apoptotic effect of nitric oxide. Cancer Res. 2001; 61: 5289–5294.

39 Ferry-Dumazet H, Mammani-Matsuda M, Dupouy M, Belloc F, Thiolat D, Marit G, Arock M, Reiffers J, Mossalayi MD. Nitric oxide induces the apoptosis of human BCR-ABL-positive myeloid leukemia cells: evidence for the chelation of intracellular iron. Leukemia 2002; 16: 708–715.

40 Roy B, Lepoivre M, Henry Y, Fontecave M. Inhibition of ribonucleotide reductase by nitric oxide derived from thionitrites. Biochemistry 1995; 34: 5411–5420.

41 Stupakova MV, Lobysheva II, Mikoyan VD, Vanin AF, Vasilèva SV. A role of iron ions in the SOS DNA repair response induced by nitric oxide in *Escherichia coli*. Biochemistry (Moscow) 2000; 65: 690–695.

42 Gorbunov NV, Yalowich JC, Gaddam A, Thampatty P, Rinov VB, Kisin ER, Elsayed NM, Kagan VE. Nitric oxide prevents oxidative damage produced by *tert*-butyl hydroperoxide in erythroleukemia cells via nitrosylation of heme and non-heme iron. J. Biol. Chem. 1997; 272: 12328–12341.

43 Vasilèva SV, Stupakova MV, Lobysheva II, Mikoyan VD, Vanin AF. Activation of the Escherichia coli SoxRs-regulon by nitric oxide and its physiological donors. Biochemistry (Moscow) 2001; 66: 984–988.

44 Lu C, Koppenol WH. Inhibition of the Fenton reaction by nitrogen monoxide. J. Biol. Inorg. Chem. 2005; 10: 732–738.

Radicals for Life: The Various Forms of Nitric Oxide
E. van Faassen and A.F. Vanin (Eds.)

CHAPTER 12

Cellular non-heme iron modulates apoptosis and caspase 3 activity

Detcho A. Stoyanovsky and Timothy R. Billiar

Department of Surgery, University of Pittsburgh, Pittsburgh, PA 15213

INTRODUCTION

Caspases are a family of cysteine proteases that play an essential role in the signaling cascade leading to apoptosis. Apoptosis, or programmed cell death, is distinguished from lytic or necrotic cell death by specific biochemical and structural events. Apogenic signals trigger specific signaling pathways, including activation of proteases, which are followed by the appearance of specific morphologic changes such as condensation of nuclei and cytoplasm, blebbing of cytoplasmic membranes, and finally fragmentation into apoptotic bodies that are phagocytosed by neighboring cells [1]. Apoptosis is important to physiologic processes such as cell selection in development and immunologic responses [2], control of organ size in maturation and regeneration [3], and normal cell turnover throughout the organism [4]. Dysregulated apoptosis may contribute to pathologic states such as autoimmune disease [5] and malignancy [6]. Upon exposure to a proapoptotic signal, zymogen forms of caspases constitutively present in cells are proteolytically cleaved and activated. Initiator caspases such as caspase 8, 9, and 10 can cleave other caspases, while executioner caspases, including caspase 3, 6, and 7, cleave death substrates [7,8].

There is a considerable body of literature indicating that NO, produced either extracellularly by low-molecular-weight *S*-nitrosothiols (LMW RSNOs) or endogenously by nitric oxide synthase (NOS) prevents caspase 3-dependent apoptosis in various cell types. These effects were suggested to reflect the *S*-nitrosation of caspase 3 and procaspase 9, respectively [9–13]. However, the nature of the nitrosating species that may selectively interact with caspases remains poorly understood. The lack of a mechanism that correlates the chemical properties of NO and its derivatives (e.g. iron- and copper-nitrosyl complexes, N_2O_3, and peroxynitrite) with substrate-selective reactions of *S*-nitrosation has proven to be a major obstacle in studies aimed at elucidating the role of NO as a signaling molecule. The basic chemistry of *S*-nitrosation reactions was discussed in Chapters 1 and 9 of this volume. Measurements of NO_2^-, a stable end product of NO oxidation, have demonstrated that cells produce considerable amounts of this species. Primary mouse macrophages stimulated with inflammatory agents produce NO_2^- at a rate of 70 nmol/mg protein/hour, limited only by L-arginine availability [14]. Similarly, vascular smooth muscle cells treated with the peptide hormone relaxin generate NO_2^- at a rate of 50 nmol/mg protein/hour for a period of 24 h [15]. In the presence of oxygen, NO undergoes oxidation to N_2O_3, which is a nitrosating species with poor substrate selectivity. In parallel, NO forms metal-nitrosyl complexes that can also act as *S*-nitrosating agents [16,17]. Since GSH is the most abundant cellular thiol (15–30 nmol/mg protein), formation of GSNO is expected to parallel the activity of NOS. In fact, several studies have confirmed the formation of GSNO in biological systems: GSNO has been detected in rat brain (15 pmol of GSNO/mg protein) [18], rat liver cytosol and mitochondria (in Refs. [19] and [20]; ~3 nM GSNO and 34 pmol of GSNO/mg protein, respectively). However, GSNO accounts for only a small fraction of the endogenously generated NO_2^-. Furthermore, GSNO levels appear to be insufficient to induce trans-*S*-nitrosation of protein thiols. Often, a concentration of at least 100 μM of GSNO is required for the *in vitro* trans-*S*-nitrosation of proteins. This suggests that the nitrosation of critical thiols on enzymes may be GSNO-independent, unless there exist cellular catalytic pathways for transfer of NO from GSNO and/or other LMW RSNOs to protein thiols. The transport of *S*-nitrosothiols across membranes is discussed in Chapter 9. One plausible pathway for modulation of the redox properties of NO includes the interaction of this radical species with metal ions. To verify the hypothesis that iron-nitrosyl complexes modulate the activity of thiol-containing enzymes, we studied the effects of NO on the activity of caspase 3 in control and iron-preloaded RAW264.7 cells.

SUMMARY OF OBSERVATIONS

Formation of iron-nitrosyl complexes parallels the inhibition of caspase 3 in NO-treated RAW264.7 cells

The effects of $FeSO_4$ and NO exposure on the formation of intracellular RSNOs, iron-nitrosyl complexes and activation of caspase 3 were assessed in RAW264.7 cells, a murine macrophage cell line that is susceptible to NO-induced apoptosis. Incubation of cells with $FeSO_4$ resulted in a marked increase in the levels of non-heme iron (Fig. 1A), while exposure of control and iron-preloaded RAW264.7 cells to NO led to the formation of RSNOs and iron-nitrosyl complexes that was more robust in iron-treated cells (Fig. 1B,C). The ESR spectra presented in Panel B indicate that exogenous NO led to the formation of intracellular iron-nitrosyl complexes, presumably *via* the interaction of NO with bis(cysteinato)Fe^{2+} and bis(glutathionato)Fe^{2+} complexes [21].

It could be further speculated that the increased content of RSNOs in iron-loaded cells reflected the nitrosation of cellular thiols by iron-nitrosyl complexes. In cell-free systems, the bis(cysteinato)dinitrosyliron^{2+} complex (DNIC) impeded the activity of purified caspase 3 in a dose dependent manner [16]. Treatment of RAW264.7 cells with *S*-nitroso-*N*-acetyl penicillamine (SNAP) led to caspase 3 activation (Fig. 1D) and apoptosis [16]. Control RAW264.7 cells exhibited higher caspase 3 activity (Fig. 1D) and, as DTT did not change the activity of caspase 3 to any significant extent, we concluded that the protease was present

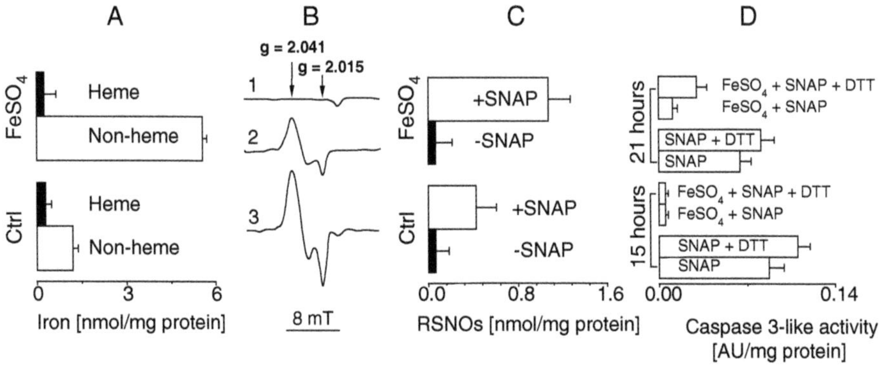

Fig. 1. Relationship between non-heme iron, iron-nitrosyl complexes, RSNOs and caspase 3 activity in RAW264.7 cells. Cells were precultured with 80 μM $FeSO_4$ for 24 h, washed with fresh medium and thereafter treated with 0.75 mM SNAP. Cellular levels of iron (A), thiolato-nitrosyl-iron$^{(2+)}$ complexes (B), and RSNOs (C) were measured as described in Ref. [16]. (B) Spectrum (1) control cells; (2) control cells plus SNAP; (3) iron-loaded cells plus SNAP. (D) RAW264.7 cells were preincubated with $FeSO_4$, washed with fresh medium, and then treated with SNAP. After 15 and 21 h, caspase 3 activity was determined by colorimetric assay using the peptide-based substrate Ac-DEVD-*p*-nitroanilide in the absence and the presence of DTT (10 mM). Results are given as means ± S.D.

in a reduced, SH-state. In contrast, the activity of caspase 3 in iron-pretreated cells was lower but markedly increased by DTT, indicating that the ratio between SH and SS (or SNO) functions on this protein was shifted in favor of the latter. Importantly, elevation of non-heme iron in RAW264.7 cells delayed the cell death, which then appeared to occur *via* necrosis instead of apoptosis [16]. The delay in cell death could be associated with the inhibition of caspase 3 thus suggesting that the level of non-heme iron is an important factor in determining the consequence of NO exposure on cell viability.

MECHANISTIC CONSIDERATIONS

Redox regulation of caspase 3 activity

Release of cytochrome c from mitochondria into cytosol [22] activates the initiator procaspase 9 [23]. In the presence of cytochrome c and dATP [24], the apoptotic protease-activating factor 1 (Apaf-1) binds to procaspase 9 to form the apoptosome complex [24]. In the apoptosome, caspase 9 is activated to process other downstream caspases, including caspase 3 [25]. Active executioner caspase 3 can further cleave downstream substrates involved in apoptotic changes [26].

Caspase 3 is a relatively small protein that consists of 2 subunits, a 12- and 17-kDa subunit that contains 3 and 5 thiol functions, respectively. Activation of caspase 3 is dependent on its dimerization to a heterotetramer, where the histidine-activated Cys-285 in the active site of the p17 subunit is conserved in the caspase superfamily and is required for enzymatic activity [27]. Caspase 3 proteinase activity is lost in the presence of thiol-oxidizing agents such as diamide, oxidized glutathione, and dithiocarbamate disulfide [28–30]. Furthermore, caspase 3 spontaneously loses its enzymatic activity [31], even in chelex-100-treated buffers containing chelators of metal ions (data not shown), suggesting that low pK_a thiol(s) that are deprotonated at physiological pH auto-oxidize with concomitant generation of superoxide anion radical [32,33]. In contrast, thiol-containing compounds such as GSH, thioredoxin (Trxn), albumin, and DTT fully restore caspase 3 activity, presumably *via* reduction of disulfide functions on the protein [34].

Several research groups have reported that LMW RSNOs impede caspase 3 activity *via* reactions of trans-*S*-nitrosation [9,10,12,13,31,35,36]. Detailed mechanistic studies by Zech et al. have established that caspase 3 can undergo poly-*S*-nitrosation, whereby all *S*-nitroso functions in the p12 subunit are cleaved with the release of NO and partial formation of protein-mixed disulfides with GSH; however, a single SNO function in the p17 subunit remained stable [31]. Since this SNO function was not observed in a mutant form of caspase 3 lacking the active site cysteine, the authors concluded that NO nitrosates the active site cysteine of caspase 3 to form *S*-nitrosocaspase 3 that is inert to reduction by GSH. In a follow-up study, Zech et al. have found that NO donors block Fas- and etoposide-induced caspase activation and apoptosis in Jurkat cells [35]. However, caspase activity was not restored by DTT, as predicted for S-nitrosation reactions. Rather, the processing of procaspases 9, 3, and 8 has been found to be decreased due to ineffective formation of the Apaf 1/caspase 9 apoptosome [35]. Recently, Kim and Tannenbaum applied the biotin-switch assay to directly detect endogenous *S*-nitroso(pro)caspases in apoptotic HT-29 human colon carcinoma cells [37]. The authors have acquired experimental proof for the formation of *S*-nitrosoprocaspase 9 but not *S*-nitrosocaspase 3, thus suggesting that NO may impede the caspase 9-catalyzed release of caspase 3 (Scheme 1). These observations are in agreement with the data presented in Fig. 1 and Ref. [16]. In control RAW264.7 cells, SNAP-activated caspase 3 in a *threo*-1,4-dimercapto-2,3-butanediol (DTT)-independent manner. On the other

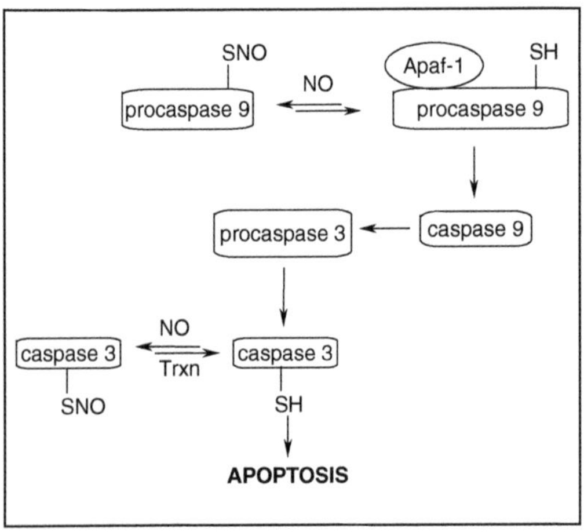

Scheme 1.

hand, the activity of caspase 3 in iron-loaded cells was less pronounced, perhaps due to the formation of *S*-nitrosoprocaspase 9; however, the activity of caspase 3 was markedly increased by DTT.

NO inhibits apoptosis by preventing increases in caspase 3-like activity *via* two distinct mechanisms

Kim et al. have reported that when lysates from apoptotic hepatocytes stimulated to express iNOS or hepatocytes exposed to NO donors were incubated with DTT, caspase 3 activity increased to about 55% of cells not exposed to a source of NO. TNF-alpha-induced apoptosis and caspase 3-like activity were also reduced in cultured hepatocytes exposed to 8-bromo-cGMP, and both effects were inhibited by the cGMP-dependent kinase inhibitor KT5823. The suppression in caspase 3-like activity in hepatocytes exposed to an NO donor was partially blocked by inhibitors of soluble guanylyl cyclase, while the incubation of these lysates with DT T almost completely restored caspase 3 activity to the level of TNF-alpha-treated controls. These data suggests that NO prevents apoptosis in hepatocytes by either directly or indirectly inhibiting caspase 3 *via* a cGMP-dependent mechanism and by direct inhibition of caspase 3 activity through protein *S*-nitrosation [13].

Enzymatic denitrosation of caspase 3

While virtually all enzymes contain critical cysteine residues whose *S*-nitrosation acts as an activity-switch, *S*-nitrosation and denitrosation of cellular thiols have been suggested to be a fundamental post-translational protein modification that is similar to protein phosphorylation and dephosphorylation, respectively [38]. However, the identification of specific reaction pathways of *S*-nitrosation has proven to be difficult in part because most RSNOs are unstable metabolites with a half-life of minutes to hours [19,39]. Three enzymatic systems are known thus far to catabolize LMW RSNOs: Trxn [40], protein disulfide isomerase (PDI) [41] and alcohol dehydrogenase class III (ADH) [42] have been shown to catalyze the denitrosation of *S*-nitrosoglutathione (GSNO) but not that of *S*-nitrosocysteine, *S*-nitrosohomocysteine and *S*-nitrosoalbumin. Recently, we reported that Trxn catalyzes the denitrosation of a series of *S*-nitrosoproteins, including that of poly-*S*-nitrosocaspase 3.

Trxn is a ubiquitous protein whose activity has been linked to cell growth, transcription factor regulation, DNA synthesis, protein binding [43–45], detoxification of free radicals [33], and regeneration of antioxidant compounds (ascorbic acid, selenium-containing substances, and ubiquinones) [46]. Trxn isozymes of all organisms contain a conserved –Cys-Gly-Pro-Cys- active site that is essential for the function of this class of proteins as general protein disulfide reductases [43,44], Trxn-$(SH)_2$ regulates the activity of thiol-containing proteins *via* reduction of their S–S bonds to protein-$(SH)_2$ at the expense of its own oxidation to Trxn-$(S)_2$. Trxn is maintained in its active, reduced form by thioredoxin reductase (TrxnR) that uses NADPH to reduce its S–S bridge between Cysteines 32 and 35, respectively. Haendeler et al. reported that overexpression of Trxn in endothelial cells activates eNOS, increases basal levels of endogenous *S*-nitrosothiols, and inhibits TNF-α-induced apoptosis [47]. Experiments with genetically manipulated cells that express cysteine 69-lacking Trxn have suggested that

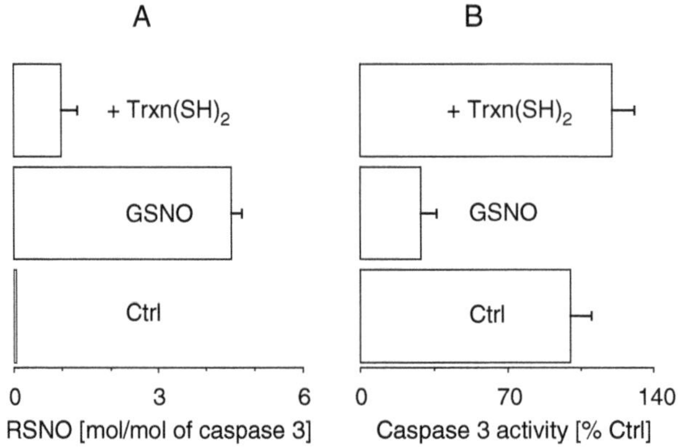

Fig. 2. Trxn denitrosates caspase 3. Experiments were carried out in 0.1 M phosphate buffer containing 0.1 mM desferrioxamine. *S*-nitrosation of caspase 3 (5 μM) by GSNO (0.1 mM; A) and denitrosation by Trxn (10 μM)/TrxnR (15 U/mL)/NADPH (0.2 mM; B) were carried out as described in Ref. [49]. Caspase activity was determined by using Enzcheck Caspase 3 assay kit (Molecular Probes, OR). Data are presented as mean values ± SE ($n = 3$).

these effects may reflect Trxn nitrosation in its redox inactive cysteine 69 to ONS-$Cys_{(69)}$-Trxn-S_2, which, in turn, inhibited activators of apoptosis by delivering NO to their active sites [47].

However, direct experiments with Trxn and caspases that support this mechanism have not been presented. This hypothesis was challenged by model experiments of Mitchell and Marletta who reported that GSNO preferentially nitrosates oxidized Trxn (Trxn-S_2) in cysteine 73 without affecting cysteine 69. The authors proposed that ONS-$Cys_{(73)}$-Trxn-S_2 may act as an inhibitor of caspase 3 *via* trans-*S*-nitrosation of this protease to ONS-$Cys_{(163)}$-caspase 3 [48]. Given these controversial data, it is important to elucidate whether (a) other nitrosating species (e.g. metal-nitrosyl complexes and *S*-nitroso(homo)cysteine) interact preferentially with Trxn-cysteine 69, and (b) ONS-Cys-Trxn-S_2 undergoes auto-denitrosation to HS-Cys-Trxn-$(SH)_2$ in physiologically relevant conditions that include the presence of TrxnR and NADPH. In our studies, the complete Trxn/TrxnR/NADPH system was able to fully reconstitute the activity of a purified poly-*S*-nitrosated caspase 3 (Fig. 2A,B; Ref. [49]). These observations may be extended to the studies of Zhang et al. who reported that exposure of lung endothelial cells (LEC) to NO results in diminished eNOS activity and decreased expression of Trxn and TrxnR, whereas overexpression of Trxn prevents eNOS inhibition in intact cells [50]. Interestingly, resistance to LPS-induced apoptosis in LEC is manifested by inhibition of caspase 3 but is not apparent until 96 h of endogenous NO generation [51,52]. Hence, it could be speculated that exhaustion of TrxnR activity may be required for the nitrosation of caspase 3 (Scheme 1).

Chelation of metal ions by caspase 3

Recently, Perry et al. have reported that caspase 3 could be inhibited by submicromolar concentrations of Zn^{2+}, which suggests a regulatory role for Zn^{2+} in modulating the upstream apoptotic machinery; mechanistically, Zn^{2+} has been proposed to form a complex with caspase 3 [53]. This hypothesis is in agreement with the studies of Tang et al. and Kondo et al. who have demonstrated that the release of Zn^{2+} from metallothioneins leads to inhibition of caspase 3 in mouse embryonic cells and cultured sheep pulmonary artery endothelial cells, respectively [54,55]. More recently, Sliskovic and Mutus have demonstrated that caspase 3 chelates iron ions with concomitant loss of activity, whereby both EDTA and DTT could reverse this effect [56]. The latter indicates that the chelation of iron ions by caspase 3 was not followed by oxidation of its thiol functions.

Iron complexes containing thiyl and NO ligands

Interactions between iron complexes and NO have been the focus of much research that dates back to the studies of Priestley [57–62]. Recently, Dobry-Duclaux has shown that submicromolar concentrations of the black Roussin's salt ($K[Fe_4S_3(NO)_7]$) impede the enzyme alcohol dehydrogenase [63,64], while Gordy and Rexroad demonstrated that complexes of nitric oxide with hemoglobin and cytochrome c exhibit ESR spectra that reflect the electronic structures of the iron atoms in these biologically important molecules [65]. Despite their notorious instability, numerous iron-nitrosyl complexes have been isolated in crystal form and characterized by x-ray analysis and IR, NMR, ESR, and MS spectrometry. Experimental proof has been provided for the formation of iron-nitrosyl complexes in neutral aqueous solutions of Fe^{2+} containing anionic ligands such as pyrophosphate, adenosine triphosphate (ATP), creatine phosphate, carbonate, maleate, and mercaptans (2-mercaptoethanol, thiourea, dithiols, and cysteine). However, it is noteworthy that most studies have focused on the geometrical and electronic structures of iron-nitrosyl complexes, whereas the (bio)chemical properties of this class of compounds have not been well-characterized.

In recent years, the formation and interactions of iron-nitrosyl complexes in biological matrices have been of particular interest with respect to the transduction of NO signals [16,17,66–69]. In weakly alkaline milieu, cysteine interacts with ferrous ions to form a bis(cysteinato)iron^{2+} complex [59]. The formation of this complex is paralleled by the appearance of a violet color that gradually fades with the consumption of O_2. Simultaneously, cysteine is partially transformed to cystine, but, as long as there remains any thiol acid, the color may be regenerated by O_2. This catalytic process has been proposed to occur *via* three steps: first, formation of bis(cysteinato)iron^{2+} complex; second, oxidation of this complex by O_2 to a Fe^{3+} complex, which is the cause of the violet color observed in the catalytic process; and third, an autoreduction of the ferric complex in which the iron is reduced to a divalent state while the thiol sulfur of cysteine is oxidized to thiyl radical cation ($LS^{+\bullet}$) with simultaneous breakdown of the complex and fading of the color. In cellular systems, non-heme iron is expected to form complexes with GSH and to exhibit even higher affinity for compounds containing vicinal thiol functions. Bonomi et al. have characterized a series of ferric and ferrous complexes of 6,8-dimercapto-octanoic acid (dihydrolipoic acid; DHLA) and dihydrolipoamide as potential substrates for enzymatic synthesis of iron–sulfur clusters, whereas Kijima et al. have reported that these complexes are strong reductants that can convert aromatic NO_2 functions to NH_2 and *N,O*-dialkylhydroxylamines to alkylamines and alcohols [70].

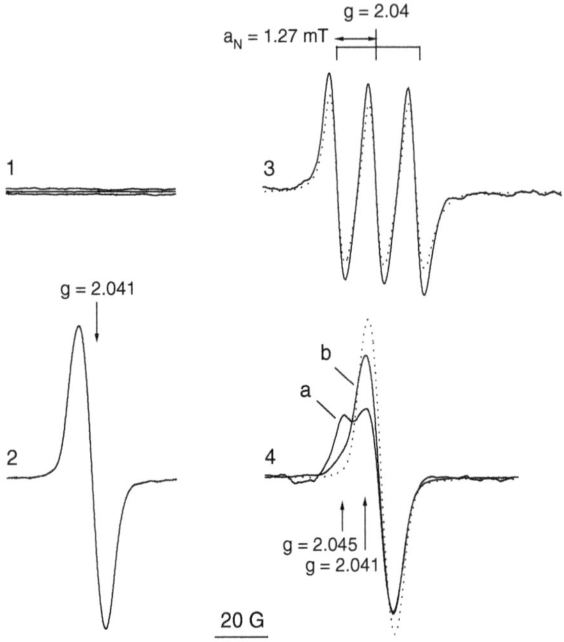

Fig. 3. ESR spectra of selected DNIC. Spectra were recorded in 0.1 M phosphate buffer saturated with N_2 (20°C). A stock solution of 0.1 M phosphate buffer (pH 7.20) saturated with NO gas (~1.8 mM) was prepared by first deoxygenating the medium with N_2 for 20 min and then gassing with NO gas that was passed through a gas trap containing KOH. NO was produced by dropping sulfuric acid to sodium nitrite. Spectra 1 – $Fe(NH_4)_2(SO_4)_2$ (0.2 mM) in the presence of cysteine (1 mM), DTT (5 mM), or *N*-methyl-D-glucamine dithio-carbamate (MGD) 0.5 mM. 2 – $Fe(NH_4)_2(SO_4)_2$, cysteine, and NO (0.2 mM). 3 – $Fe(NH_4)_2(SO_4)_2$, MGD, and NO in the absence (dashed lines) and the presence (solid lines) of cysteine. 4 – $Fe(NH_4)_2(SO_4)_2$, DTT, and NO in the absence (a) and the presence (b) of cysteine. With dashed line is presented spectrum 2.

Thiolato-nitrosyl-iron complexes are formed spontaneously in solutions of a thiol, Fe^{2+}, and NO (Fig. 3). In the case of cysteine (Fig. 3, spectrum 2), the unstable bis(cysteinato) dinitrosyliron$^{(2+)}$ complex is formed, which is then converted to cystinatodinitrosyliron$^{(2+)}$ [59]. Similarly, compounds with vicinal thiol functions form iron-nitrosyl complexes [60,61], even in the presence of equimolar concentrations of cysteine (Fig. 3, spectra 3 and 4, respectively) or GSH (data not shown).

In iron-nitrosyl complexes, the charge on the nitrogen atom is presumed to be a major determinant in the reactivity of the ligated NO. The principal bonding scheme of Fe^{2+}–NO complexes has been described by Enemark and Fetham [71]. These complexes were classified as [FeNO] [72], and it has been proposed that they show bent Fe–NO units with radical character on the nitrosyl ligand. However, this description leaves room for a large variation of the electronic structure mediated by metal-ligand covalency. A charge transfer between NO and Fe may lead to the formation of either NO^+ or NO^- as a ligand, where NO^+ is expected to interact with thiols to form RSNOs; in contrast, NO^- oxidizes thiols to disulfides [17,49,73]. The charge transfer in transition metal-nitrosyl complexes are often classified on the basis of their NO stretching frequencies in the infra red spectrum as coordination compounds containing either the NO^+(ν_{NO+} = 1500–2000 cm^{-1}) or the NO^- ligand (ν_{NO-} = 1080–1500 cm^{-1}) [74]. However, concerns have been raised that this classification may be misleading, as a series of iron-nitrosyl complexes with stretching frequencies between 1080 and 1500 cm^{-1} have been shown to contain hyponitrite, nitrito, or nitro function, rather than NO ([62] and the references therein). If the assumption that stretching frequencies of 1600–1640 cm^{-1} are associated with double bonds, then the presence of NO^- as a ligand in many iron-nitrosyl complexes cannot be ruled out [75,76]. In support of the latter hypothesis, Pearsall and Bonner have reported that Fe^{2+} reduces NO to nitroxyl (HNO) and forms iron-dinitrosyl complexes bearing both NO^+ and NO^- as *cis* positioned ligands [77]. Granozzi et al. have provided quantum mechanical calculations indicating that in the bis-cyclopentadienyl-bis(μ-nitrosyl)iron^{2+} complex, a considerable amount of charge is withdrawn from the cyclopentadienyl ion; a small part of this charge is retained by the metal ion while the remaining charge is "channeled" through the metal to the bridging NO ligand [78]. Hence, iron-nitrosyl complexes may either trans-*S*-nitrosate cellular thiols *via* transfer of NO^+ [17,79] or cause oxidation of SH functions *via* the intermediate formation of *S*-derived hydroxylamines [73]. An alternative mechanism is suggested by the studies of Kijima et al. [70], whereby the oxidation of thiols by Fe^{2+} and NO could be envisaged without the intermediate formation of *S*-nitrosothiols (Scheme 2). This mechanism is similar to the oxidation of the bis(cysteinato)Fe^{2+} complex by O_2, except that NO plays the role of an oxidant (**2** → **3**); in turn, Fe^{3+} could be reduced by one of the ligated SH functions, which would set the stage for an intramolecular disulfide ring closure (**3** → **4**). In the presence of oxygen, **5** readily undergoes oxidation to **6** with release of superoxide anion radical [32,33].

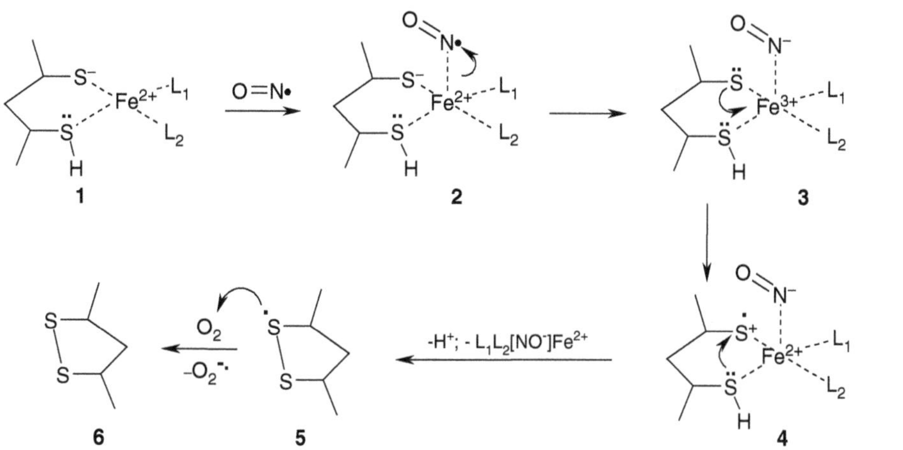

Scheme 2.

Copper-nitrosyl complexes

The interactions of copper complexes with NO and RSNOs have both synthetic and biological implications. Copper nitrosyls of the type $CuX_2 \cdot NO$ (X = halide) have been shown to interact with primary amines [80] and alcohols to form dihalides and nitrosooxyalkanes (RONO) [81,82], respectively, whereby both reactions have been proposed to occur *via* the intermediate formation of NO^+. Similarly, the $(CH_3)_3CSNO/CuCl_2$ system converts aromatic amines to the corresponding halides, presumably *via* a copper-catalyzed breakdown of $(CH_3)_3CSNO$ to $CuX_2 \cdot NO$ and $(CH_3)_3\,CS^\bullet$ [83].

The biological relevance of copper-catalyzed decomposition of RSNOs has been addressed by several research groups [84–87]. While it has been established that reduction of Cu^{2+} to Cu^+ is required for the breakdown of RSNOs to NO, the exact mechanism of this reaction is not fully understood. In the presence of Cu^+, the SNO function can undergo a reductive homolysis ($RSNO + Cu^+ \rightarrow RS^- + Cu^{2+} + NO$), where a catalytic cycle could be closed *via* RS^--dependent reduction of Cu^{2+} back to Cu^+ ($RS^- + Cu^{2+} \rightarrow RS^\bullet + Cu^+$) [87]. Alternatively, the decomposition of RSNOs may proceed *via* the intermediate chelation of Cu^+ by the S and N atoms of the SNO function [88,89]. In biological systems, however, the existence of free (aquated) Cu^+ can be ruled out. Depending on the nature of the ligand in Cu^{n+}–L_m, the breakdown of RSNOs is either partially or completely impeded [87,90,91].

Recently, Inoue et al. have reported that $0.1-1\,\mu M$ Cu^{2+} (but not Fe^{2+}, Fe^{3+}, Co^{2+}, and Mn^{2+}) and/or ceruloplasmin (CP) catalyze the *S*-nitrosation of biological thiols by NO with yields ranging from 60 to 80% [92]. Similarly, copper ions catalyze the S-(de)nitrosation of albumin [93]. In CP, a blue copper-containing protein, three different types of copper can be distinguished: type 1 with a characteristic absorption maximum at 610 nm and ESR signal with a small hyperfine splitting, and type 2 with a larger hyperfine splitting and most likely anti-ferromagnetic spin-paired Cu^{2+} couples [94]. NO forms a diamagnetic charge-transfer complex with type 1 copper whereas type 2 is unaffected [95]. Depending on the redox state of type 1 copper, NO binding to CP leads to a charge transfer with concomitant formation of either Cu^{2+}–NO^- or Cu^+–NO^+ [95,96]. Notably, Inoue et al. have found that HepG2 cells produce and release CP in amounts that are sufficient to catalyze the *S*-nitrosation of extracellular GSH. The concentrations of CP and GSH used in these experiments, as well as the amounts of GSNO generated, are physiologically conceivable. For example, the concentration of CP in the bronchoalveolar lavage fluid of adult respiratory distress syndrome patients increases up to 1 μM [97]. It is also known that a high concentration of GSH (~0.5 mM) exists in the alveolar spaces of the lung [98,99], although it is significantly lower (30 μM) in adult respiratory distress syndrome [98]. On the other hand, Zhang and Hogg have demonstrated that L-cystine enhances GSNO-dependent LMW RSNOs uptake into cells, thus increasing intracellular RSNOs levels from ~ 60 pmol/mg of protein to ~3 nmol/mg of protein. The latter process depends on the reduction of cystine to cysteine, which involves the xc-amino acid transport system, followed by formation and uptake of *S*-nitrosocysteine *via* amino acid transport system L [100,101]. Hence, it is tempting to speculate that these processes are intrinsic steps of a catalytic pathway for biosynthesis and transport of RSNOs.

CONCLUSIONS

Current research on caspases includes attempts to impede the action of these proteases in order to better understand the apoptotic pathways, as well as to develop drugs that can regulate apoptosis. Observations that NO impedes caspases and apoptosis provided a foundation for the synthesis of prodrugs that could be used for liver-specific delivery of NO. Saavedra et al. have synthesized a series of prodrugs with structural motifs that mimic specific cytochrome P450 substrates. Upon cytochrome P450 metabolism, these compounds release NO and block TNF-α-induced apoptosis and toxicity in liver, presumably *via* inhibition of caspases [102].

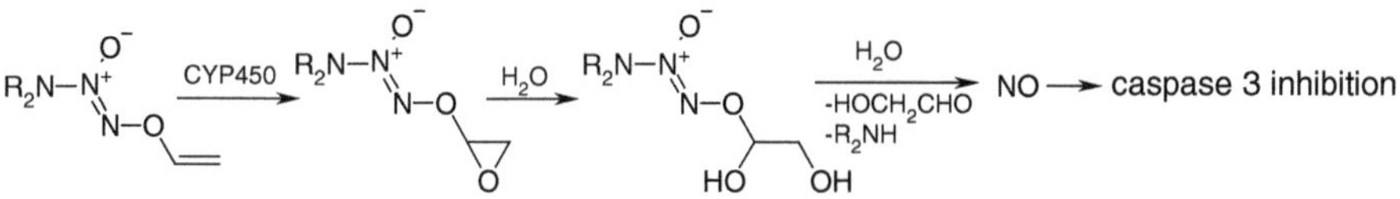

Scheme 3.

Another mechanism for regulation of apoptosis is suggested by the data presented in Fig. 1 and Ref. [16]. Elevation of non-heme iron in RAW264.7 cells delayed NO-induced cell death, which then appeared to occur *via* necrosis instead of apoptosis. The delay in cell death could be associated with *S*-nitrosation and inhibition of caspase 3, thus suggesting that the level of non-heme iron is an important factor in determining the consequence of NO exposure on cell viability. Notably, the capacity of non-heme iron and NO to impede caspases could dictate whether cells will undergo apoptosis *vs.* necrosis. However, it should be pointed out that the mechanism whereby metal-nitrosyl complexes trans-*S*-nitrosate critical cellular thiols is not deprived of intrinsic limitations: first, it is difficult to assess the distribution of ferric and ferrous ions among cellular ligands; second, thiolato-iron complexes have been shown to catalyze both the synthesis and the decomposition of RSNOs; and third, ferric and copper ions can catalyze the oxidation of thiol functions in the absence of NO. It is noteworthy that the ESR spectra presented in Fig. 3 could be observed after addition of either NO or SNAP to thiolato-iron complexes (data not shown). Nevertheless, it is interesting to speculate that LMW ligands that modulate the properties of iron- and/or copper-nitrosyl complexes may be used to control apoptotic processes.

REFERENCES

1 Steller H. Mechanisms and genes of cellular suicide. Science 1995; 267: 1445–1449.

2 Pospisil R, Young-Cooper GO, Mage RG. Preferential expansion and survival of B lymphocytes based on VH framework 1 and framework 3 expression: "positive" selection in appendix of normal and VH-mutant rabbits. Proc. Natl. Acad. Sci. USA 1995; 92: 6961–6965.

3 Raff MC. Size control: the regulation of cell numbers in animal development. Cell 1996; 86: 173–175.

4 Thompson CB. Apoptosis in the pathogenesis and treatment of disease. Science 1995; 267: 1456–1462.

5 Reap EA, Leslie D, Abrahams M, Eisenberg RA, Cohen PL. Apoptosis abnormalities of splenic lymphocytes in autoimmune lpr and gld mice. J. Immunol. 1995; 154: 936–943.

6 Liebermann DA, Hoffman B, Steinman RA. Molecular controls of growth arrest and apoptosis: p53-dependent and independent pathways. Oncogene 1995; 11: 199–210.

7 Alnemri ES, Livingston DJ, Nicholson DW, Salvesen G, Thornberry NA, Wong WW, Yuan J. Human ICE/CED-3 protease nomenclature. Cell 1996; 87: 171.

8 Casciola-Rosen L, Nicholson DW, Chong T, Rowan KR, Thornberry NA, Miller DK, Rosen A. Apopain/CPP32 cleaves proteins that are essential for cellular repair: a fundamental principle of apoptotic death. J. Exp. Med. 1996; 183: 1957–1964.

9 Dimmeler S, Haendeler J, Nehls M, Zeiher AM. Suppression of apoptosis by nitric oxide via inhibition of interleukin-1beta-converting enzyme (ICE)-like and cysteine protease protein (CPP)-32-like proteases. J. Exp. Med. 1997; 185: 601–607.
10 Melino G, Bernassola F, Knight RA, Corasaniti MT, Nistico G, Finazzi-Agro A. S-nitrosylation regulates apoptosis. Nature 1997; 388: 432–433.
11 Haendeler J, Weiland U, Zeiher AM, Dimmeler S. Effects of redox-related congeners of NO on apoptosis and caspase-3 activity. Nitric Oxide 1997; 1: 282–293.
12 Li J, Billiar TR, Talanian RV, Kim YM. Nitric oxide reversibly inhibits seven members of the caspase family via S-nitrosylation. Biochem. Biophys. Res. Commun. 1997; 240: 419–424.
13 Kim YM, Talanian RV, Billiar TR. Nitric oxide inhibits apoptosis by preventing increases in caspase-3-like activity via two distinct mechanisms. J. Biol. Chem. 1997; 272: 31138–31148.
14 Vodovotz Y, Kwon NS, Pospischil M, Manning J, Paik J, Nathan C. Inactivation of nitric oxide synthase after prolonged incubation of mouse macrophages with IFN-gamma and bacterial lipopolysaccharide. J. Immunol. 1994; 152: 4110–4118.
15 Bani D, Failli P, Bello MG, Thiemermann C, Bani Sacchi T, Bigazzi M, Masini E. Relaxin activates the L-arginine-nitric oxide pathway in vascular smooth muscle cells in culture. Hypertension 1998; 31: 1240–1247.
16 Kim YM, Chung HT, Simmons RL, Billiar TR. Cellular non-heme iron content is a determinant of nitric oxide-mediated apoptosis, necrosis, and caspase inhibition. J. Biol. Chem. 2000; 275: 10954–10961.
17 Boese M, Mordvintcev PI, Vanin AF, Busse R, Mulsch A. S-nitrosation of serum albumin by dinitrosyl-iron complex. J. Biol. Chem. 1995; 270: 29244–29249.
18 Kluge I, Gutteck-Amsler U, Zollinger M, Do KQ. S-nitrosoglutathione in rat cerebellum: identification and quantification by liquid chromatography-mass spectrometry. J. Neurochem. 1997; 69: 2599–2607.
19 Janero DR, Bryan NS, Saijo F, Dhawan V, Schwalb DJ, Warren MC, Feelisch M. Differential nitros(yl)ation of blood and tissue constituents during glyceryl trinitrate biotransformation in vivo. Proc. Natl. Acad. Sci. USA 2004; 101: 16958–16963.
20 Steffen M, Sarkela TM, Gybina AA, Steele TW, Trasseth NJ, Kuehl D, Giulivi C. Metabolism of S-nitrosoglutathione in intact mitochondria. Biochem. J. 2001; 356: 395–402.
21 Tomita X, Hirai H, Makishima S. Optical rotatory dispersion study on the iron complex with L-cysteine and its reaction with carbon monoxide and nitric oxide. Inorg. Chem. 1967; 6: 1746–1749.
22 Gross A, Yin XM, Wang K, Wei MC, Jockel J, Milliman C, Erdjument-Bromage H, Tempst P, Korsmeyer SJ. Caspase cleaved BID targets mitochondria and is required for cytochrome c release, while BCL-XL prevents this release but not tumor necrosis factor-R1/Fas death. J. Biol. Chem. 1999; 274: 1156–1163.
23 Bossy-Wetzel E, Newmeyer DD, Green DR. Mitochondrial cytochrome c release in apoptosis occurs upstream of DEVD-specific caspase activation and independently of mitochondrial transmembrane depolarization. Embo. J. 1998; 17: 37–49.
24 Li P, Nijhawan D, Budihardjo I, Srinivasula SM, Ahmad M, Alnemri ES, Wang X. Cytochrome c and dATP-dependent formation of Apaf-1/caspase-9 complex initiates an apoptotic protease cascade. Cell 1997; 91: 479–489.
25 Slee EA, Harte MT, Kluck RM, Wolf BB, Casiano CA, Newmeyer DD, Wang HG, Reed JC, Nicholson DW, Alnemri ES, Green DR, Martin SJ. Ordering the cytochrome c-initiated caspase cascade: hierarchical activation of caspases-2, -3, -6, -7, -8, and -10 in a caspase-9-dependent manner. J. Cell Biol. 1999; 144: 281–292.
26 Janicke RU, Ng P, Sprengart ML, Porter AG. Caspase-3 is required for alpha-fodrin cleavage but dispensable for cleavage of other death substrates in apoptosis. J. Biol. Chem. 1998; 273: 15540–15545.
27 Cohen GM. Caspases: the executioners of apoptosis. Biochem. J. 1997; 326 (Pt 1): 1–16.
28 Nobel CS, Kimland M, Nicholson DW, Orrenius S, Slater AF. Disulfiram is a potent inhibitor of proteases of the caspase family. Chem. Res. Toxicol. 1997; 10: 1319–1324.
29 Nobel CS, Burgess DH, Zhivotovsky B, Burkitt MJ, Orrenius S, Slater AF. Mechanism of dithiocarbamate inhibition of apoptosis: thiol oxidation by dithiocarbamate disulfides directly inhibits processing of the caspase-3 proenzyme. Chem. Res. Toxicol. 1997; 10: 636–643.

30 Sadakata N, Oda T, Komatsu N, Muramatsu T. Effects of glutathione-related compounds on increased caspase-3 and caspase-6-like activities in ricin-treated U937 cells. Biosci. Biotechnol. Biochem. 2000; 64: 202–205.
31 Zech B, Wilm M, van Eldik R, Brune B. Mass spectrometric analysis of nitric oxide-modified caspase-3. J. Biol. Chem. 1999; 274: 20931–20936.
32 Stoyanovsky DA, Goldman R, Claycamp HG, Kagan VE. Phenoxyl radical-induced thiol-dependent generation of reactive oxygen species: implications for benzene toxicity. Arch. Biochem. Biophys. 1995; 317: 315–323.
33 Goldman R, Stoyanovsky DA, Day BW, Kagan VE. Reduction of phenoxyl radicals by thioredoxin results in selective oxidation of its SH-groups to disulfides. An antioxidant function of thioredoxin. Biochemistry 1995; 34: 4765–4772.
34 Baker A, Santos BD, Powis G. Redox control of caspase-3 activity by thioredoxin and other reduced proteins. Biochem. Biophys. Res. Commun. 2000; 268: 78–81.
35 Zech B, Kohl R, von Knethen A, Brune B. Nitric oxide donors inhibit formation of the Apaf-1/caspase-9 apoptosome and activation of caspases. Biochem. J. 2003; 371: 1055–1064.
36 Mohr S, Zech B, Lapetina EG, Brune B. Inhibition of caspase-3 by S-nitrosation and oxidation caused by nitric oxide. Biochem. Biophys. Res. Commun. 1997; 238: 387–391.
37 Kim JE, Tannenbaum SR. S-nitrosation regulates the activation of endogenous procaspase-9 in HT-29 human colon carcinoma cells. J. Biol. Chem. 2004; 279: 9758–9764.
38 Lane P, Hao G, Gross SS. S-nitrosylation is emerging as a specific and fundamental posttranslational protein modification: head-to-head comparison with O-phosphorylation. Sci. STKE. 2001; 2001: RE1.
39 Kashiba-Iwatsuki M, Kitoh K, Kasahara E, Yu H, Nisikawa M, Matsuo M, Inoue M. Ascorbic acid and reducing agents regulate the fates and functions of S-nitrosothiols. J. Biochem. (Tokyo) 1997; 122: 1208–1214.
40 Nikitovic D, Holmgren A. S-nitrosoglutathione is cleaved by the thioredoxin system with liberation of glutathione and redox regulating nitric oxide. J. Biol. Chem. 1996; 271: 19180–19185.
41 Sliskovic I, Raturi A, Mutus B. Characterization of the S-denitrosation activity of protein disulfide isomerase. J. Biol. Chem. 2005; 280: 8733–8741.
42 Jensen DE, Belka GK, Du Bois GC. S-nitrosoglutathione is a substrate for rat alcohol dehydrogenase class III isoenzyme. Biochem. J. 1998; 331 (Pt 2): 659–668.
43 Laurent TC, Moore EC, Reichard P. Enzymatic synthesis of deoxyribonucleotides. IV. isolation and characterization of thioredoxin, the hydrogen donor from *Escherichia coli B*. J. Biol. Chem. 1964; 239: 3436–3444.
44 Holmgren A. Thioredoxin. Annu. Rev. Biochem. 1985; 54: 237–271.
45 Follmann H, Haberlein I. Thioredoxins: universal, yet specific thiol-disulfide redox cofactors. Biofactors 1995; 5: 147–156.
46 Nordberg J, Arner ES. Reactive oxygen species, antioxidants, and the mammalian thioredoxin system. Free Radic. Biol. Med. 2001; 31: 1287–1312.
47 Haendeler J, Hoffmann J, Tischler V, Berk BC, Zeiher AM, Dimmeler S. Redox regulatory and anti-apoptotic functions of thioredoxin depend on S-nitrosylation at cysteine 69. Nat. Cell Biol. 2002; 4: 743–749.
48 Mitchell D, Marletta M. Thioredoxin catalyzes the S-nitrosation of the caspase-3 active site cysteine. Nat. Chem. Biol. 2005; 1: 53–59.
49 Stoyanovsky DA, Tyurina YY, Tyurin VA, Anand D, Mandavia DN, Gius D, Ivanova J, Pitt B, Billiar TR, Kagan VE. Thioredoxin and lipoic acid catalyze the denitrosation of low molecular weight and protein S-nitrosothiols. J. Am. Chem. Soc. 2005; 127: 15815–15823.
50 Zhang J, Li YD, Patel JM, Block ER. Thioredoxin overexpression prevents NO-induced reduction of NO synthase activity in lung endothelial cells. Am. J. Physiol. 1998; 275: L288–L293.
51 Ceneviva GD, Tzeng E, Hoyt DG, Yee E, Gallagher A, Engelhardt JF, Kim YM, Billiar TR, Watkins SA, Pitt BR. Nitric oxide inhibits lipopolysaccharide-induced apoptosis in pulmonary artery endothelial cells. Am. J. Physiol. 1998; 275: L717–L728.

52 Tzeng E, Billiar TR, Williams DL, Li J, Lizonova A, Kovesdi I, Kim YM. Adenovirus-mediated inducible nitric oxide synthase gene transfer inhibits hepatocyte apoptosis. Surgery 1998; 124: 278–283.

53 Perry DK, Smyth MJ, Stennicke HR, Salvesen GS, Duriez P, Poirier GG, Hannun YA. Zinc is a potent inhibitor of the apoptotic protease, caspase-3. A novel target for zinc in the inhibition of apoptosis. J. Biol. Chem. 1997; 272: 18530–18533.

54 Tang ZL, Wasserloos K, St Croix CM, Pitt BR. Role of zinc in pulmonary endothelial cell response to oxidative stress. Am. J. Physiol. Lung Cell Mol. Physiol. 2001; 281: L243–L249.

55 Kondo Y, Rusnak JM, Hoyt DG, Settineri CE, Pitt BR, Lazo JS. Enhanced apoptosis in metallothionein null cells. Mol. Pharmacol. 1997; 52: 195–201.

56 Sliskovic I, Mutus B. Reversible inhibition of caspase-3 activity by iron(III): potential role in physiological control of apoptosis. FEBS Lett. 2006; 580: 2233–2237.

57 Priestley J. Experiments and Observations on Different Kinds of Air, Vol. 1, Thomas Pearson, Birmingham, 1790, p. 328.

58 Schlesinger HI, Sagathe A. Absorption spectra of nitrosylsulfuric acid and of the complex compounds of copper sulfate and of ferrous sulfate with nitric oxide. J. Am. Chem. Soc. 1923; 45: 1863–1878.

59 Tomita A, Hirai H, Makishima S. Optical rotatory dispersion study on the iron complex with L-cysteine and its reaction with carbon monoxide and nitric oxide. Inorg. Chem. 1967; 6: 1746–1749.

60 McCleverty JA, Atherton NM, Locke J, Wharthon EJ, Winscom CJ. Transition metal-dithiolene complexes. III. nitrosyl complexes of iron and cobalt. J. Am. Chem. Soc. 1967; 89: 6082–6092.

61 McDonald CC, Phillips WD, Mower HF. An electron spin resonance study of some complexes of iron, nitric oxide, and anionic ligands. J. Am. Chem. Soc. 1965; 87: 3319–3326.

62 Silverthorn W, Feltham RD. Metal nitrosyls. VII. stabilization of mononitrosyl complexes of iron. Inorg. Chem. 1967; 6: 1662–1666.

63 Dobry-Duclaux A. On the determination of the active sites of certain enzymes by means of a new specific reagent, Roussin's salt. I. Biochim. Biophys. Acta 1960; 39: 33–44.

64 Dobry-Duclaux A. On the determination of the active sites of certain enzymes by means of a new specific reagent, Roussin's salt. II. Biochim. Biophys. Acta 1960; 39: 44–52.

65 Gordy W, Rexroad HN. Free Radicals in Biological Systems. Academic Press Inc., New York, 1958, pp. 268–273.

66 Mordvintsev PI, Rudneva VG, Vanin AF, Shimkevich LL, Khodorov BI. Inhibition of platelet aggregation by dinitrosyl iron complexes with low molecular weight ligands. Biokhimiia 1986; 51: 1851–1857.

67 Vanin AF. Endothelium-derived relaxing factor is a nitrosyl iron complex with thiol ligands. FEBS Lett. 1991; 289: 1–3.

68 Mulsch A, Mordvintcev P, Vanin AF, Busse R. The potent vasodilating and guanylyl cyclase activating dinitrosyl-iron(II) complex is stored in a protein-bound form in vascular tissue and is released by thiols. FEBS Lett. 1991; 294: 252–256.

69 Vanin AF, Malenkova IV, Serezhenkov VA. Iron catalyzes both decomposition and synthesis of S-nitrosothiols: optical and electron paramagnetic resonance studies. Nitric Oxide 1997; 1: 191–203.

70 Kijima M, Nambu Y, Endo T. Electrochemical study on dihydrolipoamide-iron(II) complex and its chemical reactivity. J. Org. Chem. 1985; 50: 2522–2524.

71 Enemark JH, Feltham RD. Principles of structure, bonding, and reactivity for metal nitrosyl complexes. Coordination Chem. Rev. 1974; 13: 339–406.

72 Moller JK, Skibsted LH. Nitric oxide and myoglobins. Chem. Rev. 2002; 102: 1167–1178.

73 Doyle MP, Mahapatro SN, Broene RD, Guy JK. Oxidation and reduction of hemoproteins by trioxodinitrate(II). The role of nitrosyl hydride and nitrite. J. Am. Chem. Soc. 1988; 110: 593–599.

74 Lewis J, Irwing RJ, Wilkinson G. Infra-red spectra of transition metal-nitric oxide complexes-I. J. Inorg. Nucl. Chem. 1958; 7: 32–37.

75 Bellamy LJ. The infrared spectra of complex molecules, John Wiley and Sons, Inc., New York, 1964.

76 Brown C, Pavlosky M, Westre T, Zhang Y, Hedman B, Hodgson K, Solomon E. Spectroscopic and theoretical description of the electronic structure of S = 3/2 iron-nitrosyl complexes and their relation to O_2 activation by non-heme iron enzyme active sites. J. Am. Chem. Soc. 1995; 117: 715–732.

77 Pearsall KA, Bonner FT. Aqueous nitrosylirin(II) chemistry. 2. Kinetics and mechanism of nitric oxide reduction. The dinitrosyl complex. J. Inorg. Chem. 1982; 21: 1978–1985.
78 Granozzi G, Mougenot P, Demuynck J, Benard M. UV photoelectron spectrum and electronic structure of {Fe(n5 $-$ C_2H_5)(u-NO)}: an interpretation by means of ab initio CI calculations. Inorg. Chem. 1987; 26: 2588–2594.
79 Vanin AF, Papina AA, Serezhenkov VA, Koppenol WH. The mechanisms of S-nitrosothiol decomposition catalyzed by iron. Nitric Oxide 2004; 10: 60–73.
80 Doyle M, Siegfried B, Hammond J. Oxidative deamination of primary amines by copper halide nitrosyls. The formation of geminal dihalides. J. Am. Chem. Soc. 1976; 98: 1627.
81 Tsuge K, DeRosa F, Lim MD, Ford PC. Intramolecular reductive nitrosylation: reaction of nitric oxide and a copper(II) complex of a cyclam derivative with pendant luminescent chromophores. J. Am. Chem. Soc. 2004; 126: 6564–6565.
82 Tran D, Skelton B, White A, Laverman L, Ford P. Investigation of the nitric oxide reduction of the bis(2,9-dimethyl-1,10-phenanthroline) complex of copper(II) and the structure of $[Cu(dmp)2(H_2O)](CF_3SO_3)_2$. Inorg. Chem. 1998; 37: 2505–2511.
83 Kim Y, Shinhama K, Oae S. Direct conversion of amines to the corresponding halides by deamination with t-butyl thionitrite or t-butyl thionitrate and copper(II) halides. Tetrahedron Lett. 1978; 46: 4519–4522.
84 Askew S, Barnett D, McAninly J, Williams D. Catalysis by $Cu^{2+}$of nitric oxide release from S-nitrosothiols (RSNO). J. Chem. Soc. Perkin Trans. 2: Phy. Org. Chem. 1995; 4: 741–745.
85 Dicks AP, Williams DL. Generation of nitric oxide from S-nitrosothiols using protein-bound Cu^{2+}sources. Chem. Biol. 1996; 3: 655–659.
86 McAninly J, Williams D, Askew S, Butler A, Russell C. Metal ion catalysis in nitrosothiol (RSNO) decomposition. J. Chem. Soc. Chem. Commun. 1993; 23: 1758–1759.
87 Williams D. The mechanism of nitric oxide formation from S-nitrosothiols (thionitrites). Chem. Commun. 1996; 10: 1085–1091.
88 Smith JN, Dasgupta TP. Kinetics and mechanism of the decomposition of S-nitrosoglutathione by l-ascorbic acid and copper ions in aqueous solution to produce nitric oxide. Nitric Oxide 2000; 4: 57–66.
89 Toubin C, Yeung DY, English AM, Peslherbe GH. Theoretical evidence that Cu(I) complexation promotes degradation of S-nitrosothiols. J. Am. Chem. Soc. 2002; 124: 14816–14817.
90 Clancy R, Cederbaum AI, Stoyanovsky DA. Preparation and properties of S-nitroso-L-cysteine ethyl ester, an intracellular nitrosating agent. J. Med. Chem. 2001; 44: 2035–2038.
91 Gordge MP, Meyer DJ, Hothersall J, Neild GH, Payne NN, Noronha-Dutra A. Copper chelation-induced reduction of the biological activity of S-nitrosothiols. Br. J. Pharmacol. 1995; 114: 1083–1089.
92 Inoue K, Akaike T, Miyamoto Y, Okamoto T, Sawa T, Otagiri M, Suzuki S, Yoshimura T, Maeda H. Nitrosothiol formation catalyzed by ceruloplasmin. Implication for cytoprotective mechanism in vivo. J. Biol. Chem. 1999; 274: 27069–27075.
93 Stubauer G, Giuffre A, Sarti P. Mechanism of S-nitrosothiol formation and degradation mediated by copper ions. J. Biol. Chem. 1999; 274: 28128–28133.
94 Malkin R, Malmstrom BG. The state and function of copper in biological systems. Adv. Enzymol. Relat. Areas Mol. Biol. 1970; 33: 177–244.
95 Wever R, van Leeuwen FX, van Gelder BF. The reaction of nitric oxide with ceruloplasmin. Biochim. Biophys. Acta 1973; 302: 236–239.
96 van Leeuwen FX, Wever R, van Gelder BF. EPR study of nitric oxide-treated reduced ceruloplasmin. Biochim. Biophys. Acta 1973; 315: 200–203.
97 Krsek-Staples JA, Kew RR, Webster RO. Ceruloplasmin and transferrin levels are altered in serum and bronchoalveolar lavage fluid of patients with the adult respiratory distress syndrome. Am. Rev. Respir. Dis. 1992; 145: 1009–1015.
98 Bunnell E, Pacht ER. Oxidized glutathione is increased in the alveolar fluid of patients with the adult respiratory distress syndrome. Am. Rev. Respir. Dis. 1993; 148: 1174–1178.
99 Cantin AM, North SL, Hubbard RC, Crystal RG. Normal alveolar epithelial lining fluid contains high levels of glutathione. J. Appl. Physiol. 1987; 63: 152–157.

100 Zhang Y, Hogg N. S-nitrosothiols: cellular formation and transport. Free Radic. Biol. Med. 2005; 38: 831–838.

101 Zhang Y, Hogg N. The mechanism of transmembrane S-nitrosothiol transport. Proc. Natl. Acad. Sci. USA 2004; 101: 7891–7896.

102 Saavedra JE, Billiar TR, Williams DL, Kim YM, Watkins SC, Keefer LK. Targeting nitric oxide (NO) delivery in vivo. Design of a liver-selective NO donor prodrug that blocks tumor necrosis factor-alpha-induced apoptosis and toxicity in the liver. J. Med. Chem. 1997; 40: 1947–1954.

CHAPTER 13

Nitrite and nitrosospecies in blood and tissue: approaching the gap between bench and bedside

Tienush Rassaf* and Malte Kelm

University Hospital Aachen, Medical Clinic I, Department of Medicine, Division of Cardiology and Pulmonary Diseases, Germany

The continuous production and release of endothelium-derived nitric oxide (NO) plays an important role in vascular homeostasis and cardiac function [1]. For a long time, the signaling actions of NO in the vasculature have been thought to be short-lived as a result of its rapid reaction with oxy-hemoglobin and the immense concentration of the latter in blood. However, a series of studies within the last years have unraveled mechanisms by which NO bioactivity in blood might be sustained. Nitrite [2], *S*-nitrosothiols (RSNO) [3–5], N-nitrosamines (RNNO) [6], nitrosylhemoglobin (NOHb) [7], and S-nitrosohemoglobin (SNOHb) [8] have all been described to preserve bioactivity of NO in blood and in tissue. The mechanisms by which these agents affect the NO status of the vasculature have been discussed in other chapters of this book (nitrite in Chapters 14–16; RSNO in 9–10; RNNO in 17; the hemoglobins in 9–10). Therefore, it has been suggested that NO itself may remain active in the blood stream for longer than originally assumed [4]. Whatever mechanisms may turn out to best describe how the effects of NO are conserved in the circulation, it is clear that besides the local actions it elicits NO can also be transported throughout the body to function in a paracrine fashion, much like a hormone [9]. Although under physiological conditions this process is likely to be limited, it may well occur at pharmacological levels of NO as well as during pathological states such as in sepsis, a condition associated with high NO synthase (NOS) activity. Such mechanisms of NO conservation may explain some, if not all, of the systemic effects of NO that occur distal to the site of delivery following its administration by inhalation or infusion. Experimentally, inhaled NO has been shown to prevent lung injury from hemodialysis [10], endotoxemia [11], and ischemia/reperfusion injury [10]. Moreover, inhaled NO has been reported to reduce systemic vascular resistance [12], increase glomerular filtration rates [13] and aortic cGMP levels [14], and improve blood flow during NOS inhibition in the intestine [15] as well as in the human forearm circulation [16]. In the following, the major routes

* Author for Correspondence. E-mail: trassaf@ukaachen.de

of NO metabolism and transport in the mammalian circulation and in tissue will be described considering the plasmatic compartment and the red blood cells (RBCs) separately. A focus is laid on the role of the respective species in physiology and the opportunities they open for future diagnostic and therapeutic approaches.

NITRIC OXIDE AND RED BLOOD CELLS

By regulating blood vessel tone, inhibiting smooth muscle cell proliferation, blood cell adhesion and lipid oxidation, NO becomes fundamental for vascular homeostasis. Since NO is not only released abluminally to exert its effects on cells of the vascular wall, but also into the vessel lumen, a significant part of the NO produced by the endothelium is believed to come into direct contact with blood. Given the extremely short half-life of NO in blood and its rapid reaction with hemoglobin *in vitro*, the fate of this fraction of NO is thought to be dictated largely by its interaction with red blood cells. RBCs are believed to be a major sink for NO by virtue of the rapid co-oxidation reaction of NO with oxyhemoglobin to form methemoglobin and nitrate, the second-order rate constant of which approaches $3–4 \times 10^7\ M^{-1}s^{-1}$ [17]. Although this reaction has appreciated widespread recognition as the major inactivation pathway of NO *in vivo*, recent results obtained in humans had suggested that this may not be the predominant route under all conditions [18]. Alternatively, NO may bind to the heme group of deoxyhemoglobin to form NOHb [19]. NOHb has been detected in the blood of patients receiving nitroglycerin or inhaled NO [7,16]. A third possibility is the reaction of NO, or a higher oxidation product such as NO_2 or N_2O_3, with cysteine-93 of the β-globin chains (β-Cys93) of hemoglobin, leading to the formation of an S-nitrosated derivative of oxyhemoglobin (SNOHb). SNOHb has been suggested to participate in the regulation of blood flow [20] and platelet aggregability [21,22]. According to this theory, hemoglobin in RBCs undergoes S-nitrosation during passage through the lungs and subsequently releases part of its bound NO during arterial–venous transit in order to enhance blood flow and aid in the delivery of oxygen in the microcirculation. In the venous circulation, deoxygenated hemoglobin preferentially binds NO at the heme group to form NOHb. This proposed dynamic cycle has had a profound impact on the way we see NO today, ascribing it a most important new regulatory role in the circulation. The central observations supporting this “SNOHb hypothesis” were two reports showing arterial–venous gradients of micromolar concentrations of SNOHb and NOHb in RBCs of rats and humans [8,23]. However, numerous reports from different groups on the basal levels of intracellular SNOHb in arterial and venous blood have cast serious doubt as to the existence of such a dynamic cycle [2,24–27]. This discrepancy may have its origin in the different methodological approaches used to determine NO adducts in RBCs and the technical difficulties inherent to trace level analysis of nitroso compounds, including artifactual SNOHb/NOHb formation during sample processing.

Of particular importance in this context is the finding that the reaction rate of NO with oxyhemoglobin within erythrocytes is limited by its diffusion into the cell and occurs ~650 times slower compared to the reaction with free oxyhemoglobin [28]. Consumption of plasma NO by RBCs is reduced by increased flow, an unstirred plasma layer surrounding the RBCs, the cell-free zone near the vascular wall, and a reduced diffusion rate over the cellular

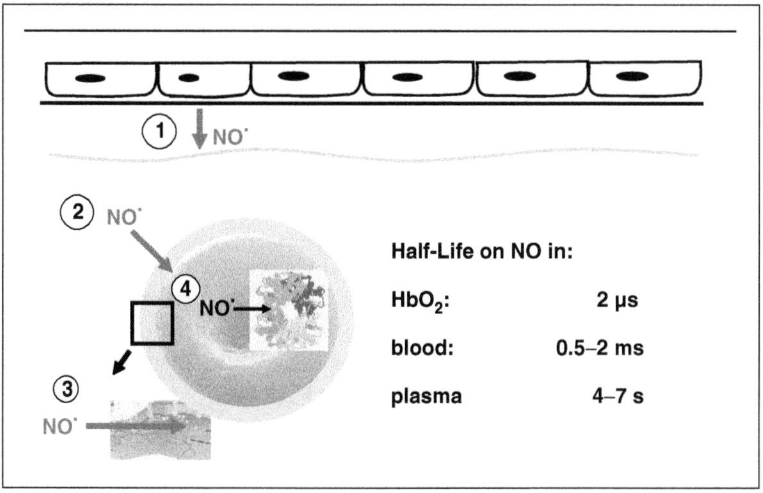

Fig. 1. Barriers of NO consumption. Consumption of plasma NO by RBCs is reduced by (1) the cell-free zone near the vascular wall, (2) an unstirred plasma layer surrounding the RBCs, (3) a reduced diffusion rate over the cellular membrane, and (4) the reaction rate with hemoglobin.

membrane [29–31]. This makes clear that at some stage endothelial NO has to react with components other than RBCs (Fig. 1).

RBC: A NOVEL SOURCE FOR NITRIC OXIDE

As shown above, RBC not only inactivate but also preserve NO bioactivity. We broaden this view by demonstrating that RBC constitutively synthesize NO [32]. This NO-formation is modulated by supplementation of eNOS-substrate in RBC. Preliminary data pointed towards the possibility that RBC might carry NOS protein. The origin of NO-syntheses within the blood compartment could not yet be reliably differentiated between the blood cell fractions due to the respective purity of preparation [33–35]. Other groups postulated either a basal or total inactive NOS isoform [36–38] or a origin by a non-enzymatic NO-syntheses [39] within RBC. Using biochemical analyses of NO and bioassays for NOS activity, we provided unequivocal evidence that RBC constitutively carry an active NOS. The specificity of RBC NOS being an eNOS isoform is supported by the lack of NO-formation in RBC from $eNOS^{-/-}$ mice. Thus RBC not simply scavenge NO but instead represent an important source of vascular NO-formation.

RBC NOS resembles a variety of specific regulatory pathways of eNOS, in that it is stereospecifically stimulated by the substrate L-arginine, it is sensitive to common NOS inhibitors, and its regulation depends on the intracellular calcium level and the phosphorylation at serine 1177 regulated by the PI3K [40–42]. Further studies are warranted to investigate the potential role of shear stress, pH, pO_2, and pCO_2 in the regulation of RBC NOS activity. In contrast to the vascular endothelium, the kininergic and muscarinergic receptors are quiescent in

RBC [43], readily explaining the lack of increases in NOS activity after exposure to acetylcholine, carbachol, metacholine, or bradykinin. The lack of NOS protein and activity in RBC from $eNOS^{-/-}$ mice—in contrast to those from wild-type (WT) mice—further support the notion that RBC carry an active eNOS isoform. Erythrocytes carry important enzymes of L-arginine metabolism, such as arginase degrading the eNOS substrate, dimethylarginine dimethylaminohydrolase (DDAH), an enzyme metabolizing endogenous NOS inhibitors [44], and cationic amino acid transporters [45]. Although admittedly speculative, RBC might fine tune their NO production *via* control of substrate availability.

The RBC NOS may substantially increase the local NO concentration at the immediate vicinity of the outer membrane thus contributing to an intrinsic barrier preventing consumption of NO derived from other sources than RBC themselves. Alternatively, intrinsic NO-formation may alter the electromechanical properties of the RBC membrane such as proteins and lipoproteins preventing consumption of NO by RBC [46].

NITRIC OXIDE AND PLASMA

Blood constitutes approximately 60% of plasma. The latter thus represents a potentially significant compartment for NO metabolism [47]. The major immediate breakdown product of NO in human plasma is nitrite. A physiologically important component of the metabolism of NO involves the generation of thionitrite esters with cysteine (Cys), leading to the formation of S-nitrosocysteine [48]. RSNOs may be generated by reaction with NO_2 or N_2O_3 produced during the oxidation of NO with dissolved oxygen [49], by reaction with nitrosonium ions (NO^+) formed from dinitrosyl-iron complexes [11] or peroxynitrite ($ONOO^-$) derived from the reaction of NO with O_2^- [50]. Alternatively, under anaerobic conditions RSNOs may be formed by direct interaction of NO with thiols in the presence of electron acceptors [51] or by reaction with nitrite [52]. The biological relevance of covalent attachement of an NO moiety to the sulfhydryl group of a plasma protein is exemplified by the occurrence of S-nitrosoalbumin (SNOAlb) in human plasma. SNOAlb is thought to represent the major reaction product of NO with plasma thiols. Albumin is the principle plasma protein and fulfills a plethora of different functions, including the maintenance of colloid osmotic pressure, acid/base buffering, and antioxidative actions, as well as binding and transport of bilirubin, hormones, fatty acids, and other endogenous ligands [53]. Human albumin contains a single free sulfhydryl group in Cys34, which is responsible for many of *its* properties. Apart from a change in its binding characteristics following reaction with NO [54], little is known about whether or not other physiological functions of albumin are altered upon nitrosation of its free SH-group. What is clear, however, is that nitrosation of albumin confers NO^+ and NO-donating properties to the molecule. Transnitrosation reactions, i.e. the transfer of an NO^+-equivalent from one molecule to another, are a common feature of all RSNOs and may account for the inhibition of cysteine-dependent enzymes by SNOAlb. Whereas reaction with glutathione and ascorbate stimulates NO release from SNOAlb [55], spontaneous decomposition to NO is rather slow, which is compatible with the notion that SNOAlb acts as a buffer and transport system for NO [5]. In spite of the fact that the physiological concentration of this circulating NO-adduct is the subject of much debate [56,57], a number of recent clinical investigations have indirectly implicated the involvement of RSNOs in disease processes [58–60].

However, whether these species represent reporter molecules suitable to monitor the progression of illnesses associated with an increased or decreased NO production, play a protective role *in vivo*, or rather represent bioinactive detoxification products that is far from being clear.

NITROSOSPECIES OTHER THAN RSNOs IN PLASMA

RSNOs have been the major focus of most recent studies, and many pathways have been proposed to be under regulatory control by S-nitrosation [61]. However, little is known about reaction sites of NO other than thiols. Recently, it has been shown that under physiological conditions, human plasma contains an approximately 5-fold higher concentration of RNNOs than RSNOs [6]. It has been known that RNNOs are generated endogenously during infections and inflammatory processes [62] either *via* NO-mediated nitrosation, intermediate formation of peroxynitrite, or bacterial action. In the acidic environment of the stomach, RNNOs are formed as a result of the reaction of nitrite with amino groups of food constituents [63]. Irrespective of whether they are taken up during occupational exposure or CONTACT with rubber products, ingested with the diet, or formed endogenously, low-molecular-weight RNNOs are potentially mutagenic and have traditionally been associated with an increased risk for cancer.

Due to the association of both the S- and N-nitroso components to serum albumin, we hypothesized that the major nitroso species in human plasma is a *S*-nitrosothiol/N-nitrosamine derivative of albumin. Whether such a species originates from competing S- and N-nitrosation reactions or is a result of S $\rightarrow$ N transnitrosation remains to be investigated. What makes this finding particularly intriguing is that it demonstrates that endogenous RNNOs are present already under non-inflammatory conditions, suggesting that they either serve a physiological role or are the result of the body's response to the continuous exposure to foreign material. The finding that the basal concentration of RNNOs exceeds that of RSNOs is important and may suggest a novel storage and/or delivery form of NO that is differentially regulated from RSNOs. However, further studies are required to elucidate whether RNNOs play a role in cardiovascular physiology and/or pathology.

NITRIC OXIDE SOLUTIONS

Intravenous application of an NO-saturated saline solution exerts systemic hemodynamic effects, as judged by a significant dilation of resistance and conduit arteries in the contralateral arm and a transient and modest decrease in arterial blood pressure. The formation of RSNO and the dilatory response lasted several orders of magnitude longer than the previously suggested half-life of NO *in vivo*, thus ruling out that the transport of free NO contributes significantly to the sustained dilation seen in both conduit and resistance vessels. In support of the conclusion that plasma RSNOs are involved in the systemic vascular effects of intravenously applied aqueous NO solution, its hemodynamic response was mimicked by intravenous application of S-nitrosoglutathione [3].

In the future, systemic application of authentic NO may substitute NO deficiency as seen in various cardiovascular diseases associated with endothelial dysfunction. The route of administration and the chosen dose may critically determine NO transport and biological effects. This includes uptake, transport, and liberation of NO along the vascular tree by physiological blood-borne carriers, the intercellular contact of blood cells during passage through the microvasculature, and the interaction of plasmatic NO stores with the endothelium and vascular smooth muscle. The metabolism and distribution of NO and thus its range of action may exhibit significant spatial heterogeneity *in vivo*, depending on vessel size, hematocrit, flow velocity, and shear rates [64] as well as oxygen saturation and tissue pO_2. These variables are expected to affect NO consumption in the RBC-free plasma zone near the vascular wall, and within RBCs [65,66]. Matters are further complicated by the fact that the RBC membrane represents a heterogenous sink [30,67], which may critically determine the effects of NO on the microcirculation [68]. These denominators might be affected differently by intraarterial, intravenous, or inhalative NO substitution therapy. Whether the application of aqueous NO solution may offer therapeutic advantages over existing NO-related therapies and furthermore be suitable to exert systemic biological effects *in vivo* remains to be investigated.

NITRIC OXIDE CAN BE TRANSPORTED IN ITS FREE FORM ALONG THE VASCULAR TREE

Beside the reactions of NO with thiols and amines, a third route has been proposed for the transport of NO in plasma. *In-vivo* investigations with authentic NO shed light on the capability of plasma to transport NO in its free form along the vascular tree. Infusions of NO solutions into the brachial artery of human volunteers led to an increase in the diameter of the downstream artery and to an increase in forearm blood flow [4]. In these experiments, the immediate effects appeared to be mediated by bioactive NO itself and the delayed effects by plasma RSNOs. The effective lifetime of NO in blood is thought to be limited by numerous reactions in plasma and, importantly, with intraerythrocytic hemoglobin (see above). However, in flowing blood, these degradative reactions are markedly attenuated for several reasons: because of the properties of the RBC membrane, entry of NO into RBCs occurs at a rate up to three orders of magnitude slower than would be expected from simple diffusion [30,31]. Moreover, in a perfused blood vessel, the endothelial surface is in contact with an RBC-free plasma zone, which has been estimated to reach up to 25% of the luminal diameter in thickness [29]. Within this RBC-free zone of laminar flowing blood, NO has a surprisingly long half-life. Considering the concentration of dissolved oxygen in blood ($\sim$150–250 μM) and assuming that physiological NO concentrations are in the nanomolar range, the biochemical lifetime of NO in such a plasma layer has been calculated to reach 100–500 s [69], provided reactions with other plasma constituents are negligible. Such a long half-life would allow NO to be transported as such from its site of synthesis to many other tissues.

PLASMA RSNOs AS POTENTIAL DISEASE MARKERS

A number of recent clinical investigations have indirectly implicated the involvement of plasma RSNOs in disease processes [58,60]. Whether these species represent markers suitable to monitor the progression of illnesses associated with an increased or decreased NO production, play a protective role *in vivo*, or rather represent bio-inactive products is not clear. In this context, many attempts have been made to quantify plasma RSNOs in humans. However, nearly 15 years after the initial report [5] there seems to be no consensus about the true physiological levels of RSNOs in human plasma. Whereas the initially proposed concentration of 7 μM RSNOs was confirmed in a study by Tyurin et al. [60], who measured 9.2 ± 1.6 μM in plasma of non-pregnant women, other studies suggested that basal RSNO levels in plasma are considerably lower, and in fact in the low nanomolar range [3,6,16,70–72]. As pointed out by others [56,57], these differences may have their origin in the different analytical approaches used to detect RSNOs and the technical difficulties associated with trace level analysis, including artifactual RSNO formation.

The question arises whether knowledge of the true plasma concentration would actually be of significance and help further our understanding of the physiological role of RSNOs. Arguably more interesting to know is to what extent RSNO concentrations change under various physiological conditions (e.g. compare physical exercise with rest, levels after food intake to fasting conditions, etc.) and in disease states such as heart failure, septic shock, hypertension, or arteriosclerosis. All this will still not help much without a clear idea about the mechanisms of bio-activation and -degradation, which are critical in determining the flux rate of NO that can be released from a given concentration of RSNO, and hence its biological effect. These mechanisms may be quite complex and under control of a number of different factors some of which are presently unknown. Thus, one may argue that knowledge of absolute concentrations is less important than that of relative changes in concentration experienced under specific conditions. Such RSNO determinations could well be carried out using different analytical methods, even if they produced different absolute concentrations, as long as their results were proportional to the true plasma concentration. However, this is true only if the technique is specific and not influenced by changes in concentration of other nitroso species, nitrite, or nitrate. Otherwise, it would invite misinterpretations about the involvement of plasma RSNOs by falsely attributing these differences to changes in RSNO levels.

ANALYSIS OF NITRIC OXIDE SPECIES: FINDING THE RIGHT APPROACH

A major obstacle in assessing the significance of formation of oxidative and nitrosative NO species *in vivo* is the limited reliability of available techniques for measurements in complex biological matrices. Investigations into the validity of the emerging hypotheses about the physiological or pathophysiological role of specific nitros(yl)ated biomolecules are often hampered by the availability of analytical methods suited for reliable quantification of the extremely low levels of such species *in vivo*. Although RSNOs have been detected in various samples and are the focus of most recent studies, NO does interact either directly or indirectly with other biological targets, including amines and heme moieties. Older colorimetric methods for nitrite and RSNO determination are subject to various interferences by proteins, suspended materials, and colored species, often lack sensitivity and require elaborate clean-up procedures when working with biological media. EPR spectroscopy for determination of NO in metal complexes or hemeproteins has the advantage to be applicable *in vivo*.

However, its sensitivity appears to be insufficient for nitrosyl detection under basal conditions [19]. Several analytical techniques for the determination of RSNOs and RNNOs have been described. Though it has been demonstrated that NO can be released from RSNOs and RNNOs by high-intensity UV-light, most analytical methods are based on a chemical denitrosation step. This procedure leads to reductive cleavage of the nitroso compounds by e.g. hydrogen bromide, iodine/triiodide mixtures, or vanadium chloride, and the released NO is subsequently detected by gas-phase chemiluminescence reaction with ozone. As the reducing properties of these solutions vary greatly, so do efficiency of reduction and selectivity. To complicate matters further, some sampling processing techniques may even produce the species under investigation, leading to false positive results.

For the determination of nitrite levels in biological samples, we favor the use of either one of the following analytical techniques: (1) flow injection analysis (FIA) [73], (2) reductive gas-phase chemiluminescence (CLD; e.g. CLD 77am sp or model 88am, Eco Physics; Duernten, Switzerland) [27], and (3) a dedicated HPLC system employing ion chromatography with post-column derivatization (e.g. ENO-20, EiCom, Kyoto, Japan) [26]. All three methods offer detection of nitrite (and nitrate, either directly or after reduction) in biological samples in the low nanomolar range with a high linearity, recovery, and a low coefficient of variation, and do not require complicated sample processing procedures.

For the determination of RSNOs, RNNOs, NO-heme compounds in plasma, RBCs, and tissues, we suggest to employ reductive cleavage by an iodine/triiodide-containing reaction mixture and subsequent determination of the released NO by chemiluminescence [27]. With larger injection volumina, this method can quantify as little as 100 fmol bound NO and has been validated extensively for use in different biological matrices. To differentiate between compound classes without having to change reaction solutions or conditions, samples can be pretreated with group-specific reagents before analysis. In detail, biological samples are typically divided into three aliquots: one used for direct injection (nitrite + nitroso compounds), one for preincubation with sulfanilamide, which efficiently removes nitrite (total nitroso species), and a third one for preincubation with $HgCl_2$/sulfanilamide (mercury-resistant nitroso compounds).

Not less important than the use of the appropriate analytical technique is the actual sample processing: blood should be collected in (chilled) tubes containing *N*-ethylmaleimide (NEM)/ethylenediaminetetraacetic acid (EDTA) (10/2 mmol/l) in order to block SH-groups and inhibit transnitrosation reactions, preventing artificial nitrosation, as well as thiolate- and ascorbate-mediated RSNO degradation. Plasma is readily obtained by centrifugation for 10 min at 800x*g* and 4°C; RBCs are lysed 1:4 in water containing NEM/EDTA (10/2 mmol/l) and kept on ice in the dark for <15 min. Cells and organs have to be homogenized in the dark on ice, using PBS with the addition of NEM/EDTA (10/2 mmol/l). We caution that many commercially available chemicals and solutions are contaminated with varying amounts of nitrite, which may result into artifactual nitrosation and therefore lead to false positive results.

PLASMA NITROSOSPECIES AND CARDIOVASCULAR DISEASE

In order to examine the role of nitrosospecies in relation to cardiovascular risk factors, we determined RXNO levels (the sum of all nitrosospecies) of plasma in patients with one or more cardiovascular risk factors [74]. Levels of RXNOs were lower in such individuals compared to controls. It can therefore be concluded that endothelial dysfunction is associated with a depletion of circulating nitroso/nitrosyl species in plasma, which is likely to contribute to the increased risk for major cardiovascular events in individuals with endothelial dysfunction. Measuring circulating nitroso/nitrosyl species may help identifying individuals at risk and serve as a therapeutic surrogate marker in the future. Further studies are needed to establish whether the plasmatic RXNO pool represents a valuable parameter that allows optimal dose titration of therapeutic agents aimed at targeting endothelial dysfunction.

EFFECTS OF NITRIC OXIDE AND RSNOs ON LEFT VENTRICULAR FUNCTION

In contrast to its well-defined vasodilator effect, the role of NO on myocardial function is still a subject of significant controversy, and experimental studies suggest both positive and negative effects on left ventricular function. Thus, the fundamental question as to whether NO contributes to the maintenance of myocardial function in humans still remains unanswered.

Initial *in-vitro* experiments in isolated ventricular myocytes as well as in the isolated guinea pig heart suggested a negative contractile effect of NO [75–77]. A bimodal effect of NO [78], with a positive contractile effect at low concentrations but a negative one at higher concentrations was postulated. In humans, the results are limited and partly divergent [79–81]. The direct impact of NO levels on left ventricular function has never been determined in humans. We therefore applied the *S*-nitrosothiol *S*-nitrosoglutathione (GSNO) on top of inhibited NO-synthase (NOS)-activity [by infusion of N^{G}-monomethyl-L-arginine (L-NMMA)], and the mean arterial blood pressure at the beginning of the experiments served as the target value. This led to an increase in stroke volume index (SVI), which was measured by MRI-technology, and was accompanied by a decrease in arterial blood pressure compared to L-NMMA. To investigate whether the decrease in afterload and /or a sympathetic activation is responsible for the increase in SVI, we applied the vasodilator dihydralazine (a vasodilator not influencing the NO-pathway) to decrease arterial pressure to a level comparable to that resulting from GSNO. We observed no significant change in SVI after dihydralazine, suggesting that the effect of GSNO on cardiac function cannot be explained solely by the decrease in arterial blood pressure. The lack of a significant difference in plasma noradrenaline after application of GSNO or hydralazine ruled out a major sympathetic influence of either substance on SVI. Therefore, NO (at low concentration) improves left ventricular function even after matching for afterload and sympathetic tone pointing towards a direct NO-related effect on the myocardium [82]. Because of the bimodal effects of NO [78], with a positive effect on LV function at low amounts and a negative one at high amounts, the threshold level for the switch from a positive to a negative effect of NO seems to be of importance.

PLASMA NITRITE AND eNOS-ACTIVITY

The majority of intravascular NO is inactivated by its reaction with hemoglobin to form nitrate [17] (Fig. 2). Nitrate concentrations are influenced by a variety of NO-synthase (NOS)

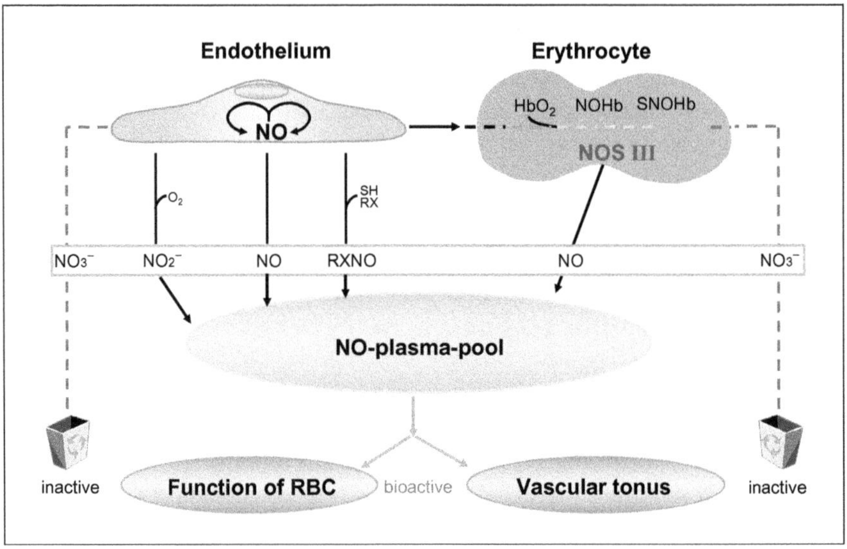

Fig. 2. Constitutive NO synthesis and vascular homeostasis. The eNOS constitutively synthesizes NO in RBC and endothelial cells. Endothelium-derived NO may act in an autocrine fashion within the vascular wall and may then be inactivated to nitrate (NO_3^-) representing a potential salvage pool for NO or oxidized to nitrite (NO_2^-). Alternatively, it may undergo nitros(yl)ation thus contributing to circulating plasma pool of conserved NO bioactivity yielding *S*-nitrosothiols and nitrosoamines (RXNO). Within RBC, NO may be inactivated *via* the co-oxidation with oxyhemoglobin (HbO_2) forming nitrate and methemoglobin. Alternatively, its bioactivity may be conserved through nitros(yl)ation of Hb yielding nitrosylhemoglobin (NOHb) and *S*-nitrosohemoglobin (SNOHb). Enzymatic NO-formation in RBC contributes also to plasma NO pool. Both, RBC and endothelium, export bioactive NO exerting an array of functions important for vascular homeostasis.

independent factors, including dietary nitrate intake, formation of saliva, bacterial nitrate synthesis within the bowel, denitrifying liver enzymes, inhalation of atmospheric gaseous nitrogen oxides, and renal function [47,83]. Due to these factors and the high background level, small changes in plasma nitrate concentrations may not sensitively reflect acute changes in NOS-activity [84]. Stimulation of eNOS has been shown to result in an acute change in plasma nitrite concentration in the human forearm vasculature [84]. The reported basal nitrite concentrations in plasma of mammals range from "non-detectable" [85] over nanomolar (450 nmol/l) [86] to micromolar (26 μmol/l) [87] levels. These enormous differences in nitrite levels can be rationalized considering the emerging confounding factors and variations in blood sampling and sample processing as well as the methodological problems inherent to the analytical procedures used. Some methods simply do not possess the sensitivity to allow a precise measurement of nitrite in the proposed physiological concentration range. In addition, the analysis might be affected by proteins, varying redox conditions, and trace contamination with nitrite during sample processing [73].

Using multiple distinct analytical methods, the uniform occurrence of nanomolar levels of plasma nitrite in different mammalian species has been demonstrated, indicative of comparable constitutive NOS-activity. Nitrite was measurable in plasma samples of humans (305 ± 23 nmol/l), monkeys (367 ± 62 nmol/l), minipigs (319 ± 24 nmol/l), dogs (305 ± 50 nmol/l), rabbits (502 ± 21 nmol/l), guinea pigs (412 ± 44 nmol/l), rats (191 ± 43 nmol/l), and mice (457 ± 51 nmol/l). Nitrite levels were in the nanomolar range (100–600 nmol/l) in spite of subtle differences in sample preparation and analytical methods used [88]. Gas-phase chemiluminescence, a method that depends on the reductive conversion of nitrite to NO and subsequent detection of the liberated NO by its reaction with ozone [27], an HPLC technique which employs ion chromatography to separate nitrite and nitrate from other plasma constituents, online reduction and subsequent postcolumn derivatization with the Griess reagent [6], and a flow injection analysis in combination with the Griess reagent, a method that employs colorimetric reaction with nitrite without prior separation [73] from other plasma constituents were applied. The accuracy of the reported ranges of plasma nitrite concentrations is supported by the fact that these different and independent analytical approaches revealed very similar absolute levels of plasma nitrite.

Experiments with NOS-inhibitors showed that the majority of the basal nitrite concentration in all tested mammals was derived from NOS-activity, corroborating earlier results by Rhodes et al. [89]. In contrast to nitrite, plasma nitrate did not change significantly during acute NOS inhibition. However, changes in nitrate concentration due to a diminished NO:oxyHb interaction during NOS-inhibition would be expected to be in the nanomolar range which would represent only a very small fraction of total plasma nitrate concentration [47,90]. The detection of such small changes in plasma nitrate might be hampered by the analytical difficulty to discriminate concentration differences in the nanomolar range against a micromolar background concentration of nitrate. In addition, nitrate levels are influenced by a variety of NOS-independent factors such as the dietary nitrate intake, saliva formation, bacterial synthesis in the bowl, denitrifying liver enzymes, inhalation of atmospheric gaseous nitrogen compounds, and the renal function. Thus, nitrate is a less sensitive marker for *acute* changes of NOS-activity. Plasma nitrate, which is eliminated *via* the kidneys, has a significant longer half-life than nitrite, and therefore requires more time to decrease in response to changes of NOS-activity. Only long-lasting (5 days) and chronic changes in NOS-activity in mice caused a significant reduction of plasma nitrate concentration. If NOS-independent factors which influence plasma nitrate are controlled, nitrate may have the potential for clinical balancing studies of the L-arginine/NO-pathway reflecting chronic changes of NOS-activity.

Mice deficient in eNOS showed similar plasma nitrite levels under basal conditions as WT mice during NOS inhibition. Two-thirds of basal plasma nitrite are derived from basal eNOS activity. These observations highlight the uniformity of constitutive NOS-activity across mammalian species and the significance of eNOS for the generation of plasma nitrite. The remaining plasma nitrite in the eNOS(−/−) mice did not diminish significantly under NOS-inhibition. Major compensatory upregulation of nNOS-isoform seems unlikely but iNOS upregulation cannot be excluded. The remaining nitrite in eNOS(−/−) mice may be of alimentary origin, inhalation of gaseous N-oxides, be derived from nitrate reduction by xanthine oxidase [91], result from the decay of bound NO in the mammalian circulation or be due to an upregulation of the iNOS-isoform.

These findings may have important implications for future *in-vitro* or *in-vivo* investigations and the potential development of a diagnostic marker for atherosclerosis. The latter is associated with a diminished bioavailability of NO due to secondary endothelial dysfunction to reduced NOS-activity [92–94]. Further investigations are required to clarify whether plasma nitrite may be a useful index of regional NOS-activity in such pathological disease processes.

ORIGIN OF PLASMA NITRITE

The net concentration of nitrite in plasma is a result of its formation and consumption. Several routes of formation of nitrite exist in mammals. Nitrite is an oxidation product of endothelium-derived NO. Moreover, nitrite is present in food, especially in processed meat, in which nitrite is used to prevent botulism [95]. Furthermore, plasma nitrite increases after ingestion of large amounts of inorganic nitrate. This increase is entirely due to enterosalivary circulation of nitrate (as much as 25% is actively taken up by the salivary glands) and reduction to nitrite by commensal bacteria [95]. This nitrite enters the circulation when saliva is swallowed [96]. The formation of nitrite is counterbalanced by several pathways of elimination. Nitrite can be oxidized to nitrate by oxyhemoglobin [97] in a reaction that is by far slower than the oxidation of NO to nitrite. In addition, nitrite can be reduced to NO under acidic conditions [98–101]. However, this will only occur at a pH less than 7, which is seen in tissues during ischemia [102]. Moreover, it has been shown that xanthine oxidase may reduce nitrite to NO [103]. Nitrite reduction by xanthine oxidase is greatly enhanced at low oxygen tensions and acidic conditions such as those seen during ischemia. Furthermore, nitrite is recycled back into bioactive NO *via* reduction by desoxyhemoglobin [98]. It is suggested that this mechanism ensures an autoregulated NO generation in regions of poor oxygenation where desoxyhemoglobin predominates [104]. The different routes of formation and metabolism result into a biological relevant steady-state concentration of nitrite.

NITRITE SIGNALING

In an experimental study, the capacity of nitrite to modulate multiple signaling pathways known to be mediated by nitrosylation and /or S-nitrosation was demonstrated [52]. Nitrite can nitrosylate hemes and nitrosate thiols even at low doses; in rats, physiological levels of nitrite account for basal levels of heme nitrosylation and S-nitrosation. Thus nitrite, not NO, may turn out to be the substance that determines the level of these products under physiological conditions. The results further show several aspects of heme nitrosylation and S-nitrosation from nitrite that indicate these signaling mechanisms follows a chemical path distinct from that of NO. This leads to the conclusion that nitrite must perform a unique signaling role in mammalian tissues. The nitrite-based signaling pathway must serve a crucial function that the seemingly more sophisticated formation of NO cannot fulfill. It has been hypothesized that nitrite plays a role not only as a back-up system for NO production under hypoxic conditions (which are incompatible with oxygen-dependent NO production from NOS), but that it serves important but yet unrecognized co-ordinating/synchronizing functions already under normal physiological conditions. This assumption is based on the following observations: by virtue of its radical nature, NO readily reacts with a number of reactive biological species, which on the one hand offers plenty of opportunities to interact with multiple regulatory and signaling pathways, but on the other hand makes it suited only for local/regional control of cellular functions. As a result, NO acts largely in a paracrine fashion. In contrast, nitrite is considerably less reactive, which translates into a much longer lifetime, allowing nitrite to reach targets that locally produced NO would not be able to interact with. Since conversion of nitrite into nitroso/nitrosyl species does not rely on the presence of oxygen, nitrite is ideally suited to serve as an endocrine messenger.

NITRITE AND DIAGNOSTICS

Taking into account that endothelial dysfunction is at least in part reversible [105], an early diagnosis of this disorder by assessing eNOS activity may have prognostic and therapeutic consequences. While direct biochemical evidence for an impaired eNOS activity has been obtained in experimental models, this approach is difficult in humans so far. It has recently been shown that plasma nitrite levels decrease with increasing numbers of cardiovascular risk factors [106]. In order to improve sensitivity and to get a higher discriminatory power, in the next step stimulation of eNOS activity has been chosen.

An increase in shear stress, i.e. the tangential force exerted by the flow of blood over the surface of the endothelium, is the strongest physiological stimulus of eNOS activity and leads to increases in NO formation. Shear stress induced NO-dependent dilation of the brachial artery can be measured non-invasively as flow-mediated dilation (FMD) using high-resolution ultrasound and is commonly used to characterize endothelial function [107]. This ultrasonographic method quantifies the dilation of conduit arteries, e.g. brachial artery, in response to physiologically relevant increases in laminar shear stress induced by ischemic dilation of the downstream microvasculature. Increases in shear stress lead to a rapid activation of endothelial NO-synthase with consecutive increases in NO formation. Accordingly, FMD is largely abolished following NOS inhibition [105], and therefore provides a valuable "read-out" of local vascular NO availability. A biochemical approach to determine eNOS activity is still missing. In order to attempt to establish an index to assess eNOS capacity biochemically in humans, the functional reserve of eNOS activity was determined and local plasma nitrite concentrations in the antecubital vein at baseline as well as during reactive hyperemia following 5 min of forearm ischemia in young healthy subjects were measured. Plasma nitrite increased by factor 2 and the time course of nitrite mirrored the increase in brachial artery diameter. Individuals with endothelial dysfunction, as defined by reduced flow-mediated dilation, did not exhibit an increase in nitrite [108]. The correlation between local nitrite increase and the degree of vasodilation suggest that nitrite changes may reflect endothelial function.

TISSUE STORES OF NITRIC OXIDE

Most literature on the transport and metabolism of NO focused almost exclusively on the reaction of NO with proteins circulating in blood. However, with the abundance of hemeproteins such as cytochrome P_{450}, cyclooxygenase, and peroxidases in tissues and their known high affinity for NO, heme moieties in tissue appears to be obvious acceptors for endogenous NO *in vivo*. It has been shown that endothelial NO production results in local formation of NO-adducts that may act as storage forms of NO. Biochemical analyses revealed that rat aortic tissue contains equimolar concentrations of S- and N-nitroso compounds as well as nitrite and nitrate [109]. Using a functional approach, RSNOs and nitrite showed the capability to release NO upon illumination with light and dilate blood vessels (photorelaxation). Whether these stores can be bioactivated *in vivo* to the extent they are optically, what activation mechanisms are involved and whether circulating NO does exchange with tissue acceptors at the level of the microcirculation is not clear at present and requires further investigation. Similarly, the major targets of cellular nitrosation are unclear. Whereas nitrosation is expected to occur predominantly in hydrophobic over hydrophilic compartments of the cell [110], one target of obvious relevance is the mitochondrion.

NO reacts with free radicals, hemeproteins, and—following reaction with molecular oxygen—thiols, all of which are abundant at high concentration in mitochondria. Therefore, it comes as no surprise that NO interaction with mitochondria would have significance for cell function [111,112]. The discovery of mitochondrial NO formation in 1997 [113] suggested the existence of a highly localized source of NO in the mitochondrial inner membrane [114]. The basal formation of NO in mitochondria seems to be one of the main regulators of cellular respiration, mitochondrial transmembrane potential, and transmembrane pH gradient [115]. Moreover, mitochondrially derived NO plays an important role as an antioxidant by reacting with potential harmful ROS. However, an overproduction of NO may result in the generation of RNOS (reactive oxygen and nitrogen species) that cannot be neutralized by the mitochondrial barriers, which results in oxidative stress and cell death.

SUMMARY

Nitric Oxide activity in blood and tissue can be preserved for much longer than originally believed over the formation of SNOHb, NOHb, RXNO, nitrite, and free NO. The sum of all these, the NO pool can be characterized and offers an important diagnostic and therapeutic tool. Together with the discovery that nitrite can be recycled into bioactive NO again *in vivo*, these new insights may stimulate the development of new approaches in the diagnosis and treatment of NO deficiency states inherent to many vascular diseases. This may help closing the gap between bench and bedside.

REFERENCES

1 Ignarro LJ. Biological actions and properties of endothelium-derived nitric oxide formed and released from artery and vein. Circ. Res. 1989; 65: 1–21.

2 Gladwin MT, Shelhamer JH, Schechter AN, Pease-Fye ME, Waclawiw MA, Panza JA, Ognibene FP, Cannon III RO. Role of circulating nitrite and S-nitrosohemoglobin in the regulation of regional blood flow in humans. Proc. Natl. Acad. Sci. USA 2000; 97: 11482–11487.

3 Rassaf T, Kleinbongard P, Preik M, Dejam A, Gharini P, Lauer T, Erckenbrecht J, Duschin A, Schulz R, Heusch G, Feelisch M, Kelm M. Plasma nitrosothiols contribute to the systemic vasodilator effects of intravenously applied NO: Experimental and clinical study on the fate of NO in human blood. Circ. Res. 2002; 91: 470–477.

4 Rassaf T, Preik M, Kleinbongard P, Lauer T, Heiß C, Strauer BE, Feelisch M, Kelm M. Evidence for in vivo transport of bioactive nitric oxide in human plasma. J. Clin. Invest. 2002; 109: 1241–1248.
5 Stamler JS, Jaraki O, Osborne J, Simon DI, Keaney J, Vita J, Singel D, Valeri CR, Loscalzo J. Nitric oxide circulates in mammalian plasma primarily as an S-nitroso adduct of serum albumin. Proc. Natl. Acad. Sci. USA 1992; 89: 7674–7677.
6 Rassaf T, Bryan NS, Kelm M, Feelisch M. Concomitant presence of N-nitroso and S-nitroso proteins in human plasma. Free Radic. Biol. Med. 2002; 33: 1590-1596.
7 Gladwin MT, Ognibene FP, Pannell LK, Nichols JS, Pease-Fye ME, Shelhamer JH, Schechter AN. Relative role of heme nitrosylation and β-cysteine 93 nitrosation in the transport and metabolism of nitric oxide by hemoglobin in the human circulation. Proc. Natl. Acad. Sci. USA 2000; 97: 9943–9948.
8 Jia L, Bonaventura C, Bonaventura J, Stamler JS. S-nitrosohaemoglobin: a dynamic activity of blood involved in vascular control. Nature 1996; 380: 221–226.
9 Schechter AN, Gladwin MT. Hemoglobin and the paracrine and endocrine functions of nitric oxide. N. Engl. J. Med. 2003; 348: 1483–1485.
10 Barbotin-Larrieu F, Mazmanian M, Baudet B, Detruit H, Chapelier A, Libert JM, Dartevelle P, Herve P. Prevention of ischemia-reperfusion lung injury by inhaled nitric oxide in neonatal piglets. J. Appl. Physiol. 1996; 80: 782–788.
11 Boese M, Mordvintcev PI, Vanin AF, Busse R, Mulsch A. S-nitrosation of serum albumin by dinitrosyl-iron complex. J. Biol. Chem. 1995; 270: 29244–29249.
12 Takahashi Y, Kobayashi H, Tanaka N, Sato T, Takizawa N, Tomita T. Nitrosyl hemoglobin in blood of normoxic and hypoxic sheep during nitric oxide inhalation. Am. J. Physiol. Heart Circ. Physiol. 1998; 274: H349–H357.
13 Troncy E, Francoeur M, Salazkin I, Yang F, Charbonneau M, Leclerc G, Vinay P, Blaise G. Extra-pulmonary effects of inhaled nitric oxide in swine with and without phenylephrine. Br. J. Anaesth. 1997; 79: 631–640.
14 Kermarrec N, Zunic P, Beloucif S, Benessiano J, Drouet L, Payen D. Impact of inhaled nitric oxide on platelet aggregation and fibrinolysis in rats with endotoxic lung injury. Role of cyclic guanosine 5′-monophosphate. Am. J. Respir. Crit. Care. Med. 1998; 158: 833–839.
15 Fox-Robichaud A, Payne D, Hasan SU, Ostrovsky L, Fairhead T, Reinhardt P, Kubes P. Inhaled NO as a viable antiadhesive therapy for ischemia/reperfusion injury of distal microvascular beds. J. Clin. Invest. 1998; 101: 2497–2505.
16 Cannon III RO, Schechter AN, Panza JA, Ognibene FP, Pease-Fye ME, Waclawiw MA, Shelhamer JH, Gladwin MT. Effects of inhaled nitric oxide on regional blood flow are consistent with intravascular nitric oxide delivery. J. Clin. Invest. 2001; 108: 279–287.
17 Doyle MP, Hoekstra JW. Oxidation of nitrogen oxides by bound dioxygen in hemoproteins. J. Inorg. Biochem. 1981; 14: 351–358.
18 McMahon T, Moon RE, Luschinger BP, Carraway MS, Stone AE, Stolp BW, Gow AJ, Pawloski JR, Watke P, Singel DJ, Piantadosi CA, Stamler J. Nitric oxide in the human respiratory cycle. Nat. Med. 2002; 8: 711–717.
19 Wennmalm A, Benthin G, Edlund A, Jungersten L, Kieler-Jensen N, Lundin S, Westfelt UN, Petersson A-S, Waagstein F. Metabolism and excretion of nitric oxide in humans. An experimental and clinical study. Circ. Res. 1993; 73: 1121–1127.
20 Stamler JS, Jia L, Eu JP, McMahon TJ, Demchenko IT, Bonaventura J, Gernert K, Piantadosi CA. Blood flow regulation by S-nitrosohemoglobin in the physiological oxygen gradient. Science 1997; 276: 2034–2037.
21 Pawloski JR, Swaminathan RV, Stamler JS. Cell-free and erythrocytic *S*-nitrosohemoglobin inhibits human platelet aggregation. Circulation 1998; 97: 263–267.
22 Stamler JS, Slivka A. Biological chemistry of thiols in the vasculature and in vascular-related disease. Nutr. Rev. 1996; 54: 1–30.
23 McMahon TJ, Moon RE, Luschinger BP, Carraway MS, Stone AE, Stolp BW, Gow AJ, Pawloski JR, Watke P, Singel DJ, Piantadosi CA, Stamler JS. Nitric oxide in the human respiratory cycle. Nat. Med. 2002; 8: 711–717.

24 Gladwin MT, Wang X, Reiter CD, Yang BK, Vivas EX, Bonaventura C, Schechter AN. S-nitrosohemoglobin is unstable in the reductive red cell environment and lacks O2/NO-linked allosteric function. J. Biol. Chem. 2002; 277: 27818–27828.

25 Hobbs AJ, Gladwin MT, Patel RP, Williams DLH, Butler AR. Haemoglobin: NO transporter, NO inactivator or None of the above? Trends Pharmacol. Sci. 2002; 23: 406–411.

26 Rassaf T, Bryan NS, Maloney RE, Specian V, Kelm M, Kalyanaraman B, Rodriguez J, Feelisch M. NO adducts in mammalian red blood cells: too much or too little? Nat. Med. 2003; 9: 481–482.

27 Feelisch M, Rassaf T, Mnaimneh S, Singh N, Bryan NS, Jourd'heuil D, Kelm M. Concomitant S-, N-, and heme-nitros(yl)ation in biological tissues and fluids: implications for the fate of NO in vivo. FASEB J. 2002; 16: 1775–1785.

28 Joshi MS, Ferguson Jr. TB, Han TH, Hyduke DR, Liao JC, Rassaf T, Bryan N, Feelisch M, Lancaster Jr. JR. Nitric oxide is consumed, rather than conserved, by reaction with oxyhemoglobin under physiological conditions. Proc. Natl. Acad. Sci. USA 2002; 99: 10341–10346.

29 Liao JC, Hein TW, Vaughn MW, Huang K-T, Kuo L. Intravascular flow decreases erythrocyte consumption of nitric oxide. Proc. Natl. Acad. Sci. USA 1999; 96: 8757–8761.

30 Liu X, Miller MJS, Joshi MS, Sadowska-Krowicka H, Clark DA, Lancaster Jr. JR. Diffusion-limited reaction of free nitric oxide with erythrocytes. J. Biol. Chem. 1998; 273: 18709–18713.

31 Vaughn MW, Huang K-T, Kuo L, Liao JC. Erythrocytes possess an intrinsic barrier to nitric oxide consumption. J. Biol. Chem. 2000; 275: 2342–2348.

32 Kleinbongard P, Schulz R, Rassaf T, Lauer T, Dejam A, Jax T, Kumara I, Gharini P, Kabanova S, Ozuyaman B, Schnurch HG, Godecke A, Weber AA, Robenek M, Robenek H, Bloch W, Rosen P, Kelm M. Red blood cells express a functional endothelial nitric oxide synthase. Blood 2006; 107: 2943–2951.

33 Deliconstantinos G, Villiotou V, Stavrides JC, Salemes N, Gogas J. Nitric oxide and peroxynitrite production by human erythrocytes: a causative factor of toxic anemia in breast cancer patients. Anticancer Res. 1995; 15: 1435–1446.

34 Jubelin BC, Gierman JL. Erythrocytes may synthesize their own nitric oxide. Am. J. Hypertens. 1996; 9: 1214–1219.

35 Chen LY, Mehta JL. Evidence for the presence of L-arginine-nitric oxide pathway in human red blood cells: relevance in the effects of red blood cells on platelet function. J. Cardiovasc. Pharmacol. 1998; 32: 57–61.

36 Ghigo D, Todde R, Ginsburg H, Costamagna C, Gautret P, Bussolino F, Ulliers D, Giribaldi G, Deharo E, Gabrielli G, Pescarmona G, Bosia A. Erythrocyte stages of *Plasmodium falciparum* exhibit a high nitric oxide synthase (NOS) activity and release an NOS-inducing soluble factor. J. Exp. Med. 1995; 182: 677–688.

37 Kang ES, Ford K, Grokulsky G, Wang Y-B, Chiang TM, Acchiardo SR. Normal circulating adult human red blood cells contain inactive NOS proteins. J. Lab. Clin. Med. 2000; 135: 444–451.

38 Bhattacharya S, Patra SC, Roy SB, Kahn NN, Sinha AK. Purification and properties of insulin-activated nitric oxide synthase from human erythrocyte membranes. Arch. Physiol. Biochem. 2001; 109: 441–449.

39 Metha JL, Metha P, Li D. Nitric oxide synthase in adult red blood cells: vestige of an earlier age or a biologically active enzyme? J. Lab. Clin. Med. 2000; 135: 430–431.

40 Dimmeler S, Fleming I, Fisslthaler B, Hermann C, Busse R, Zeiher AM. Activation of nitric oxide synthase in endothelial cells by Akt-dependent phosphorylation. Nature 1999; 399: 601–605.

41 Sessa WC. The nitric oxide synthase family of proteins. J. Vasc. Res. 1994; 31: 131–143.

42 Fulton D, Gratton J-P, Sessa WC. Post-translational control of endothelial nitric oxide synthase: Why isn't calcium/calmodulin enough? J. Pharmacol. Exp. Ther. 2001; 299: 818–824.

43 Sekar MC, Hokin LE. Phosphoinositide metabolism and cGMP levels are not coupled to the muscarinic-cholinergic receptor in human erythrocyte. Life Sci. 1986; 39: 1257–1262.

44 Kang ES, Cates TB, Harper DN, Chiang TM, Myers LK, Acchiardo SR, Kimoto M. An enzyme hydrolyzing methylated inhibitors of nitric oxide synthase is present in circulating human red blood cells. Free Radic. Res. 2001; 35: 693–707.

45 Angelo S, Irarrázabal C, Devés R. The binding specificity of amino acid transport system $y^{+}L$ in human erythrocytes is altered by monovalent cations. J. Membr. Biol. 1996; 153: 37–44.

46 Tsuda K, Kimura K, Nishio I, Masuyama Y. Nitric oxide improves membrane fluidity of erythrocytes in essential hypertension: an electron paramagnetic resonance investigation. Biochem. Biophys. Res. Commun. 2000; 275: 946–954.
47 Kelm M. Nitric oxide metabolism and breakdown. Biochim. Biophys. Acta 1999; 1411: 273–289.
48 Baumgartner I, Pieczek A, Manor O, Blair R, Kearney M, Walsh K, Isner JM. Constitutive expression of phVEGF$_{165}$ after intramuscular gene transfer promotes collateral vessel development in patients with critical limb ischemia. Circulation 1998; 97: 1114–1123.
49 Wink DA, Miranda KM, Mitchell JB, Grisham MB, Fukuto JM, Feelisch M. The chemical biology of nitric oxide. Balancing nitric oxide with oxidative and nitrosative stress. In Mayer B, editor. Handbook of Experimental Pharmacology. Berlin, Germany, 2000, pp. 7–29.
50 Moro MA, Darley-Usmar VM, Goodwin DA, Read NG, Zamora-Pino R, Feelisch M, Radomski MW, Moncada S. Paradoxical fate and biological action of peroxynitrite on human platelets. Proc. Natl. Acad. Sci. USA 1994; 91: 6702–6706.
51 Gow AJ, Buerk DG, Ischiropoulos H. A novel reaction mechanism for the formation of *S*-nitrosothiol *in vivo*. J. Biol. Chem. 1997; 272: 2841–2845.
52 Bryan NS, Fernandez BO, Bauer SM, Garcia-Saura MF, Milsom AB, Rassaf T, Maloney RE, Bharti A, Rodriguez J, Feelisch M. Nitrite is a signaling molecule and regulator of gene expression in mammalian tissues. Nat. Chem. Biol. 2005; 1: 290–297.
53 Peters TJR. All about albumin. Biochemistry, genetics and medical applications. Academic Press, 1996.
54 Kashiba-Iwatsuki M, Miyamoto M, Inoue M. Effect of nitric oxide on the ligand-binding activity of albumin. Arch. Biochem. Biophys. 1997; 345: 237–242.
55 Scorza G, Pietraforte D, Minetti M. Role of ascorbate and protein thiols in the release of nitric oxide from S-nitroso-albumin and S-nitroso-glutathione in human plasma. Free Radic. Biol. Med. 1997; 22: 633–642.
56 Rossi R, Giustarini D, Milzani A, Colombo R, Dalle-Donne I, di Simplicio P. Physiological levels of s-nitrosothiols in human plasma. Circ. Res. 2001; 89: E47.
57 Tsikas D, Frölich JC. S-Nitrosoalbumin plasma levels in health and disease: facts or artifacts? Value of analytical chemistry in nitric oxide, clinical research. Circ. Res. 2002; 90: e39.
58 Gaston B, Sears S, Woods J, Hunt J, Ponaman M, McMahon T, Stamler JS. Bronchodilator s-nitrosothiol deficiency in asthmatic respiratory failure. Lancet 1998; 351: 1317–1319.
59 Corradi M, Montuschi P, Donnelly LE, Pesci A, Kharitonov SA, Barnes PJ. Increased nitrosothiols in exhaled breath condensate in inflammatory airway diseases. Am. J. Respir. Crit. Care Med. 2001; 163: 854–858.
60 Tyurin VA, Liu S-X, Tyurina YY, Sussman NB, Hubel CA, Roberts JM, Taylor RN, Kagan VE. Elevated levels of s-nitrosoalbumin in preeclampsia plasma. Circ. Res. 2001; 88: 1210–1215.
61 Stamler JS, Lamas S, Fang FC. Nitrosylation: the prototypic redox-based signaling mechanism. Cell 2001; 106: 675–683.
62 Ohshima H, Bartsch H. Chronic infections and inflammatory processes as cancer risk factors: possible role of nitric oxide in carcinogenesis. Mutat. Res. 1994; 305: 253–264.
63 Lijinsky W. Significance of in vivo formation of N-nitroso compounds. Oncology 1980; 37: 223–226.
64 Pearson MJ, Lipowsky HH. Influence of erythrocyte aggregation on leukocyte margination in postcapillary venules of rat mesentery. Am. J. Physiol. Heart Circ. Physiol. 2000; 279: H1460–H1471.
65 Unger RH, Foster DW. Diabetes mellitus. In Wilson JD, Foster DW, Kronenberg HM, Larsen PR, editors. Williams Textbook of Endocrinology. W.B. Saunders Co., Philadelphia, 1998: 973–1059.
66 Butler AR, Megson IL, Wright PG. Diffusion of nitric oxide and scavenging by blood in the vasculature. Biochim. Biophys. Acta 1998; 1425: 168–176.
67 Liu X, Miller MJS, Joshi MS, Thomas DD, Lancaster JR. Accelerated reaction of nitric oxide with O_2 within the hydrophobic interior of biological membranes. Proc. Natl. Acad. Sci. USA 1998; 95: 2175–2179.
68 Vaughn MW, Kuo L, Liao JC. Effective diffusion distance of nitric oxide in the microcirculation. Am. J. Physiol. Heart Circ. Physiol. 1998; 274: H1705–H1714.
69 Ford PC, Wink DA, Stanbury DM. Autoxidation kinetics of aqueous nitric oxide. FEBS Lett. 1993; 326: 1–3.

70 Goldman RK, Vlessis AA, Trunkey DD. Nitrosothiol quantification in human plasma. Anal. Biochem. 1998; 259: 98–103.

71 Marley R, Feelisch M, Holt S, Moore K. A chemiluminescense-based assay for s-nitrosoalbumin and other plasma s-nitrosothiols. Free Radic. Res. 2000; 32: 1–9.

72 Moriel P, Pereira IRO, Bertolami MC, Abdalla DSP. Is ceruloplasmin an important catalyst for S-Nitrosothiol generation in hypercholesterolemia? Free Radic. Biol. Med. 2001; 30: 318–326.

73 Kleinbongard P, Rassaf T, Dejam A, Kerber S, Kelm M. Griess method for nitrite measurement of aqueous and protein containing sample. Methods Enzymol. 2002; 359: 158–168.

74 Heiss C, Lauer T, Dejam A, Kleinbongard P, Hamada S, Rassaf T, Matern S, Feelisch M, Kelm M. Plasma nitroso compounds are decreased in patients with endothelial dysfunction. J. Am. Coll. Cardiol. 2006; 47: 573–579.

75 Grocott-Mason R, Anning P, Evans H, Lewis MJ, Shah AM. Modulation of left ventricular relaxation in isolated ejecting heart by endogenous nitric oxide. Am. J. Physiol. Heart Circ. Physiol. 1994; 267: H1804–H1813.

76 Brady AJB, Warren JB, Poole-Wilson PA, Williams TJ, Harding SE. Nitric oxide attenuates cardiac myocyte contraction. Am. J. Physiol. 1993; 176–182.

77 Kelm M, Schäfer S, Dahmann R, Dolu B, Perings S, Decking U, Schrader J, Strauer BE. Nitric oxide induced contractile dysfunction is related to a reduction in myocardial energy generation. Cardiovasc. Res. 1997; 36: 185–194.

78 Massion PB, Feron O, Dessy C, Balligand JL. Nitric oxide and cardiac function. Ten years after, and continuing. Circ. Res. 2003; 93: 388–398.

79 Paulus WJ, Vantrimpont PJ, Shah AM. Acute effects of nitric oxide on left ventricular relaxation and diastolic distensibility in humans. Assessment by bicoronary sodium nitroprusside infusion. Circulation 1994; 89: 2070–2078.

80 Stamler JS, Loh E, Roddy MA, Currie KE, Creager MA. Nitric oxide regulates basal systemic and pulmonary vascular resistance in healthy humans. Circulation 1994; 89: 2035–2044.

81 Cotton JM, Kearney MT, MacCarthy PA, Grocott-Mason RM, McClean DR, Heymes C, Richardson PJ, Shah AM. Effects of nitric oxide synthase inhibition on basal function and the force-frequence relationship in the normal and failing human heart in vivo. Circulation 2001; 104: 2318–2323.

82 Rassaf T, Poll LW, Brouzos P, Lauer T, Totzeck M, Kleinbongard P, Gharini P, Andersen K, Schulz R, Heusch G, Modder U, Kelm M. Positive effects of nitric oxide on left ventricular function in humans. Eur. Heart J. 2006; 27: 1699–1705.

83 Tannenbaum SR, Witter JP, Gatley SJ, Balish E. Nitrate and nitrite: origin in humans. Science 1979; 205: 1333–1337.

84 Lauer T, Preik M, Rassaf T, Strauer BE, Deussen A, Feelisch M, Kelm M. Plasma nitrite rather than nitrate reflects regional endothelial nitric oxide synthase activity but lacks intrinsic vasodilator action. Proc. Natl. Acad. Sci. USA 2001; 98: 12814–12819.

85 Meulemans A, Delsenne F. Measurement of nitrite and nitrate levels in biological samples by capillary electrophoresis. J. Chromatogr B. Biomed. Appl. 1994; 660: 401–404.

86 Leone AM, Francis PL, Rhodes P, Moncada S. A rapid and simple method for the measurement of nitrite and nitrate in plasma by high performance capillary electrophoresis. Biochem. Biophys. Res. Commun. 1994; 200: 951–957.

87 Gorenflo M, Zheng C, Poge A, Bettendorf M, Werle E, Fiehn W, Ulmer HE. Metabolites of the L-arginine-NO pathway in patients with left-to-right shunt. Clin. Lab. 2001; 47: 441–447.

88 Kleinbongard P, Dejam A, Lauer T, Rassaf T, Schindler A, Picker O, Scheeren T, Godecke A, Schrader J, Schulz R, Heusch G, Schaub GA, Bryan NS, Feelisch M, Kelm M. Plasma nitrite reflects constitutive nitric oxide synthase activity in mammals. Free. Radic. Biol. Med. 2003; 35: 790–796.

89 Rhodes PM, Leone AM, Francis PL, Struthers AD, Moncada S. The L-arginine:nitric oxide pathway is the major source of plasma nitrite in fasted humans. Biochem. Biophys. Res. Commun. 1995; 209: 590–596.

90 Williams DA, Lewis DA, Krause WJ, Flood MH, Miller VM. Species characterization of plasma nitrite/nitrate (NOx) concentration. Comp. Med. 2003; 53: 21–28.

91 Li H, Samouilov A, Liu X, Zweier JL. Characterization of the magnitude and kinetics of xanthine oxidase-catalyzed nitrate reduction: evaluation of its role in nitrite and nitric oxide generation in anoxic tissues. Biochemistry (Mosc) 2003; 42: 1150–1159.

92 Ignarro LJ. Biosynthesis and metabolism of endothelium-derived nitric oxide. Annu. Rev. Pharmacol. Toxicol. 1990; 30: 535–560.

93 Busse R, Fleming I. Regulation and functional consequences of endothelial nitric oxide formation. Ann. Med.1995; 27: 331–340.

94 Moncada S, Higgs A. The L-arginine-nitric oxide pathway. N. Engl. J. Med. 1993; 329: 2002–2012.

95 Lundberg JO, Govoni M. Inorganic nitrate is a possible source for systemic generation of nitric oxide. Free Radic. Biol. Med. 2004; 37: 395–400.

96 Lundberg JO, Govoni M. Inorganic nitrate is a possible source for systemic generation of nitric oxide. Free Radic. Biol. Med. 2004; 37: 395–400.

97 Doyle MP, Herman JG, Dykstra RL. Autocatalytic oxidation of hemoglobin induced by nitrite: activation and chemical inhibition. Free Radic. Biol. Med. 1985; 1: 145–153.

98 Doyle MP, Pickering RA, DeWeert TM, Hoekstra JW, Pater D. Kinetics and mechanism of the oxidation of human deoxyhemoglobin by nitrites. J. Biol. Chem.1981; 256: 12393–12398.

99 Lundberg JO, Weitzberg E, Lundberg JM, Alving K. Intragastric nitric oxide production in humans: measurements in expelled air. Gut 1994; 35: 1543–1546.

100 Benjamin N, O'Driscoll F, Dougall H, Duncan C, Smith L, Golden M, McKenzie H. Stomach NO synthesis. Nature 1994; 368: 502.

101 Zweier JL, Wang P, Samouilov A, Kuppusamy P. Enzyme-independent formation of nitric oxide in biological tissues. Nat. Med. 1995; 1: 804–809.

102 Modin A, Bjorne H, Herulf M, Alving K, Weitzberg E, Lundberg JO. Nitrite-derived nitric oxide: a possible mediator of 'acidic-metabolic' vasodilation. Acta Physiol. Scand. 2001; 171: 9–16.

103 Millar TM, Stevens CR, Benjamin N, Eisenthal R, Harrison R, Blake DR. Xanthine oxidoreductase catalyses the reduction of nitrates and nitrite to nitric oxide under hypoxic conditions. FEBS Lett. 1998; 427: 225–228.

104 Cosby K, Partovi KS, Crawford JH, Patel RP, Reiter CD, Martyr S, Yang BK, Waclawiw MA, Zalos G, Xu X, Huang KT, Shields H, Kim-Shapiro DB, Schechter AN, Cannon III RO, Gladwin MT. Nitrite reduction to nitric oxide by deoxyhemoglobin vasodilates the human circulation. Nat. Med. 2003;.9: 1498–1505.

105 Heiss, C., Lauer, T., Dejam, A., Kleinbongard, P., Hamada, S, Rassaf, T., Matern, S, and Kelm, M. Plasma Nitrosos compounds are Drecreased in Patients with Endothelial Dysfunction. J. Am. Coll. Cardiol. 2006; 47: 573–579.

106 Kleinbongard P, Dejam A, Lauer T, Jax T, Kerber S, Gharini P, Balzer J, Zotz RB, Scharf RE, Willers R, Schechter AN, Feelisch M, Kelm M. Plasma nitrite concentrations reflect the degree of endothelial dysfunction in humans. Free Radic. Biol. Med. 2006; 40: 295–302.

107 Coretti MC, Anderson TJ, Benjamin EJ, Celermajer D, Charbonneau F, Creager M, Deanfield J, Drexler H, Gerhard-Herman M, Herrington D, Vallance P, Vita J, Vogel R. Guidelines for the ultrasound assessment of endothelial-dependent flow-mediated vasodilation of the brachial artery. J. Am. Coll. Cardiol. 2002; 39: 257–265.

108 Rassaf T, Heiss C, Hendgen-Cotta U, Balzer J, Matern S, Kleinbongard P, Lee A, Lauer T, Kelm M. Plasma nitrite reserve and endothelial function in the human forearm circulation. Free Radic. Biol. Med. 2006; 41: 295–301.

109 Rodriguez J, Maloney RE, Rassaf T, Bryan NS, Feelisch M. Chemical nature of nitric oxide storage forms in rat vascular tissue. Proc. Natl. Acad. Sci. USA 2003; 100: 336–341.

110 Rafikova O, Rafikov R, Nudler E. Catalysis of S-nitrosothiols formation by serum albumin: the mechanism and implication in vascular control. Proc. Natl. Acad. Sci. USA 2002; 99: 5913–5918.

111 Ghafourifar P, Colton CA. Mitochondria and nitric oxide. Antioxidants and redox signaling 2003; 5: 249–250.

112 Lacza Z, Pankotai E, Csordas A, Gero D, Kiss L, Horvath E, Kollai M, Busija D, Szabo C. Mitochondrial NO and reactive nitrogen species production: does mtNOS exist? Nitric Oxide 2006; 14: 162–168.

113 Ghafourifar P, Richter C. Nitric oxide synthase activity in mitochondria. FEBS Lett. 1997; 418: 291–296.
114 Ghafourifar P, Cadenas E. Mitochondrial nitric oxide synthase. Trends Pharmacol. Sci. 2005; 26: 190–195.
115 Brown GC. Nitric oxide regulates mitochondrial respiration and cell functions by inhibiting cytochrome oxidase. FEBS Lett. 1995; 369: 136–139.

PART IV

Nitrites and Nitrates as a NO Source in Cells and Tissues

Radicals for Life: The Various Forms of Nitric Oxide
E. van Faassen and A.F. Vanin (Eds.)

CHAPTER 14

Nitrite as endothelial NO donor under anoxia

Ernst van Faassen[1], Anatoly F. Vanin[2] and Anny Slama-Schwok[3]

[1]*Debye Institute, Section Interface Physics, Ornstein Laboratory, Utrecht University, 3508 TA, Utrecht, The Netherlands*
[2]*Semenov Institute of Chemical Physics, Russian Academy of Sciences, Moscow, 119991, Russian Federation*
[3]*Laboratory for Optics and Biosciences, INSERM U696, CNRS UMR 7645, Ecole Polytechnique, Palaiseau, France*

Nitrite anions (NO_2^-) appear as metastable intermediates in the oxidation cascade [1] leading from nitric oxide (NO) radicals to the stable metabolite nitrate (NO_3^-). The occurrence of nitrite in living tissues reflects the metabolic fate of oxidized nitrogen species in organisms. Nitrite anions share some characteristics of free NO but lack others. They have low molecular weight and show rapid diffusive transport in water but cannot cross lipid cell membranes. They may induce relaxation of precontracted vessel rings [2] but fail to prevent platelet aggregation [3]. Although excessive nitrite concentrations are known to elicit vasodilatory response [3–5], nitrite as such was considered for a long time biologically inert at the physiological concentrations found in plasma or tissues. However, recent observations in animals and humans have greatly increased the interest in possible biological functions of nitrite, especially with respect to blood flow in the vascular system. The observation that nitrite may be converted to NO under conditions of hypoxia attracted special interest. As such, nitrite is now considered to act as a hypoxic buffer and assist in maintaining vasodilation and mitochondrial respiration under hypoxia. Its importance stems from the abundance of nitrite in tissues and the blood circulation. Typical nitrite levels [1,6–9]. are 0.1–1.0 μM in plasma, 0.5–2 μM in normoxic tissues, with still higher levels of up to 20 μM reported in vascular tissue. These quantities should be compared with levels of other NO-related compounds. Table 1 lists various concentrations as determined in rats. For aortic rat tissue, comparable numbers were reported in Ref. [10,11]. It should be noted that significantly higher levels of metabolites arise in human plasma [12].

These nitrite concentrations are at 2–3 orders of magnitude higher than the basal concentrations of nitrosothiols found in tissues. Therefore, nitrite represents the largest pool of nitrogen oxides in humans from which the NO radicals may be recovered.

In humans, three dominant sources contribute to this endogenous nitrite pool: (1) Dietary intake of nitrite from food like cured meat or certain vegetables. These dietary sources of nitrite are discussed in detail in Chapter 16. (2) Endogenous reduction of dietary nitrate to

Table 1 Magnitude of various pools of NO-metabolites in Wistar rats (adapted from Ref [9]). Basal levels of *O*-Nitroso compounds [10] and DNIC remain below detection limits. The [GSH]/[GSSG] ratio is a marker for the redox state of the tissue. The rightmost column gives NO-metabolite values from human plasma for Ref. [12]

NO-metabolite	Aorta	Brain	Heart	Plasma	Erythrocytes	Human plasma¶
Nitrite (μM)	23 ± 9	1.7 ± 0.3	0.80 ± 0.08	0.29 ± 0.05	0.68 ± 0.06	0.20 ± 0.02
Nitrate (μM)	49 ± 7	6.1 ± 1.1	5.9 ± 1.7	5.7 ± 0.6	10.2 ± 1.2	14.4 ± 1.7
S-nitroso moieties (nM)	96 ± 24	22 ± 6	13 ± 2	1.4 ± 0.5	246 ± 32	7.2 ± 1.1
N-nitroso moieties (nM)	19 ± 11	61 ± 8	14 ± 2	3.5 ± 0.4	95 ± 14	32.3 ± 5.0
Nitrosyl-heme (nM)	Below detection*	160 ± 30	15 ± 1	Below detection*	10.8 ± 1.8	n.d.
GSH (mM)	0.34 ± 0.08	1.28 ± 0.06	0.93 ± 0.12	0.02†	0.8 mM † (whole blood)	n.d.
[GSH]/[GSSG]	6.8	24	14	5†	> 10† (whole blood)	n.d

* Detection limit 1 nM.
†[13].
¶[12].

nitrite by certain bacterial strains in the gastrointestinal tract. (3) Finally, the oxidation of endogenous NO to nitrite. Some additional nitrite may also originate from exogenous sources like NO donors, organic nitrates or nitrites (cf Chapter 17).

Nitrite levels in the vascular system are clearly strongly affected by the enzymatic activity of endothelial nitric oxide synthase (eNOS) [1,5]. Plasma nitrite levels were reduced from 1.6 μM in wild-type mice to 0.74 μM in eNOS knockout mutants [14] and it was verified that almost all circulating nitrite in the blood of fasting humans derives from the L-arginine-NO pathway [15]. Studies of human vascular flow have suggested that a significant fraction of infused NO is rapidly oxidized to nitrite and transported in this form for considerable distances along the vascular tree [16]. Such convective transport is facilitated by nitrite having a fairly long lifetime of several minutes in human blood [1]. Interestingly, the nitrite concentrations in oxygen-rich arterial blood were found to be higher than in venous blood [6,17]. This arterial–venous gradient was interpreted as a manifestation of nitrite delivery to the perivascular tissues and suggested that nitrite actually plays an active physiological role in control of the vascular flow. However, infusion studies did not show vasodilatory capacity of nitrite as such in humans under normoxia [5]. Rather, low oxygen tension appeared to be a crucial aspect of *in vivo* studies of vasodilation by direct infusion of nitrite [18].

The physiological effects of nitrite can be roughly classified into four categories: First, as a potential contributor to methemoglobinemia [19] by reacting rapidly with oxy-hemoglobin to nitrate and methemoglobin which hinders to transport oxygen. Second, a direct signaling function of nitrite itself in normoxic vasodilation [3,4]. This phenomenon will not be discussed further as the dilatory threshold of ca 10–15 μM plasma nitrite exceeds physiological nitrite levels. Third, an indirect regulatory role by coupling to the pool of nitrosylated or nitrosated compounds circulating in the body. Specific examples are the *S*-nitrosothiols

(Chapters 9 and 10) and the dinitrosyl iron complexes (Chapters 2 and 3). There is increasing evidence from mammalian tissues that nitrite couples to these nitrosylated and nitrosated compounds *via* an equilibrium with a timescale of several minutes and which does not involve the formation of freely diffusing NO intermediates [2,3,9,20]. This balance depends on oxygen status, since application of hypoxia enhanced the nitrosylated Hb and nitrosothiols in tissues of rats [9]. Although *S*-nitrosothiols themselves are inefficient as vasodilators, they may potentially have high significance for the vasculature as intermediates for the subsequent formation of long-lived and very potent vasodilators like dinitrosothiol iron complexes (DNIC). This might seem paradoxical at first, because the formation of significant quantities of DNIC from free NO is possible *in vitro*, but improbable *in vivo* because second order NO has low physiological concentration. However, in the presence of non-transferrin-bound iron (NTBI), such *S*-nitrosothiols are known to establish a reversible equilibrium with DNIC. The micromolar nitrosothiol concentrations could provide a more efficient reaction pathway for the *in vivo* formation of DNIC. A detailed discussion of this pathway was given in Chapters 2 and 3. As the fourth category, a rapidly increasing body of clinical data attests to the protective action of nitrite against ischemic damage to vascular and other tissues [8,21–23]. In this case, the beneficial effect of nitrite seems mediated by the release of free-NO radicals or NO-like species under hypoxia.

Apart from the above physiological functions, it should be mentioned that irradiation with intense blue light will photolyze nitrite and release NO radicals. It shares this property with the photo-labile nitrosothiols, and it has been demonstrated [10] that endogenous nitrite contributes significantly to the photo-induced NO release from aortic tissues of rats.

The first three mechanisms have been discussed in previous chapters, whereas the fourth will be the topic of this chapter. It concentrates on the mechanism behind the intriguing beneficial effects of nitrite when living tissue is subjected to conditions of low oxygen tension as may arise in ischemia, hypoxia or anoxia.

When discussing the physiological relevance of nitrite under low oxygen tension, we should clearly distinguish three states of oxygenation for vascular tissues: Normoxia, hypoxia and real anoxia. Normoxic blood has a $pO_2 \sim 80$–100 Torr. The boundary between the hypoxic and anoxic regimes is taken as the critical oxygen concentration below which NOS ceases to consume arginine. From Fig. 11 of Ref. [24], we define anoxia as $[O_2] < 15\,\mu M$, corresponding to $pO_2 \sim 12$ Torr. Hypoxia refers to the interval 12–80 Torr.

Hypoxia affects tissue cells in a variety of ways. Chronic hypoxia can affect cellular proliferation and induce adaptation to the conditions of low oxygen, for example, by promoting the expression of NOS enzyme in a variety of tissues [25]. Endothelial cells react to oxygen deficiency by release of hypoxia-inducible factors (HIFs) [26] that initiate the transcription of a range of genes and by vascular endothelial growth factors (VEGFs) [27] that promote neovascularization and homeostasis. On shorter timescales, hypoxia elicits acute responses from cells that affect the bioavailability of NO. For example, it was reported before [28–30] that imposition of hypoxia enhances NO release from endothelial cultures. The hypoxic enhancement was attributed to upregulation of the arginine pathway by increases in the cytosolic calcium levels. Under the conditions chosen in these studies, the oxygen levels remained comfortable above the critical level of $15\,\mu M$, and the arginine pathway unimpeded. In this chapter, we consider the role of nitrite in the truly anoxic regime where oxygen is carefully excluded from the reaction vessels, culture media and atmospheres in the head space.

Our study shows that NO release from eNOS is enhanced even under true anoxia where arginine pathway is blocked by the absence of oxygen. The enzyme starts to function as a nitrite reductase.

The undisputed beneficial effects of nitrite under ischemia are attributed to the reduction of nitrite back to NO, but the dominant mechanism for this reduction has been controversial. The loss of vasodilation upon removal of the endothelial monolayer has localized the agent for the nitrite reduction in the endothelium [2]. Direct uncatalyzed reduction is very slow [31] except at extreme acidic conditions as may arise in the stomach [32], urine [33] or during acidosis [34]. Therefore enzymatic mediators for the reduction have been proposed. Dedicated nitrite reductases are known in plants, algae and cyanobacteria [35] but are lacking in mammals. However, certain mammalian enzymes show some nitrite reductase capacity supplementary to their normal physiological function, such as glutathione-*S*-transferase [36], xanthine oxidoreductase (XOR) [37,38], deoxy-hemoglobin [8,39–41] and cytochrome P-450 [42]. Finally, a mitochondrial mechanism was reported [43–45] involving ubiquinol and complex III of the respiratory chain. So far, most attention in the literature has been devoted to deoxy-hemoglobin and XOR as mediators of anoxic nitrite reduction. Both pathways are discussed in detail in Chapter 15. In addition, we recently discovered that endothelial nitric oxide synthase (eNOS) also has the capacity to reduce nitrite to NO under anoxia [46]. The remaining part of this chapter will be devoted to the possible role of endothelial NOS as the enzymatic mediator for anoxic reduction of nitrite.

AQUEOUS REDUCTION PATHWAYS OF NITRITE

At physiological pH ~7.0–7.4, nitrite anions are fairly stable in the absence of compounds like oxy-hemoglobin. In view of this stability, nitrite is often taken as an indirect measure for NO. Nitrite easily binds to many transition metal ions and decelerates their participation in redox reactions. This property is often used to cure meats or sausages by addition of nitrite. However, the presumed stability of nitrite should be approached with caution. As mentioned above, nitrite rapidly reacts with oxy-hemoglobin to methemoglobin which is unable to bind and transport oxygen. Therefore, excessive nitrite levels may contribute to methemoglobinemia, and excessive consumption of nitrite from canned spinach has been known to cause death in this way.

At acidic pH, nitrite is protonized to nitrous acid (HNO_2, pK = 3.37) which is very unstable and prone to spontaneous decay *via* secondary reactions. In the dark, acidic nitrite solutions are in equilibrium with a very small quantity of nitrosonium cations by heterolytic fission of the bond

$$NO_2^- + H^+ \leftrightarrow HO\text{—}NO \leftrightarrow HO^- \ldots N^+ \equiv O \leftrightarrow OH^- + NO^+ \quad (1)$$

In water, this reaction requires [47] a Gibbs energy of $\Delta G = +94$ kJ/mol $= -RT \ln K$ with $K = [NO^+]/[NO_2^-] \sim 4 \cdot 10^{-17}$ at pH ~7.0.

Therefore, the release of free nitrosonium from nitrite may be considered insignificant in water at physiological pH.

Nitrite anions have a single broad absorption peak at 354 nm ($\varepsilon_{354} \sim 24\ \mathrm{Mcm}^{-1}$) which changes into a well-resolved multiplet upon protonation to nitrous acid HNO_2 (cf Fig. 2 of Chapter 1). Nitrous acid is rather unstable by itself and its decomposition depends on its concentration. At low concentrations, illumination with blue light may induce homolytic fission of nitrous acid into hydroxyl radicals and nitric oxide radicals:

$$\mathrm{ON{-}OH} \xrightarrow{h\nu} \mathrm{NO}\cdot + \mathrm{OH}\cdot \tag{2}$$

Atmospheric research has shown that photolytic release of OH· affects the ozone budget in the troposphere and is one of the negative effects of the nitrous smog gases in the atmosphere. In aqueous environment also, photolysis was demonstrated. Endogenous nitrite was found to be the major source of nitric oxide radicals that are released when aortic tissue was illuminated with UV light [10].

At higher concentrations, nitrous acid dismutates into nitric oxide, nitrogen dioxide and water

$$2\mathrm{HNO_2} \rightleftarrows \mathrm{N_2O_3} + \mathrm{H_2O} \rightleftarrows \mathrm{NO}\cdot + \mathrm{NO_2}\cdot + \mathrm{H_2O}$$

with $NO_2\cdot$ subsequently dimerizing into N_2O_4 and reacting irreversibly with water to nitrite, nitrate and protons [for this reaction cf Chapter 1, Eq. (8)].

IN VITRO EXPERIMENTS ON eNOS IN BUFFERED SOLUTION

eNOS is the endothelial enzyme that catalyzes the synthesis of NO from L-arginine *via* an oxygen-consuming pathway. This arginine pathway is regulated in a complicated way by a combination of local calcium concentration, phosphorylation of serine moieties and the redox state of intrinsic protein thiols [48]. eNOS is unique among the NOS isoforms to show irreversible myristoylation of the glycine residue at its N terminus as a prerequisite for subsequent reversible palmitoylation and localization at the cell membrane [49,50]. The arginine pathway is blocked upon depletion of the substrate but also if local oxygen levels fall below a threshold value of ca 15 μM [24]. In our experiments we studied the anoxic reduction of nitrite by buffered eNOS solutions. The anoxia was induced by removing oxygen from the solutions by bubbling with argon and conducting the experiments under an argon atmosphere.

The anoxic activity of eNOS was studied in buffered solution in an optical quartz cuvette that was sealed by a teflon stopper with three holes for the NO electrode, an inflow argon purge and an outflow exhaust, respectively. The setup allows simultaneous monitoring of the optical absorption spectrum of the enzyme and the local NO concentration inside the cuvette. Additionally, the NO content of the exhaust gas flow was determined by feeding the exhaust into a vial containing a solution of NO traps like iron-dithiocarbamate complexes. The protein was either full-length eNOS holoenzyme or only its oxygenase subdomain (eNOSoxy) and obtained by overexpression in *E. coli* [51]. The enzymatic reaction was started by injecting reductants into the cuvette *via* the inflow argon purge. The full-length protein contains the flavins of the reductase domain and was started by adding NADPH. The eNOSoxy domain does not contain any flavins and the electron flow was initiated with dithionite. In the presence of oxygen, calmodulin, BH_4, L-arginine and NADPH, the holoenzyme produced copious amounts of NO as detected by the electrode. As expected, the electrode failed to detect NO release in the absence of both oxygen and nitrite.

Interestingly, the electrode showed release of significant quantities of NO under anoxic conditions if nitrite anions were present in the solution (cf Fig. 1). This formation of NO could be confirmed by two other independent methods: First, the optical absorbance showed the appearance of a shoulder near 440 nm and a resolved red-shifted band at 560 nm (cf Fig. 2). These bands indicate the formation of nitrosylated ferrous heme [52]. Second, NO radicals could also be detected in the exhaust gas flow after NO-trapping with Fe–DETC (Fe-diethyldithiocarbamate) complexes and detection of the paramagnetic NO—Fe^{2+}–DETC adducts with EPR (cf Chapter 18). Monitoring the NADPH concentration at 340 nm showed that NO was released from nitrite under consumption of NADPH. This new nitrite reduction pathway of eNOS proceeded in Tris buffer (50 mM, pH = 7.5).

Three aspects distinguish the nitrite reduction pathway from the regular NO release *via* the arginine pathway: First, the nitrite reduction proceeds in the absence of oxygen. Second, the reduction is slowed but not blocked by the removal of BH_4. And finally, using ^{15}N-labeled nitrite proved that the NO was generated from nitrite anions rather than the unlabeled arginine. The ^{15}N and ^{14}N isotopes carry nuclear spin ½ and 1, respectively. As explained in Chapter 18, the difference in nuclear spin determines the multiplicity of the hyperfine splitting observed in the EPR spectra of the paramagnetic NO—Fe^{2+}–DETC complexes formed upon NO trapping (Fig. 3).

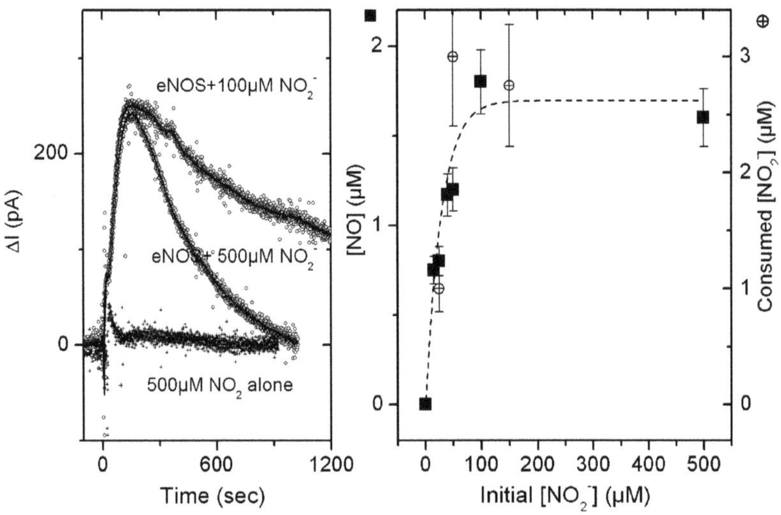

Fig. 1. Left panel: Electrode traces obtained after injection of 70 μM NADPH into a solution containing 500 μM nitrite in the absence (bottom) or presence of 2 μM eNOS (intermediate), and eNOS with 100 μM nitrite (upper trace), 50 mM Tris buffer at pH = 7.6, 1 mM arginine, 12 μM BH_4, 8 μM calmodulin. An electrode current of 195 ± 20 pA corresponds to 1 μM NO; Right panel: Correlation between the NO concentration (black squares, electrode current) and the nitrite consumption (crossed circles, Griess reaction) from eNOS catalyzed nitrite reduction as a function of the initial nitrite concentration; the data have been normalized to 4–5 μM eNOS; same conditions as in the left panel.

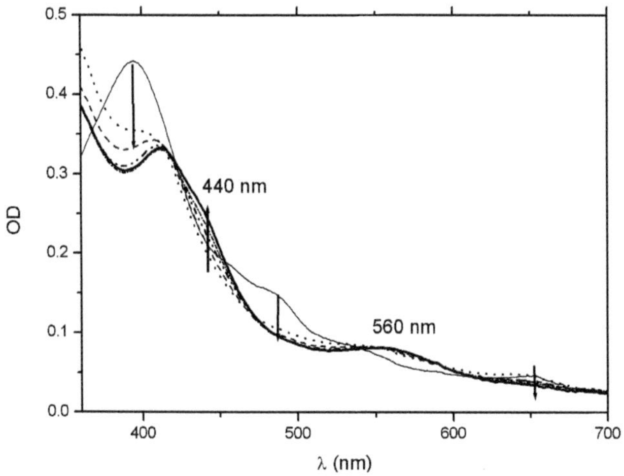

Fig. 2. Absorption spectra of eNOS (4.0 μM) in the presence of nitrite (500 μM). Thin solid line: before injection of NADPH. Dotted line: $t = 30$ s after injection of 100 μM of NADPH. Dashed line: $t = 2$ min. Dash-dotted line: $t = 5$ min. Fine dots: $t = 11$ min. Thick solid line: $t = 25$ min. The arrows indicate the direction of time. The reaction proceeded at room temperature in 50 mM Tris buffer (pH = 7.5) containing 150 mM NaCl and 5% glycerol, 1 mM arginine, 40 μM BH_4, 10 μM calmodulin and 1 mM Ca^{2+}.

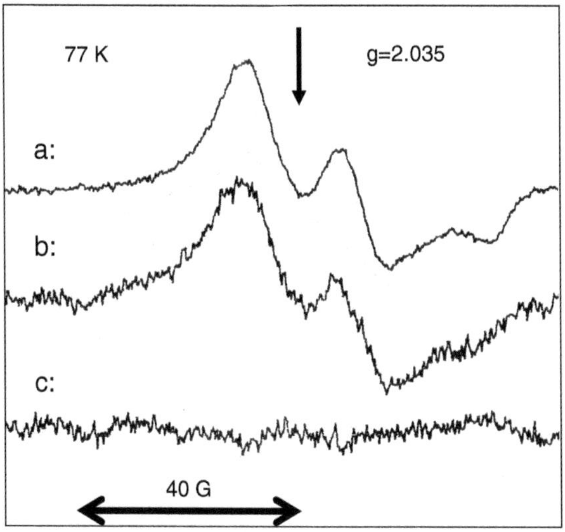

Fig. 3. EPR spectra at 77 K from trapping of traces of ^{15}NO carried by the argon purge of the reaction cuvette. This cuvette contained ^{15}N-labeled nitrite anions (500 μM) and eNOS = 5 μM. The hyperfine doublet near $g = 2.035$ is the ^{15}N-labeled ferrous MNIC adduct. Trapping was for 15 min. Curve a: Reference EPR spectrum from 2.7 nmol $^{15}NO–Fe^{2+}–(MGD)_2$ in frozen solution. The intensity is scaled down by a factor of 20. Curve b: 130 ± 13 pmol ^{15}N-MNIC adducts after trapping of ^{15}NO carried by the argon purge of the cuvette containing reduced full-length eNOS. Curve c: Absence of adducts after trapping of ^{15}NO from the argon purge when the cuvette did not contain any eNOS.

The site of this nitrite reduction is presumably localized in the oxygenase domain of eNOS, since eNOSoxy proved capable of nitrite reduction as well, though at a slower rate. The reaction utilizes the flavins from the reductase domain to channel electrons from NADPH towards the heme. Arginine was found to play an interesting role in this reaction. In the absence of arginine, optical spectroscopy showed extensive nitrosylation of the ferrous heme, but no release of free NO could be detected electrochemically and the NADPH consumption came to a halt by nitrosylation of the heme moiety.

The main interest in this new pathway comes from its potential relevance for physiology. Since the conventional arginine pathway is blocked under conditions of low oxygen tension [24], we speculated that the newly discovered anoxic nitrite reductase pathway might provide a significant alternative source of NO for tissues under acute hypoxia. We therefore proceeded to test this hypothesis in a more complex model, namely endothelial cell cultures.

EXPERIMENTS ON ENDOTHELIAL CELL CULTURES

The cell experiments were done with immortalized murine microvascular brain endothelial bEND.3 cells which have a good expression of endothelial nitric oxide synthase and good reproducibility from generation to generation. These cells were cultivated to confluence in 75 cm^2 flasks at 37°C under a controlled atmosphere containing 5% CO_2 and 20% O_2. At confluence, the bottom of the flask was covered by a cellular monolayer consisting of ca $7.5 \pm 0.5 \times 10^6$ endothelial cells. Confluence and cell count are important parameters for NO synthesis [49] and are always verified by optical inspection *via* a stereomicroscope. The NO production is monitored *via* spin trapping of the NO radicals with iron-dithiocarbamate complexes and quantifying the yield of paramagnetic NO—Fe^{2+}–DETC adducts with electron paramagnetic resonance spectroscopy.

Anoxia was imposed by replacing the growth medium containing 2.5 mM DETC, 10% fetal calf serum, 2 mM L-glutamine, 10 IU/ml penicillin and 100 mg/L streptomycin by the same argon-bubbled medium, then subsequently adding 10 μM ferrous sulfate and flushing the flask with argon before closing it with an airtight top. After 20 min of NO trapping at 37°C, enzymatic activity was terminated by placing the flask on ice. The cellular fraction containing the Fe–DETC complexes was harvested by ultracentrifugation and snap-frozen in liquid nitrogen until EPR assay. Intracellular nitrite concentrations were determined by the colorimetric Griess assay of the lysated cellular fraction. Imposition of anoxia did not affect the viability of the cells (cell death as determined by trypan blue staining was below 0.1% for normoxic as well as anoxic cultures).

Upon NO trapping for 20 min in the controlled atmosphere with 5% CO_2 and 20% O_2, the cellular fraction from ca $7.5 \pm 0.5 \times 10^6$ endothelial cells had acquired a yield of 110 ± 8 pmol paramagnetic MNIC (NO—Fe^{2+}–DETC) as detected by EPR. This basal yield was obtained without any stimulus of the NO production from the cells. A typical EPR spectrum (Fig. 4A) showed a clear triplet hyperfine structure centered at $g = 2.035$ characteristic for MNIC. As expected for biological samples [53], a small contribution from paramagnetic Cu^{2+}–DETC complexes was superposed. The most intense hyperfine line of this copper complex is visibl at $g = 2.01$ in Fig. 4. Preincubation with N_ω-nitro-L-arginine (NLA) diminished the MNIC yield in a dose-dependent manner (Fig. 4B).

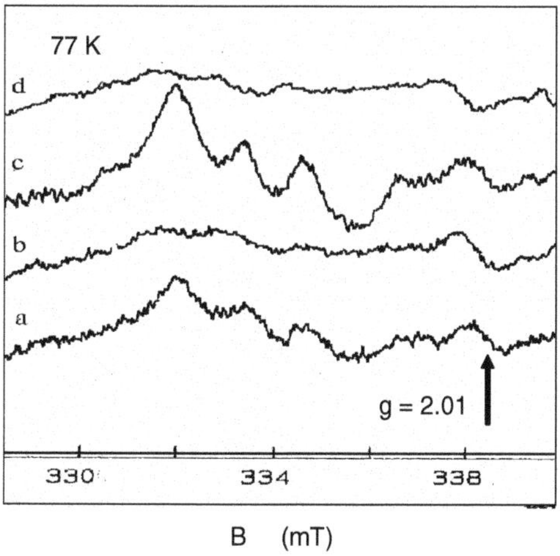

Fig. 4. EPR spectra at 77 K from cellular fractions of ca $7.5 \pm 0.5 \times 10^6$ endothelial cells after 20 min NO trapping at 37°C with iron-dithiocarbamate complexes.The arrow indicates one of the following. Curve a: 110 pmol MNIC formed under a controlled atmosphere containing 5% CO_2 and 20% O_2. Curve b: ca 25 pmol MNIC formed under the controlled atmosphere in the presence of 5 μM NOS inhibitor NLA. Curve c: 160 pmol MNIC formed under anoxia. Curve d: ca 33 pmol MNIC formed under anoxia in the presence of 5 μM NOS inhibitor NLA.

These basal unstimulated yield of 110 ± 8 pmol compares favorably with a total MNIC yield of 400 ± 30 pmol as obtained when the cellular NOS production was stimulated by administration of 5 μM Ca-ionophore A23187.

Interestingly, when anoxia was applied by argon, MNIC yields increased significantly over basal. In the absence of inhibitors, the anoxic yields from ca $7.5 \pm 0.5 \times 10^6$ endothelial cells were ca 160 ± 10 pmol MNIC, these values being typically some 50% higher than basal in the presence of oxygen (Fig 4C). In the presence of specific NOS inhibitors like N_ω-nitro-L-arginine (NLA) and N_ω-nitro-L-arginine-methylester (L-NAME), the MNIC yield under anoxia was also diminished in a dose-dependent manner. The intensity of the EPR absorption from paramagnetic Cu^{2+}–DETC complexes was not affected by anoxia. The yields of the trapping experiments are compiled in Table 2.

Preincubation with 10 mM of the heme-binding inhibitor imidazole diminished both oxic and anoxic yields to 80 ± 10 pmol. In contrast, preincubation for 20 min with the molybdenum-binding xanthine oxidase inhibitor oxypurinol did not affect the anoxic MNIC yield.

The kinetics of the anoxic NO release from the cells showed a linear increase with time up to ca 30 min after induction of anoxia, when the signal intensity saturated at an asymptotic value of ca 200 ± 20 pmol MNIC (cf Fig. 5).

The main and most prominent result from these cell culture experiments is the clear and large increase of MNIC yield upon introduction of anoxia. The anoxic NO production had a surprising duration of 30 min and its magnitude exceeded that of the conventional enzymatic pathway for NO production from arginine. In view of our earlier *in vitro* experiments [40] we attribute the anoxic NO production to the reduction of intracellular nitrite by

Table 2 Yields of MNIC adducts (in pmole) in the cellular fractions of a single 75 cm^2 flask containing ca 7.5×10^6 endothelial cells. Trapping proceeded for 20 min at 37°C. The second row gives the preincubation time τ_{inc} (min) of the supplements. During preincubation the cells were kept at 37°C in an atmosphere with 5% CO_2 and 20% O_2. (With permission from Ref. [54])

	Basal unstimulated	CaI (5 μM)	NLA (5 μM)	NLA (57 μM)	L-NAME (5 μM)	L-NAME (50 μM)	Imidazole (10 mM)	Oxypurinol (100 μM)	$NaNO_2$ (250 μM)
τ_{inc}	–	–	20′	20′	2′	2′	1′	20′	20′
oxia	110 ± 8	400 ± 30	25 ± 2	†	93 ± 10	51 ± 8	85 ± 10	n.d.	116 ± 8
anoxia	160 ± 10	n.d.	33 ± 2	†	87 ± 8¶	61 ± 8	80 ± 10	170 ± 10	154 ± 8

† Below detection limit of ca 10 pmol.
¶ 85 ± 8 pmol with $\tau_{inc} = 20'$.

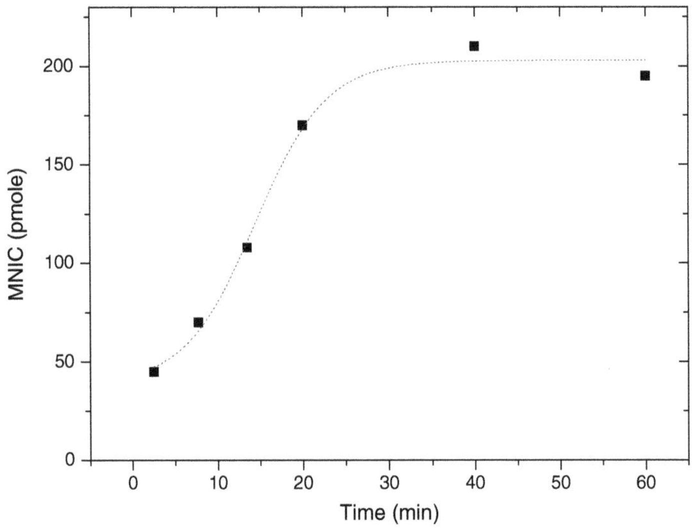

Fig. 5. Kinetics of the formation of nitrosylated MNIC adducts in ca $7.5 \pm 0.5 \times 10^6$ endothelial cells. Anoxia is applied at 0 min. The detection limit is ca 10 pmol MNIC, and the experimental error ca 10%.

nitric oxide synthase. The enzymatic mediator of the reaction could be unambiguously as a form of NOS. The significant suppression of the MNIC yields by imidazole suggests that a hemeprotein be involved in the observed effect. The dose-dependent inhibition by NLA and L-NAME inhibitors is very specific for nitric oxide synthases. We did not quantify the IC_{50} dosages as these depend on the (unknown) intracellular L-arginine concentration, but we noted that the anoxic pathway and the arginine pathways were inhibited by comparable dosages. It suggests that both pathways are inhibited *via* the same mechanism, namely direct competition between the nitro-arginine-inhibitors with arginine for the arginine binding site of NOS.

The anoxic NO yields in our endothelial cell cultures were unaffected by addition of 100 μM oxypurinol. This observation rules out xanthine oxidoreductase (XOR) as a mediator of the anoxic nitrite reduction, since oxypurinol is an effective inhibitor of the catalytic molybdenum site in XOR. This flavoenzyme finds expression in endothelial cells and acts as a nitrite reductase *in vitro* under anoxia [37].

In addition to enzymatic mediators of nitrite reduction, alternative reaction pathways for the release of NO should be considered. First and foremost, acidic reduction of intracellular nitrite could be ruled out. Although anoxia may stimulate acidification of the intracellular compartment (acidosis), in our case the observed NO release was completely canceled by very specific NOS inhibitors. Artificial NO release from spurious quantities of specific catalytic metal ions can be ruled out as well for the same reason.

Next, the substrate for the release of NO should be discussed. Five potential sources should be considered: Intracellular nitrite, extracellular nitrite, nitrate, arginine and endogenous nitrosothiols. Extracellular nitrite could be ruled out as the source of anoxic NO since the yield did not increase upon administration of massive 250 μM of extracellular nitrite prior to anoxia. Nitrate can be ruled out as its reduction requires extreme reductive conditions far beyond those found in cell cultures, or the presence of specific metal ions that are lacking in our assay. Arginine can be ruled out as its oxidation requires the presence of dioxygen. Although some residual oxygen could remain after imposition of anoxia, the quantity is not enough to keep the arginine pathway active for half an hour, let alone enhance its activity by 50%. It is known [24] that the synthesis of NO from L-arginine is blocked if dioxygen concentrations fall below ca 15 μM. Nitrosothiols can also be ruled out as their NO release is known to be catalyzed by the presence of reduced transition metal ions and remains unaffected by the presence of NOS inhibitors. Since the observed NO yields did not depend on extracellular nitrite levels, only intracellular nitrite remains as a plausible source of the NO released under hypoxia. The consumption of intracellular nitrite was directly confirmed by the colorimetric Griess assay on the lysates of the cell fractions that had been exposed to anoxia. While the total nitrite content of a single flask of normoxic cells was ca 1.1 nmol, the nitrite content had fallen to ca 0.15 nmol after 10 min of anoxia. These observations are in full agreement with our previous observations [46] that eNOS is capable of nitrite reduction under hypoxia *in vitro*. Since 7.5×10^6 cells have a total intracellular volume of ca 50 μl, we estimate that the intracellular nitrite concentration of normoxic cells is 1.1 nmol/50 μl $\sim$22 μM. This is a reasonable value in agreement with nitrite concentrations reported [9,10] for aortic tissues.

The magnitude of the anoxic NO release from the cell cultures suggests that the anoxic pathway be relevant to counter the effects of oxygen deficiency in vascular tissues. The kinetics of anoxic NO release showed that the MNIC yield increased linearly with time for up to about 30 min with a steady rate of formation of ca 160 pmol/20 min $\sim$8 pmol MNIC/min. This value is a lower limit on the total NO production since some NO will be bound in the form of diamagnetic NO—Fe^{3+}–DETC complexes and some will be lost from observation *via* other reaction pathways. It seems reasonable to estimate the total NO formation in a 75 cm^2 flask as $\sim$20–50 pmol NO/min, with an intracellular volume of ca 50 μl. This brings the anoxic NO formation to 0.4–1.0 pmol NO/min/mg endothelial cells. The normoxic NO formation is lower with ca 5 pmol MNIC/min and total 0.2–0.6 pmol NO/min/mg endothelial cells. This rough estimate corresponds remarkably well with the estimate [16,55] of 0.8 pmol NO/min/mg endothelial cells as basal yield in humans.

After 30 min, the MNIC yield became stationary. This saturation was definitely not caused by depletion of Fe–DETC traps. After all, the addition of exogenous nitrite and subsequent reduction with dithionite proved the formation of at least 4 nmol Fe–DETC traps in a cell culture. This quantity is at least an order of magnitude larger than the asymptotic yield of 200 pmol MNIC and proves that only a small fraction of traps actually binds NO under the conditions used. Therefore, the observed saturation of the MNIC yield indicates that the formation of NO under anoxia ceases after ca 30 min. Several reasons should be considered. First, the sustained anoxia causes irreversible damage to the cells and changes in the chemical composition of the interior compartment. Such ischemic damage is well documented for vascular tissue and endothelial cells in particular, and is manifested from the induction of massive apoptosis in the first 12 h after reoxygenation. However, our experiments are restricted to much shorter timescales of up to an hour, and trypan-blue staining of the cell cultures showed that the imposition of anoxia did not enhance cell death. A second reason could be inhibition of eNOS itself by free NO through nitrosylation of the heme moiety [24]. Such self-inhibition of eNOS was previously observed by us [46] in *in vitro* experiments with anoxic nitrite-reduction. The self-inhibition becomes manifested at higher nitrite concentration as can be seen in the left panel of Fig. 1. In our trapping experiments described here, however, the abundance of Fe–DETC traps acts as an efficient NO sink and should keep the concentration of free-NO radicals to a very low value and prevent significant nitrosylation of the heme. In a separate *in vitro* assay, the nitrosylation of eNOS was indeed prevented by the presence of albumin with Fe–DETC complexes in the solution. This makes self-inhibition of the cellular eNOS by free NO highly improbable. The third, and in our opinion most plausible reason could be depletion of the intracellular nitrite. The Griess assay showed that the endothelial cells contain ca 1.1 nmol intracellular nitrite per 75 cm^2 flask before anoxia is imposed. From the anoxic NO formation of ~20–50 pmol NO/min (see above), we expect depletion of intracellular nitrite after ca 55–22 min. This estimate suggests that nitrite depletion is the reason that MNIC yields become stationary after ca 30 min. We did not investigate further the reason for the saturation behavior since we expect that 30 min of anoxia has caused many changes inside the cells and made their physiology less and less representative for actual tissue endothelium as time proceeds. Instead, the magnitude of the anoxic NO release and particularly its surprising duration indicates that the nitrite reductase capacity of eNOS is a remarkably robust reaction mechanism.

The strong and sustained release of free NO in anoxic endothelial cell cultures suggests that the reduction of nitrite by eNOS has physiological relevance for NO levels near the endothelium under acute hypoxia.

MECHANISTIC HYPOTHESIS

Our experiments have shown that the nitrite reductase pathway of eNOS utilizes a significant stretch of the normal electron transport chain of the enzyme. The electrons are received from NADPH at the flavins of the reductase domain and optical spectroscopy shows that the electrons are channeled towards the heme just like with the normal arginine pathway. Therefore, both oxic and anoxic pathways should be similarly affected by the binding of calmodulin. Binding of this coenzyme facilitates the passage of electrons from the reductase to the oxygenase domain. The binding of calmodulin is controlled by the local Ca^{2+} concentration and this binding/unbinding of calmodulin represents one of the dominant regulatory mechanisms of the enzymatic activity of eNOS [43].

The site of the nitrite reduction is clearly located in the oxygenase domain, since the latter domain is by itself capable of reducing nitrite if electrons are provided to the heme, in our case by administration of dithionite. The significant inhibition of nitrite reduction by heme inhibitors like imidazole shows that the heme moiety of the oxygenase domain is crucial. Additionally, the effect of the arginine analogs NLA and L-NAME proved that steric hindrance of access to the axial heme position is strongly inhibiting the nitrite reduction. Finally, we observed that anoxic nitrite reduction leads to nitrosylation of the heme without the release of any free NO if no arginine is added to the solution. These observations suggest that the anoxic reduction may take place at the heme itself. It is known that ferrous heme may reduce nitrite in a proton-consuming reaction

$$NO_2^- + Fe^{2+} + H^+ \rightarrow Fe^{3+} + NO + OH^- \tag{3}$$

A prominent and well-studied example of this reaction is afforded by deoxy-hemoglobin [18]. The ferrous heme reduces nitrite to nitric oxide and ferric methemoglobin, consuming a proton in the process. A similar reaction is documented for the so-called NIR-cd_1 bacterial nitrite reductases [35,56]. The functional enzyme is a homodimer containing one heme c and one heme d_1 per subunit. The catalytic action of the nitrite reductase takes place at this d_1 heme and entails three successive stages: First, binding of the nitrite anion to the heme, with the nitrogen atom coordinating to the iron. Second, the reduction to NO under consumption of a proton and dehydration, with the nitrosyl remaining bound to the heme. The final step is the release of NO from the enzyme. In NIR-cd_1, the two final stages are mediated by nearby amino acid residues (cf Fig. 6 of Ref. [35]).

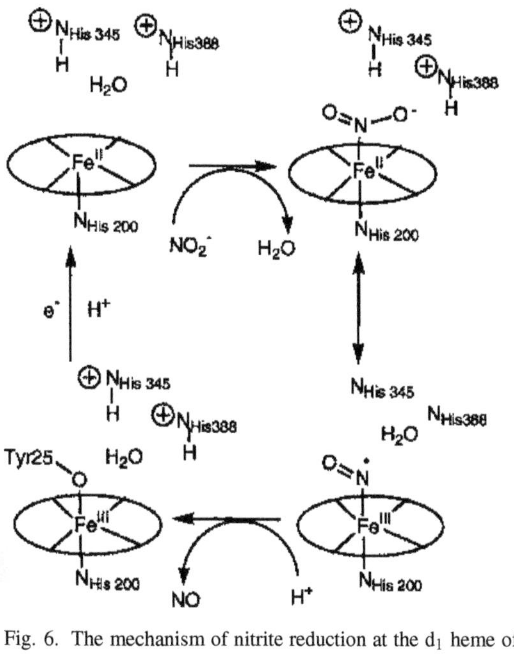

Fig. 6. The mechanism of nitrite reduction at the d_1 heme of NIR-cd_1. (With permission from Ref. [35].)

The reduction step is a dehydration reaction where the proton is provided by one of the pairs of nearby distal histidines and the NO remains liganded to the ferric iron. In Enemark–Feltham notation the electronic configuration is that of an $\{FeNO\}^6$ species. The final step is the subsequent replacement of the axial NO ligand on the ferric heme by a tyrosine residue, followed by the release of free NO from the enzyme. Its heme is left in ferric d^5 configuration. The next cycle is initiated by reduction of the d_1 heme to ferrous state after receiving an electron from the nearby c-heme that acts as the electron entry into the enzyme. For this particular enzyme, the binding of nitrite to the heme does not depend only on the charge state of the iron and the geometry of the heme pocket. Theoretical considerations suggest that protonation of these nearby histidines greatly enhances the binding of nitrite to d_1 heme [56].

Possibly, anoxic nitrite reduction by eNOS may proceed in an analogous manner, *via* a reaction like Eq. (3) that takes place at the heme of the oxygenase domain. The first step is binding of the nitrite anion at the axial position of the heme, with the nitrogen atom coordinating to the iron, followed by reduction of the heme to ferrous state by an electron channeled from the flavins of the reductase domain in the usual way. The subsequent step is acidic reduction of the nitrite according to Eq. (3), with the proton presumably being provided by a nearby amino acid. The final step is the release of the NO ligand from the enzyme, and is a complex kinetic process depending on factors like the charge state of the heme and the presence of cofactors like arginine and tetrahydrobiopterin. We tried to confirm direct binding of nitrite to the heme by optical spectroscopy. However, we failed to detect any spectroscopic changes from such binding at the relevant nitrite concentrations in the submillimolar range.

In our enzymatic assay, arginine was found to play a decisive role in the release of NO from the complex. In the absence of arginine, optical spectroscopy showed extensive heme nitrosylation by the characteristic shoulder around 440 nm. It is direct proof of formation of NO. However, this NO was not released from the heme since free NO levels remained below the detection threshold of the NO electrode. In the presence of arginine, copious quantities of free NO were detected. The binding site of arginine is in close proximity to the heme, and kinetic studies [57,58] have shown that binding of arginine affects the geminate recombination of NO to heme. The kinetic constants show that the presence of arginine facilitates the release of the NO ligand from the heme. When eNOS functions as nitrite reductase, arginine itself is not consumed. Rather, its binding to the nearby site promotes the release of free NO from the enzyme. EPR spectroscopy on nitrosylated eNOS heme has shown that the binding of L-arginine significantly reduces the motional freedom of the NO ligand with respect to the heme plane [59]. The effect was attributed to a weakening of the Fe—NO bond due to the electrostatic attraction between negatively charged oxygen of the NO ligand and the delocalized positive charge on the protonated guanidine group of the L-arginine.

NITRIC OXIDE IS RELEASED FROM FULL-LENGTH eNOS BUT NOT FROM nNOS UNDER ANOXIA

In an anaerobic 50 mM Tris buffer containing BH_4, arginine and 0.5 mM nitrite at $pH = 7.6$, eNOS is able to release NO after the administration of NADPH. The NO-specific electrode

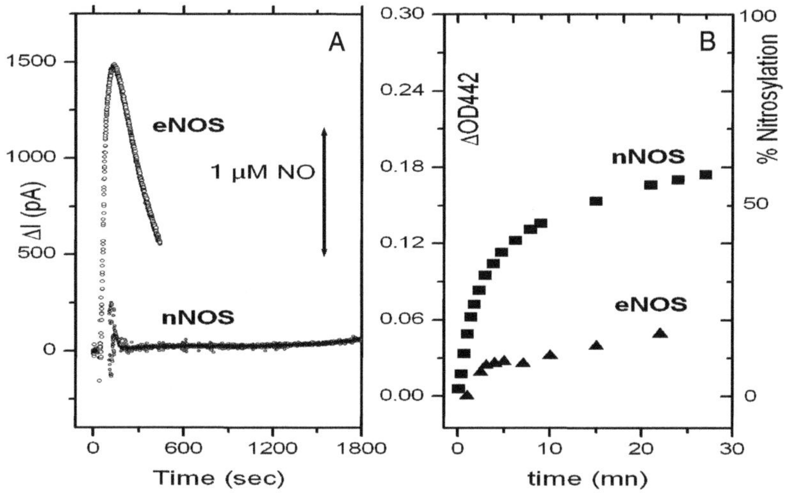

Fig. 7. NO formation from nitrite reduction by eNOS and nNOS at 500 μM nitrite. The reaction is initiated by injection of 100 μM of NADPH.The electrode current and optical density are studied simultaneously in a single experiment. (A) NO release as monitored by the NO electrode current. Upper trace is 5 μM eNOS. Lower trace is 7 μM nNOS. (B) Formation of ferrous heme-nitrosyl complex after injection of NADPH. Triangles represent 5 μM eNOS, the squares 7 μM nNOS. The right scale gives the degree of nitrosylation as computed from $\Delta\varepsilon_{442} = [\varepsilon_{442}(\text{NO} - \text{heme}) - \varepsilon_{442}(\text{heme})] = 4 \times 10^4\,\text{M}^{-1}\,\text{cm}^{-1}$. The experiments are performed under a constant argon flow in 50 mM Tris buffer at pH = 7.6 containing 150 mM NaCl and 5% glycerol, 1 mM arginine, 40 μM BH_4, 10 μM calmodulin, 1 mM Ca^{2+}.

(Fig. 7A) showed that the NO was released from the enzyme into the solution. In contrast, the electrode current (Fig. 7A) showed that anoxic nNOS failed to release significant quantities of free NO. When the measurement of nNOS was continued for 90 min, a small signal corresponding to 0.32 ± 0.07 μM NO could be observed.

Fig. 7B shows the change in absorbance at 442 nm, simultaneously recorded from the same solution, which is indicative of heme nitrosylation. The kinetics of Fig. 7A shows that free NO was released in a fast burst within a few seconds required for mixing NADPH. This release of NO is clearly faster than the nitrosylation of eNOS heme, attesting some NO rebinding to the protein after being released.

It is interesting to note that the eNOS and nNOS exhibit significantly different degrees of nitrosylation (cf Fig. 7 and Table 3). The optical density at 442 nm showed that the degree of nitrosylation of eNOS reached 9% after 5 min and increased to 18% after 25 min. For nNOS the numbers are far higher with 25% at 5 min, 55% at 25 min, and seemed to saturate at an asymptotic value of 62% nitrosylation.

Although the electrode detected the release of small quantities of free NO, the extensive nitrosylation of nNOS shows that the dominant fraction of NO is retained as a nitrosyl ligand by nNOS. Clearly, the NO was formed but was unable to diffuse out of the protein.

Table 3 Percentage of nitrosylated heme at given time points after injection of NADPH. Data are taken from Fig. 7

	prior	5 min	10 min	25 min	∞
eNOS	0	9	11	18	20
nNOS	0	25	45	55	62

COMPARISON OF THE RATES OF NITRIC OXIDE GEMINATE RECOMBINATION TO THE HEME OF eNOS AND OF nNOS

The above differences between eNOS and nNOS in terms of NO release and heme nitrosylation are related to the stability of their ferrous nitrosyl complexes. For nNOS, this ferrous nitrosyl complex of nNOS is very stable. The rates for spontaneous binding and dissociation of NO to nNOS were reported as $k_{on} = 2 \times 10^7\ M^{-1}s^{-1}$ and $k_{off} = 1 \times 10^{-4}s^{-1}$ [60]. This is consistent with the majority of the nNOS protein that exist as a ferrous heme-NO complex in steady-state (aerobic) catalysis, since the nitrosylation governs and limits the rate of NO synthesis [61]. In contrast, in eNOS, little heme-NO complex forms during normal turnover when arginine is the substrate. The nitrosylation rate of ferrous eNOS was reported as $k_{on} = 1.1 \times 10^6\ M^{-1}s^{-1}$ whereas conflicting values were published for the dissociation rate: $k_{off} = 70\ s^{-1}$ and $6 \times 10^{-4}\ s^{-1}$ [23,54]. These dissociation rates are at least an order of magnitude larger than for nNOS. These numbers explain the experimental fact that eNOS has a significantly lower degree of nitrosylation than nNOS, once NO is formed *via* bimolecular recombination of NO to the proteins. The kinetics of photo-induced release and rebinding of free NO from the ferrous nitrosylated eNOS and nNOS were measured by ultra-fast absorption spectroscopy (geminate rebinding of the nitrosyl ligand after photodissociation). The buffered solution had a high 50 μM concentration of ferrous protein that was nitrosylated by prolonged bubbling of the solution with argon containing 0.1% NO. Optical absorption confirmed that bubbling achieved full nitrosylation of the heme within the experimental error of ca 5%. We used a 30 fs pump pulse of 565 nm to selectively cleave the NO-heme bond of the ferrous nitrosyl complex. This bond cleavage generated the unliganded Fe^{2+} heme at the expense of the ferrous nitrosyl complex. The kinetics of the absorption spectrum after bond cleavage was monitored with a wideband probe beam (between 390 and 550 nm) with high time resolution. A typical trace of the kinetics is shown in Fig. 8. The amplitude ΔOD represents the change in optical density by application of the pump pulse. At given optical pathlength d, this observable value is proportional to the concentration of non-nitrosyl heme $[Fe^{2+}]$ released by the pump pulse

$$\Delta \mathrm{OD}\,(t = 0) = d\{\varepsilon_{410}(\mathrm{heme}) - \varepsilon_{410}(\mathrm{NO{-}heme})\} \cdot [\mathrm{Fe}^{2+}]$$

Significantly, the optical density does not return to its original value just prior to the pump pulse, but it approaches a positive asymptotic value. This phenomenon is particularly manifest for the eNOS isoform, but applies to nNOS as well. It proves that not all cleaved NO-heme bonds are restored by recapture of the nitrosyl ligand. Phrased otherwise, a certain fraction of NO ligands was actually released from the enzyme as free NO.

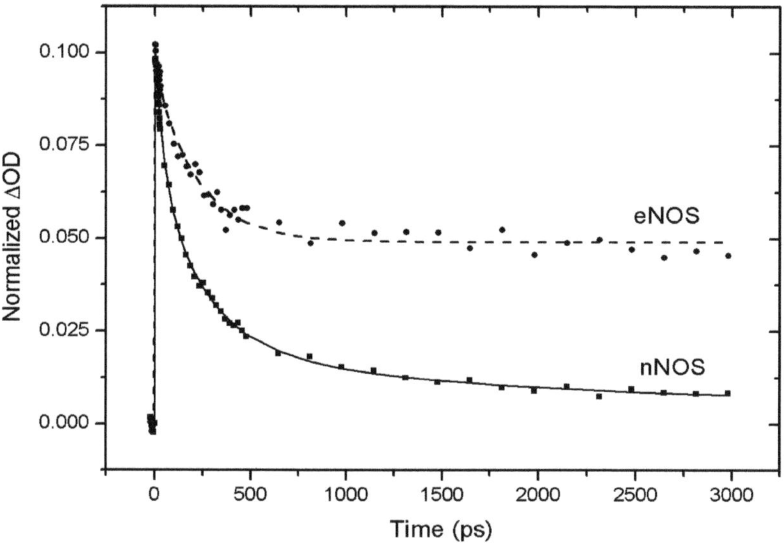

Fig. 8. Kinetics of the optical density at 410 nm due to geminate recombination of NO ligands to the ferrous heme of eNOS and nNOS. The reaction is initiated by a short pump pulse at 565 nm. The solutions contain ca 50 μM nitrosylated eNOS after bubbling with argon containing 0.1% NO. The kinetic constants are given in Table 4. The decay curves are fit according to Eq. (4).

Table 4 Kinetic constants of the geminate recombination of NO ligands to ferrous heme of eNOS and nNOS. The parameter A_4 represents the probability that the photodissociated NO is released from the protein and escapes recombination with the heme

	eNOS	nNOS
τ_1 (ps)	44 ± 9	42 ± 9
τ_2 (ps)	390 ± 70	230 ± 50
τ_3 (ns)	4.1 ± 0.5	1.3 ± 0.2
A_1	0.35	0.29
A_2	0.20	0.48
A_3	0.25	0.17
$A_4 = 1 - A_3 - A_2 - A_1$	0.20	0.06

As often found for the recombination kinetics of hemeproteins, NO rebinding kinetics could be described as the sum of three exponentials with clearly distinct timescales $\tau_1 < \tau_2 < \tau_3$ where $A_1 + A_2 + A_3 + A_4 = 1$

$$\frac{D(t)}{D(t=0)} = A_1 e^{-t/\tau_1} + A_2 e^{-t/\tau_2} + A_3 e^{-t/\tau_3} + A_4 \tag{4}$$

The first exponential (A_1, τ_1) accounts for rapid NO rebinding from the immediate vicinity to the heme iron, and is activationless. The second and third exponentials account for much slower kinetics where the NO rebinding takes place from larger distances to the heme, but still within the heme pocket or its vicinity. These phases are usually required to overcome internal energy barriers for NO recombination to the heme. The constant term A_4 is the probability of escape of free NO from the protein. It is clearly seen from Fig. 8 that this escape probability A_4 is much higher in eNOS than in nNOS (0.20 *versus* 0.06). It confirms that very little NO is able to escape from the reduced nNOS. This result is in complete agreement with the absence of spontaneous (thermolytic) release of NO from nitrosylated nNOS and the high degree of nitrosylation of nNOS in the presence of nitrite (cf Fig. 7). This NO-bound ferrous complex probably arises from NO generated by nitrite reduction, but is unable to escape from the ferrous nNOS. In contrast, once oxygen is able to oxidize the heme to its ferric state, this NO should be released immediately in a "burst." This NO burst could be very toxic since it happens as the oxygen is freshly supplied, for example, at reperfusion. Moreover, it can combine with the superoxide ions formed under decoupling conditions to generate peroxynitrite. This could be the basis of the reported evidence of nNOS promoting inflammation in the cerebral microcirculation whereas eNOS blunts the extent of this response to episodic hypoxia and exerts neuroprotective effects [62].

IMPLICATIONS FOR HUMAN PHYSIOLOGY

In recent years, clinical and research evidence has accumulated that nitrite anions are the dominant endogenous store of nitric oxide in the human body. A range of enzymatic pathways for the conversion of nitrite to NO have been identified. Nitric reductase activities have been well documented for deoxy-hemoglobin, xanthine-oxidoreductase, myoglobin and endothelial NOS as described above. In these enzymatic pathways, the presence of oxygen had decisive inhibiting effects. Therefore the nitrite pool can sustain the local NO levels only during physiological or pathological hypoxia. It was suggested [63] that nitrite should be regarded as a "critical hypoxic buffer" that couples to the NO pool and supports vasodilation and mitochondrial respiration under the strain of hypoxia, with beneficial protection against ischemia-reperfusion injuries and infarction.

Our findings are in-line with a plethora of animal studies showing a protective role of eNOS in the early stages of ischemia [64–71]. Interestingly, it appears that the protective role cannot be attributed simply to all NOS isoforms indiscriminately but seems to be associated with eNOS only. Ischemic insult resulting in reduced blood flow and diminished myocyte viability in eNOS-KO mice can be rescued by local delivery of eNOS and underlines a critical role of eNOS in ischemic remodeling [67,71]. The cerebrovascular inflammation after brief episodic hypoxia was tested on eNOS or nNOS-KO mice by imaging leukocyte dynamics [62]. nNOS was reported to promote inflammation in the cerebral microcirculation whereas eNOS blunts the extent of this response to episodic hypoxia. The anti-inflammatory effect is neuroprotective [62]. These observations were in agreement with previous studies in KO animals [64–66]. Conversely, overexpression of endothelial nitric oxide synthase in endothelial cells was found to protect against ischemia-reperfusion injury in mouse [72]. Recent work observed a cytoprotective effect of nitrite in ischemic hepatic injury in eNOS-KO animals [63], but the addition of nitrite still reduced the hepatic enzyme aspartate aminotransferase (AST) without a contribution from eNOS. Activation of hepatic eNOS activity and enhanced NO production protected female mice against ischemia-reperfusion injury of the liver [73]. Wild-type mice have far better chance than eNOS-KO mutants to survive the pulmonary edema after pulmonary ischemia reperfusion [74]. Taken together, these works suggest that eNOS protects against ischemic damage to various organs. The protection appears as an immediate but short-term response [30,69,70]. Taken together, these animal studies show that eNOS and nNOS respond very differently in early stages of hypoxia. Experiments to study nitrite reduction in cells and animal tissues are currently in progress to determine the relevance of this newly discovered pathway to physiology and pathology.

REFERENCES

1 Kelm M. Nitric oxide metabolism and breakdown. Biochim. Biophys. Acta 1999; 1411: 273–289.

2 Demoncheaux EA, Higenbottam T, Foster P, Borland C, Smith H, Marriott H, Bee D, Akamire S, Davies M. Circulating nitrite anions are a directly acting vasodilator and are donors for nitric oxide. Clin. Sci. 2002; 102: 77–83.

3 Bryan N, Fernandez B, Bauer S, Garcia-Saura MT, Milsom AR, Rassaf T, Maloney RE, Bharti A, Rodriguez J, Feelisch M. Nitrite is a signaling molecule and regulator of gene expression in mammalian tissues. Nature Chem. Biol. 2005; 1: 290–297.

4 Vleeming W, van de Kuil A, te Biesebeek J, Meulenbelt J, Boink A. Effect of nitrite on blood pressure in anaesthetized and free-moving rats. Food. Chem. Toxicol. 1997; 35: 615–619.

5 Lauer T, Preik M, Rassaf T, Stauer B, Deussen A, Feelisch M, Kelm M. Plasma nitrite rather than nitrate reflects regional endothelial nitric oxide synthase activity but lacks intrinsic vasodilator action. Proc. Natl. Acad. Sci. USA 2001; 98: 12814–12819.

6 Gladwin MT, Shelhamer JH, Schechter AN, Pease-Fye M, Waclaweiw M, Panza J, Oguibene F, Cannon R. Role of circulating nitrite and S-nitrosohemoglobin in the regulation of the regional blood flow in humans. Proc. Natl. Acad. Sci. USA 2000; 97: 11482–11487.

7 Kleinbongard P, Dejam A, Rassaf T, Lauer T, Rassaf T, Schindler A, Piecker O, Schreen T, Godecke A, Schrader J, Schulz R, Hersch G, Schaub GA, Bryan NS, Feelisch M, Kelm M. Plasma nitrite reflects constitutive nitric oxide synthase activity in mammals. Free Rad. Biol. Med. 2003; 35: 790–796.

8 Duranski MR, Greer J, Dejam A, Jaganmohan S, Hogg N, Langston W, Patel RP, Yet S, Wang Y, Kevil C, Gladwin M, Lefer D. Cytoprotective effects of nitrite during in vivo ischemia-reperfusion of the heart and liver. J. Clin. Invest. 2005; 115: 1232–1240.

9 Bryan N, Rassaf T, Maloney R, Rodriguez CM, Saijo F, Rodriguez JR, Feelisch M. Cellular targets and mechanisms of nitros(yl)ation: an insight into their nature and kinetics in vivo. Proc. Natl. Acad. Sci. USA 2004; 101: 4308–4313.

10 Rodriguez J, Maloney R, Rassaf T, Bryan N, Feelisch M. Chemical nature of nitric oxide storage forms in rat vascular tissue. Proc. Natl. Acad. Sci. 2003; 100 (1): 336–341.

11 Dhawan V, Schwalb D, Shumway M, Warren MC, Wexler RS, Zemtseva IS, Zifcak BM, Janero DR. Selective nitros(yl)ation induced in vivo by a nitric oxide-donating cyclooxygenase-2 inhibitor: A NObonomic analysis. Free Rad. Biol. Med. 2005; 39: 1191–1207.

12 Rassaf T, Bryan N, Kelm M, Feelisch M. Concomitant presence of N-nitroso and S-nitroso proteins in human plasma. Free Rad. Biol. Med. 2002; 33: 1590–1596.

13 Halliwell B, Gutteridge J. Free radicals in biology and medicine, Clarendon Press, Oxford, 1989.

14 Godecke A, Decking Z, Ding J, Hirshenhain J, Bidmon HJ, Godecke S, Schrader J. Coronary hemodynamics in endothelial NO synthase knockout mice. Circ. Res. 1998; 82: 186–194.

15 Rhodes P, Leone A, Francis P, Struthers A, Moncada S. The L-arginine:NO pathway is the major source of plasma nitrite in fasted humans. Biochem. Biophys. Res. Comm. 1995; 209: 590–596.

16 Rassaf T, Preik M, Kleinbongard P, Lauer T, Heiss C, Strauer BE, Feelisch M, Kelm M. Evidence for in vivo transport of bioactive nitric oxide in human plasma. J. Clin. Invest. 2002; 109: 1241–1248.

17 Kelm M, Preik-Steinhoff H, Preik M, Strauer B. Serum nitrite sensitivity reflects endothelial NO formation in human forearm vasculature; evidence for biochemical assessment of the endothelial L-arginine pathway. Cardiovasc. Res. 1999; 41: 765–772.
18 Cosby K, Partovi K, Crawford JH, Patel RP, Reiter CD, Martyr S, Yang BK, Waclawin M, Zalos G, Xu X, Huang KT, Shields H, Kim-Shapiro DB, Schechter AN, Cannon R, Gladwin MT. Nitrite reduction to nitric oxide by deoxyhemoglobin vasodilates the human circulation. Nat. Med. 2003; 9: 1498–1505.
19 Kohn M, Melnick R, Frank Y, Portier C. Pharmacokinetics of sodium nitrite-induced methemoglobinemia in the rat. Drug Metab. Disp. 2002; 30: 676–683.
20 Gladwin M, Schechter A, Kim-Shapiro D, Rakesh RP, Hogg N, Shiva S, Cannon RV, Kelm M, Wink DA, Espey MG, Oldfield EH, Pluta RM, Freeman BA, Lancaster JR, Feelisch M, Lundberg JO. The emerging biology of the nitrite anion. Nature Chem. Biol. 2005; 1: 308–314.
21 Webb A, Bond R, McLean P, Uppal R, Benjamin N, Ahluwalia A. Reduction of nitrite to nitric oxide protects against myocardial ischemia-reperfusion damage. Proc. Natl. Acad. Sci. USA 2004; 101: 13683–13688.
22 Gladwin M, Haldane T. Hot dogs, halitosis and hypoxic vasodilation. The emerging biology of the nitrite anion. J. Clin. Invest. 2004; 113: 19–21.
23 Agvald P, Adding L, Gustafsson L. Influence of oxygen, temperature and carbon dioxide on NO generation from nitrite as measured in expired gas from in situ perfused rabbit lungs. Vascular Pharmacology 2005; 43: 441–448.
24 Abu-Soud HM, Ichimori K, Presta A, Stuehr DJ. Electron transfer, oxygen binding, and nitric oxide feedback inhibition in endothelial nitric-oxide synthase. J. Biol. Chem. 2000; 275: 17349–17357.
25 Xue C, Rengasamy A, Le Cras T, Koberna P, Daily G, Johns R. Distribution of NOS in normoxic vs. hypoxic rat lungs: Upregulation of NOS by chronic hypoxia. Am. J. Physiol. 1994: L667–L678.
26 Bracken C, Whitelaw M, Peet D. The hypoxia-inducible factors; key transcriptional regulators of hypoxic responses. Cell. Mol. Life Sci. 2003; 60: 1376–1393.
27 Cebe-Suarez S, Zehnder-Fjallman A, Ballmer-Hofer K. The role of VEGF receptors in angiogenesis; complex partnership. Cell. Mol. Life Sci. 2006; 63: 601–615.
28 Hampl V, Cornfield D, Cowan N, Archer S. Hypoxia potentiates nitric oxide synthesis and transiently increases cytosolic calcium levels in pulmonary artery endothelial cells. Eur. Respir. J. 1995; 8: 515–522.
29 Whorton A, Simonds D, Piantadosi C. Regulation of nitric oxide synthesis by oxygen in vascular endothelial cells. Am. J. Physiol. 1997; L1161–L1166.
30 Beleslin-Cokic B, Cokic V, Yu X, Weksler B, Schechter A, Noguchi C. Erythropoietin and hypoxia stimulate erythropoietin receptor and nitric oxide production by endothelial cells. Blood 2004; 104: 2073–2080.
31 Samouilov A, Zweier J. Development of chemiluminescence-based methods of specific quantitation of nitrosylated thiols. Anal. Biochem. 1998; 258: 322–330.
32 McKnight G, Smith L, Drummond R, Duncan CW, Golden M, Benjamin N. Chemical synthesis of nitric oxide in the stomach from dietary nitrate in humans. Gut 1997; 40: 211–214.
33 Lundberg J, Carlsson S, Engstrand L, Morcos E, Wiklund NP, Weitzberg E. Urinary nitrite: more than a marker of infection. Urology 1997; 50: 187–191.
34 Zweier J, Wang P, Samouilov A, Kuppusamy P. Enzyme-independent formation of nitric oxide in biological tissues. Nature, Med. 1995; 1: 804–809.
35 Wasser I, de Vries S, Loccoz P, Schroder I, Karlin K. Nitric oxide in biological denitrification: Fe/Cu metalloenzyme and metal complex Nox redox chemistry. Chem. Rev. 2002; 102: 1201–1234.
36 Hill KE, Hunt Jr RW, Jones R, Hoover RL, Burk RF. Metabolism of nitroglycerine by smooth muscle cells. Involvement of glutathione-S-transferase. Biochem. Pharmacol. 1992; 43: 561–566.
37 Millar TM, Stevens C, Benjamin M, Eisenthal R, Harrison P, Blake DR. Xanthine oxidoreductase catalyses the reduction of nitrates and nitrite to nitric oxide under hypoxic conditions. FEBS Lett. 1998; 427: 225–228.
38 Godber BL, Doel JL, Sapkota GP, Blake DR, Stevens CR, Eisenthal R, Harrison R. Reduction of nitrite to nitric oxide catalyzed by xanthine oxidoreductase. J. Biol. Chem. 2000; 275: 7757–7763.
39 Doyle PM, Pickering RA, DeWeert TM, Hoekstra D, Pater J. Kinetics and mechanism of the oxidation of human deoxyhemoglobin by nitrites, J. Biol. Chem. 1981; 256: 12393–12398.

40 Luchsinger BP, Rich E, Yan Y, Williams E, Stamler J, Singel D. Assessments of the chemistry and vasodilatory activity of nitrite with emoglobin uder physiologically relevant conditions. J. Inorg. Biochem. 2005; 99: 912–921.
41 Crawford J, Scott Isbell T, Huang Z, Shiva S, Chacko BK, Schechter AN, Darley-Usmar VM, Kerhy JD, Lang JD, Kraus D, Ho C, Gladwin MT, Patel RP. Hypoxia, red blood cells, and nitrite regulate NO dependent hypoxic vasodilation. Blood 2006; 107: 566–574.
42 Delafoge M, Servent D, Wirsta P, Ducrocq C, Mansye D, Lenfant M. Particular ability of cytochrome P-450 CYP3A to reduce glycerol trinitrate in rat liver microsomes: subsequent formation of nitric oxide. Chem. Biol. Interact. 1993; 86: 103–117.
43 Kozlov A, Staniek K, Nohl H. Mitochondrial NO-generation from nitrite, FEBS lett. 1999; 454: 127–130.
44 Nohl H, Staniek K, Kozlov A. Involvement of mammalian mitochondria in recycling of the NO-metabolite nitrite to nitric monoxide. Free Radical Biol. Med. 1999; 27: 245–252.
45 Nohl H, Staniek K, Kozlov A, The existence and significance of a mitochondrial nitrite reductase. Redox. Report 2005; 10: 281–286.
46 Gautier C, van Faassen E, Mikula I, Martasek P, Slama-Schwok A. Endothelial nitric oxide synthase reduces nitrite anions to NO under hypoxia. Biochem. Biophys. Res. Commun. 2006; 341: 816–821.
47 Koppenol W. Thermodynamics of reactions involving nitrogen-oxide compounds. Meth. Enzymol. 1996; 268: 7–12.
48 Hofmann H, Schmidt H. Thiol dependence of nitric oxide synthase. Biochemistry 1995; 34: 13443–13452.
49 Govers R, Rabelink T. Cellular regulation of endothelial nitric oxide synthase. Am. J. Physiol. Renal Physiol. 2001; 280: F193–F206.
50 Liu J, Garcia-Gardena G, Sessa W. Palmitoylation of endothelial nitric oxide synthase is necessary for optimal stimulated release of nitric oxide: implications for caveolae localization. Biochemistry 1996; 35: 13277–13281.
51 Martasek P, Liu Q, Liu J, Roman LJ, Gross SS, Sessa WC, Masters BS. Characterization of bovine endothelial nitric oxide synthase expressed in E.coli. Biochem. Biophys. Res. Comm. 1996; 219: 359–366.
52 Wang J, Rousseau D, Abu-Soud H, Stuehr D. Heme coordination of NO in NO synthase. Proc. Natl. Acad. Sci. 1994; 91: 10512–10516.
53 Suzuki Y, Fujii S, Tominaga T, Yoshimoto T, Yoshimura T, Kamada H. The origin of an EPR signal observed in dithiocarbamate-loaded tissues: Copper(II)-dithiocarbamate complexes account for the narrow hyperfine lines. Biochim. Biophys. Acta-Gen Subj. 1997; 1335: 242–245.
54 Vanin A, Bevers L, Slama-Schwok A, van Faassen E. Cell. Mol. Life Sci. 2007; 64: 96–103.
55 Kelm M, Feelisch M, Deussen A, Strauer B, Schrader J. Release of endothelium derived nitric oxide in relation to pressure and flow. Cardiovasc. Res. 1991; 25: 831–836.
56 Marti M, Crespo A, Bari S, Doctorovich F, Estrin D. QM-MM study of nitrite reduction by nitrite reductase of Pseudmonas Aeruginosa. J. Phys. Chem. 2004; B108: 18073–18080.
57 Negrerie M, Berka V, Vos MH, Liebl U, Lambry JC, Tsai AL, Martin JL. Geminate recombination of nitric oxide to endothelial nitric-oxide synthase and mechanistic implications. J. Biol. Chem. 1999; 274: 24694–24702.
58 Gautier C, Mikula I, Nioche P, Martasek P, Raman CS, Slama-Schwok A. Dynamics of NO rebinding to the heme domain of NO synthase- like proteins from bacterial pathogens, nitric oxide. Biochem. Biophys. Res. Comm. 2006; 341: 816–821.
59 Migita C, Salerno J, Siler Masters B-S, Martasek P, McMillan K, Ikeda-Saito M. Substrate binding-induced changes in the EPR spectra of the ferrous nitric oxide complexes of neuronal nitric oxide synthase. Biochemistry 1997; 36: 10987–10992.
60 Scheele JS, Bruner E, Kharitonov VG, Martasek P, Roman LJ, Masters BS, Sharma VS, Magde D. Kinetics of NO ligation with nitric-oxide synthase by flash photolysis and stopped-flow spectrophotometry. J. Biol. Chem. 1999; 274: 13105–13110.
61 Santolini J, Meade AL, Stuehr DJ. Differences in Three Kinetic Parameters Underpin the Unique Catalytic Profiles of Nitric-oxide Synthases I, II, and III*. J. Biol. Chem. 2001; 276: 48887–48898.

62 Altay T, Gonzales ER, Park TS, Gidday JM. Cerebrovascular inflammation after brief episodic hypoxia: modulation by neuronal and endothelial NOS J. Appl. Physiol. 2004; 96: 1223–1230.
63 Duranski M, Elrod J, Calwert J, Bryan NS, Feelisch M, Lefer JM. Genetic overexpression of eNOS attenuates hepatic ischemia-reperfusion injury. Am. J. Physiol. Heart Circ. Physiol. 2006; 291: H2980–H2986.
64 Bolanos J, Almeida A. Roles of nitric oxide in brain hypoxia-ischemia. BBA. 1999; 1411: 415–436.
65 Wei G, Dawson VL, Zweier JL. Roles of neuronal and endothelial NOS in NO generation in the brain following cerebral ischemia BBA. 1999; 1411: 23–34.
66 Huang PL. Neuronal and endothelial NOS gene knock-out mice. Braz. J. Med. Biol. Res. 1999; 32: 1353–1359.
67 Yu J, Demuinck ED, Zhuang Z, Drinane M, Kauser K, Rubanyi GM, Qian HS, Murata T, Escalante B, Sessa WC. Endothelial nitric oxide synthase is critical for ischemic remodeling, mural cell recruitment, and blood flow reserve. Proc. Natl. Acad. Sci. USA 2005; 102: 10999–11004.
68 Moro M, Cardenas A, Hurtado J, Leza C, Lizasoain I. Role of nitric oxide after brain ischaemia. Cell Calcium 2004; 36: 265–275.
69 Bolli R. Cardioprotective function of inducible NO synthase and role of nitric oxide in myocardial ischemia and preconditioning: an overview of a decade of research. J. Mol. Cell. Cardiol. 2001; 33: 1897–1918.
70 Endres M, Laufs U, Lia JK, Moskowitz MA. Targeting eNOS for stroke protection. Trends Neurosci. 2004; 27: 283–289.
71 Murohara T, Asahara T, Silver M, Bauter C, Masuda H, Kalka C, Kearny M, Chen D, Chen D, Symes JF, Fishman MC, Huang PL, Isner JM. Nitric oxide synthase modulates angiogenesis in response to tissue ischemia. J. Clin. Invest. 1998; 101: 2567–2578.
72 Ozaki M, Kawashima S, Hirase T, Yamashita T, Namiki M, Inoue N, Hirata Ki K, Yokoyama M. Overexpression of endothelial nitric oxide synthase in endothelial cells is protective against ischemia-reperfusion injury in mouse skeletal muscle. Am. J. Pathol. 2002; 160: 1335–1344
73 Harada H, Pavlick KP, Hines IN, Lefer DJ, Hoffman JM, Bharwani S, Wolf RE, Grisham MB. Sexual dimorphism in reduced-size liver ischemia and reperfusion injury in mice: role of the endothelial cell nitric oxide synthase. Proc. Natl. Acad. Sci. USA 2003; 100: 739–744.
74 Kaminski A, Pohl CB, Sponholz C, Ma N, Stamm C, Vollmar B, Steinhoff G. Up-regulation of endothelial nitric oxide synthase inhibits pulmonary leukocyte migration following lung ischemia-reperfusion in mice. Am. J. Pathol. 2004; 164(6) 2241–2249.

Radicals for Life: The Various Forms of Nitric Oxide
E. van Faassen and A.F. Vanin (Eds.)

CHAPTER 15

Nitrite as NO donor in cells and tissues

Alexandre Samouilov, Haitao Li and Jay L. Zweier

From the Center for Biomedical EPR Spectroscopy and Imaging, the Davis Heart and Lung Research Institute, and the Division of Cardiovascular Medicine, Department of Internal Medicine, Ohio State University College of Medicine, Columbus, Ohio 43210

INTRODUCTION

The metabolism of L-arginine by NO synthase (NOS), to produce $NO^{\bullet}$ and citrulline is widely accepted as the primary source of NO from biological tissues. However, a growing body of evidence supports alternative NOS-independent mechanisms of NO synthesis that operate in situations in which conventional NO production in cells is impaired [1–7]. Inorganic nitrite (NO_2^-), an endogenous substance produced by the oxidation of NO in aerobic conditions, can be detected in biological tissues that are able to generate nitric oxide. *In vivo* plasma levels of nitrite are in the range from 0.15 to 1 μM [8,9], and the concentration in aortic ring tissue is above 10 μM [8,10,11]. Tissue nitrite levels vary sharply and can be in the range of 0.5–50 μM [3,7,8,12–14]. Nitrite levels are controlled by a number of factors including diet (cf Chapter 16) and concentration of ambient NO in inhaled air as well as by production from NOS or other enzymes [1,15–17]. Many pathological conditions that are accompanied by high levels of NOS induction markedly increase nitrite levels [18–20]. Therefore in many experimental settings nitrite levels have been used as a marker of NOS activity. However, these nitrite levels can be also raised by high dietary ingestion, pharmacological administration of organic nitrates, or other NO-donating therapeutics. For that reason, in many settings there may be a lack of direct correlation between nitrite levels and NO synthase (NOS) activity. Nitrite, when administrated at high concentrations, demonstrates vasodilating activity *in vitro* [21–27]. This vasodilating activity, in general, is attributed to the conversion of nitrite to NO. Thus, provided that physiological mechanisms exist to reduce nitrite to NO, it can comprise a significant storage pool of NO. Nitrite, in aqueous solutions, spontaneously disproportionates to NO. This reaction is facilitated by acidic conditions (cf Chapters 1 and 9) and can be evoked in biological systems. It was originally described in the acidic conditions of the human stomach, in which NO_2^- derived from the sequential reduction of dietary nitrate (NO_3^-) [28–31]. This reaction has important functional effects whereby human saliva rich in NO_2^- improves gastric mucosal blood flow and mucus thickness when applied to the rat stomach [32]. Also, this alternative mechanism of NO synthesis may be particularly important in ischemic conditions, because the generation of NO from L-arginine by NOS enzymes depends on oxygen, which is rapidly depleted in ischemia. Our studies in ischemic rat myocardium clearly demonstrate NOS-independent generation of NO, detected with EPR spectroscopy [7].

Initially, we [7] and then Ferrari et al. [33] and Gabel et al. [34] proposed that myocardial $NO^{\bullet}$ is derived from a simple disproportionation of nitrite, which is accelerated by the acidic media. Indeed, our quantitative chemiluminescence analysis of this reaction [35] showed that ischemia-induced acidosis, associated with a drop in pH to 5.5 in the isolated rat heart preparation after 20 min of global ischemia, is able to generate $NO^{\bullet}$ at the rate up to 100 pM/s which corresponds to as much as 10% of the maximum production that would occur from NOS under normal physiological conditions. At these low pH values, NOS loses its catalytic activity [36], so that NO formation from nitrite could be of particular importance. However, measurements of anaerobic $NO^{\bullet}$ production from 1 mM nitrite at pH 5.5 after addition of an aliquot of ischemic heart tissue homogenate demonstrated a 40–50-fold increase in the rate of NO generation compared with disproportionation (Fig. 1). Aliquots of tissue alone did not release measurable NO under these conditions, suggesting that this nitrite reduction depends on enzymatic catalysis and/or reducing equivalents or substrates.

XANTHINE OXIDASE-CATALYZED NITRITE REDUCTION

Xanthine oxidase (XO) is a ubiquitous enzyme in mammalian cells that plays important roles in both physiological and pathological conditions. It is involved in the catabolism of purine and pyrimidines, oxidizing hypoxanthine to xanthine and xanthine to uric acid. XO also reduces oxygen, to superoxide and hydrogen peroxide production, and is one of the key enzymes responsible for superoxide-mediated cellular injury.

Although it has been established that XO can reduce nitrite to NO [4,5,37], questions remained regarding the biological importance of this pathway of NO production, as well as the mechanism, magnitude, and substrate specificity of this process. It was first reported that NADH, but not xanthine, can act as an electron donor to XO and catalyze nitrite reduction [4,37]. Xanthine or hypoxanthine was found to inhibit this NO formation from XO [37]. However, later it was reported that xanthine can serve as a reducing substrate to stimulate nitrite reduction [38]. In contrast to this, other investigators reported that XO in the presence of xanthine does not reduce nitrite to NO [39]. In addition, in these studies, there were large differences regarding the rates of NO formation and K_m values of XO for nitrite or the requisite reducing substrate. In view of these uncertainties, it had not been possible to ascertain the biological relevance and importance of this pathway of NO generation. Therefore, studies using EPR, chemiluminescence NO analyzer, and NO electrode techniques were performed to measure the magnitude and kinetics of NO formation that arise due to XO-mediated nitrite reduction [40]. Data obtained using each of these three methods confirmed that XO does reduce nitrite to NO under anaerobic conditions. Each of the typical reducing substrates xanthine, DBA, and NADH can act as electron donors to support this XO-mediated nitrite reduction (Fig. 2). The results of these studies, along with the inhibition seen with oxypurinol, suggested that reduced XO was the direct electron donor to nitrite, with nitrite binding and reduction occurring at the molybdenum site. Whereas NADH-stimulated NO generation was inhibited by the flavin modifier DPI, NO generation stimulated by xanthine

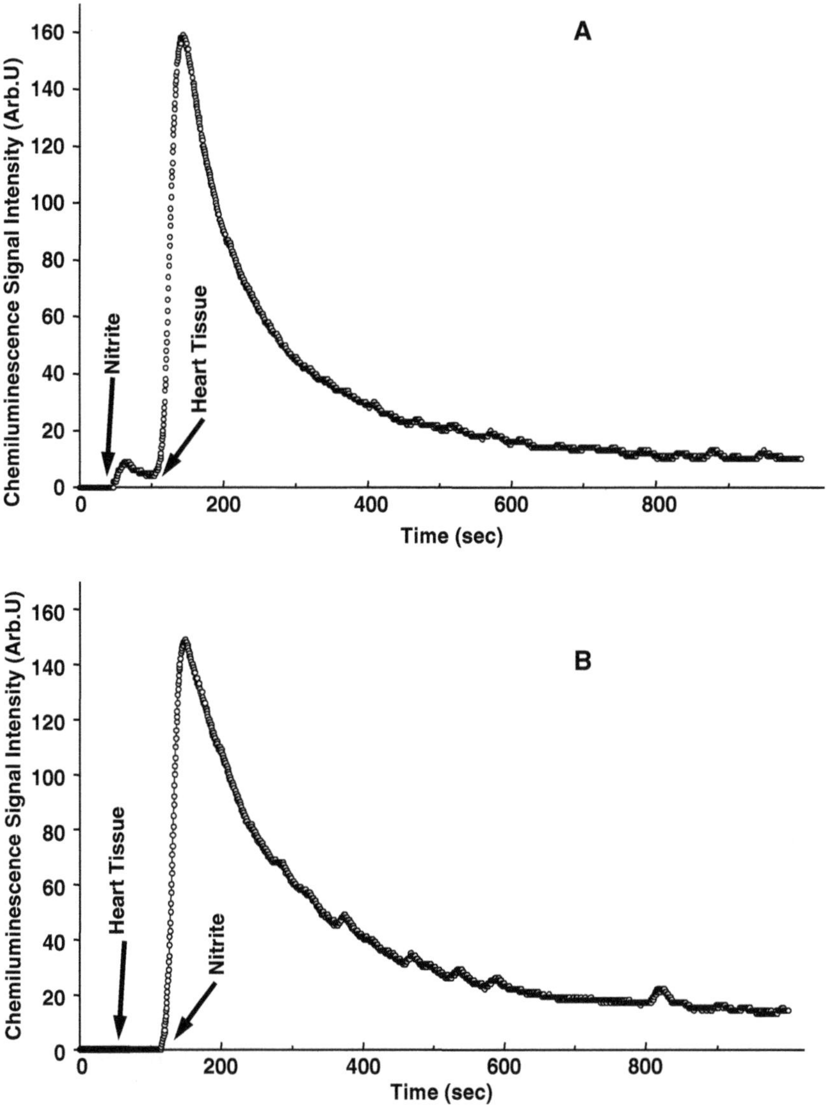

Fig. 1. Effect of ischemic heart tissue homogenate on the magnitude of NO production from nitrite. (A) Chemiluminescence measurements were performed with addition of nitrite 1 mM (first arrow) and then 0.1 ml ischemic heart tissue homogenate (second arrow) to 2 ml of phosphate buffer at pH 5.5. Addition of heart homogenate stimulated a marked increase in the rate of NO formation. (B) With addition of heart tissue homogenate prior to nitrite (first arrow), NO generation was not observed; however, after addition of nitrite (second arrow), NO formation occurred identical to that in (A).

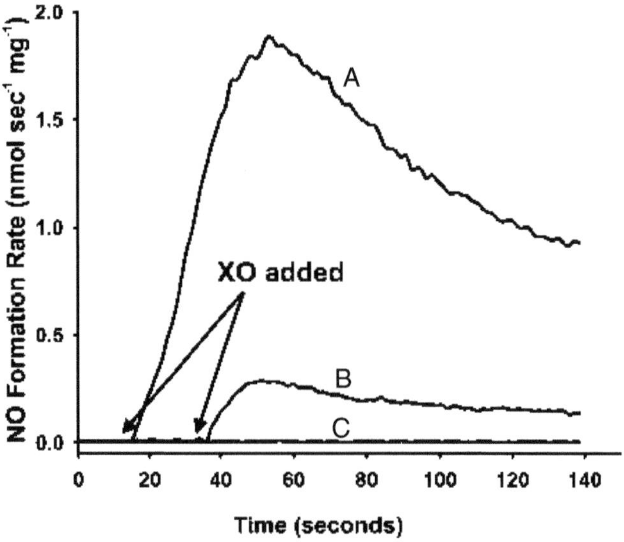

Fig. 2. Measurement of the rate of NO generation from XO-catalyzed nitrite reduction. Measurements were performed using a chemiluminescence NO analyzer under anaerobic conditions at 37°C in PBS, pH 7.4. The *arrows* show the time at which XO (0.02 mg/ml) was added to A and B. Tracing A shows the data for 1.0 mM nitrite and 5 μM xanthine, B shows 1.0 mM nitrite in the presence of 1.0 mM NADH, and C shows 1.0 mM nitrite without XO.

or DBA was unaffected. Thus, whereas xanthine or DBA directly reduces the molybdenum center, NADH initially reduces the flavin, which subsequently transfers electrons to the molybdenum.

Initial studies reported a K_m of 22.9 mM for nitrite reduction in the presence of NADH [37]. A subsequent study reported a K_m value of 2.4 mM [4]. However, the assessment of the K_m by measurement of NADH depletion rather than NO generation reported a K_m of 16 mM [4]. We observed that the K_m for nitrite was consistently about 2.3 mM in the presence of 1.0 mM NADH. This observation agrees closely with the report of Zhang et al. [4]. In contrast with the studies that report only enzyme reduction by NADH and the study reporting a K_m of 36 mM for nitrite in the presence of xanthine, we measured the K_m for nitrite is 2.4 ± 0.2 mM for each of the three types of substrates studied, NADH (1 mM), xanthine (5 mM), and 2,3-dihydroxybenz-aldehyde (DBA 40 mM). Possible reasons for variable results could relate to the conditions used for NO purging from the solutions, leak of oxygen into the measurement system, or partial enzyme inactivation or denaturation. Indeed, in the report, Godber et al. [41] acknowledge that phase equilibration and gas flow factors led them to delay their measurements for 2 min after initiation of the reaction. In their system, they measure the spontaneous liberation of NO into the gas headspace and follow steady state conditions with equilibration of the XO in the presence of NO that is generated. Because only a small fraction of the NO generated is used for detection, this approach limits the sensitivity that can be obtained, and therefore, these studies were performed with high nitrite concentrations in the range of 5–120 mM, about 1000 times above typical tissue levels. In our study, the entire

NO generated were rapidly purged into the gas phase and this enabled efficient NO detection and limited NO-mediated effects on the enzyme, enabling detection of the initial rate of the enzyme with physiological/pharmacological nitrite concentrations of 5 μM to 1 mM. Of note, our parallel measurements of the rates of NO generation performed by NO electrode yielded similar values.

Although xanthine was the highest efficiency reducing substrate of XO-catalyzed nitrite reduction, excessive xanthine exhibited inhibition of NO production. Previous studies reported that enzyme inactivation resulted from NO-induced conversion of XO to its relatively inactive desulfo-form [4]. However, this could not explain why XO kept its activity (95%) when NADH acted as reducing substrate [41]. Also, presence of excess NADH (0.10 mM) or DBA (0.2 mM) had no inhibitive effect on XO-catalyzed NO generation. $ONOO^-$ markedly inhibits XO activity in dose-dependent manner, whereas NO from NO gas in concentrations up to 200 mM had no effect [42]. So inactivation of XO [41] could be caused by $ONOO^-$ formation triggered upon exposure to oxygen at the time of spectrophotometric activity assay.

The mechanism of NO formation occurs due to nitrite reduction at the molybdenum site, with either NADH or xanthine serving as reducing substrates. Since oxypurinol inhibits substrate binding at the molybdenum site of the enzyme, this suggests that nitrite binds to the reduced Mo site. Diphenyleneiodonium (DPI), which acts at the flavin adenine dinucleotide (FAD) reaction site, inhibited XO-dependent nitrite reduction only when NADH was used as the reducing substrate and it did not inhibit NO generation when xanthine was used (Fig. 3). This suggests that NADH donates electrons to FAD, and then electrons are transported back to

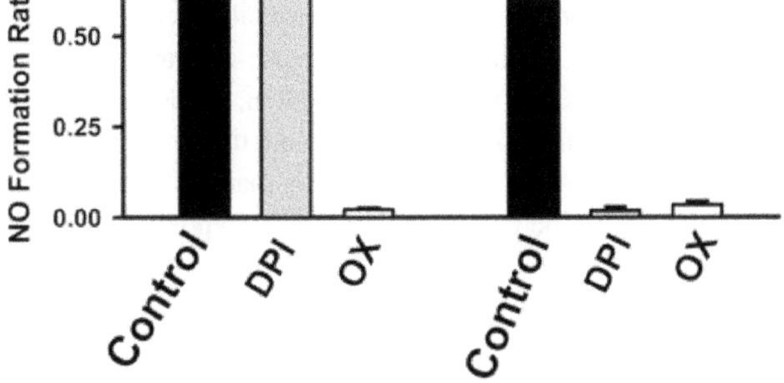

Fig. 3. Effect of site-specific inhibitors on XO-mediated NO formation. The inhibitive effect of oxypurinol, which binds to the molybdenum site, and DPI, which modifies the flavin, were determined for xanthine (X) or NADH-mediated NO generation. For the *left set of bars*, experiments were performed with 0.5 mM nitrite, 5 μM xanthine, and 0.02 mg/ml XO, and for the *right set of bars*, experiments were performed with 1.0 mM nitrite, 1.0 mM NADH, and 0.04 mg/ml XO. Control, without inhibitor; DPI, 20 μM; oxypurinol, 20 μM.

reduce the Mo that in turn reduces nitrite to NO. When xanthine or aldehydes are the electron donors, both XO reduction (by xanthine or aldehydes) and oxidation (by nitrite) takes place at the Mo site so that only oxypurinol could inhibit XO-dependent NO formation. Although xanthine is the highest efficiency reducing substrate of XO-catalyzed nitrite reduction, excessive xanthine exhibits inhibition of NO production by binding to the molybdenum site of the reduced enzyme [43,44], thus blocking the binding of nitrite at this enzyme site. This xanthine-mediated inhibition, which has also been demonstrated by Godber et al. [41], may explain the prior failure to detect XO-mediated NO generation from nitrite in studies in which 150 mM xanthine were used [37].

It has been reported that purine and aldehyde substrate hydroxylation takes place *via* a base-catalyzed mechanism and that substrate must be protonated for hydroxylation [41]. The rate of XO reduction by purine and aldehydes greatly increases when the pH value is increased from 6.0 to 8.0, and this increased rate of XO reduction will lead to an increased rate of nitrite reduction. However, experiments showed that acidic conditions promote XO-catalyzed nitrite reduction. NO generation increased as the pH was decreased from 8.0 down to 7.4 or from 7.4 down to 6.0 (see Table 1), suggesting that nitrite reduction takes place *via* an acid-catalyzed mechanism, presumably due to nitrite protonation. HNO_2 concentration increases when the pH decreases, and it could be the direct binding substrate of XO. Although the decrease of pH would decrease the rate of XO reduction by reducing substrates, it would greatly increase the speed of XO oxidation by nitrite/HNO_2.

From the studies performed, it is clear that XO can catalyze the process of NO generation from nitrite under anaerobic or markedly hypoxic conditions similar to those occurring in ischemic tissues. The key questions are, what is the magnitude of this process, and whether the levels of NO produced are likely to have functional significance.

The activity of XO in the postischemic rat heart is 16.8 milliunits/g of protein [17], which corresponds to 0.013 mg of XO/g of protein or ~3.4 mg/g of cell water. The total XO and xanthine dehydrogenase (XDH) activity, however, is 10-fold above this value. In the ischemic heart, xanthine levels rise from near zero to values in the order of 10–100 mM, and nitrite levels are ~10 mM [1,2,23]. At normal pH values of 7.4, the rate of nitrite degradation due to simple chemical disproportionation is ~0.05 pM/s, as previously reported [35], whereas the rate of XO-catalyzed nitrite reduction would be ~100 pM/s. It has been previously reported in studies from rat heart homogenates that maximally activated nitric oxide synthase produces 1.5 nM/s of NO. Thus xanthine oxidoreductase-mediated NO generation could approach that of the maximal NO production from NOS. Under conditions with increased tissue nitrite concentrations, the magnitude of NO production from this pathway would be further increased; however, it is also clear that with marked elevations in xanthine, it would be inhibited. Because under the acidic and markedly hypoxic conditions

Table 1 Effect of pH on NO generation rate ($nmol \cdot mg^{-1} \cdot s^{-1}$) from 1 mM nitrite in the presence of 0.02 mg/ml XO

pH	6.0	7.4	8.0
Xanthine (10 μM)	2.15 ± 0.10	1.87 ± 0.09	0.34 ± 0.03
NADH (1 mM)	0.7 ± 0.05	0.3 ± 0.03	0.11 ± 0.01
DBA (0.1 mM)	1.96 ± 0.10	0.76 ± 0.05	0.45 ± 0.04

occurring during ischemia, NOS does not function to synthesize NO, this NOS-independent NO generation could be of particular importance. Indeed, it has been shown that the acidosis occurring during ischemia results in reversible denaturation of NOS, which progresses to irreversible denaturation and enzyme degradation [36].

Overall, it is clear that XO-mediated NO generation can potentially be an important source of NO under ischemic conditions in biological tissues that contain substantial levels of the enzyme along with nitrite and reducing substrates. In tissues such as the liver and gastrointestinal tract, which contain high levels of the enzyme, this could be even more pronounced than for the example of the heart considered above [45]. Beyond the obligatory need for the enzyme, the levels of tissue nitrite and enzyme-reducing substrates have a critical role in controlling this process. Nitrite is required, and overall it is the most limiting substrate, because its K_m is ~2.5 mM, whereas typical tissue levels of nitrite are at least 2 orders of magnitude below this value. A number of factors that increase tissue nitrite levels, such as prior activation of constitutive or inducible NOS in inflammatory conditions, dietary sources, pharmacological sources, or bacterial sources, could all modulate this pathway of NO generation [19,46–52]. This pathway also requires a reducing substrate, such as NADH or xanthine. Xanthine was the most effective substrate, triggering NO generation under anaerobic conditions with a V_{max} 4-fold higher than that of NADH. Although only low xanthine concentrations are required, because its K_m value is about 1.5 mM, high levels of xanthine, above 20 mM, resulted in prominent substrate-mediated inhibition. If particularly high levels of xanthine accumulate, this pathway would be inhibited, and perhaps this may serve as a regulatory role to prevent overproduction of NO. Thus, XO can be an important source of NOS-independent NO generation. Under anaerobic conditions, XO reduces nitrite to NO at the molybdenum site of the enzyme with xanthine, NADH, or aldehyde substrates serving to provide the requisite reducing equivalents. The substrate-dependent rate relationship for anaerobic nitrite reduction by XO was determined, and it was demonstrated that under conditions of tissue ischemia, the rate of NO generation is greatly increased above the rate of nitrite disproportionation. This NO production from the enzyme could serve as an alternative source of NO under ischemic conditions in which NO production from NOS is impaired. NO derived from nitrite would accumulate during ischemia. Initially, it could serve to provide protection *via* compensatory vasodilation, whereas upon reperfusion it would react with superoxide, forming peroxynitrite, which can result in protein nitration and cellular injury [53,54].

EFFECT OF OXYGEN ON XO-MEDIATED NITRIC OXIDE GENERATION FROM NITRITE [2]

It is clear that XO-mediated nitrite and nitrate reduction occurs and can be an important source of NO, particularly under conditions of limited tissue perfusion and resulting acidosis. However, questions remained regarding whether XO-mediated NO generation also occurs in the presence of oxygen. In mammalian organs under normoxic conditions, O_2 concentration ranges from 10 to 0.5%, with values of ~14% in arterial blood and ~5% in the myocardium. During mild hypoxia, myocardial O_2 levels drop to ~1–3% or lower [55]. Therefore, combined studies using EPR, chemiluminescence NO analyzer, and NO electrode technique were performed to measure the magnitude and kinetics of XO-mediated NO formation under different oxygen tensions. NO chemiluminescence detection from the reaction mixtures was done in a glass-purging vessel equipped with pressure-monitoring device, which allowed the maintenance of atmospheric pressure inside the purging vessel. Because purging of the released NO was performed using gas mixtures containing variable concentrations of oxygen, studies were performed to estimate the applicability of chemiluminescence measurements in the presence of oxygen. Compared with measurements performed with argon (100%), the efficiency of NO measurement with air, 10%, 5%, or 2% oxygen/nitrogen, was 91.9%, 92.5%, 96.3%, or 98.6%, respectively. Chemiluminescence linearly increased with NO generation when purging with any of these gas mixtures. Alternatively, the reaction solution (or heart tissue) for nitrite reduction was placed in purging vessel 1, and the NO released was purged out by flow of inert gas (argon) or oxygen mixtures of known concentrations to purging vessel 2 (Fig. 4). Purging vessel 2 (trap vessel) was filled with aqueous solutions containing the ferrous iron complex of MGD (*N*-methyl-D-glucamine dithiocarbamate), Fe–MGD, which forms the stable, water-soluble mononitrosyl adduct $(MGD)_2$–Fe–NO, that exhibits a characteristic triplet EPR spectrum at $g = 2.04$ and $aN = 12.8$. This setup was designed to isolate the reaction solution in the purging vessel from the spin trap and thus avoid any possible perturbation caused by the reaction of $(MGD)_2$–Fe with nitrite or with the enzyme [56,57]. Samples from purging vessel 2 were aliquoted at desired times for EPR quantitative measurements.

All three typical reducing substrates of XO triggered NO generation from XO-mediated nitrite reduction; however, their kinetics are quite different in the presence of molybdenum-site-binding substrates xanthine or DBA, compared with that in the presence of the FAD-site-binding substrate NADH. With xanthine or DBA as reducing substrates that donate electrons to XO at the molybdenum site of enzyme, the rate of NO production followed typical Michaelis–Menten kinetics, and kinetic studies show that oxygen acts as a strong competitive inhibitor of nitrite reduction.

Under aerobic conditions, with xanthine or DBA as reducing substrates, XO-mediated NO production is less than 10% of NO production under anaerobic conditions [2]. With the FAD site binding reducing substrate, NADH, as electron donor, XO-mediated NO production is maintained at more than 70% of the anaerobic levels, and the XO-catalyzed NO generation rate only changes from $\sim$0.30 nmol·mg^{-1}s^{-1} under anaerobic conditions to $\sim$0.22 nmol·mg^{-1}s^{-1} under aerobic conditions in the presence of the same enzyme and substrate concentrations. With NADH, under aerobic conditions, XO-mediated nitrite reduction did not follow Michaelis–Menten kinetics. NADH serves as electron donor to XO at the FAD site, the same site as that for oxygen binding, whereas nitrite reduction takes place at the molybdenum site of the enzyme [2]. With NADH as reducing substrate, the possible XO-mediated NO generation may occur through two possible processes as shown on Scheme 1.

In Process I, XO is in the reduced state. With FAD site free, both oxygen and nitrite can accept electrons from reduced XO. Thus, under aerobic conditions, oxygen is a strong competitive inhibitor to XO-mediated nitrite reduction in Process I. NO generation through Process I would decrease greatly in the presence of oxygen. However, in Process II, the FAD site is occupied by the binding of NADH, thus oxygen reduction is totally blocked; meanwhile, at the molybdenum site, XO-mediated nitrite reduction is unaffected. Thus, the rate of XO-mediated nitrite reduction would be similar in the presence or absence of oxygen.

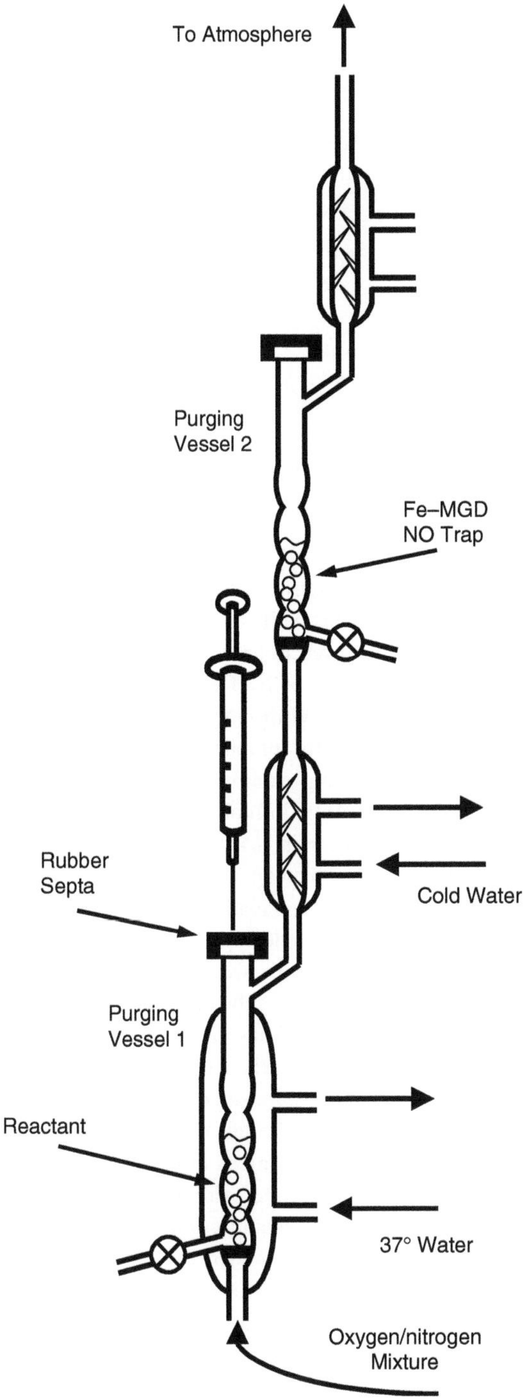

Fig. 4. Experimental setup used for EPR and chemiluminescence NO measurements. NO generated from XO under aerobic conditions in purging vessel 1 was trapped in purging vessel 2, which contained NO spin-trap Fe–MGD; this spin-trap solution was then measured by EPR spectroscopy.

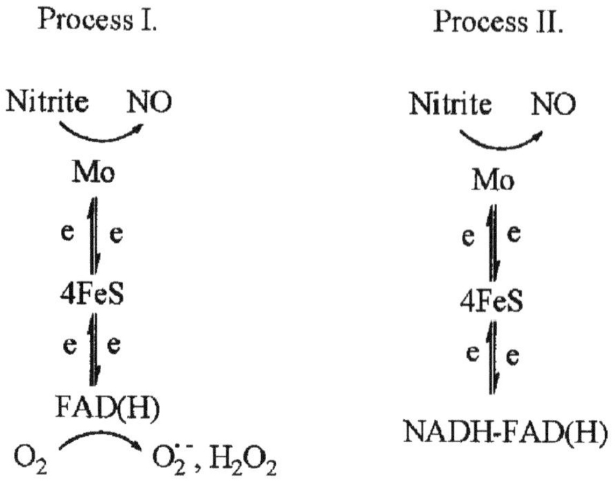

Scheme 1. Two possible processes for XO-mediated NO generation with NADH as reducing substrate. (Explanation in the text.)

In Process II, under aerobic conditions, less than 30% of the nitrite reductase activity of XO is inhibited, which suggests that most nitrite reduction happens while the FAD site is occupied with NADH. NADH is necessary for many biochemical reactions within the body and is found in every living cell. Brain cells contain about 50 μg of NADH per gram of tissue, and heart cells contain 90 μg of NADH per gram of tissue. With molybdenum-site binding electron donors xanthine or DBA, nitrite reduction is greatly inhibited by the presence of oxygen, whereas with NADH, XO-mediated NO generation remains at more than 70% of anaerobic levels. The relatively high concentration of NADH in biological systems and its inhibitive effects on the binding of oxygen strongly suggest that NADH would be the major electron donor for XO-catalyzed NO production under aerobic conditions.

Interestingly, DPI, the inhibitor of FAD site-related function, greatly increased NO generation under aerobic conditions with xanthine or DBA used as reducing substrate. It is known that oxypurinol blocks the binding of xanthine, DBA, and nitrite, whereas DPI inhibits the reduction of XO by NADH. With xanthine or DBA as reducing substrates, the presence of DPI inhibits XO-mediated oxygen reduction at the FAD site and thus increases the capability of the enzyme for nitrite reduction at the molybdenum site. Both the reduction of nitrite and the oxidation of xanthine and DBA take place on the molybdenum site of XO. The potential effects of DPI in stimulating NO generation from XO should be taken into account when DPI is used in biological systems, especially when high concentrations of nitrite are present.

In contrast to the superoxide generation, where maximum superoxide production occurs at alkali conditions (pH 8–9), the aerobic XO-mediated NO generation rates increase more than 10 times when pH values fall from 8.0 to 6.0, and further increase about 3-fold as pH values decrease from 6.0 to 5.0. With lower pH, a more rapid increase of XO-mediated NO generation rate was observed under aerobic conditions than under anaerobic conditions. This would be expected, because under aerobic conditions, the acidosis would significantly increase XO-mediated nitrite reduction and simultaneously inhibit the competitive reaction of oxygen reduction, thus facilitating NO generation under aerobic conditions. The simultaneous production of NO and superoxide can form the potent oxidant peroxynitrite. In the setting of inflammatory disease or pharmacological treatment with organic nitrates or NO-donating compounds, nitrite concentrations can rise by more than an order of magnitude [19,46–52]. Without the protection of antioxidants or antioxidant enzymes, accumulated nitrite can become an important source of peroxynitrite production that can damage cells or tissues. Superoxide dismutase in biological systems is an extremely potent antioxidant enzyme that is responsible for the elimination of cytotoxic active oxygen by catalyzing the dismutation of the superoxide radical to oxygen and hydrogen peroxide. Because NO is readily inactivated by superoxide, the bioactivity of NO is dependent upon the local activity of SOD [58]. Also, there are numerous peroxynitrite scavengers, such as uric acid and NADH, in biological systems [42,59]. These results suggest that under aerobic conditions, NADH would be the main electron donor for XO-catalyzed NO production in mammalian cells and tissues. During ischemia, the myocardial $NADH/NAD^+$ concentration ratio can increase more than 10-fold [60], xanthine levels can rise to the level of 10–100 μM, with nitrite levels of about 10 μM [19,20]; the low oxygen pressure and acidosis greatly facilitate XO-mediated NO generation and limit superoxide production. The magnitude of XO-mediated NO generation can approach that of the maximal NO production from NOS [40]. Even with mild to moderate levels of hypoxia, as can occur with subtotal coronary lesions or regional ischemia in the presence of collateral flow, this process would be stimulated. This could allow NO to accumulate and exert a vasodilator role during ischemia. Upon reperfusion, the accumulated NO would react with XO-derived superoxide, giving a burst of peroxynitrite production that can mediate protein nitration and cellular injury [54]. Thus, provided that the environmental conditions are appropriate, it may be possible that XOR acts as a salvage pathway to maintain levels of NO in situations where conventional constitutive NOS activity may be compromised. Such situations would include inflammatory cardiovascular conditions (atherosclerosis) with associated endothelial dysfunction and particularly myocardial infarction [1–7]. Indeed, XOR activity is upregulated during hypoxia [37,61–63], with increasing acidosis [41], and with atherosclerosis. In patients with coronary artery disease, endothelium-bound XOR activity is increased twofold [64].

Thus, XO-mediated NO generation occurs under aerobic conditions as well as under anaerobic conditions. With substrates such as xanthine or DBA that bind at the molybdenum site of the enzyme, oxygen serves as a competitive inhibitor of nitrite reduction, whereas with NADH, which binds at the FAD site, oxygen exerts only a modest inhibition of nitrite reduction. This process of aerobic XO-mediated NO generation is modulated by oxygen tension, pH, nitrite levels, and reducing substrate concentrations.

MEASUREMENT OF NITRIC OXIDE FORMATION IN ISCHEMIC MYOCARDIUM [6,7,54]

The ability of NO to react and to form high affinity paramagnetic complexes with a variety of metal chelators and metalloproteins was used for quantitative measurements of NO generation [35,65]. While nitrosyl-heme formation is an intrinsic NO trap, these complexes are labile in the presence of O_2. The Fe^{2+}-diethyldithiocarbamate (DETC) complex has been proposed as a more stable and O_2-independent trap suitable for measuring NO [66]. This complex has

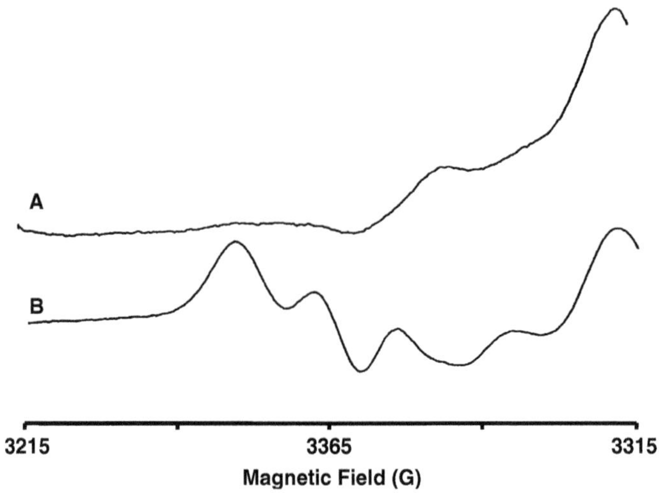

Fig. 5. EPR spectra of NO trapping in hearts labeled with Fe–MGD. (A) Frozen tissue from a normally perfused heart; (B) frozen tissue from a heart subjected to 30 min of ischemia.

limited solubility in water, therefore, the ferrous iron complex of *N*-methyl-D-glucamine-dithiocarbamate (MGD), Fe^{2+}–MGD_2 (Fe–MGD), was also developed and has been applied for measuring NO in living tissues [66]. Alterations in NO generation had been hypothesized to be a critical cause of injury in the ischemic heart; however, the alterations in NO which occur were unknown. Therefore, we performed EPR studies measuring NO in isolated rat hearts subjected to global ischemia, using Fe–MGD, which binds NO giving rise to a characteristic triplet EPR spectrum with $g = 2.04$, $aN = 13.2$ G. While only a small triplet signal was observed in normally perfused, control heart (Fig. 5A), a tenfold increase in this signal occurred after 30 min ischemia indicating an NO formation (Fig. 5B). NO formation increased as a function of the duration of the ischemia [3,65]. With short ischemic durations of 30 min or less, NO generation was decreased by the nitric oxide synthase (NOS) blocker L-NAME [6,7]. Blockade of NO generation with L-NAME also resulted in increased recovery of contractile function after reperfusion [3]. At 1 mM concentration, L-NAME, totally inhibits the enzyme. Infusion of 1 mM L-NAME into normally perfused hearts also caused maximum depression of the coronary flow. Even though L-NAME totally blocked NOS-mediated vasorelaxation, formation of the NO triplet signal seen in ischemic hearts after 30 min or longer was only partially inhibited, with a 60–80% decrease. With higher L-NAME concentrations or with L-NMMA, no further inhibition occurred. This lack of total inhibition of NO formation with blockade of NOS suggested the existence of a NOS independent pathway of NO formation [1,40,65]. To confirm this, direct measurements of NO formation *via* its binding to intrinsic met allo-heme centers within the tissue were also performed. The binding of NO to heme proteins gives rise to a unique EPR spectrum with axial symmetry and with a 5-coordinate complex, a characteristic inverted triplet is seen due to the hyperfine coupling of the nitrogen nucleus of bound NO. These complexes are inherently O_2 labile and are best observed at low O_2 tensions as occur with prolonged ischemia. In control hearts, no NO-heme signal was seen; however, in ischemic tissue after 2–12 h, increasingly prominent NO-heme signal was seen, confirming that increased NO formation occurs during myocardial ischemia. Pretreatment of hearts with L-NAME partially decreased these NO signals with only a 30–50% decrease after 4–8 h of ischemia, further supporting the existence of a NOS-independent pathway of NO generation. Myocardial ischemia results in intracellular acidosis and severe hypoxia leading to a highly reduced state that could cause nitrite reduction to NO. To determine if nitrite, NO_2^-, was reduced during ischemia to form NO, experiments were performed measuring NO with Fe–MGD in hearts subjected to ischemia in the presence of isotopically labeled $^{15}NO_2^-$ [7]. Since ^{15}N has a nuclear spin of 1/2, doublet hyperfine splitting will be observed in the EPR spectra of NO complexes instead of the triplet splitting observed for the natural abundance ^{14}N that has a nuclear spin of 1. In the normally perfused control hearts, $^{15}NO_2^-$ did not result in significant NO formation, however in labeled hearts which were subjected to 30 min of ischemia, marked ^{15}NO formation was seen. In matched experiments with natural abundance $^{14}NO_2^-$, large ^{14}NO triplet signals were seen. Thus, NO_2^- is reduced to NO in the ischemic heart. Nitrosyl-heme formation was also measured in ischemic hearts labeled with 1 mM $^{15}NO_2^-$. A prominent doublet nitrosyl-heme signal was seen due to the formation and binding of ^{15}NO to these proteins, further confirming that NO is generated from NO_2^- (Fig. 6B). With $^{14}NO_2^-$, a similar magnitude triplet ^{14}NO signal was observed. Further experiments performed measuring the concentration of nitrite within the heart prior

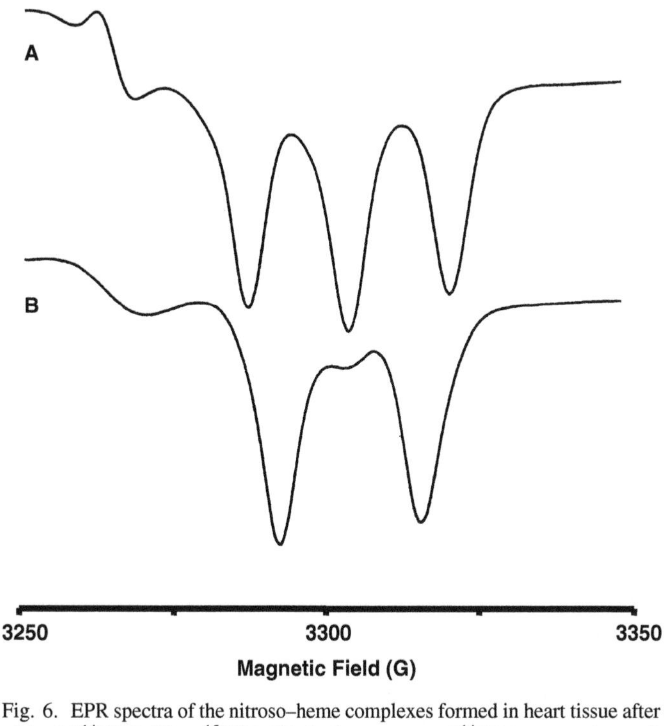

Fig. 6. EPR spectra of the nitroso–heme complexes formed in heart tissue after 8 h of ischemia in the presence of 1 mM $^{14}NO_2^-$ (A) or $^{15}NO_2^-$ (B). In the presence of $^{14}NO_2^-$, a prominent triplet nitroso-heme signal is seen (A), while in the presence of $^{15}NO_2^-$, a doublet signal is observed (B), indicating that the NO formation was directly derived from nitrite.

to ischemia, using an NO analyzer [7,35], showed that relatively large nitrite concentrations of 12 μM, were present.

EVALUATION OF THE ROLE OF NITRITE-DERIVED NITRIC OXIDE IN POSTISCHEMIC INJURY

As was previously shown [67], NO generated from nitric oxide synthase can result in a loss of recovery of contractile function after reperfusion. To evaluate the relative importance of enzyme-dependent *vs.* enzyme-independent NO generation in the pathogenesis of postischemic injury, hemodynamic studies were performed in hearts which were subjected to 30 min of global 37°C ischemia followed by 45 min of reperfusion with measurement of the recovery of contractile function. As reported previously [67], L-NAME-treated hearts exhibited significantly higher recovery of contractile function than untreated control hearts. Oxyhemoglobin-treated hearts (oxyhemoglobin is an efficient scavenger of NO), however, exhibited an even higher recovery of contractile function than the L-NAME-treated hearts. While, L-NAME only blocks NO formation from nitric oxide synthase, oxyhemoglobin scavenges NO formed from either pathway. Thus, this data suggests that enzyme independent NO formation contributes to the process of postischemic injury in the reperfused heart. To determine if further enhancing NO production from nitrite would reverse the protection afforded by L-NAME, hearts were also infused with 10 μM nitrite during the 5 min immediately prior to ischemia. It was observed that this nitrite loading further increased postischemic injury and almost totally blocked the protective action of L-NAME (Fig. 7).

Amount of NO generated at the end of ischemia, measured using Fe–MGD, showed invert correlation with the final recovery of contractile function observed after reperfusion. Preischemic infusion with arginine did not significantly increase NO generation and did not alter the recovery of contractile function. This indicates that tissue arginine concentrations remained well above the K_m of nitric oxide synthase even in the absence of arginine infusion. L-NAME pretreatment decreased nitric oxide generation by about 65% and resulted in approximately a twofold increase in the recovery of contractile function. With preischemic infusion of nitrite in addition to L-NAME, NO generation markedly increased and functional injury was restored. With preischemic loading of nitrite, NO generation was increased by 2–3-fold above levels seen in the untreated controls, and marked impairment of contractile function was seen; however, the function was only slightly lower than in the untreated controls. In both of these groups, very severe functional injury was present with more than a 90% loss of contractile function.

Because of the severe, near-complete, extent of this injury, further increases in NO generation from nitrite may only result in further small decreases in functional recovery which were seen. Alternatively, since NO-mediated injury can be due to its reaction with superoxide to form the potent oxidant peroxynitrite [53], it is possible that the magnitude of tissue injury could reach a maximum value when the concentrations of NO approach those of superoxide. With oxyhemoglobin treatment, NO was almost totally quenched with more than a 3-fold increase in the recovery of contractile function above the values observed in untreated controls. Thus, this data further confirms the presence and functional importance of enzyme-independent NO generation. In general, it is clear from these studies that both

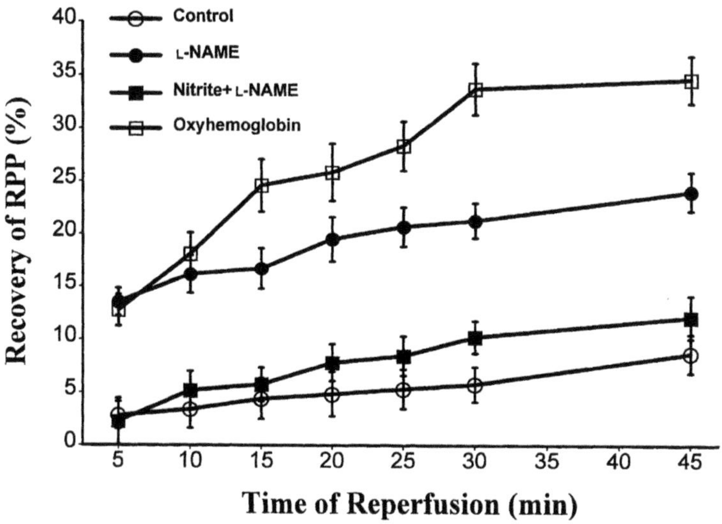

Fig. 7. Assessment of the importance of enzyme-dependent *vs.* enzyme-independent NO formation in the pathogenesis of postischemic injury. Four groups of hearts were subjected to 30 min of global 37°C ischemia followed by 45 min of reperfusion. The four groups were as follows: (1) untreated control; (2) pretreated with 1.0 mM L-NAME for the 5 min prior to the onset of ischemia to block nitric oxide synthase; (3) treated with 10 μM oxyhemoglobin 1 min prior to ischemia as well as the first 5 min of reflow, to scavenge NO generated by either pathway; and (4) pretreated with both 10 μM nitrite and 1.0 mM L-NAME, to determine if NO_2^- can modulate the protective effects of enzyme inhibition. Left ventricular pressures and heart rate were measured using a left ventricular balloon. The graph shows the recovery of the product of left ventricular developed pressure and heart rate (RPP), an index of cardiac contractile work, expressed as % of baseline preischemic values. While inhibition of enzyme-dependent NO formation resulted in cardioprotection, further protection was seen in the presence of oxyhemoglobin which scavenges NO from both pathways. It was observed that supplementation of nitrite within the heart abolished the protection seen with enzyme inhibition.

enzyme-dependent and enzyme-independent NO formation is increased during ischemia and that the magnitude of this NO generation correlated with the severity of functional injury observed upon reperfusion.

These results demonstrate that NOS-independent NO formation not only contributes to postischemic injury, but it may also reverse the protective effects of NOS blockers.

NITROSYL-HEME FORMATION AND NITRIC OXIDE SIGNALING DURING BRIEF MYOCARDIAL ISCHEMIA

In addition to generation from specific nitric oxide (NO) synthases, we have observed that NO formation from nitrite occurs in the ischemic heart. While NO binding to heme-centers is the basis for NO-mediated signaling, as occurs through guanylate cyclase (GC), it was not known if this process is triggered with physiologically relevant periods of sublethal ischemia and if nitrite serves as a critical substrate. Therefore, EPR studies to measure nitrosyl-heme formation during the time course of myocardial ischemia and reperfusion and

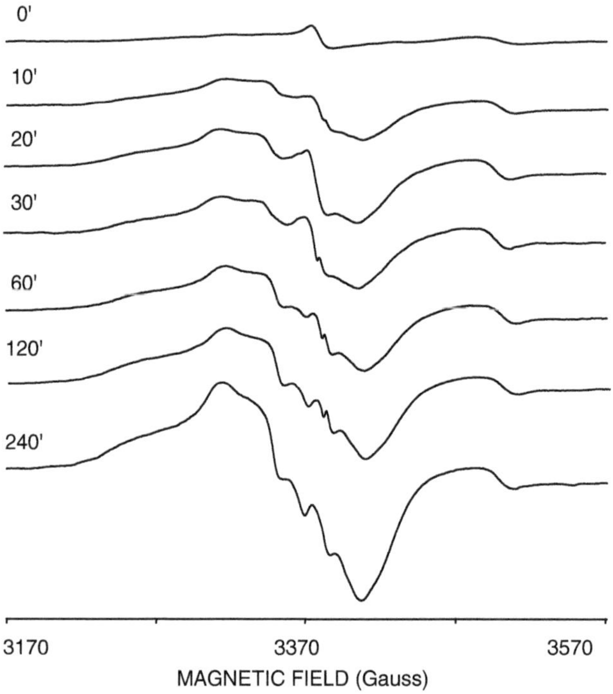

Fig. 8. EPR spectra of NO-heme formation in hearts during ischemia. Top to bottom, ischemic durations of 0–240 min. Spectra were recorded at 77 K with a frequency of 9.786 GHz; 10.0 mW power.

the role of nitrite in this process were performed [3]. Ischemic hearts loaded with nitrite 50 μM showed prominent spectra consisting of six-coordinate nitrosyl-heme complexes, primarily NO–myoglobin, that increased as a function of ischemic duration. In nonischemic-controls these signals were not seen (Fig. 8). Quantitative analysis of EPR spectra showed that total NO-heme concentrations within the heart were 6.6 $\pm$ 0.7 μM after 30 min of ischemia, and 12.7 μM at 240 min. With the increasing nitrite-loading-dose concentrations a linear correlation was observed between dose and amount of Mb–NO formed from 3.04 $\pm$ 0.01 μM in hearts loaded with 25 μM nitrite to 27.0 $\pm$ 0.3 μM in hearts with 400 μM nitrite. NO–Mb in controls (without nitrite load) also progressively increased as a function of ischemic duration but was about 8–10-fold lower than the nitrite-loaded hearts with concentrations of 0.5 or 1.6 μM after 30 or 240 min, respectively.

The levels of NO-heme complexes formed in ischemic myocardium are quite high compared with the concentrations of free NO in cells of 10–100 nM or less [68]. To prove that the observed NO-heme complex formation was derived from nitrite, isotope tracer experiments were performed measuring NO-heme formation in hearts infused with isotopically labeled ^{15}N nitrite. With ^{15}N nitrite labeling, the characteristic centrally split ^{15}NO–Mb spectrum was seen, while with ^{14}N nitrite the typical ^{14}NO–Mb spectrum was observed [3]. Thus, under ischemic conditions myoglobin binds and stabilizes nitrite-derived NO in the form of NO–Mb complexes, and these complexes serve as a store of NO.

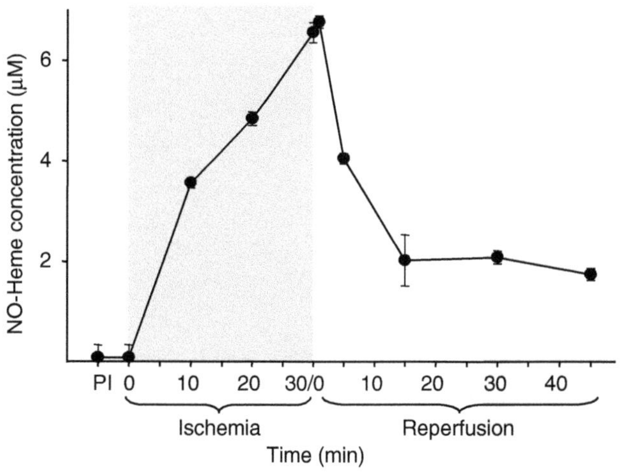

Fig. 9. Time course of NO-heme levels before, during, and following ischemia. From EPR measurements on a series of three nitrite loaded hearts at each preischemic (PI), ischemic, and reperfusion time.

Upon reperfusion the concentration of NO-heme complexes decrease, as would be expected with the reoxygenation that accompanies reperfusion. Initially a rapid decrease was seen in the observed NO-heme EPR spectra over the first 15 min of reperfusion followed by a slow decrease thereafter, with 26% of the signal intensity persisting even after 45 min of reperfusion (Fig. 9). The time course of the early decrease parallels the oxidant burst accompanied by superoxide and superoxide-derived radical generation that occurs in ischemic myocardium. It is possible that the early rapid decrease in NO-heme complexes is due to superoxide or other radical reaction with the bound NO. The subsequent slow decrease could be due to a slow rate of spontaneous oxidation secondary to the levels of oxygen present in the reperfused heart. The decrease in NO-heme signal seen upon reperfusion could be due to oxidation or facilitated release/exchange of the NO bound at the heme site of Mb. With oxidation, nitrate formation would be observed, whereas for released NO, nitrite formation would be expected. In the hearts preloaded with nitrite, nitrite levels were elevated only over the first minute of reperfusion, which can also be explained by simple nitrite washout. However, nitrate levels remained elevated over the first 5 min of reperfusion and exhibited a time course of decrease that paralleled the observed decrease in NO-heme concentrations within the heart. As described above, the observed NO-heme formation is primarily due to the formation of NO–Mb complexes. However, trace amount of five coordinate NO-heme complexes were also seen, as would arise from NO bound to guanylate cyclase [69,70]. Although these could also arise from other heme centers [71,72].

NO binding to the heme of sGC (soluble guanylate cyclase) is the critical event triggering the activation of sGC that results in cyclic guanosine monophosphate (cGMP) formation in the presence of guanosine triphosphate (GTP). It has been previously demonstrated that NO–Mb effectively donates NO-heme to sGC, resulting in activation of the enzyme with cGMP formation [73]. In view of this, we performed experiments to determine whether the observed NO–Mb formation in the ischemic heart was associated with activation of sGC.

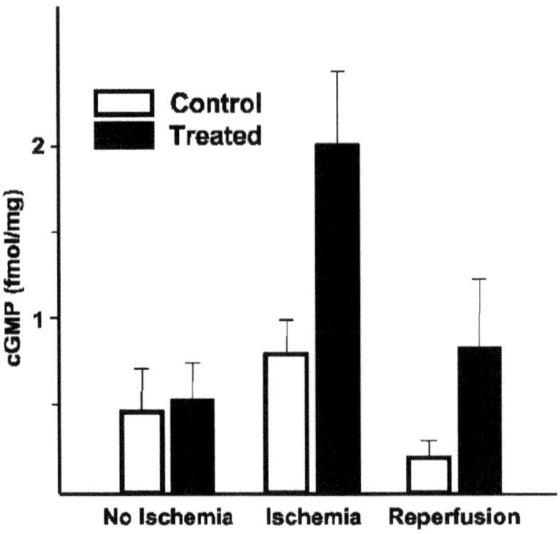

Fig. 10. Levels of cGMP in control and 50 μM nitrite treated hearts. Tissue cGMP levels were measured in normally perfused nonischemic hearts, after 30 min of 37°C global ischemia, or after reperfusion for 15 min following after 30 min of ischemia. Control untreated hearts (white bars) and nitrite (50 μM)-treated hearts (black bars) were studied. Values are expressed in units of fmol/mg of heart tissue, wet weight.

In nonischemic hearts that were either nitrite-treated or untreated, only low levels of cGMP were seen. However, after 30 min of ischemia, a modest increase in cGMP levels was present in untreated hearts, whereas a large 4-fold increase was seen in nitrite-treated hearts compared with preischemic levels. Furthermore, in ischemic hearts the levels of cGMP was 2.5-fold higher with nitrite treatment than in otherwise identical untreated control hearts (Fig. 10). Upon reperfusion the levels of cGMP declined, and after 15 min of reperfusion more than a 2.4-fold decrease in cGMP levels was observed from ischemic values. These changes in cGMP levels paralleled the changes seen for NO heme levels. This observed increase in cGMP levels was independent of NOS, since 1 min of pre-infusion with 1.0 mM *N*–G-monomethyl-L-arginine (NOS inhibitor) before the onset of ischemia did not significantly change the levels of cGMP (compared with 1.6 ± 0.3 fmol/mg wet weight for treated hearts compared with 2.0 ± 0.4 fmol/mg wet weight for hearts without NOS inhibition). Activation of sGC accompanied the formation of NO–Mb complexes and thus, the observed nitrite-derived NO-heme formation is paralleled by activation of myocardial-signaling pathways.

Nitrite-mediated NO-heme formation during ischemia is paralleled by subsequent depressed recovery of contractile function upon reperfusion. In nitrite-treated hearts, the recovery of contractile function as measured by left ventricular-developed pressure and rate-pressure product was significantly decreased, compared with that in untreated control hearts, with final recovery at 30 min diminished by more than 3-fold. In contrast to the impaired recovery of contractile function in nitrite-treated hearts, the recovery of coronary flow was paradoxically mild elevated.

Thus, nitrite-mediated nitrosyl heme formation occurs in ischemic myocardium and can be an important regulator of myocardial signaling and injury in the postischemic heart.

CONCLUSION

Studies from many laboratories including ours have shown that NOS enzyme-independent NO formation occurs in biological systems. The NO formation is a generalized phenomenon which can occur in many biological tissues. In many disease states, such as ischemia or shock, where acidosis and marked hypoxia occur, this pathway becomes the major source of NO and the magnitude of NO generation can be much greater than that which would be formed by normal tissue concentrations of NOS. The substrate source of this NO formation is nitrite, rather than the NOS substrate and a number of pathways of nitrite reduction to NO was proposed, which include nitrite disproportionation, nitrite reduction by myoglobin [3,74–80], reduction by anoxic mitochondria [81,82], by xanthine oxidoreductase [2,4,37,40,41,83], by hemoglobin [84–90]. For in-depth critical review of hemoglobin/RBC mechanism NO production, look at Ref. [91]. Thus, these pathways may “substitute” NO production from NOS in pathophysiological conditions, when NOS function is impaired and conditions might be favorable for nitrite reduction. From first glance, nitrite should display cytoprotective effect, since knockout of endothelial NOS leaves the hearts of mice more sensitive to ischemic damage [92,93]. Protection should be similar to that observed with $NO^{\bullet}$ donors [94–97]. Indeed, in models of infarction, nitrite has been reported to protect against ischemia–reperfusion damage [98–101] in a way similar to NO donors. However, there is contradiction in literature. Studies including ours, in isolated rat hearts, suggest that nitrite-derived cellular NO (or its derivatives) may be contributing to damage [3,7,54,67,102,103] and, therefore, the effects of NO, derived from nitrite in myocardial injury are uncertain [95,96].

Potentially, there are great therapeutical benefits in control of the process of nitrite reduction; however present state of understanding of nitrite biochemistry does allow this control, thus demanding further efforts of research in this field.

REFERENCES

1 Li H, Samouilov A, Liu X, Zweier JL. Characterization of the magnitude and kinetics of xanthine oxidase-catalyzed nitrate reduction: evaluation of its role in nitrite and nitric oxide generation in anoxic tissues. Biochemistry 2003; 42: 1150–1159.

2 Li H, Samouilov A, Liu X, Zweier JL. Characterization of the effects of oxygen on xanthine oxidase-mediated nitric oxide formation. J. Biol. Chem. 2004; 279: 16939–16946.

3 Tiravanti E, Samouilov A, Zweier JL. Nitrosyl-heme complexes are formed in the ischemic heart: evidence of nitrite-derived nitric oxide formation, storage, and signaling in post-ischemic tissues. J. Biol. Chem. 2004; 279: 11065–11073.

4 Zhang Z, Naughton D, Winyard PG, Benjamin N, Blake DR, Symons MC. Generation of nitric oxide by a nitrite reductase activity of xanthine oxidase: a potential pathway for nitric oxide formation in the absence of nitric oxide synthase activity. [erratum appears in Biochemistry Biophys. Res. Commun. 1998; 251(2): 667] Biochemistry Biophys. Res. Comms. 1998; 249: 767–772.

5 Zhang Z, Naughton DP, Blake DR, Benjamin N, Stevens CR, Winyard PG, Symons MC, Harrison R. Human xanthine oxidase converts nitrite ions into nitric oxide (NO). Biochemistry Soc. Trans. 1997; 25: 524S.

6 Zweier JL, Samouilov A, Kuppusamy P. Non-enzymatic nitric oxide synthesis in biological systems. Biochim. Biophys. Acta 1999; 1411: 250–262.

7 Zweier JL, Wang P, Samouilov A, Kuppusamy P. Enzyme-independent formation of nitric oxide in biological tissues. [see comment] Nat. Med. 1995; 1: 804–809.

8 Gladwin MT, Shelhamer JH, Schechter AN, Pease-Fye ME, Waclawiw MA, Panza JA, Ognibene FP, Cannon RO 3rd. Role of circulating nitrite and S-nitrosohemoglobin in the regulation of regional blood flow in humans. Proc. Natl. Acad. Sci. USA 2000; 97: 11482–11487.

9 Lauer T, Preik M, Rassaf T, Strauer BE, Deussen A, Feelisch M, Kelm M. Plasma nitrite rather than nitrate reflects regional endothelial nitric oxide synthase activity but lacks intrinsic vasodilator action. Proc. Natl. Acad. Sci. USA 2001; 98: 12814–12819.

10 Rodriguez J, Maloney RE, Rassaf T, Bryan NS, Feelisch M. Chemical nature of nitric oxide storage forms in rat vascular tissue. Proc. Natl. Acad. Sci. USA 2003; 100: 336–341.

11 Rassaf T, Bryan NS, Maloney RE, Specian V, Kelm M, Kalyanaraman B, Rodriguez J, Feelisch M. NO adducts in mammalian red blood cells: too much or too little? [comment]. Nat. Med. 2003; 9: 481–482; author reply 482–483.

12 Sastry KV, Moudgal RP, Mohan J, Tyagi JS, Rao GS. Spectrophotometric determination of serum nitrite and nitrate by copper-cadmium alloy. Anal. Biochemistry 2002; 306: 79–82.

13 Tari A, Kodama K, Kurihara K, Fujihara M, Sumii K, Kajiyama G. Does serum nitrite concentration reflect gastric carcinogenesis in Japanese Helicobacter pylori-infected patients? Dig. Dis. Sci. 2002; 47: 100–106.

14 Tari A, Kodama K, Kitadai Y, Ohta M, Sumii K, Kajiyama G. Is apoptosis in antral mucosa correlated with serum nitrite concentration in Japanese Helicobacter pylori-infected patients? [see comment] J. Gastroenterol. Hepatol. 2003; 18: 498–504.

15 Oldreive C, Bradley N, Bruckdorfer R, Rice-Evans C. Lack of influence of dietary nitrate/nitrite on plasma nitrotyrosine levels measured using a competitive inhibition of binding ELISA assay. Free Radic. Res. 2001; 35: 377–386.

16 Mcknight GM, Duncan CW, Leifert C, Golden MH. Dietary nitrate in man: friend or foe? Br. J. Nutr. 1999; 81: 349–358.

17 Xia Y, Zweier JL. Direct measurement of nitric oxide generation from nitric oxide synthase. Proc. Natl. Acad. Sci. USA 1997; 94: 12705–12710.

18 Nava E, Palmer RM, Moncada S. The role of nitric oxide in endotoxic shock: effects of NG-monomethyl-L-arginine. J. Cardiovasc. Pharmacol. 1992; 20 (Suppl 12): S132–S134.

19 Evans T, Carpenter A, Silva A, Cohen J. Inhibition of nitric oxide synthase in experimental gram-negative sepsis. J. Infect. Dis. 1994; 169: 343–349.

20 Tracey WR, Tse J, Carter G. Lipopolysaccharide-induced changes in plasma nitrite and nitrate concentrations in rats and mice: pharmacological evaluation of nitric oxide synthase inhibitors. J. Pharmacol. Exp. Ther. 1995; 272: 1011–1015.

21 Ignarro LJ, Gruetter CA. Requirement of thiols for activation of coronary arterial guanylate cyclase by glyceryl trinitrate and sodium nitrite: possible involvement of S-nitrosothiols. Biochim. Biophys. Acta 1980; 631: 221–231.

22 Ignarro LJ, Lippton H, Edwards JC, Baricos WH, Hyman AL, Kadowitz PJ, Gruetter CA. Mechanism of vascular smooth muscle relaxation by organic nitrates, nitrites, nitroprusside and nitric oxide: evidence for the involvement of S-nitrosothiols as active intermediates. J. Pharmacol. Exp. Ther. 1981; 218: 739–749.

23 Moulds RF, Jauernig RA, Shaw J. A comparison of the effects of hydrallazine, diazoxide, sodium nitrite and sodium nitroprusside on human isolated arteries and veins. Br. J. Clin. Pharmacol. 1981; 11: 57–61.

24 Gruetter CA, Kadowitz PJ, Ignarro LJ. Methylene blue inhibits coronary arterial relaxation and guanylate cyclase activation by nitroglycerin, sodium nitrite, and amyl nitrite. Can. J. Physiol. Pharmacol. 1981; 59: 150–156.

25 Gruetter CA, Gruetter DY, Lyon JE, Kadowitz PJ, Ignarro LJ. Relationship between cyclic guanosine 3′:5′-monophosphate formation and relaxation of coronary arterial smooth muscle by glyceryl trinitrate, nitroprusside, nitrite and nitric oxide: effects of methylene blue and methemoglobin. J. Pharmacol. Exp. Ther. 1981; 219: 181–186.

26 Matsunaga K, Furchgott RF. Interactions of light and sodium nitrite in producing relaxation of rabbit aorta. J. Pharmacol. Exp. Ther. 1989; 248: 687–695.

27 Laustiola KE, Vuorinen P, Porsti I, Metsa-Ketela T, Manninen V, Vapaatalo H. Exogenous GTP enhances the effects of sodium nitrite on cyclic GMP accumulation, vascular smooth muscle relaxation and platelet aggregation. Pharmacol. Toxicol. 1991; 68: 60–63.

28 Duncan C, Dougall H, Johnston P, Green S, Brogan R, Leifert C, Smith L, Golden M, Benjamin N. Chemical generation of nitric oxide in the mouth from the enterosalivary circulation of dietary nitrate [see comment] Nat. Med. 1995; 1: 546–551.

29 Dougall HT, Smith L, Duncan C, Benjamin N. The effect of amoxycillin on salivary nitrite concentrations: an important mechanism of adverse reactions? Br. J. Clin. Pharmacol. 1995; 39: 460–462.

30 Benjamin N, O'Driscoll F, Dougall H, Duncan C, Smith L, Golden M, Mckenzie H. Stomach NO synthesis. [see comment] Nature 1994; 368: 502.

31 Lundberg JO, Weitzberg E, Lundberg JM, Alving K. Intragastric nitric oxide production in humans: measurements in expelled air. Gut 1994; 35: 1543–1546.

32 Bjorne HH, Petersson J, Phillipson M, Weitzberg E, Holm L, Lundberg JO. Nitrite in saliva increases gastric mucosal blood flow and mucus thickness. [see comment][erratum appears in J. Clin. Invest. 2004; 113(3): 490] J. Clin. Invest. 2004; 113: 106–114.

33 Ferrari R, Cargnoni A, Bernocchi P, Pasini E, Curello S, Ceconi C, Ruigrok TJ. Metabolic adaptation during a sequence of no-flow and low-flow ischemia. A possible trigger for hibernation. Circulation 1996; 94: 2587–2596.

34 Gabel SA, Cross HR, London RE, Steenbergen C, Murphy E. Decreased intracellular pH is not due to increased H+ extrusion in preconditioned rat hearts. Am. J. Physiol. 1997; 273: H2257–H2262.

35 Samouilov A, Kuppusamy P, Zweier JL. Evaluation of the magnitude and rate of nitric oxide production from nitrite in biological systems. Arch. Biochem. Biophys. 1998; 357: 1–7.

36 Giraldez RR, Panda A, Xia Y, Sanders SP, Zweier JL. Decreased nitric-oxide synthase activity causes impaired endothelium-dependent relaxation in the postischemic heart. J. Biol. Chem. 1997; 272: 21420–21426.

37 Millar TM, Stevens CR, Benjamin N, Eisenthal R, Harrison R, Blake DR. Xanthine oxidoreductase catalyses the reduction of nitrates and nitrite to nitric oxide under hypoxic conditions. FEBS Lett. 1998; 427: 225–228.

38 Godber BL, Doel JJ, Sapkota GP, Blake DR, Stevens CR, Eisenthal R, Harrison R. Reduction of nitrite to nitric oxide catalyzed by xanthine oxidoreductase. J. Biol. Chem. 2000; 275: 7757–7763.

39 Trujillo M, Alvarez MN, Peluffo G, Freeman BA, Radi R. Xanthine oxidase-mediated decomposition of S-nitrosothiols. J. Biol. Chem. 1998; 273: 7828–7834.

40 Li H, Samouilov A, Liu X, Zweier JL. Characterization of the magnitude and kinetics of xanthine oxidase-catalyzed nitrite reduction. Evaluation of its role in niitric oxide generation in anoxic tissues. J. Biol. Chem. 2001; 276: 24482–24489.

41 Godber BLJ, Doel JJ, Sapkota GP, Blake DR, Stevens CR, Eisenthal R, Harrison R. Reduction of nitrite to nitric oxide catalyzed by xanthine oxidoreductase. J. Biol. Chem. 2000; 275: 7757–7763.

42 Lee CI, Liu X, Zweier JL. Regulation of xanthine oxidase by nitric oxide and peroxynitrite. J. Biol. Chem. 2000; 275: 9369–9376.

43 Hille R, Stewart RC. The inhibition of xanthine oxidase by 8-bromoxanthine. J. Biol. Chem. 1984; 259: 1570–1576.

44 Rubbo H, Radi R, Prodanov E. Substrate inhibition of xanthine oxidase and its influence on superoxide radical production. Biochim. Biophys. Acta 1991; 1074: 386–391.

45 Sarnesto A, Linder N, Raivio KO. Organ distribution and molecular forms of human xanthine dehydrogenase/xanthine oxidase protein. Lab. Invest. 1996; 74: 48–56.

46 Evans TJ, Buttery LD, Carpenter A, Springall DR, Polak JM, Cohen J. Cytokine-treated human neutrophils contain inducible nitric oxide synthase that produces nitration of ingested bacteria. Proc. Natl. Acad. Sci. USA 1996; 93: 9553–9558.

47 Farrell AJ, Blake DR. Nitric oxide. Ann. Rheum. Dis. 1996; 55: 7–20.

48 Farrell AJ, Blake DR, Palmer RM, Moncada S. Increased concentrations of nitrite in synovial fluid and serum samples suggest increased nitric oxide synthesis in rheumatic diseases. Ann. Rheum. Dis. 1992; 51: 1219–1222.

49 Hukkanen M, Hughes FJ, Buttery LD, Gross SS, Evans TJ, Seddon S, Riveros-Moreno V, Macintyre I, Polak JM. Cytokine-stimulated expression of inducible nitric oxide synthase by mouse, rat, and human osteoblast-like cells and its functional role in osteoblast metabolic activity. Endocrinology 1995; 136: 5445–5453.

50 Torre D, Ferrario G, Speranza F, Martegani R, Zeroli C. Increased levels of nitrite in the sera of children infected with human immunodeficiency virus type 1. Clin. Infect. Dis. 1996; 22: 650–653.
51 Torre D, Ferrario G, Speranza F, Orani A, Fiori GP, Zeroli C. Serum concentrations of nitrite in patients with HIV-1 infection. J. Clin. Pathol. 1996; 49: 574–576.
52 Torre D, Zeroli C, Ferrario G, Pugliese A, Speranza F, Orani A, Casari S, Bassi P, Poggio A, Carosi GP, Fiori GP. Levels of nitric oxide, gamma interferon and interleukin-12 in AIDS patients with toxoplasmic encephalitis. Infection 1999; 27: 218–220.
53 Beckman JS, Beckman TW, Chen J, Marshall PA, Freeman BA. Apparent hydroxyl radical production by peroxynitrite: implications for endothelial injury from nitric oxide and superoxide. Proc. Natl. Acad. Sci. USA 1990; 87: 1620–1624.
54 Wang P, Zweier JL. Measurement of nitric oxide and peroxynitrite generation in the postischemic heart. Evidence for peroxynitrite-mediated reperfusion injury. J. Biol. Chem. 1996; 271: 29223–29230.
55 Roy S, Khanna S, Bickerstaff AA, Subramanian SV, Atalay M, Bierl M, Pendyala S, Levy D, Sharma N, Venojarvi M, Strauch A, Orosz CG, Sen CK. Oxygen sensing by primary cardiac fibroblasts: a key role of p21(Waf1/Cip1/Sdi1). Circ. Res. 2003; 92: 264–271.
56 Tsuchiya K, Jiang JJ, Yoshizumi M, Tamaki T, Houchi H, Minakuchi K, Fukuzawa K, Mason RP. Nitric oxide-forming reactions of the water-soluble nitric oxide spin-trapping agent, MGD. Free Radi. Biol. Med. 1999; 27: 347–355.
57 Tsuchiya K, Yoshizumi M, Houchi H, Mason RP. Nitric oxide-forming reaction between the iron-N-methyl-D-glucamine dithiocarbamate complex and nitrite. J. Biol. Chem. 2000; 275: 1551–1556.
58 Lynch SM, Frei B, Morrow JD, Roberts LJ 2nd, Xu A, Jackson T, Reyna R, Klevay LM, Vita JA, Keaney Jr. JF. Vascular superoxide dismutase deficiency impairs endothelial vasodilator function through direct inactivation of nitric oxide and increased lipid peroxidation. Arterioscler. Thromb. Vasc. Biol. 1997; 17: 2975–2981.
59 Kirsch M, De Groot H. Reaction of peroxynitrite with reduced nicotinamide nucleotides, the formation of hydrogen peroxide. J. Biol. Chem. 1999; 274: 24664–24670.
60 Salem JE, Saidel GM, Stanley WC, Cabrera ME. Mechanistic model of myocardial energy metabolism under normal and ischemic conditions. Ann. Biomed. Eng. 2002; 30: 202–216.
61 Hassoun PM, Yu FS, Shedd AL, Zulueta JJ, Thannickal VJ, Lanzillo JJ, Fanburg BL. Regulation of endothelial cell xanthine dehydrogenase xanthine oxidase gene expression by oxygen tension. Am. J. Physiol. 1994; 266: L163–L171.
62 Kayyali US, Donaldson C, Huang H, Abdelnour R, Hassoun PM. Phosphorylation of xanthine dehydrogenase/oxidase in hypoxia. J. Biol. Chem. 2001; 276: 14359–14365.
63 Poss WB, Huecksteadt TP, Panus PC, Freeman BA, Hoidal JR. Regulation of xanthine dehydrogenase and xanthine oxidase activity by hypoxia. Am. J. Physiol. 1996; 270: L941–L946.
64 Spiekermann S, Landmesser U, Dikalov S, Bredt M, Gamez G, Tatge H, Reepschlager N, Hornig B, Drexler H, Harrison DG. Electron spin resonance characterization of vascular xanthine and NAD(P)H oxidase activity in patients with coronary artery disease: relation to endothelium-dependent vasodilation. Circulation 2003; 107: 1383–1389.
65 Kuppusamy P, Wang P, Samouilov A, Zweier JL. Spatial mapping of nitric oxide generation in the ischemic heart using electron paramagnetic resonance imaging. Magn. Reson. Med. 1996; 36: 212–218.
66 Vanin AF, Liu X, Samouilov A, Stukan RA, Zweier JL. Redox properties of iron-dithiocarbamates and their nitrosyl derivatives: implications for their use as traps of nitric oxide in biological systems. Biochim. Biophys. Acta 2000; 1474: 365–377.
67 Zweier JL, Wang P, Kuppusamy P. Direct measurement of nitric oxide generation in the ischemic heart using electron paramagnetic resonance spectroscopy. J. Biol. Chem. 1995; 270: 304–307.
68 Moncada S. Nitric oxide in the vasculature: physiology and pathophysiology. Ann. N.Y. Acad. Sci. 1997; 811: 60–67; Discussion 67–69.
69 Stone JR, Sands RH, Dunham WR, Marletta MA. Electron paramagnetic resonance spectral evidence for the formation of a pentacoordinate nitrosyl-heme complex on soluble guanylate cyclase. Biochem. Biophys. Res. Commun. 1995; 207: 572–577.
70 Zhao Y, Hoganson C, Babcock GT, Marletta MA. Structural changes in the heme proximal pocket induced by nitric oxide binding to soluble guanylate cyclase. Biochemistry 1998; 37: 12458–12464.

71 Henry Y, Lepoivre M, Drapier JC, Ducrocq C, Boucher JL, Guissani A. EPR characterization of molecular targets for NO in mammalian cells and organelles. FASEB J. 1993; 7: 1124–1134.

72 Henry YA, Guissani A, Ducastel B. Nitric Oxide Research from Chemistry to Biology: EPR Spectroscopy of Nitrosylated Compounds. Springer Verlag, Austin, TX, 1997, pp. 205–233.

73 Ignarro LJ, Adams JB, Horwitz PM, Wood KS. Activation of soluble guanylate cyclase by NO-hemoproteins involves NO-heme exchange. Comparison of heme-containing and heme-deficient enzyme forms. J. Biol. Chem. 1986; 261: 4997–5002.

74 Laverman LE, Wanat A, Oszajca J, Stochel G, Ford PC, Van Eldik R. Mechanistic studies on the reversible binding of nitric oxide to metmyoglobin. J. Am. Chem. Soc. 2001; 123: 285–293.

75 Wanat A, Gdula-Argasinska J, Rutkowska-Zbik D, Witko M, Stochel G, Van Eldik R. Nitrite binding to metmyoglobin and methemoglobin in comparison to nitric oxide binding. J. Biol. Inorg. Chem. 2002; 7: 165–176.

76 Wanat A, Schneppensieper T, Stochel G, Van Eldik R, Bill E, Wieghardt K. Kinetics, mechanism, and spectroscopy of the reversible binding of nitric oxide to aquated iron(II). An undergraduate text book reaction revisited. Inorg. Chem. 2002; 41: 4–10.

77 Nakamura M, Nakamura S. Conversion of metmyoglobin to NO myoglobin in the presence of nitrite and reductants. Biochim. Biophys. Acta 1996; 1289: 329–335.

78 Ishida Y, Tomoda A, Momose K. NO production through catalase and myoglobin, hemeproteins, in vascular smooth muscle. Nippon Yakurigaku Zasshi 1999; 114 (Suppl 1): 27P–32P.

79 Reutov VP. Biochemical predetermination of the NO synthase and nitrite reductase components of the nitric oxide cycle. Biochemistry-Russia 1999; 64: 528–542.

80 Reutov VP. Mediko-Biologicheskie Aspekty Tsiklov Oksida Azota i Subperoksidnogo Anion-Radikala. Vestn. Akad. Med. Nauk. SSSR. 2000; 35–41.

81 Kozlov AV, Staniek K, Nohl H. Nitrite reductase activity is a novel function of mammalian mitochondria. FEBS Lett. 1999; 454: 127–130.

82 Nohl H, Staniek K, Sobhian B, Bahrami S, Redl H, Kozlov AV. Mitochondria recycle nitrite back to the bioregulator nitric monoxide. Acta Biochim. Pol. 2000; 47: 913–921.

83 Millar TM, Stevens CR, Blake DR. Xanthine oxidase can generate nitric oxide from nitrate in ischaemia. Biochem. Soc. Trans. 1997; 25: 528S.

84 Cosby K, Partovi KS, Crawford JH, Patel RP, Reiter CD, Martyr S, Yang BK, Waclawiw MA, Zalos G, Xu X, Huang KT, Shields H, Kim-Shapiro DB, Schechter AN, Cannon III RO, Gladwin MT. Nitrite reduction to nitric oxide by deoxyhemoglobin vasodilates the human circulation. [see comment] Nat. Med. 2003; 9: 1498–1505.

85 Huang KT, Keszler A, Patel N, Patel RP, Gladwin MT, Kim-Shapiro DB, Hogg N. The reaction between nitrite and deoxyhemoglobin. Reassessment of reaction kinetics and stoichiometry. J. Biol. Chem. 2005; 280: 31126–31131.

86 Huang Z, Shiva S, Kim-Shapiro DB, Patel RP, Ringwood LA, Irby CE, Huang KT, Ho C, Hogg N, Schechter AN, Gladwin MT. Enzymatic function of hemoglobin as a nitrite reductase that produces NO under allosteric control. J. Clin. Invest. 2005; 115: 2099–2107.

87 Xu X, Cho M, Spencer NY, Patel N, Huang Z, Shields H, King SB, Gladwin MT, Hogg N, Kim-Shapiro DB. Measurements of nitric oxide on the heme iron and beta-93 thiol of human hemoglobin during cycles of oxygenation and deoxygenation. Proc. Natl. Acad. Sci. USA 2003; 100: 11303–11308.

88 Nagababu E, Ramasamy S, Abernethy DR, Rifkind JM. Active nitric oxide produced in the red cell under hypoxic conditions by deoxyhemoglobin-mediated nitrite reduction. J. Biol. Chem. 2003; 278: 46349–46356.

89 Rifkind JM, Ramasamy S, Manoharan PT, Nagababu E, Mohanty JG. Redox reactions of hemoglobin. Antioxid. Redox Signal. 2004; 6: 657–666.

90 Rifkind JM, Nagababu E, Ramasamy S, Ravi LB. Hemoglobin redox reactions and oxidative stress. Redox Rep. 2003; 8: 234–237.

91 Crawford JH, Isbell TS, Huang Z, Shiva S, Chacko BK, Schechter AN, Darley-Usmar VM, Kerby JD, Lang Jr. JD, Kraus D, Ho C, Gladwin MT, Patel RP. Hypoxia, red blood cells, and nitrite regulate NO-dependent hypoxic vasodilation. Blood 2006; 107: 566–574.

92 Jones SP, Girod WG, Palazzo AJ, Granger DN, Grisham MB, Jourd'Heuil D, Huang PL, Lefer DJ. Myocardial ischemia-reperfusion injury is exacerbated in absence of endothelial cell nitric oxide synthase. Am. J. Physiol. 1999; 276: H1567–H1573.

93 Sumeray MS, Rees DD, Yellon DM. Infarct size and nitric oxide synthase in murine myocardium. J. Mol. Cell. Cardiol. 2000; 32: 35–42.

94 Du Toit EF, Meiring J, Opie LH. Relation of cyclic nucleotide ratios to ischemic and reperfusion injury in nitric oxide-donor treated rat hearts. J. Cardiovasc. Pharmacol. 2001; 38: 529–538.

95 Brunner F, Leonhard B, Kukovetz WR, Mayer B. Role of endothelin, nitric oxide and L-Arginine release in ischaemia/reperfusion injury of rat heart. Cardiovasc. Res. 1997; 36: 60–66.

96 Ma XL, Gao F, Liu GL, Lopez BL, Christopher TA, Fukuto JM, Wink DA, Feelisch M. Opposite effects of nitric oxide and nitroxyl on postischemic myocardial injury. Proc. Natl. Acad. Sci. USA 1999; 96: 14617–14622.

97 Bolli R. Cardioprotective function of inducible nitric oxide synthase and role of nitric oxide in myocardial ischemia and preconditioning: an overview of a decade of research. J. Mol. Cell. Cardiol. 2001; 33: 1897–1918.

98 Duranski MR, Greer JJ, Dejam A, Jaganmohan S, Hogg N, Langston W, Patel RP, Yet SF, Wang X, Kevil CG, Gladwin MT, Lefer DJ. Cytoprotective effects of nitrite during in vivo ischemia-reperfusion of the heart and liver. J. Clin. Invest. 2005; 115: 1232–1240.

99 Webb A, Bond R, Mclean P, Uppal R, Benjamin N, Ahluwalia A. Reduction of nitrite to nitric oxide during ischemia protects against myocardial ischemia-reperfusion damage. PNAS 2004; 101: 13683–13688.

100 Johnson G, IIIrd, Tsao PS, Mulloy D, Lefer AM. Cardioprotective effects of acidified sodium nitrite in myocardial ischemia with reperfusion. J. Pharmacol. Exp. Ther. 1990; 252: 35–41.

101 Johnson G, IIIrd, Tsao PS, Lefer AM. Cardioprotective effects of authentic nitric oxide in myocardial ischemia with reperfusion. Crit. Care Med. 1991; 19: 244–252.

102 Woolfson RG, Patel VC, Neild GH, Yellon DM. Inhibition of nitric oxide synthesis reduces infarct size by an adenosine-dependent mechanism. Circulation 1995; 91: 1545–1551.

103 Wink DA, Miranda KM, Katori T, Mancardi D, Thomas DD, Ridnour L, Espey MG, Feelisch M, Colton CA, Fukuto JM, Pagliaro P, Kass DA, Paolocci N. Orthogonal properties of the redox siblings nitroxyl and nitric oxide in the cardiovascular system: a novel redox paradigm. Am. J. Physiol. Heart Circ. Physiol. 2003; 285: H2264–H2276.

Radicals for Life: The Various Forms of Nitric Oxide
E. van Faassen and A.F. Vanin (Eds.)

CHAPTER 16

The anti-microbial and cytotoxic actions of nitrite, and the use of DNIC as a marker for these actions

Tetsuhiko Yoshimura

Research Project of Biofunctional Reactive Species, Yamagata Promotional Organization for Industrial Technology, Yamagata 990-2473, JAPAN

INTRODUCTION

Biological systems can produce nitrite ions *via* enzymatic and non-enzymatic processes. Enzymatically, nitrite can be produced by a two-electron reduction of nitrate during denitrification and nitrate assimilation processes in various plants and microbes, or alternatively *via* the four-electron oxidation of hydroxylamine by nitrifying bacteria [1–3]. The nitrite ions formed by these enzymatic pathways are involved in the nitrogen redox cycle in the biosphere [4]. Non-enzymatically, nitrite may be produced by the chemical oxidation of nitric oxide (NO) with dissolved oxygen in bio-fluids.

Historically, the food industries have used inorganic nitrate for preserving food and curing meat [5,6]. Nitrate is relatively non-reactive in chemical and biological systems, but its enzymatic reduction product, nitrite, shows anti-microbial activity. The effectiveness of nitrite as an anti-microbial agent is closely associated with its ability to trigger formation of NO, which reacts with microbial proteins that contain iron–sulfur clusters, resulting in their destruction to form dinitrosyl dithiolato iron complex (DNIC) [7–9].

The nitrate content of fresh vegetables is relatively high compared to that of nitrite, and this content ratio varies across different foods. Thus, humans are exposed to different proportions of food-derived nitrates and nitrites. The chemical reactivity of nitrite toward biological constituents in body fluid is considered an important factor in maintaining the desired low steady-state nitrite concentrations. In healthy humans, serum nitrite and nitrate levels are usually maintained at 6.6 ± 11 μM and 34 ± 18 μM, respectively, while cerebrospinal fluid nitrite and nitrate levels are maintained at 3.4 ± 3.1 μM and 7.6 ± 4 μM, respectively [10].

In contrast, the plasma nitrite and nitrate levels in plasma from endotoxin-treated rats were found to be ~6 μM and ~260 μM, respectively, indicating a sharp increase in nitrate [11]. In humans, resting salivary nitrite levels are approximately 50 μM [12], and may rise as high as 2 mM following ingestion of foods with high nitrate content [13]. When salivary nitrite enters the stomach, the acidity and ascorbic acid present in gastric juice combines to convert nitrite to NO. Thus, the physiological significance of high NO concentrations in body fluids has attracted considerable attention. This chapter will focus on the anti-microbial and cytotoxic activities of nitrite, and the significance of DNIC as a biomarker for this activity. The roles of nitrite in the vasculature will not be covered in this chapter. For this, the readers are referred to Chapters 14 and 15.

THE ANTI-MICROBIAL AND CYTOTOXIC ACTIVITIES OF NITRITE

Nitrite acts as a colorant, flavorant, antioxidant and anti-botulinal agent in cured meat

In ancient times, meat was preserved with saline desert sands and sea salts, both of which contain nitrate [5]. The reddening effect of nitrate in preserved meat was mentioned as far back as the late Roman era, but it wasn't until the early twentieth century that the bacterial reduction product, nitrite, was identified as the agent responsible for coloring and curing meat [5]. In meat curing, nitrate functions as a reservoir for nitrite, which acts not only to color meat, but also acts as a flavorant, an antioxidant and an anti-microbial agent [14]. These days, it is generally accepted that most of these effects are due to the action of NO, which is generated by the reduction of nitrite [15]. However, in recent years the use of nitrite has faced considerable negative pressure, because its reaction with amines in meat has been shown to produce nitrosamines, which are known carcinogens and possibly mutagens [16–18].

The characteristic red pigment in cured meat has been attributed to nitrosylmyoglobin, which is formed by the reaction of muscle tissue myoglobin with nitrites in the curing agent [18,19]. However, although nitrite and/or NO are generally accepted as playing a crucial role in this process, the reaction mechanism of myoglobin with nitrite and/or the reduction mechanism of nitrite to NO in cured meat remain under some debate. In fresh minces, nitrite has been proposed to be enzymatically reduced to NO by the cell respiratory system [20]. In addition, various reductants that are present in or added to fresh meat (e.g. ascorbate and cysteine) can reduce nitrite to NO *via* non-enzymatic routes [21]. In cell-free systems, the addition of nitrite to oxyhemoglobin has been demonstrated to transform proteins to their met-forms [22], indicating that nitrites can be considered causative agents for methemoglobinemia, particularly in infants [23–25]. In contrast, metmyoglobin has been shown to react with excess nitrite at pH < 7 to yield a nitrimyoglobin with an unusual nitrovinyl group on its heme-side chain [26,27]. Nitrimyoglobin is a green pigment involved in the phenomenon called "nitrite burning" or "nitrite greening" in improperly cured meat. Thus, the color of nitrite-treated meat appears to be controlled by multiple factors, including the utilized agents, pH, temperature and the presence/absence of other additives.

In nitrite-cured meat, NO appears to play a crucial antioxidant role. For example, undesirable lipid oxidation in nitrite-cured meat is inhibited by *S*-nitrosothiol and nitrosylmyoglobin [28,29], both of which are formed *via* nitrite-derived NO [30]. In addition, the cysteine-containing nitrosyl iron complex, which is produced by the reaction of cysteine, iron(II) salt and NO, showed inhibitory activities against lipid peroxidation [31]. *S*-nitrosothiol, nitrosylmyoglobin and nitrosyl iron complex with cysteine all have reactive NO groups, are able to release an NO molecule and may undergo a transnitrosation reaction under appropriate physiological conditions [31–34]. Further, NO can undergo rapid radical–radical combination reactions with other radical species, leading to quenching of those species; thus it can act as a terminating species for radical chain reactions during lipid oxidation [35,36]. Thus, the chemical reactivity of nitrite-derived NO are largely responsible for the antioxidant activity of nitrite in cured meat.

Nitrite has also been shown to exert significant anti-botulinal effects in cured meat [37–42]. Initially, this anti-botulinal activity was thought to be due to the interaction of nitrite as nitrous acid with thiol-containing constituents of the bacterial cell [43]. However, in clostridia, nitrite was found to directly interact with pyruvate-ferredoxin oxidoreductase, an iron–sulfur cluster that contains enzymes necessary for botulinal energy production in some clostridial vegetative cells. The interaction of nitrite with pyruvate-ferredoxin oxidoreductase was found to inhibit the phosphoroclastic system in *Clostridium sporogenes* and *Clostridium botulinum* [44,45]. In terms of an action mechanism for this effect, Reddy et al. (1983) reported that iron–sulfur proteins in vegetative cells of *C. botulinum* reacted with nitrite to form iron–NO complexes, resulting in destruction of the iron–sulfur clusters. Inactivation of iron–sulfur enzymes (e.g. ferredoxin) by NO binding was shown to inhibit the growth of *C. botulinum* [7]. Additional studies identified the active iron–NO complex in this mechanism as a dinitrosyl dithiolato iron complex [46,47]. However, although iron-thiol-nitrosyl complexes are extremely inhibitory toward botulinal growth, Payne et al. (1990) reported that there was no correlation between botulinal growth inhibition and the content of iron-thiol-nitrosyl complexes, suggesting that direct inhibition and/or destruction of iron–sulfur enzymes may not be the principal basis of the anti-botulinal activity of nitrite [8,9].

Ingested nitrate may be reduced to nitrite, which plays numerous roles in the stomach

Nitrate, an essential plant nutrient, is ultimately metabolized to form plant proteins. High concentrations of nitrate are found in fresh, green, leafy vegetables such as spinach and lettuce [23], while most other non-vegetable food products contain relatively low nitrate contents. When nitrate-containing food is ingested, the nitrate is absorbed from the small intestine, and nearly a quarter of it is transported to the salivary glands and re-secreted into the mouth [48–52]. In the mouth, bacteria on the dorsum of the tongue reduce about 30% of the salivary nitrate to nitrite [48–53], and may further reduce this nitrite to NO under anaerobic conditions [51]. Under a fasting condition, the salivary nitrite concentration is approximately 50 μM [12], and this rises to as high as 2 mM after ingesting food with a high nitrate content such as green lettuce [13], as described above. When salivary nitrite enters the stomach, it is reduced to NO in the acidic media (pH 1–2) through following reactions [54]:

$$NO_2^- + H^+ \rightarrow HNO_2 (pKa = 3.4)$$

$$2HNO_2 \rightarrow N_2O_3 + H_2O$$

$$N_2O_3 \rightarrow NO + NO_2$$

$$3NO_2 + H_2O \rightarrow NO + 2H^+ + 2NO_3^-$$

In the stomach, nitrite can be totally converted to nitrous acid because its pKa is lower than the pH in the gastric juice, subsequently leading to the generation of reactive nitrogen oxide species (RNOS) such as N_2O_3, NO and NO_2 (N_2O_4). In addition, ascorbic acid is actively secreted within the gastric juice of the healthy stomach and is stabilized under acidic conditions [55]. Since ascorbic acid rapidly reduces nitrite to NO, salivary nitrite can be efficiently converted to NO in the stomach [56–59]. High concentrations of NO resulting from enterosalivary recirculation of dietary nitrate have been detected within the lumen of the stomach by employing a variety of methods [60–63]. NO and other RNOS derived from salivary nitrite have been demonstrated to play critical roles in the physiology of the normal stomach [64].

However, these molecules may also play pathophysiological roles. For example, salivary nitrite-derived N_2O_3 and NO_2 may act as nitrosating agents in the stomach, reacting with a variety of dietary amines to yield N-nitrosamines, which are chemical carcinogens known to cause gastric cancer, likely *via* their ability to deaminate DNA bases and inactivate DNA repair enzymes [65,66]. Recently, particularly high levels of NO were reported in the gastroesophageal junction (GEJ) and cardia, where salivary nitrite first encounters gastric acid [63]. These NO molecules diffuse into the adjacent gastric tissues, increasing local glutathione consumption in the tissue [67] and enhancing the formation of DNIC [68]. These findings suggest that high levels of NO may be involved in the high prevalence of mutagenesis and neoplasia at the GEJ.

In contrast to their pathophysiological roles, salivary-derived nitrite and the resulting NO are likely to play a protective role in the stomach, guarding against ingested pathogens and maintaining gastric mucosal integrity by improving mucosal blood flow and mucus secretion [69]. Although nitrite has limited anti-microbial activity at neutral pH, this activity is profoundly enhanced in acidic media such as gastric juice. Acidified nitrite exhibits strong bactericidal activity against *Candida albicans*, *Salmonella enteritidis*, *Salmonella typhimurium*, *Yersinia enterocolitica*, *Shigella sonney*, *Escherichia coli* O157:H7 [61,70] and *E. coli* CM120 [71]. In contrast, *Helicobacter pylori* and five lactobacilli species have been shown to be relatively resistant to acidified nitrite [71,72]. The anti-microbial activity of acidified nitrite appears to be influenced by many local environmental factors, including the presence of ascorbic acid, thiocyanate and chloride, the oxygen concentration and the culture medium [64,71]. In the stomach, other constituents of gastric juice such as amino acids, peptides and proteins are also likely to affect the level of anti-microbial activity. Although the antimicrobial mechanisms of salivary-derived nitrite are not well understood, several studies have suggested the involvement of RNOS arising from nitrite under acidic conditions [69,71,73].

DNIC MAY BE USED AS A BIOMARKER FOR THE ANTI-MICROBIAL AND CYTOTOXIC ACTIONS OF NITRITE

The interaction of nitrite or NO with the iron–sulfur protein/enzyme is closely implicated in nitrite treatment of biological systems as described in the previous sections. Dinitrosyl dithiolato iron complex (DNIC) formed through the interaction has been shown to be relatively stable paramagnetic molecule that exhibits a characteristic electron paramagnetic resonance (EPR) signal both in solution and the frozen state [46,74–78]. In nitrite-treated cells of clostridia [7–9] and *E. coli* O157:H7 [79], DNIC can be detected at 77 K as an axially symmetric EPR signal ($g_{\|} = 2.04$, $g_{\perp} = 2.015$). In cured meat, the DNIC signal was found to overlap with that of nitrosyl myoglobin [80], suggesting that both nitrosyl iron complexes were simultaneously formed. At the rat GEJ, the DNIC signal was found in tissues where salivary nitrite is likely to encounter highly acidic gastric juices, yielding high concentrations of NO [68], as details shown below. The presence of DNIC suggests that synchronous inactivation of iron–sulfur proteins/enzymes is involved in vital-activity as well as the release of intracellular iron after the reaction of free iron with thiol-containing proteins [47,76], further indicating that detection of DNIC in nitrite-treated specimens could be a biomarker for the anti-microbial and cytotoxic actions of nitrite. In what follows, two good examples illustrating the utility of DNIC as a biomarker will be described.

Infection with *E. coli* O157:H7 causes hemorrhagic diarrhea and hemolytic uremic syndrome [81,82]. This gram-negative bacterial species produces a large quantity of verotoxins during the course of the infection. Verotoxins are released from bacterial cells in the event of bacteriolysis, exerting a disastrous effect on the patient. Although some antibiotics and bacteriostatic sodium chloride can inhibit or kill *E. coli* O157:H7, these bactericidal agents cause bacteriolysis, releasing large quantities of verotoxins contained within these cells [78]. In contrast, nitrite treatment effectively inhibited the growth of this bacterium without triggering increased verotoxin release. EPR spectroscopy of frozen *E. coli* suspensions treated with nitrite in the presence or absence of ascorbate revealed that nitrite treatment abrogated the iron–sulfur protein signal at $g = 1.94$ and initiated the DNIC marker signals at $g = 2.036$ and 2.011, and that the additional treatment with ascorbate intensified the DNIC signals (Fig. 1) [78]. The addition of NO donor reagents exerts the same effect. In addition, nitrite treatment inhibited the synthesis of ATP in *E. coli* O157:H7 cells. These results indicate that nitrite-derived NO can inhibit bacterial growth through the inactivation of iron–sulfur enzymes in the respiratory chain. Thus, nitrite confers antibacterial activity against *E. coli* O157:H7 without increasing verotoxin release, because the action mechanism does not include bacteriolysis. Furthermore, DNIC acts as a valuable marker for assessing this effect.

High concentrations of NO are luminally generated at the GEJ through the enterosalivary recirculation of dietary nitrate in humans [63]. In the GEJ gastric tissues of nitrite-treated rats, the DNIC marker signal increased time- and dose-dependently, whereas no signal was observed at the distal stomach in the same rats [68]. In humans after nitrate ingestion, a low level of DNIC was detected in biopsy specimens from the cardia, but not the antrum [68]. The aconitase activity of GEJ tissues was significantly lower in nitrite-treated rats *vs.* control rats, while that in the distal stomach was similar between the two groups [68]. These results collectively suggest that salivary-derived NO diffuses from the stomach lumen into adjacent

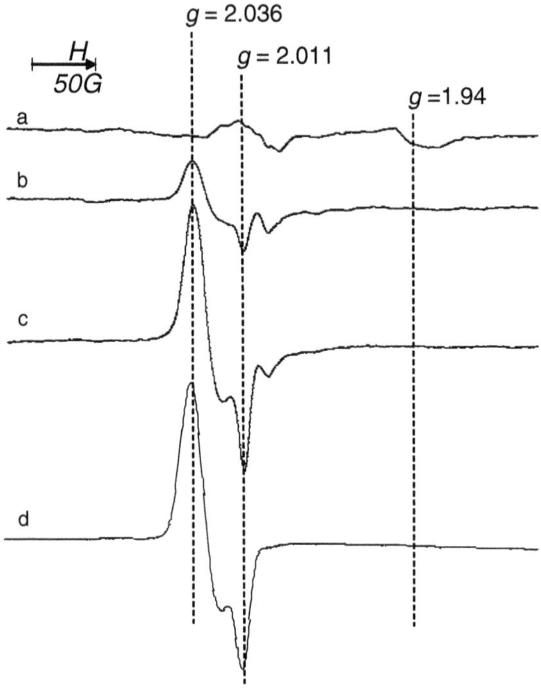

Fig. 1. EPR spectra of cell suspensions of *E. coli* O157:H7 at 77 K. (Curve a) Untreated; (Curve b) treated with $NaNO_2$ (200 mg/l); (Curve c) treated with both $NaNO_2$ (200 mg/l) and sodium ascorbate (500 mg/l); (curve d) treated with 3-[2-hydroxy-1-(1-methylethyl) -2-nitrosohydrazino]-1-propanamine as an NO donor reagent. (Reproduced with permission from Ref. [78]. Copyright 2004 Japan Society for Bioscience, Biotechnology, and Agrochemistry.)

tissues, where it interacts with and disassembles vulnerable Fe–S cluster proteins, or reacts with intracellular free iron and thiol-containing proteins. This mechanism may be involved in the high prevalence of inflammation and intestinal metaplasia at GEJ in humans. Thus, it is valuable to consider the use of DNIC monitoring in future studies.

CONCLUDING REMARKS

Nitrite, its oxidized form, nitrate, and its reduced form, NO, are bioactive molecules involved in the life cycle of all organisms. In humans, nitrite contributes to host defense mechanism against a number of pathogenic microorganisms in the mouth, stomach and skin. The addition of nitrite to meat for preservation is responsible for the characteristic color and flavor of meat, and helps combat the decay of cured meats. DNIC, which has been found in all of these tissues, cells and bacteria, has a characteristic, readily identifiable EPR signal, suggesting that it may be used as a paramagnetic biomarker for the anti-microbial and cytotoxic actions of nitrite.

ACKNOWLEDGMENTS

The author thanks Drs. K. Iijima, T. Ueno, K. Asanuma and N. Ara of Tohoku University School of Medicine for their contributions and Dr. Y. Kotake of Oklahoma Medical Research Foundation for valuable comments and criticism of the manuscript. The work was supported in part by a Grant-in-Aid for Scientific Research (17550146) from the Japan Society for the Promotion of Science and for Scientific Research on Priority Areas "Application of Molecular Spins" (15087212) from the Ministry of Education, Culture, Sports, Science and Technology.

REFERENCES

1 Brittain T, Blackmore R, Greenwood C, Thomson AJ. Bacterial nitrite-reducing enzymes. Eur. J. Biochem. 1991; 209: 793–802.
2 Zumft WG. The biological role of nitric oxide in bacteria. Arch. Microbiol. 1993; 160: 253–364.
3 Hollocher TC. The enzymology and occurrence of nitric oxide in the biological nitrogen cycle. In Nitric Oxide: Principles and Actions, (Lancaster Jr. J, ed.), Academic Press, SanDiego, 1996, pp. 289–344.
4 Sprent JI. The Ecology of the Nitrogen Cycle. Cambridge University Press, Cambridge, 1987.
5 Binkind E, Kolari OE. The history and use of nitrate and nitrite in the curing of meat. Fd. Cosmet. Toxicol. 1975; 13: 655–661.
6 MacDougall DB, Mottram DS, Rhodes DN. Contribution of nitrite and nitrate to the colour and flavour of cured meats. J. Sci. Fd. Agric. 1975; 26: 1743–1754.
7 Reddy D, Lancaster JR, Conforth DP. Nitrite inhibition of *Clostridium botulinum*: electron spin resonance detection of iron-nitric oxide complexes. Science 1983; 221: 769–770.
8 Payne MJ, Woods LJ, Gibbs P, Cammack R. Electron paramagnetic resonance spectroscopic investigation of the inhibition of the phosphoroclastic system of *Clostridium sporogenes*. J. Gen. Microbiol. 1990; 136: 2067–2076.
9 Payne MJ, Glidewell C, Cammack R. Interactions of iron-thiol-nitrosyl compounds with the phosphoroclastic system of *Clostridium sporogenes*. J. Gen. Microbiol. 1990; 136: 2077–2087.
10 Friedberg MA, Hinsdale ME, Shihabi ZK. Analysis of nitrate in biological fluids by capillary electrophoresis. J. Chromatogr. 1997; A781: 491–496.
11 Paya D, Maupoil V, Schott C, Rochette L, Stoclet J-C. Temporal relationships between levels of circulating NO derivatives, vascular NO production and hyporeactivity to noradrenaline induced by endotoxin in rats. Cardiovasc. Res. 1995; 30: 952–959.
12 McKnight GM, Smith LM, Drummond RS, Duncan CW, Benjamin N. Chemical synthesis of nitric oxide in the stomach from dietary nitrate in humans. Gut 1997; 40: 211–214.
13 Pannala A, Mani AR, Spencer JP, Skinner V, Bruckdorfer KR, Moore KP, Rice-Evans CA. The effect of dietary nitrate on salivary, plasma, and urinary nitrate metabolism in humans. Free Radic. Biol. Med. 2003; 34: 576–584.
14 Cassens RG, Greaser ML, Ito T, Lee M. Reactions of nitrite in meat. Food Technol. 1979; 33: 46–57.
15 Conforth D. Role of nitric oxide in treatment of foods. In Nitric Oxide: Principles and Actions, (Lancaster, Jr. J, ed.), Academic Press, SanDiego, 1996, pp. 259–287.
16 Tannenbaum SR, Sinskey AJ, Weisman M, Bishop W. Nitrite in human saliva. Its possible relationship to nitrosamine formation. J. Natl. Cancer Inst. 1974; 33: 79–84.
17 O'Boyle AR, Aladin-Kassam N, Rubin LJ, Diosady LL. Encapsulated cured-meat pigment and its application in nitrite-free ham. J. Food Sci. 1992; 57: 807–812.
18 Arihara K, Kushida H, Kondo Y, Itoh M, Luchansky JB, Cassens RG. Conversion of metmyoglobin to bright red myoglobin derivatives by *Chromobacterium violaceum*, *Kurthia* sp., and *Lactobacilus fermentum* JCM 1173. J. Food Sci. 1993; 58: 38–42.
19 Giddings GG. The basis of color muscle foods. J. Food Sci. 1977; 42: 288–294.
20 Walters CL, Taylor AM. Nitrite metabolism by muscle in vitro. Biochim. Biophys. Acta 1964; 86: 448–458.

21 Fox JB, Ackerman SA. Formation of nitric oxide myoglobin: mechanisms of the reaction with various reductants. J. Food Sci. 1968; 33: 364–370.

22 Martin H, Huisman THJ. Formation of ferrihaemoglobin of isolated human haemoglobin types by sodium nitrite. Nature 1963; 200: 898–899.

23 Lee DHK. Nitrates, nitrites and methemoglobinemia. Environ. Res. 1970; 3: 484–511.

24 Magee PN. Nitrogen as health hazard. Ambio. 1977; 6: 123–125.

25 Kosaka H, Imaizumi K, Tyuma I. Mechanism of autooxidation of oxyhemoglobin by nitrite. An intermediate detected by electron spin resonance. Biochim. Biophys. Acta 1982; 702: 237–241.

26 Fox JB, Thomson JS. The formation of green heme pigments from metmyoglobin and methemoglobin by the action of nitrite. Biochemistry 1964; 3: 1323–1328.

27 Bondoc LJ, Timkovich R. Structural characterization of nitrimyoglobin. J. Biol. Chem. 1989; 264: 6134–6145.

28 Kanner J, Juven BJ. *S*-nitrosocystein as an antioxidant, color-developing, and anticlostridial agent in comminuted turkey meat. J. Food. Sci. 1980; 45: 1105–1108.

29 Kanner J, Ben-Gera I, Berman S. Nitric-oxide myoglobin as an inhibitor of lipid oxidation. Lipids 1980; 15: 944–948.

30 Emi-Miwa M, Okitani A, Fujimaki M. Comparison of the fate of nitrite added to whole meat, meat fractions and model systems. Agr. Biol. Chem. 1976; 40: 1387–1392.

31 Kanner J, Harel S, Shagalovich J, Berman S. Antioxidative effect of nitrite in cured meat products: nitric oxide-iron complexes of low molecular weight. J. Agric. Food Chem. 1984; 32: 512–515.

32 Liu Z, Rudd MA, Freedman JE, Loscalzo J. *S*-transnitrosation reactions are involved in the metabolic fate and biological actions of nitric oxide. J. Pharmacol. Exp. Ther. 1998; 284: 526–534.

33 Williams DLH. The chemistry of *S*-nitrosothiols. Acc. Chem. Res. 1999; 32: 869–876

34 Hobbs AJ, Gladwin MT, Patel RP, Williams DLH, Butler AR. Haemoglobin: NO transporter, NO inactivator or none of the above? Trends Phamacol. Sci. 2002; 23: 406–411.

35 Wink DA, Hanbauer I, Krishna MC, DeGraff W, Gamson J, Mitchell JB. Nitric oxide protects against cellular damage and cytotoxicity from reactive oxygen species. Proc. Natl. Acad. Sci. USA 1993; 90: 9813–9817.

36 Rubbo H, Darley-Usmar V, Freeman BA. Nitric oxide regulation of tissue free radical injury. Chem. Res. Toxicol. 1996; 9: 809–820.

37 Castellani AG, Niver CF. Factors affecting the bacteriostatic action of sodium nitrite. Appl. Microbiol. 1955; 3: 154–159.

38 Shank JL, Silliker JH, Harper RH. The effect of nitric oxide on bacteria. Appl. Microbiol. 1962; 10: 185–189.

39 Steinke PKW, Foster EM. Botulinum toxin formation in liver sausage. Food Res. 1951; 16: 477–484.

40 Perigo JA, Whiting E, Bashford TE. Observations on the inhibition of vegetative cells of *Clostridium sporogenes* by nitrite which has been autoclaved in a laboratory medium, discussed in a context of sub-lethally processed cured meat. J. Food Technol. 1967; 2: 377–397.

41 Benedict RC. Biochemical basis for nitrite-inhibition of *Clostridium botulinum* in cured meat. J. Food Protection 1980; 43: 877–891.

42 Pierson MD, Smoot LA. Nitrite, nitrite alternatives, and the control of *Clostridium botulinum* in cured meat. Crit. Rev. Food Sci. Nutr. 1982; 17: 141–187.

43 O'leary V, Solberg M. Effect of sodium nitrite inhibition on intracellular thiol groups and on the activity of certain glycolytic enzymes. Appl. Environ. Microbiol. 1976; 31: 208–212.

44 Woods LFJ, Wood JM, Gibbs PA. The involvement of nitric oxide in the inhibition of the phosphoroclatic system in *Clostridium sporogenes* by sodium nitrite. J. Gen. Microbiol. 1981; 125: 399–406.

45 Woods FJ, Wood JM. A note on the effect of nitrite inhibition on the metabolism of *Clostridium botulinum.* J. Appl. Bacteriol. 1982; 52: 109–110.

46 Vanin F, Kleschyov AL. EPR detection and biological implications of nitrosyl iron complexes. In Nitric Oxide in Transplant Rejection and Anti-tumor Defense, (Lukiewicz S, Zweier JL, eds.), Academic publishers, Kluwer Boston, 1999, pp. 49–82.

47 Lancaster Jr. J, Stuehr DJ. The intracellular reactions of nitric oxide in the immune system and its enzymatic synthesis. In Nitric Oxide: Principles and Actions, (Lancaster Jr. J, ed.), Academic Press, SanDiego, 1996, pp. 139–175.
48 Walker R. Nitrates, nitrites and *N*-nitroso compounds: a review of the occurrence in food and diet and the toxicological implications. Food Addit. Contam. 1990; 7: 717–768.
49 Bos PM, van den Brandt PA, Wedel M, Ockhuizen T. The reproducibility of the conversion of nitrate to nitrite in human saliva after a nitrate load. Food Chem. Toxic. 1988; 26: 93–97.
50 Wagner DA, Schultz DS, Deen WM, Young VR, Tannenbaum SR. Metabolic fate of an oral dose of ^{15}N-labeled nitrate in humans: Effect of diet supplementation with ascorbic acid. Cancer Res. 1983; 43: 1921–1925.
51 Duncan C, Dougall H, Johnston P, Green S, Brogan R, Leifert C, Smith L, Golden M, Benjamin N. Chemical generation of nitric oxide in the mouth from the enterosalivary circulation of dietary nitrate. Nat. Med. 1995; 1: 546–551.
52 Granli T, Dahl R, Brodin P, Bockman OC. Nitrate and nitrite concentrations in human saliva: variations with salivary flow-rate. Food Chem. Toxic. 1989; 27: 675–680.
53 Sasaki T, Matano K. Formation of nitrite from nitrate at the dorsum linguae. J. Food Hyg. Soc. Jpn. 1979; 20: 363–369.
54 Greenwood NN, Earnshaw A. Chemistry of the elements, Chapter 11 Nitrogen, Pergamon Press, Oxford, 1984.
55 Sobala GM, Schorah CJ, Sanderson M, Dixon MF, Tompkins DS, Godwin P, Axon ATR. Ascorbic acid in the human stomach. Gastroenterology 1989; 97: 357–363.
56 Bunton CA. Oxidation of ascorbic acid and similar reductones by nitrous acid. Nature 1959; 183: 163–164.
57 Mirvish SS, Wallcave L, Eagen M, Shubik P. Ascorbate-nitrite reaction: possible means of blocking the formation of carcinogenic N-nitroso compounds. Science 1972; 177: 65–68.
58 Mirvish SS. Blocking the formation of *N*-nitroso compounds with ascorbic acid *in vitro* and *in vivo*. Ann. N. Y. Acad. Sci. 1975; 258: 175–180.
59 Licht WR, Fox JG, Deen WM. Effect of ascorbic acid and thiocyanate on nitrosation of proline in the dog stomach. Carcinogenesis 1988; 9: 373–377.
60 Aneman A, Snygg J, Fandriks L, Pettersson A. Continuous measurement of gastric nitric oxide production. Am. J. Physiol. 1996; 27: G1039–G1042.
61 Benjamin N, O'Driscoll F, Dougall H, Duncan C, Smith L, Golden M. Stomach NO synthesis. Nature 1994; 368: 502.
62 Lundberg JON, Weitzberg E, Lundberg JM, Alving K. Intragastric nitric oxide production in humans: measurements in expelled air. Gut 1994; 35: 1543–1546.
63 Iijima K, Henry E, Moriya A, Wirz A, Kelman AW, McColl KEL. Dietary nitrate generates potentially mutagenic concentrations of nitric oxide at the gastro-esophageal junction. Gastroenterology 2002; 122: 1248–1257.
64 Benjamin N, Dykhuizen R. Nitric oxide and epithelial host defense. In Nitric Oxide and Infection, (Fang FC, ed.), Kluwer Academic, New York, 1999, pp. 215–230.
65 Tamir S, Tannenbaum SR. The role of nitric oxide ($NO^{\bullet}$) in the carcinogenic process. Biochim. Biophys. Acta 1996; 1288: F31–F36.
66 Burney S, Caufield JL, Niles JC, Wishnok JS, Tannenbaum SR. The chemistry of DNA damage from nitric oxide and peroxynitrite. Mut. Res. 1999; 424: 37–49.
67 Asanuma K, Iijima K, Sugata H, Ohara S, Shimosegawa T, Yoshimura T. Diffusion of cytotoxic concentrations of nitric oxide generated luminally at the gastro-oesophageal junction of rats. Gut 2005; 54: 1072–1077.
68 Asanuma K, Iijima K, Ara N, Yoshitake J, Koike T, Ohara S, Shimosegawa T, Yoshimura T. Fe-S cluster proteins are intracellular targets for nitric oxide generated luminally at the gastro-oesophageal junction, to be published.
69 Lundberg JO, Weitzberg E, Cole JA, Benjamin N. Nitrate, bacteria and human health. Nat. Rev. 2004; 2: 593–602.

70 Dykhuizen RS, Frazer R, Duncan C, Smith CC, Golden M, Benjamin N, Leifert C. Anitimicrobial effect of acidified nitrite on gut pathogens: importance of dietary nitrate in host defense. Antimicrob. Agents Chemother. 1996; 40: 1422–1425.

71 Xu J, Verstraete W.The bacterial effect and chemical reactions of acidified nitrite under conditions simulating the stomach. J. Appl. Microbiol. 2001; 90: 523–529.

72 Dykhuizen RS, Frazer A, McKenzie H, Golden M, Leifert C, Benjamin N. *Helicobacter pylori* is killed by nitrite under acidic conditions. Gut 1998; 42: 334–337.

73 Fang FC. Mechanisms of nitric oxide-related antimicrobial activity. J. Clin. Invest. 1997; 99: 2818–2825.

74 Vanin AF, Nalbandyan RM. Free radicals of a new type in yeast cells. Biofizika (Rus.) 1965; 11: 167–168.

75 Vithaythil J, Ternberg JL, Commoner B. Changes in electron spin resonance signals of rat liver during chemical carcinogenesis. Nature 1965; 207: 1246–1249.

76 McDonald CC, Phillips WD, Mower HF. An electron spin resonance study of some complexes of iron, nitric oxide, and anionic ligands. J. Am. Chem. Soc. 1965; 87: 3319–3326.

77 Woolum JC, Tiezzi E, Commoner B. Electron spin resonance of iron-nitric oxide complexes with amino acids, peptides and proteins. Biochim. Biophys. Acta 1968; 160: 311–320.

78 Ueno T, Yoshimura T. The physiological activity and in vitro distribution of dinitrosyl dithiolato iron complex. Jpn. J. Pharmacol. 2000, 82: 95–101.

79 Morita H, Yoshikawa H, Suzuki T, Hisamatsu S, Kato Y, Sakata R, Nagata Y, Yoshimura T. Antimicrobial action against verotoxigenic *Escherichia coli* O157:H7 of nitric oxide derived from sodium nitrite. Biosci. Biotechnol. Biochem. 2004; 68: 1027–1034

80 Bonnett R, Chandra S, Charalambides AA, Sales KD, Scourides PA. Nitrosation and nitrosylation of haemoprotein and related compounds. Part 4. Pentaco-ordinate nitrosylprotohaem as the pigment of cooked cured meat. Direct evidence from E.S.R. spectroscopy. J. Chem. Soc. Perkin I. 1980; 1706–1710.

81 Griffin PM, Ostroff SM, Tauxe RV, Greene KD, Wells JG, Lewis JH, Blake PA. Illnesses associated with *Escherichia coli* O157:H7 infections. A broad clinical spectrum. Ann. Intern. Med. 1988; 109: 705–712.

82 Tarr PI, Gordon CA, Chandler WL. Shiga-toxin-producing *Escherichia coli* and haemolytic uraemic syndrome. Lancet. 2005; 365: 1073–1086.

CHAPTER 17

Organic nitrates and nitrites as stores of NO bioactivity

Gregory R.J. Thatcher

Department of Medicinal Chemistry & Pharmacognosy, College of Pharmacy, University of Illinois at Chicago, Chicago, IL 60612

"I had obtained good results in such cases by the inhalation of nitrite of amyl, and the present seemed an admirable opportunity of testing its virtues." The words of Sherlock Holmes from his creator Sir Arthur Conan Doyle show the experimental status of nitrovasodilators in 1894. Although in clinical use, nitroglycerine (glyceryl trinitrate, GTN) was better known for its explosive qualities as Conan Doyle also wrote 20 years later: *"The nitroglycerine on one side and the gun-cotton on the other are kneaded into a sort of devil's porridge which is the next stage of manufacture. Those smiling khaki-clad girls who are swirling the stuff round in their hands would be blown to atoms in an instant if certain very small changes occurred. The changes will not occur and the girls will still smile and stir their devil's porridge, but it is a narrow margin between life and death."* The small changes that must occur in the bioactivation of nitrates are intrinsic to their therapeutic activity. The narrow margin between life and death for a cell is intimately linked with levels of cellular NO, and it is the ability of nitrates to mimic the activity of NO without releasing high levels of NO itself that continues to underlie their therapeutic benefits.

INTRODUCTION

Therapeutic agents related to NO include prodrugs that elevate NO levels, scavengers of NO, and inhibitors of endogenous NO synthesis. Beyond these, there are preclinical therapeutic agents that modulate the concentration and actions of nitrogen oxides derived from or related to NO, for example, peroxynitrite ($ONOO^-$), NO_2^- (inorganic nitrite), and HNO (nitroxyl). Even drugs such as the statins, with primary targets not directly involved in regulation of NO, might be considered NO-related therapeutics because of benefits derived from downstream regulation of NO and nitrogen oxides [1–3]. The organic nitrates ($RONO_2$) represent the major focus of this review, however, the history and chemistry of organic nitrites (RONO) are intertwined with the nitrates and therefore some discussion of nitrites is essential. In this review, the terms nitrates and nitrites will be used exclusively to refer to the organic esters of nitric and nitrous acid, respectively. The inorganic ions will be referred to as nitrate ion (NO_3^-) and nitrite ion (NO_2^-), or simply inorganic nitrate and nitrite.

The nitrates represent a drug class that comprises the most venerable NO-related therapeutic agents, the nitrate vasodilators, in clinical use for more than 130 years, of which GTN (glyceryl trinitrate) is the most notorious [4]. The compounding of nitroglycerin in the form of dynamite was the foundation of Alfred Nobel's fortune. Nobel's contemporary, Thomas Brunton, had already studied the effects of organic nitrites in angina pectoris, as illustrated by the comments of Conan Doyle on the properties of nitrite of amyl [5]. The similarities of amyl nitrite to GTN led in 1878 to William Murrell pioneering the clinical use of sublingual GTN for relief from the intense pain of angina [4]. The classical nitrates are of equal contemporary relevance: isosorbide dinitrate (ISDN) has received a new lease of life as a cardioprotective agent; and hybrid nitrates, especially the so-called NO-donating non-steroidal anti-inflammatory drugs (NO-NSAIDs), are the focus of exciting clinical trials and extensive preclinical activity. Nitrate-induced changes in hemodynamic parameters remain central to the therapeutic focus of nitrates, with cardiovascular indications still the dominant disease targets, but in NO-NSAIDs the further role in modulating activity in response to endogenous prostaglandin levels has been emphasized. Most recently, several papers have described the potential utility of NO/cGMP modulating agents in neurological disorders; paralleling the entry of nitrates into clinical trials for neurological indications.

But, how do nitrates exert their biological actions? The dogma that all nitrates are simple NO donors, regardless of their structural and chemical diversity, would justify the discussion of nitrates in a treatise on "radicals for life." Given the plethora of biological roles for NO, the NO donor doctrine might even be compatible with the broad spectrum of therapeutic targets for which nitrates are being studied. But, many contemporary studies of the archetype classical nitrate, GTN, remain focused upon metabolic bioactivation and chemical biotransformation; solving the mystery of the therapeutic action of GTN, more than 130 years after its introduction in the clinic, thus emphasizing (a) how much still remains to be discovered and (b) the importance of chemical reactivity in influencing the biological activity of nitrates. Reviews covering the pharmacology, or medicinal and biological chemistry of nitrates have appeared [6–9]. The purpose of this article is to present the most recent studies on nitrates, with additional reference to nitrites, in the context of previous research and whilst emphasizing the central role of chemical reactivity under physiological conditions.

NITRATE THERAPEUTICS

The importance of the discovery of NO biology was recognized by award of the Nobel Prize in 1998. In the first decade after the identification of NO as endothelium-derived relaxing factor and the expanding roles of NO in biology, medicinal chemists focused upon developing inhibitors of the NOS isoforms, because of the proposed contributions of NO to cellular cytotoxicity and tissue damage, most notably from inducible NOS (iNOS) in inflammation, neuronal NOS (nNOS) in CNS excitotoxicity, and the toxicity of peroxynitrite [10]. Somewhat ironically the new nitrate drugs being developed in parallel were described as NO donors and NO-enhanced medicines with cytoprotective properties [11]. Classical nitrates continue to be of clinical use for angina pectoris, despite the clinical tolerance that is well documented for GTN, and controversy surrounding the phenomenon [12]. The clinical efficacy of GTN is enhanced by its selective venodilator effect resulting in decreased cardiac preload and myocardial oxygen consumption. The classical nitrate, ISDN, in a cocktail with hydralazine entered clinical use in 2005 for cardioprotection in African–American males with heart failure, marking a move to chronic drug treatment with nitrates [13].

Hybrid nitrates that conjugate a nitrate group to a commonly used drug *via* a labile linker have been the subject of numerous preclinical and clinical studies. The companies pioneering the use of hydrid nitrates have in recent years allied with "big pharma" in the areas of nitrate anti-inflammatories, antihypertensives, and glaucoma therapeutics. The medicinal chemistry of a variety of hybrid nitrates has recently been reviewed [14], for example: HCT 3012 is a naproxen hybrid in Phase III clinical trials for pain and inflammation; NCX 4016 in various Phase II clinical studies is the most well-studied aspirin hybrid (NO-ASA); NCX 701, an acetaminophen hybrid is in Phase II trials for pain; and, NCX 1015 is a prednisolone hybrid in Phase I trials for inflammation (Fig. 1).

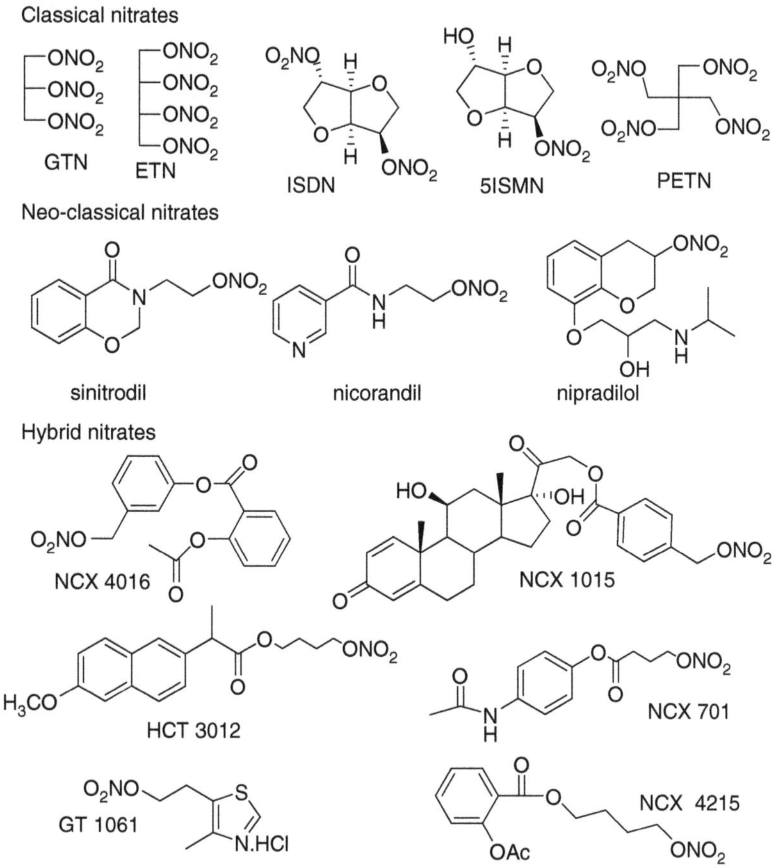

Fig. 1. Nitrate therapeutic agents in clinical use or clinical trials.

Sinitrodil, nicorandil, and nipradilol represent neo-classical nitrates, containing functionalities other than simple hydrocarbon or sugar skeletons, which influence biological activity, but which remain directed at vascular targets (Fig. 1). Nicorandil is a nitrovasodilator, also active at K_{ATP} channels, that is in preclinical and clinical studies for vascular diseases, including myocardial infarction [15,16]. Sinitrodil is a cardiovascular agent proposed to surpass classical nitrates because of reduced dilation of the smaller coronary and resistance vessels and the resultant effects on mean arterial blood pressure and heart rate [17,18]. Nipradilol, containing a pharmacophore with adrenergic antagonist activity, and used clinically as an anti-glaucoma therapeutic, was developed to exploit ocular nitrovasodilator activity, but subsequently has been shown to have neuroprotective actions [19–21].

Applications of nitrates beyond vascular targets are less common, but, supported by animal model data demonstrating cognition enhancement and neuroprotection, GT 1061 has recently entered clinical trials for Alzheimer's disease (AD) [22–24]; and, other exciting applications of nitrates as neuroprotective agents have also been described [25–27]. GT 1061 is an NO chimera, a nitrate that contains an ancillary pharmacophore designed to supplement the beneficial effects of nitrates in a specific disease state, in this case, neurodegeneration associated with AD [26,28]. In many brain regions, elevation of tissue cGMP levels and NO/cGMP signal transduction are triggered by activation of both the *N*-methyl-D-aspartate subtype of excitatory amino acid receptors (NMDAR) and cholinergic muscarinic receptor subtypes [29–31]. The NO/sGC/cGMP signal transduction system is important for modulating synaptic transmission and plasticity in brain regions such as the hippocampus and cerebral cortex, which are critical for learning and memory, and which are targeted in AD pathology [32–35]. Thus, GT 1061 was originally targeted at AD because NO plays a critical role in signal transduction cascades that are compromised in AD. Damage to cholinergic neurons has long been associated with AD, but recent evidence directly linking loss of NO/cGMP signalling, with NMDAR dysfunction, and the amyloid cascade theory of AD progression, unequivocally supports a need for supplementation of NO/cGMP in the AD brain [36,37]. GT 1061 has been studied in a variety of experimental paradigms where cognition deficits are induced, including: (a) injection of the muscarinic receptor antagonist scopolamine [38]; (b) administration of the cholinergic neurotoxin 192 IgG-saporin [39,40]; and (c) chronic, daily, bilateral, intracerebroventricular infusion of β-amyloid peptide (Aβ1-40) [41]. GT 1061 reversed cognition deficits in behavioral models including the Morris water maze, the step-through passive avoidance test, contextual memory, and a visual memory delayed matching to sample test.

There is general acceptance of the use of nitrates in indications with a vascular pathology; the field with which classical nitrates are closely associated. A more conservative attitude towards use of novel nitrates in other indications, such as neurological, may be explained by a number of factors. First, nitrates do not fit the contemporary, rational drug design paradigm: nitrates are extensively metabolized and bioactivated, and are expected to have multimeric actions and pleiotropic effects (though in most disease states this should be seen as a benefit rather than disadvantage). Second, there is a perception that cytotoxicity is inherently linked with NO, which partly derives from the study of NO as an atmospheric pollutant, before its renaissance as an essential component of life and health. The extensive drug discovery directed at inhibition of NOS can make the pursuit of nitrate drugs, that are generally almost universally described as NO donors, appear counterintuitive. Numerous reports on NOS inhibitors showing cytoprotective and chemopreventive effects are counterpoised by a large body of data showing cytoprotection and chemoprevention by nitrates and NO donors.

Several pathophysiological conditions clearly involve disruption of cellular NO homeostasis, which argues for therapeutic benefit in modulation of NO, rather than simply inhibition of its formation or action. Nitrates are NO mimetics, but poor NO donors, which is probably relevant to the safety of nitrate therapeutics: there is no evidence for quantitative bioactivation to NO; and at pharmacological concentrations the fluxes of NO detected are much lower than for genuine NO donors, such as the diazeniumdiolates (NONOates). There are numerous reports of the capacity, at higher concentrations *in vitro*, of NO and nitrosating agents (such as acidified NO_2^-) to cause DNA-modification [42]. The NO donor, diethylamine-NONOate, gaseous NO, and sodium nitrite are reported to be mutagenic in several bacterial strains in the reverse mutation assay (AMES test) [43,44]. The situation with nitrates is mixed: ISDN and several other nitrates have had negative AMES tests reported, whereas GTN was positive. There is no evidence for carcinogenicity of GTN or other clinical nitrates and several new nitrates have entered the clinic without genotoxicity concerns.

NITRITE THERAPEUTICS

Alkyl nitrites, most notably *iso*-amyl nitrite (IAN), share many of the biological activities of nitrates. They are smooth muscle relaxants used as nitrovasodilators and as recreational drugs known as poppers. IAN is also used in kits to treat cyanide poisoning, its function being to oxidize haemoglobin (Hb) to methaemoglobin (metHb) which reacts with cyanide to form the non-toxic cyano-metHb. Ethyl nitrite (EtONO) is a gas at room temperature and pressure that has been the focus of recent preclinical and clinical studies, in which its use has been proposed as a haemodynamic regulator in laparoscopic procedures [45], and as a haemodynamic regulator to alleviate pulmonary hypertension and hypoxaemia in neonates [46]. Simple alkyl nitrites are excellent nitrosating reagents, particularly towards biological thiols, such as glutathione (GSH). Inhalable EtONO therefore represents a simple and rapid means of elevating cardiopulmonary *S*-nitrosoglutathione (GSNO). Although there is no evidence for carcinogenicity, it is possible that the development of novel nitrite therapeutics will be hindered by genotoxic potential at high concentrations in cell culture assays because of high nitrosative reactivity and rapid production of NO_2^-.

NITRATES AS NITRIC OXIDE MIMETICS

In experiments in blood vessels that exploited the sensitivity of ESR (electron spin resonance) detection using a tissue permeable spin trap, NO was measured from endogenous production and from addition of an exogenous nitrosothiol [47–49]. However, using the same methodology, NO was not detected from pharmacologically relevant concentrations of GTN and ISDN. At much higher, suprapharmacological concentrations of nitrates (>10 μM), detection of NO was possible. A second recent study failed to detect NO in aortic tissue treated with GTN at pharmacological concentrations; in this study, a fluorescent indicator was used that responded to nitrosation of diaminorhodamine in the presence of NO [50,51]. In contrast, intracellular NO was detected using this fluorescent probe from a variety of agents that elevate NO: acetylcholine (that activates endothelial NOS; eNOS), and three separate NO donors (from different chemical classes). These agents were used at comparable concentrations to GTN and produced concentration-dependent responses. In addition, all four agents inhibited mitochondrial O_2 consumption, an effect associated with the interaction between NO and cytochrome c oxidase, and of potential pathophysiological significance. Again, in contrast, GTN was reported to have no effect on mitochondrial respiration. In the same study, vasorelaxation induced by all agents was inhibited by the "NO trap" oxy-haemoglobin (oxyHb), although vasorelaxation by GTN was inhibited much less by oxyHb than by all other agents. In a third study reporting measurement of coronary vascular resistance in the isolated Langendorf heart, oxyHb did not inhibit the relaxation induced by GTN [49,52].

Several researchers use phrases such as "delivery of NO bioactivity." In part, this recognizes the problematic data supporting the exclusive and simplistic sequence:

$$\text{nitrates} \rightarrow \text{NO} \rightarrow \text{bioactivity}$$

But what is NO bioactivity, or bioactive NO? Part of the problem is that NO to a chemist is a very specific chemical formula for a nitrogen-monoxide-free-radical species, whereas to some others, NO is a catholic symbol incorporating various nitrogen oxides and sometimes related thiol adducts. The biological activity of NO may be exerted by NO coordination to the Fe^{2+}-haem of soluble guanylyl cyclase (sGC) and elevation of intracellular cGMP, and also through cGMP-independent mechanisms *via* a variety of other possible reactive nitrogen oxide species (RNOS) (Fig. 2). Furthermore, the oxidative metabolism of NO and RNOS is intimately linked with a number of factors: oxygen and thiol concentrations; cellular redox state; redox active transition metal ions; and microenvironment [53]. The chemical reactions that connect true NO to the various RNOS are discussed more extensively in Chapter 1. The reactions of NO and RNOS with endogenous thiols have particular relevance (discussed in

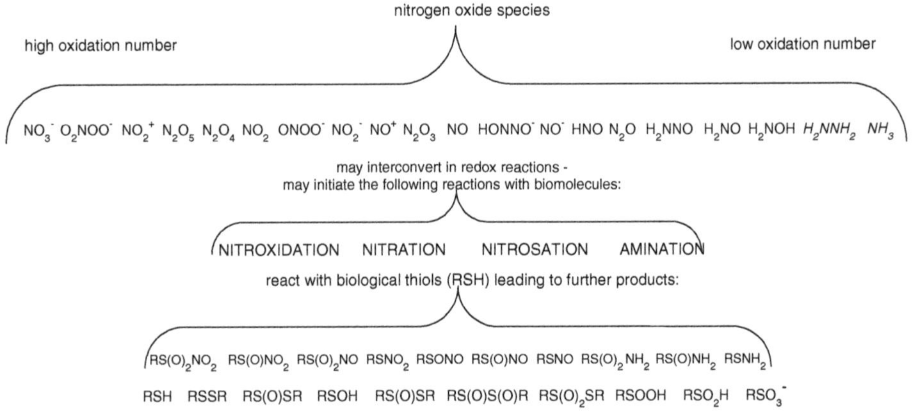

Fig. 2. Nitric oxide (NO) is only one nitrogen oxide that may form in a biological milieu. "NO bioactivity" may result from actions of NO itself or of the related nitrogen oxides and their reaction products, in particular those resulting from reaction with biological thiols.

Part 3 of this book), and are strongly affected by the presence of transition metal ions (as discussed in Chapters 9 and 11 of this book). Several metastable RNOS and their thiol adducts are difficult to quantitate and therefore the easily detected, stable end products of NO and RNOS metabolism are generally measured as surrogates: the majority of NO quantitations reported in the literature measure nitrosonium-like species ("NO^+"); in most cases acidified NO_2^-. This is of particular importance for nitrates, because inorganic nitrite (NO_2^-) is the dominant metabolic product of nitrate denitration, produced directly from nitrates without the intermediacy of NO. NO_2^- itself is well known to be biologically active at higher concentrations and able to mimic aspects of NO bioactivity. These roles for NO_2^- are discussed in detail in Chapters 14–16 of this book. Under hypoxic conditions, NO_2^- has been shown to function as a source of NO: nitrite infusions are in clinical trials and inhaled NO_2^- is in preclinical studies [54,55].

The endogenous formation of NO from arginine is an oxygen-dependent $5e^-$ oxidation requiring multiple cofactors; whereas, bioactivation of a nitrate to NO is a $3e^-$ reduction requiring oxygen atom transfer. The oxidation number of nitrogen in the various species that mediate NO bioactivity and the actions of drugs, including nitrates, span -3 to $+5$ (Fig. 3). The higher oxidation state nitrates may mediate nitrosation, nitration, and nitroxidation (also see Chapter 1 and references therein). But, we should be reminded that nitrates are not RNOS; for example, they are not comparable to the reactive nitroxidizing agents, NO_2 and peroxynitrite. It is interesting to note that nitrogen oxide species in higher oxidation states may induce physiological effects *without the requirement of prior reduction to NO*, thus accounting for the observation that *nitrates show significant bioactivity without evidence for formation of large amounts of NO*. As a simple example of this concept, consider thiol nitrosation, an important cGMP-independent bioactivity attributed to NO. Nitrate $2e^-$ reduction to organic nitrite is reasonably accompanied by nitroxidation of a metal centre or thiol [12]. Organic nitrites are excellent nitrosating agents, as is the RNOS, N_2O_3 (both $+3$ oxidation number nitrogen oxide species):

$$RONO_2 + R'SH \rightarrow RONO + R'SOH$$

$$RONO + R''SH \rightarrow ROH + R''SNO$$

Thus in two steps, nitroxidation and nitrosation, *actions associated with NO bioactivity, have been accomplished without the nitrate functioning as an NO donor*. Hence, we prefer the term NO mimetic, in place of NO donor, when referring to nitrates. The term NO mimetic is inexact in that nitrates cannot be expected to mimic all the activities attributed to NO, but is more accurate than NO donor. In order for NO itself to effect nitroxidation and nitrosation, the latter *via* the endogenous nitrosating agent N_2O_3, the following sequence is required (also see Chapter 1):

$$2NO + O_2 \rightarrow N_2O_4 \rightarrow 2NO_2$$

$$NO + NO_2 \rightarrow N_2O_3$$

$$N_2O_3 + R''SH \rightarrow R''SNO + NO_2^- + H^+$$

$$NO_2 + R'SH \rightarrow R'S^{\bullet} + NO_2^- + H^+$$

$$R'S^{\bullet} + O_2 \rightarrow R'SOH$$

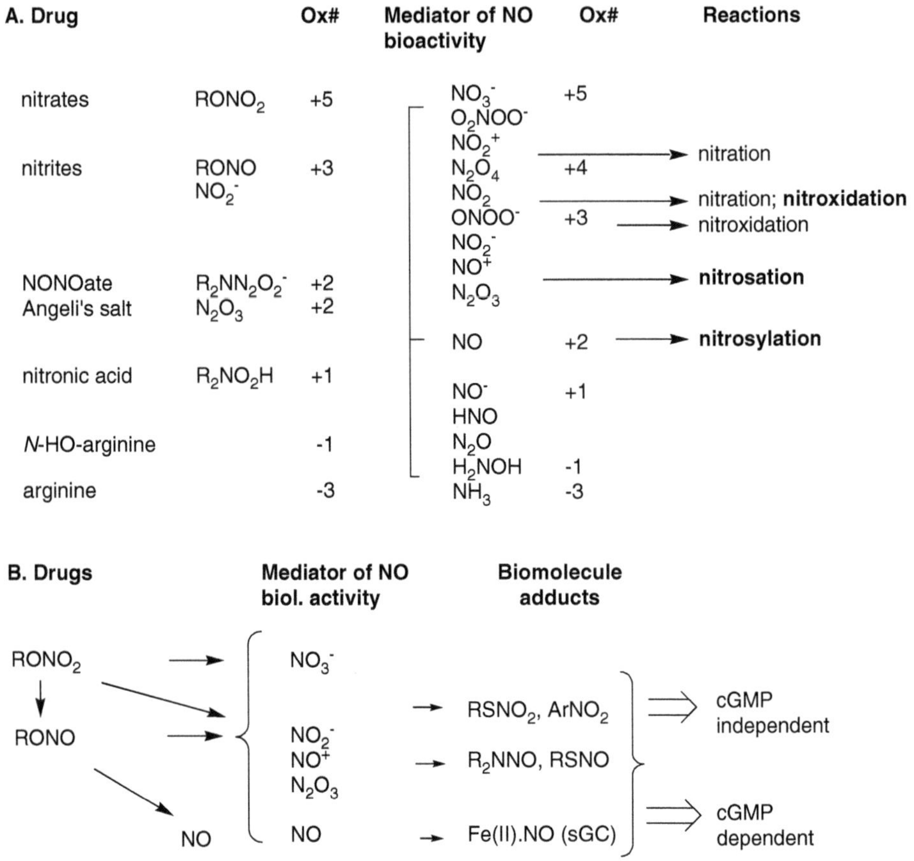

Fig. 3. (A) The listed drugs range in oxidation number at nitrogen from +5 to −3. All chemical species listed as mediators of NO bioactivity are implicated in the metabolism or activity of nitrogen oxides. Conversion of a drug to a mediator of the same oxidation number does not require redox activation. Several mediators, or RNOS, undergo reaction with biomolecules yielding a range of potential products. (B) Nitrates, in the highest oxidation state, can carry out nitroxidation and nitrosation *without* the intermediacy of NO, either directly or *via* intermediates such as nitrites. cGMP-independent activity includes that mediated *via* nitrosation, nitroxidation, and radical scavenging by NO.

This sequence requires several NO equivalents, consumes O_2, and generates RNOS: higher concentrations of NO are associated with formation of the oxidizing, cytotoxic RNOS, NO_2, and peroxynitrite:

$$NO + O_2^{\bullet -} \rightarrow ONOO^-$$

$$3ONOO^- + H^+ \rightarrow NO_2 + HO^\bullet + 2NO_3^-$$

$$\text{or } 3ONOO^- + CO_2 \rightarrow NO_2 + CO_3^{\bullet -} + 2NO_3^-$$

The corollary of this argument is that nitrates are not predicted to manifest identical physiological actions to other agents and drugs that are generally classed as true donors of free NO.

ENDOGENOUS NITRATES AND NITRITES

In mammalian systems, nitrates and nitrites theoretically may be formed as products of the reactions of RNOS: nitrates may be produced by nitrating RNOS or by the action of nitrosating and nitroxidizing RNOS; nitrites may be produced by nitrosating RNOS. Nitrates and nitrites have been proposed as endogenous products of the reactions of NO, peroxynitrite, or NO_2 with lipids, and other biomolecules, under oxidative and nitrosative stress. In particular, nitrite and nitrate derivatives of lipids were proposed to result from the antioxidant activity of NO towards lipid peroxyl radicals [56–59]. The capacity of NO_2 to react *via* H atom abstraction and *via* radical addition underlies its reactions with olefins, including polyunsaturated fatty acids, leading to formation of nitro, nitroso (nitrite), and nitrooxy (nitrate) derivatives, under both anaerobic conditions and in the presence of O_2 [60–64]. Both homolytic decomposition pathways for peroxynitrite and oxidation of NO_2^- by enzymes such as peroxidase represent physiological sources of NO_2 [60,64–66].

NO is an efficient antioxidant acting *via* lipid peroxyl radical chain termination. The potential for formation of nitrite and nitrate products as natural components of lipid membranes from the trapping of peroxyl radicals by NO is apparent [58,67–69].

$$4NO + 2ROO^{\bullet} + H_2O \rightarrow 2HNO_2 + RONO_2 + RONO$$

This possibility begs the question as to whether nitrites and nitrates would exacerbate or attenuate lipid peroxidation. Study of the effects on lipid peroxidation of a variety of nitrates, nitrites, and NONOate NO donors in rat brain synaptosomes and in model systems demonstrated that true donors of free NO inhibited lipid peroxidation [70–72]. Thus, classical nitrates such as GTN had no effect in contrast to novel NO releasing nitrates. Furthermore, the alkyl nitrites studied showed an antioxidant capacity deriving from NO release and from additional antioxidant mechanisms.

Trapping of lipid peroxyl radicals by NO is anticipated to form a number of potential nitrogen oxide derivatives, however, to date, the only characterized nitrated lipid is nitrolinoleic acid, a nitroalkane [73]. This nitrolipid is reported to contain a β-hydroxynitroalkane moiety similar to that formed on treatment of olefins with peroxynitrite (Fig. 4) [74]. This nitrolipid has been reported to be a store of NO bioactivity; the proposed mechanism of NO release involves formation of a nitronic acid intermediate followed by NO release from a nitrosoalkane (Fig. 4) [73]. Nitrosoalkanes and related *C*-nitroso compounds have recently been reviewed and several are known to be sources of NO bioactivity [75]. Interestingly, a class of nitroalkane NO-donor drugs previously developed by Fujisawa from oxidative nitrosation of microbial broths, are likely to release NO by a pathway almost identical to nitrolinoleic acid, *via* a nitronic acid intermediate and a nitrosoalkane NO progenitor (Fig. 4) [76]. These nitroalkanes are commercially available as the NOR family of NO donors.

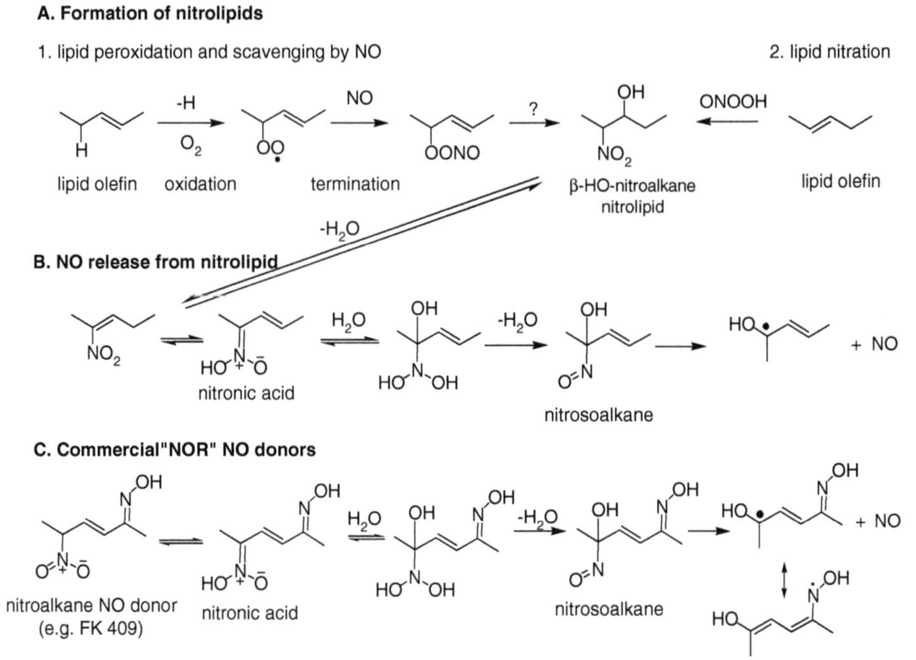

Fig. 4. Mechanistic pathways for: (A) nitration of lipids to form nitroalkane nitrolipids; (B) release of NO from nitrolipids; and (C) comparison to release of NO from semisynthetic nitroalkane NO donor drugs.

CHEMICAL REACTIVITY OF NITRATES

The reactions of nitrates are sketched in Fig. 5. These are all heterolytic processes, except for the homolysis of the sulfenyl nitrite that may be formed from initial reaction of a nitrate with thiol. The bond dissociation energy for organic nitrate homolysis, ~43 kcal/mol, is substantially higher than that calculated for homolysis of nitrosothiols (~32 kcal/mol) [8,28,72]. Nitrate homolysis to NO_2 is improbable as a thermal reaction relevant to biology, nevertheless, there is substantial evidence for photolysis of nitrates to give NO_2 [24]. Nitrates are photolabile, although simple aliphatic nitrates require extended reaction times [77–80]. In one case, detection of NO_2 as a reaction product was reported using ESR spectroscopy [77]. There is no evidence for direct photolysis leading to NO.

Two initial observations concerning the reactions shown in Fig. 5 are, first that none lead directly to NO and second that all nine reaction pathways shown are potentially viable under physiological conditions given the catalytic apparatus available in human enzymes, including Lewis acids, redox metal centres, Bronsted acids and bases, and redox cofactors.

The dominant reaction mediated by nitrates in organic chemical transformations is oxidation. Nitration of nucleophiles and arenes can also be effected by nitrates in the presence of a Lewis acid catalyst, and with Bronsted acids; nitrates are subject to ionization in concentrated sulfuric acid giving rise to the production of nitronium ion (NO_2^+) [81]. Nitrates are stable in dilute alkaline and acidic solution, allowing use as protecting groups in carbohydrate

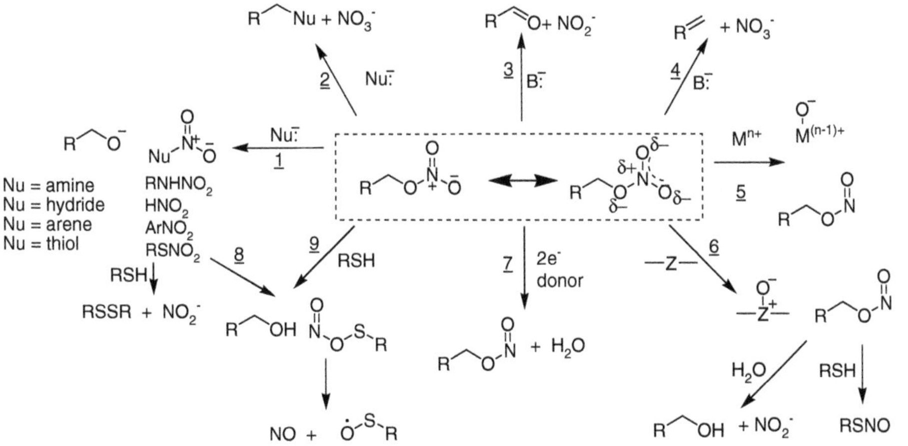

Fig. 5. Chemical reactions of nitrates of potential biological significance. (Paths 1 and 2) Reactions with nucleophiles at C or at the nitro group, leading to NO_3^- or NO_2^-. (3 and 4) Elimination reactions with strong bases. (5 and 6) Oxygen atom transfer to transition metals or reducing agents such as S, P, or N centres. (7) Reaction with $2e^-$ donor such as flavin cofactor. (8) Unimolecular rearrangement of a thionitrate to a sulfenyl nitrite. (9) Reaction of thiol at the nitro group generating a sulfenyl nitrite.

chemistry [82]. In strong alkaline solution, nitrate esters undergo solvolytic decomposition by S_N2 nucleophilic substitution yielding NO_3^-, β-elimination yielding NO_3^-, and α-elimination yielding NO_2^- and aldehyde (Fig. 5) [83–86]. Nitrate solvolysis generally proceeds *via* C—O bond fission, but evidence for O—N bond fission exists in reactions with a number of nucleophiles, including hydrazine and hydroxylamine, as evidenced by isotopic labelling and stereochemical studies [87–91]. Sulfur nucleophiles react with nitrates *via* C—O or N—O bond fission, yielding NO_3^- or NO_2^-, respectively. Benzyl nitrate was reported to undergo substitution at C with the S-nucleophile, thiophenolate, to yield NO_3^- [92,93]. Sulfides reacted with alkyl nitrates to yield NO_2^-: initial reaction at the N or O of the nitrate group would give an intermediate of the form $S_x(NO_2)^-$ [89,94]. Initial reaction at N would yield a thionitrate ($RSNO_2$), whereas a number of other reactions yield a nitrite ester (RONO) (Fig. 5).

Thionitrates

Tert-butyl thionitrate was synthesized in 1932 by oxidation of the corresponding nitrosothiol using fuming nitric acid as the oxidant: it was reported that the resulting thionitrate was more stable than the starting nitrosothiol [95]. An alternative synthesis, *via* reaction with N_2O_4 [96], gave alkyl and aryl thionitrate esters; the latter were thermolytically unstable [97]. The synthesis and crystal structure analysis of an aryl thionitrate has been reported, utilizing a sterically hindered thiol to increase stability [98]. Thionitrates are colorless in contrast to the pink or green colors of the corresponding nitrosothiols [99–101]. The isomeric sulfenyl and sulfinyl nitrites have not been isolated. Thermolytically stable thionitrates were observed to undergo reduction by phosphines and selenides to give nitrosothiol as product by oxygen atom transfer [102]. In neutral, aqueous solution, *t*-butyl thionitrate was reported to decompose with a half-life of approximately 10 min in a buffer-catalyzed reaction that gave NO, and cleanly yielded *t*-butyl sulfinyl (tBuS = O) and *tert*-butyl sulfonyl ($tBuSO_2$) products [101]. NO release from *t*-butyl thionitrate, measured electrochemically, was inhibited by addition of cysteine, and the thionitrate was observed not to activate partially purified sGC. These data imply a high reactivity of thionitrates towards free and protein-thiols-producing disulfide and NO_2^- as products.

CHEMICAL REACTIVITY OF NITRITES

The chemistry of organic nitrites includes heterolytic and homolytic reactions, in addition to oxidation of haemoglobin [103]. There is an extensive literature on the dominant reaction pathway for nitrites in solution, transnitrosation, primarily reported by Williams and co-workers [104–107]. In parallel, these researchers studied the reaction of other nitroso compounds including nitrosothiols, although the fact that many studies were not carried out in pH 7.4 phosphate buffer has led to much of this work being overlooked in contemporary research. In the mechanistic studies on nitrite hydrolysis, the sole N-containing product was assumed to be NO_2^-, and therefore there was no attempt to detect other products including NO [68,103–105]. Nitrites are excellent nitrosating agents subject to acid catalysis: the nitrosation of different substrates (alcohols, thiols, amines) by alkyl nitrites (especially tertiary alkyl nitrites) generally occurs at rates orders of magnitudes higher than competing hydrolysis [107]. In an aqueous biological compartment, the fast rate of reaction with glutathione (GSH; $k \sim 2\ M^{-1}s^{-1}$) will result in the biological activity of the nitrite largely being mediated through GSNO [106]. In lipid compartments, the identity of the nitroso acceptor is not so clearly defined.

The homolytic cleavage of the O—NO bond of alkyl nitrites produces NO, with values for the BDE quoted as 35–40 kcal/mol [72,108]. A number of synthetically useful photochemical reactions of nitrites have been reported resulting from initial O—N bond photolysis, but in general there has not been an attempt to quantify NO as a by-product [109,110]. Photolysis of nitrites in organic solvents yields aldehydes and other products, including compounds such as nitrosoalkanes and nitroxyl radicals that are radical scavenging species [111–113]. In aqueous solution, we have observed that alkyl nitrites spontaneously generate a low flux of NO, as detected electrochemically and by chemiluminescence. This behavior differentiates nitrites from nitrates, which do not spontaneously release NO, and from nitrosothiols, which generally in pure form in solution and in the dark do not spontaneously release NO [71,114,115]. NO is a minor product of the decomposition of nitrites in aqueous solution and the detailed mechanism of formation requires further study. Possible pathways include the reaction with a species formed on initial nitrite solvolysis, or weakening of the O—N bond by stabilization of the strong nitroso dipole in aqueous solution in simile with phenols [116].

Nitrites are nitrovasodilator drugs used both clinically and recreationally. The history of nitrites is interleaved with that of organic nitrates, since the vasodilatory action of IAN and its potential for use in angina was recognized before that of GTN [5]. Doyle studied the reaction of organic nitrites with Hb, reporting binding at the haem site, leading to reduction to NO and alcohol as products [117]. The molybdoenzyme, xanthine oxidoreductase (XOR) mediates the rapid reduction of nitrites to NO under anaerobic conditions in the presence of xanthine [118]. Thus there are numerous options for physiologically relevant 1-electron-reductive bioactivation of nitrites to NO, in addition to the rapid transnitrosation reaction with thiols, which will supplement levels of GSNO (a molecule extensively discussed in Chapters 9–11).

BIOLOGICAL REACTIVITY OF NITRATES AND BIOACTIVATION

Mediation of the vasodilator response to GTN by NO requires the efficiency of bioactivation to NO to be high, since GTN induces tissue relaxation with nanomolar potency and bioactive concentrations of NO are also nanomolar [119–121]. However, inspection of the chemical reactions of nitrates, depicted in Fig. 5, indicates that there is no *direct* pathway for the $3e^-$ reduction to NO. On the basis of chemical reactivity, the most facile reactions of nitrates *in vivo* are predicted to be:

1. Formation of NO_2^- as a major product; possible formation of NO_3^- (Fig. 5 Path 2).
2. Oxidation of reactive thiols, either by direct oxidation (Path 6), *via* a thionitrate (Paths 1 + 9), or *via* a sulfenyl nitrite (Path 9 or 1 + 8).
3. Oxidation of transition metal centres (Path 5).
4. Formation of nitrosating agents (RONO or RSNO; Paths 5, 6, and 7).

NO bioactivity incorporates the following cGMP-independent processes mediated by RNOS:

1. Protein nitrosation: acting to modulate protein function, in signalling, and RNOS-storage.
2. Metal ion displacement (e.g. from Zn-finger proteins [122,123]).
3. Redox signalling *via* thiol oxidation and *via* subsequent glutathiolation [124,125].

Therefore the cGMP-independent bioactivity of NO attributed to nitrates can be accounted for without recourse to reductive bioactivation of nitrates to NO itself.

Several NO-NSAIDs continue to be intensively studied and it is instructive in considering biological reactivity to discuss data on the isomeric NO-ASAs, NCX 4016, and NCX 4040, that contain the hydroxybenzyl nitrate (HBN) moiety [126–130]. In rats, NCX 4016 itself was not detected in plasma after oral administration, and metabolism was complete within 5 min in rat liver microsomes. HBN was the sole nitrate observed. In accordance with the known chemical reactivity of benzyl nitrates towards thiols [86,92], the GSH adduct, derived from nucleophilic substitution on HBN displacing NO_3^- (Fig. 5 Path 4), was also observed. After administration of NCX 4016 to rats, plasma NO_2^- levels were unchanged ($\sim$1 μM) over 6 h, but plasma NO_3^- was elevated from $\sim$20 μM to $\sim$130 μM. Biotransformation of NCX 4016 generates NO_3^-, but it is difficult to conceive that NCX 4016 acts therapeutically as an NO_3^- donor. There is little evidence for significant biological activity associated with NO_3^-, the levels of which in human plasma and rat aorta are relatively high ($\sim$25 μM and $\sim$50 μM, respectively) and fluctuate substantially with diet [131–133].

ESR spin trapping of NO has become a popular method of NO detection for nitrates because of its sensitivity, use in biological fluids, and the low levels of NO that are formed from pharmacologically relevant amounts of nitrates. Paramagnetic nitrosyl-haemoglobin,

NO-Hb, was observed in rat blood and nitrosyl-myoglobin, NO-Mb, in myocardial tissue *in vivo* and *in vitro* after administration of NCX 4016. In the study of nitrates in biological systems, care must be taken in the interpretation of ESR detection of low amounts of NO-Hb, because the primary metabolite of nitrates is generally NO_2^-; several mammalian enzymes have been shown to have the capacity for reduction of NO_2^- to NO, including XOR, deoxy-Hb and cytochrome-P450 oxidases (CYP) (details of these pathways are discussed in Chapters 14 and 15 with full references). Both oxy-Hb and deoxy-Hb react directly with nitrates and nitrites [115,134].

ESR measurements in rat blood showed that both ISMN and NCX 4016, at high concentrations (1 mM), generated similar levels of NO-Hb (~30 μM at 1 h). NCX 4016 largely absorbed in the small intestine, is subject to substantial first pass metabolism, and possesses very limited systemic bioavailability, features common in many NO-NSAIDs. NO-flurbiprofen, an example of a hydroxybutyl nitrate NO-NSAID, was also reported to undergo rapid first pass metabolism and not to be detectable in target tissue [135,136], but in this case, levels of NO_2^- in plasma were increased over 8 h [137].

NO-ASA has been the focus of clinical trials, most recently for colon cancer chemoprevention. The bioactivity of NO-ASA includes antiproliferative and pro-apoptotic activity in cancer cell lines [138], but only pro-apoptotic activity in a mouse intestinal carcinogenesis model [139]. It has been reported that *the bioactivity results from induction of oxidative stress that begins with the depletion of GSH* by the reaction of GSH with HBN forming the GSH adduct [140]. In cell culture, NCX 4040 depleted GSH (IC_{50} ~ 50 μM), elevated ROS, and activated the intrinsic apoptosis pathway, by a mechanism that incorporates β-catenin cleavage (at NCX 4040 > 50 μM). The same compound was reported to potentiate the cytotoxicity of cisplatin in resistant ovarian cancer cells, and NCX 4016 (100 μM) was shown to deplete GSH (by 50% in cells with elevated basal GSH) [141]. However, in this study it was implied that the release of sustained amounts of NO was a causative factor: NO was detected by ESR in incubations of NCX 4016 (0.5 mM) using the $Fe(MGD)_2$ spin trap at 25 min and increased up to 90 min. Interestingly, NCX 4040 was reported to inhibit iNOS upregulation and activity in colon cancer cells (IC_{50} ~ 40 μM), although these researchers expressed caution as to the applicability of these cell culture data to the clinical context [142].

The structural diversity of reported hybrid nitrates is very limited: the nitrate in most NO-NSAIDs is a benzyl mononitrate or butyl mononitrate moiety. In colon cancer cells, the metabolism and cytotoxic/apoptotic activity of NO-ASAs (including the regioisomers NCX 4016 and NCX 4040) and control compounds was reported [143]. Substantial differences in potency for apoptotic and antiproliferative activity were observed in the 3 isomers of NCX 4016. For example, under conditions in which NCX 4016 manifested an antiproliferative IC_{50} of 200 μM and 5% apoptosis, the isomers were 100–200-fold more potent and induced 100% apoptotic or atypical cell morphology: NCX 4040 is thus significantly more cytotoxic and pro-apoptotic than NCX 4016. The extent of denitration of the isomers ($NO_3^- + NO_2^-$ measured by Griess assay) was greater than for NCX 4016 by a factor of about 3 after 1 h incubation, whereas NCX 4215 released negligible amounts of $NO_3^- + NO_2^-$. The rationale for the structure–activity relationship, the mechanism of bioactivation, and the source of NO-ASA bioactivity remain undetermined. Two important observations were made: (1) the combination of ASA with genuine NO donors at high concentrations (SNAP or DETA/NO at 700 μM) gave only weak antiproliferative activity and;

(2) cyclooxygenase (COX) inhibition had practically no bearing on the biological activity of NO-ASA. The simplest interpretation of these observations is that *it is the nitrate (not ASA) that is responsible for the biological activity of NO-ASA and that the nitrate bioactivity is not identical to that of NO.*

PATHWAYS FOR REDUCTION TO NITRIC OXIDE

Reductive bioactivation of nitrates to NO must involve formation of an intermediate species, even if this is a short-lived species at the active site of an enzyme responsible for bioactivation. Before reviewing evidence for the roles of specific enzymes, the potential chemical pathways for this $3e^-$ reduction will be reviewed, with reference to Fig. 5. The overall $3e^-$ reduction may be accomplished by a $(2+1)e^-$ transfer:

$$2e^- + 1e^- : \quad RONO_2 \rightarrow NO_2^- \rightarrow NO; \text{ or } RONO_2 \rightarrow RONO \rightarrow NO$$

A route *via* initial single electron transfer has rarely been discussed or elaborated on, although nitrates are good e^- acceptors. A radical anion was postulated in the interaction of GTN with sGC, but dismissed by the same workers [144]:

$$1e^- + 2e^- : \quad RONO_2 \rightarrow RONO_2^{\bullet -} \rightarrow NO$$

The nitrate radical anion may be an intermediate in flavin-dependent reduction to a nitrite (Fig. 6):

$$FlH^- + RONO_2 + H^+ \rightarrow Fl^{\bullet -} + RONO_2^{\bullet -} + H^+ \rightarrow Fl_{ox} + RONO + H_2O$$

Model systems are of value in exploring chemical mechanisms of nitrate reductive bioactivation and predicting biomolecules able to mediate bioactivation. A catalytically active,

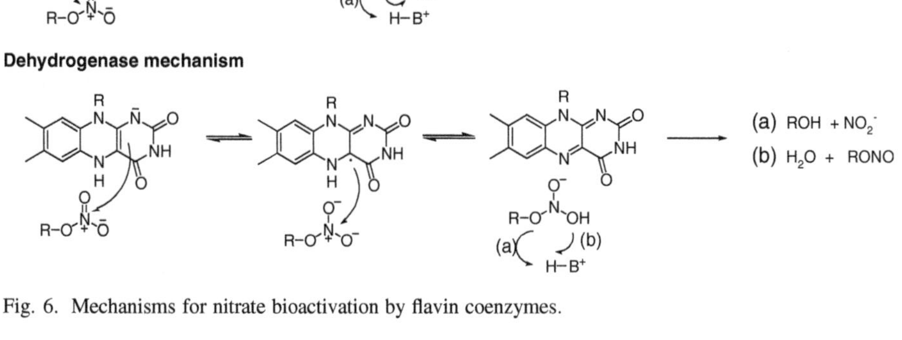

Fig. 6. Mechanisms for nitrate bioactivation by flavin coenzymes.

synthetic Mo-complex, that uses a cofactor to reduce nitrates to NO, supports the general ability of Mo-metaloenzymes to bioactivate nitrates *via* oxygen atom transfer [8,145]. In this model system, nitrates, nitrites, NO_2^- and NO_3^- were all shown to be substrates for reduction to NO, but the greatest NO fluxes were observed for nitrites. As might be expected, the 1e– reductions of nitrites and NO_2^- to NO were observed to be more facile than the $3e^-$ reductions of GTN and NO_3^- to NO; uncatalyzed NO release was detected from both NO_2^- and nitrites. In this model system, nitrites and NO_2^- were observed to be kinetically competent intermediates in reduction of nitrates to NO.

The proven concept of using intramolecular reactions to provide models for enzyme-mediated reactions [146,147] has been useful in modelling sulfhydryl-dependent nitrate reduction. Thiols, such as cysteine, possess low reactivity towards the nitrate functional group [120], therefore incorporation of a thiol group juxtaposed to a nitrate group in the same molecule allows the highly favorable intramolecular reaction. 1,2-Dinitrooxy-3-mercaptopropane (GT 150) provides a model for the reactivity of GTN bound at a protein-binding site adjacent to a cysteine thiol functionality. In aqueous solution, GT 150 is a spontaneous NO donor, generating fluxes of NO comparable to NONOates and other NO donors, and displaying the expected properties of a genuine NO donor in both activating sGC and inhibiting lipid peroxidation [70,71]. Moreover, the disulfide congener (GT 715) and related aryl disulfanyl dinitrates liberate NO in the presence of added thiols through formation of GT 150 [70]. Interestingly, GT 715 exhibits neuroprotective properties in several animal models of ischemic stroke and reverses cognition deficits induced in animal behavioral models [26,27,148]. The available mechanistic data on GT 150, GT 715, and related nitrates suggests that initial oxygen atom transfer from N to S gives a nitrite intermediate that is subsequently reduced to NO by well-known thiol-dependent mechanisms. These model systems therefore firmly establish metal-ion-dependent and sulfhydryl-dependent mechanisms for nitrate reduction to NO *via* an organic nitrite intermediate.

Via inorganic nitrite

A bioactivation route *via* NO_2^- was the earliest postulated: the 1981 Ignarro hypothesis posited intracellular conversion to NO_2^- by reaction with cysteine, followed by liberation of NO from disproportionation of nitrous acid [149]. *The corollary of this hypothesis is that nitrates are intracellular NO_2^- donors*; NO_2^- is the predominant metabolite of vascular biotransformation of GTN and of many other nitrates. There are two concerns with this pathway. First, NO_2^- has low vasodilator potency and high endogenous levels [150–153]: the NO_2^- concentration in rat aorta is reported to be between 10 and 25 μM [132], and NO_2^- is distributed evenly between the intravascular and extravascular compartments [154]. GTN is a vasodilator of nanomolar potency, and thus NO_2^- derived from GTN would only minimally increase the intracellular concentration of NO_2^-. Second, the rate of NO production from nitrous acid at physiological pH is very low. In the proton-rich environment of the mitochondrial intermembrane space, where this reaction has been implicated [155], the NO flux can be estimated from the incremental increase in NO_2^- concentration: at pH 4, the rate of NO formation from NO_2^- is approximately 0.01% per second [156], leading to production of picomolar concentrations of NO from the nanomolar concentrations of NO_2^- derived from GTN in the short time frame for GTN-induced vasodilation. The mitochondrial electron transport chain, including the bc_1 complex and cytochrome c oxidase, are theoretically capable of mediating the subsequent $1e^-$ reduction of NO_2^- to NO under anaerobic conditions [157,158]. Mitochondrial reduction of NO_2^- to NO may play a role in the activity of nitrates in myocardial ischemia and in the selective venodilation observed for nitrates [159,160], however, under normoxic, aerobic conditions, cytochrome oxidase is expected to remove NO from the mitochondria by oxidation to NO_2^- [161,162]:

$$NADH + 3H^+ + \frac{1}{2}O_2 + NO \rightarrow NAD^+ + H_2O + 2H^+ + NO_2^-$$

There is substantial current interest in the biological activity of NO_2^-, discussed in Chapters 14 and 15. NO_2^- has been shown to be oxidized by peroxidase to NO_2 [65,66], and to be reduced by ferrous Hb to yield NO-Hb or NO [133]. A contribution from NO_2^- in mediating the bioactivity of nitrates cannot be completely ruled out despite its relatively high physiological and pathophysiological concentrations [163,164] and low biological activity in several studies [131,150,153], especially if intracellular NO_2^- delivery is highly compartmentalized [133,165–167].

Via organic nitrite

Nitrites are putative intermediates in the $3e^-$ bioactivation of nitrates, and since reduction of nitrites to NO is only a $1e^-$ reduction, unsurprisingly, differences between the biological activity of nitrates and nitrites have been observed [168]. The chemical mechanism and biotransformation apparatus responsible for the reduction of neither nitrates nor nitrites is clearly defined. It has been speculated for some time that nitrites are intermediates in the bioactivation of nitrates to NO, one mechanism incorporating enzyme-mediated reaction of nitrites with glutathione to afford S-nitrosoglutathione which could subsequently release NO [169]. In another instance, nitrates were proposed to be bioactivated to a nitrosothiol *not* to NO itself [170,171]. There is no direct chemical reaction to convert a nitrate to a nitrosothiol. If nitrates are reduced to nitrites, it is known from Williams's extensive research that the reaction with thiols is very rapid: nitrites will definitely yield nitrosothiols in a biological milieu [172–174]. Comparison between the biological activity of nitrates and nitrites has been made, showing significant differences, especially in the lack of evidence in animal and tissue models for a phenomenon mirroring nitrate tolerance [174–176]. This does not rule out nitrites as intermediates in nitrate bioactivation, since it is more reasonable that tolerance would be associated with the initial $2e^-$ reduction of nitrates to nitrites. 1,2-Dinitrooxy-3-nitrosooxypropane (NOGDN; Fig. 7), the putative nitrite intermediate in GTN bioactivation, was reported in 1994 and observed to be a reactive nitrosating agent [173].

Via thionitrate

Early observations on loss of thiol and formation of NO_2^- from reaction of thiols with nitrates in neutral aqueous solution were explained by initial formation of thionitrate intermediates [177,178]. A cohesive chemical mechanism that identified an organic thionitrate as the common intermediate in a mechanism leading to either NO_2^- or NO was published in 1992,

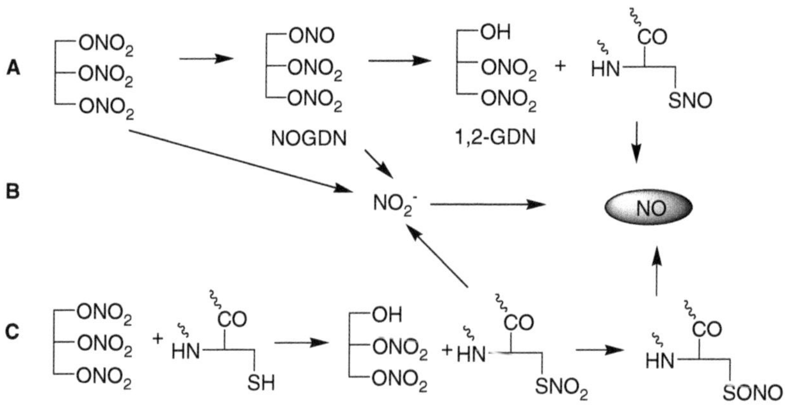

Fig. 7. Pathways for GTN bioactivation to NO. (A) *Via* nitrite (sulfhydryl-dependent). (B) *Via* NO_2^-. (C) *Via* thionitrate (sulfhydryl-dependent). The alternate GDN isomer, 1,3-glyceryl dinitrate (1,3-dinitrooxypropan-2-ol) has been omitted for clarity.

recognizing the importance of thionitrate isomerization (Fig. 7) [169]. In 1994/95, computational and experimental studies reported the spontaneous formation of NO from a synthetic thionitrate, supporting a mechanism *via* rearrangement to a sulfenyl (RS(O)NO) or sulfinyl nitrite (RSONO) [179,180]. Sulfenyl and sulfinyl nitrites are thermodynamically accessible from thionitrate isomerization and possess low bond dissociation energies for homolysis to NO and sulfinyl radical (10–15 kcal/mol) [8]:

$$RONO_2 + R'SH \rightarrow ROH + R'SNO_2 \rightarrow R'SONO \rightarrow NO + R'SO \rightarrow {}^1\!/\!_2R'S(O)_2SR'$$

An alternate, direct pathway to a sulfenyl nitrite intermediate (Fig. 5 Path 9) may be drawn if the bond forming with the sulfur nucleophile occurs at the oxygen of the nitrate.

A number of alternative pathways may be drawn, taking into consideration the reactive forms of sulfur commonly seen in biological systems: thiyl radical ($RS^{\bullet}$); disulfide radical anion ($RSSR^{\bullet-}$); and sulfenate (RSO^-; the conjugate base of sulfenic, pKa 6.1), for example:

$$RONO_2 + R'S^{\bullet} \rightarrow RON(O)OSR' \rightarrow R'SO^{\bullet} + RONO$$

However, in a photochemical study, we observed no evidence for a rapid reaction between nitrates and thiyl radical [24].

PROTEIN MEDIATORS OF BIOTRANSFORMATION

Bioactivation of nitrates both to yield NO and to yield cGMP-independent bioactivity requires reduction and oxygen atom transfer. Therefore, an enzyme or protein mediator of bioactivation requires a functional group capable of supplying electrons and/or accepting an oxygen atom; the logical candidates are metaloproteins and protein thiols and sulfides. Several proteins have been identified that are capable of mediating the denitration of GTN, including Hb, Mb, XOR, old yellow enzyme, glutathione *S*-transferase (GST), CYP,

NADPH cytochrome P450 reductase (CPR), and aldehyde dehydrogenase type-2 (ALDH2) [119,134,155,181–192]. Most of these proteins were reported in the last century to mediate nitrate biotransformation, and several have been subject to intensive further study in recent years. In all cases, a problem in unambiguous interpretation of both *in vivo* and tissue studies, of the role of specific proteins in nitrate bioactivation, is the lack of specificity of many enzyme inhibitors.

Nitrate bioactivation (sometimes termed mechanism-based biotransformation [6]) has been intimately linked with nitrate tolerance, the phenomenon that describes the decreased vasodilator and anti-anginal efficacy of nitrates, in particular GTN, after longer term clinical administration. It is reasonably postulated that tolerance occurs when the biological apparatus responsible for reductive bioactivation is exhausted, presumably through oxidation/consumption of reducing equivalents. A role in tolerance is therefore often viewed as a requirement for the nitrate bioactivation mechanism. Denitration of GTN yields the dinitrate metabolites, glyceryl-1,2-dinitrate (1,2-GDN) and glyceryl-1,3-dinitrate (1,3-GDN); regioselectivity being demonstrated in vascular biotransformation for formation of 1,2-GDN, which is lost in tolerant tissue [119,185,189,193–200]. Thus, regioselectivity has been used as a marker in the search and validation of bioactivation mechanisms. The fascinating observation of stereoselective bioactivation of nitrates has been used less often, but has been used to support a role for CPR in nitrate bioactivation [201,202]. In this work, stereoselectivity was lost on addition of the flavoprotein inhibitor DPI. However, DPI shifted GTN dose–response curves to the right in both tolerant and non-tolerant aortic tissue arguing against a dominant role for CPR in vascular bioactivation associated with tolerance [119].

More recent work has supported a function for CPR in nitrate biotransformation [203]. Under anaerobic conditions and in the presence of both NAD(P)H and added thiol, CPR was observed to generate nitrosothiol and NO, the formation of which was inhibited by DPI. The requirement for added thiols implicates the thiol-reactive nitrite, NOGDN, as the product of CPR-mediated bioactivation, and the progenitor of NO. Inhibition of GTN metabolism by DPI is compatible with a role for flavoenzymes in nitrate denitration and possibly bioactivation. Mechanisms may be drawn for the nitrate-dependent conversion of reduced flavin to oxidized flavin in both flavin-dependent dehydrogenases and in flavin-dependent oxidases (Fig. 6). In the oxidase mechanism, the nitrate substitutes for O_2, and therefore this pathway may be important in nitrate bioactivation under hypoxic conditions.

Mammalian molybdoenzymes (aldehyde oxidoreductase, sulfite oxidoreductase, and XOR) catalyze oxygen atom transfer and e^- transfer: for example, a bacterial nitrate reductase molybdoenzyme is able to catalyze the $2e^-$ reduction of NO_3^- to NO_2^-. The presence of flavin and iron–sulfur clusters in addition to molybdenum redox centres, presents a number of options for nitrate group reduction. Over 50 molybdoenzymes have been identified, however, only XOR has been examined for nitrate bioactivation [118,166,181,192,204–206].

XOR has been frequently used as a source of $O_2^{\bullet -}$ *in vitro* and therefore in aerobic solution may function as an NO scavenger. However, a body of work by Harrison's group showed that in anaerobic solutions, XOR mediated the reduction of organic and inorganic nitrates and nitrites and gave NO as product [118,166,181,192,206]. XOR was reported to mediate the NADH or xanthine dependent: (a) $2e^-$ reduction of nitrates to NO_2^- at the flavin-site; (b) $1e^-$ reduction of NO_2^- to NO at the Mo site; and (c) $1e^-$ reduction of nitrites to NO at the flavin site. GTN, ISDN, and ISMN, were observed to be substrates for reduction under anaerobic conditions, generating a low flux of NO, but XOR was inactivated in the process. A more recent study of the XOR-mediated reduction of GTN and ISDN confirmed aspects of this work, but importantly, NO was not detectable as a product without the addition of thiols or ascorbate [207]. The thiol dependence and detection of nitrosothiols argue for XOR bioactivation of nitrates to nitrites, followed by non-enzymatic trapping by thiols to generate a nitrosothiol that itself generates NO. It was remarked that the maximum rate of NO production observed was less than 10% of that of nitrite hydrolysis to NO_2^-. These observations are in accordance with the rapid trapping of nitrites by thiols and the rapid hydrolysis of NOGDN, the nitrite intermediate formed from GTN, which was first reported in 1994 [173]. XOR itself catalyzes NO_2^- reduction to NO at the Mo site [166,206], however, a leading role for XOR in vascular nitrate bioactivation is not supported by observations that the XOR inhibitor, allopurinol, has no influence on tissue relaxation [119]. A role for XOR in ischemia-reperfusion injury has often been cited and the ability of XOR to generate cytoprotective levels of NO and nitrosothiols in ischemia may be relevant.

Haemoproteins represent prime candidates for mediating nitrate bioactivation, because nitrates, nitrites, and related molecules have been demonstrated to coordinate with and be reduced at Fe-haem sites: (a) GTN reacts at the ferrous-haem of haemoglobin and myoglobin to give GDN; and (b) GTN is rapidly consumed by deoxyHb to give NO_2^- [134,208]. CYP isoforms have often been proposed to mediate GTN bioactivation to NO, but it has recently been reported that the reduction of NO_2^- to NO by CYP is a minor contributor to GTN bioactivation [203]. Mixed metaloprotein-sulfhydryl pathways have been proposed [9]. Doyle reported Hb-mediated reduction of nitrites to NO, including *via* a mixed haem-sulfhydryl pathway involving nitrosation of the Hb-β-93 cysteine thiol residue [117]. A mixed haemoprotein-sulfhydryl pathway would involve initial reduction to a nitrite, however to date, only NO_2^- production from Hb-mediated reactions of GTN has been reported.

The high affinity NO receptor, sGC, responsible for mediating the cGMP-dependent bioactivity of NO, is itself a ferrous haemoprotein [209,210]. Containing 10–24 cysteines per subunit, the various dimeric isoforms of sGC are also cysteine rich, suggesting possible sulfhydryl or mixed pathways of reaction of GTN with sGC itself. Given the oxidative chemistry of nitrates, it would be surprising if sGC was not reactive towards nitrates, especially since high concentrations of nitrates are required for sGC activation *in vitro*. Furthermore, even at these high concentrations, nitrates do not activate sGC *in vitro* unless cysteine or certain other thiols are added. In incubations with sGC, the flux of NO generated from simple reaction of GTN with cysteine is too low to account for activation of sGC by NO, unless the flux of NO is increased by reaction with sGC or other proteins in such incubations [120]. In the presence of cysteine, low potency, partial agonism of sGC activity is observed in response to GTN. The concentration–response curve for sGC activation by a true NO-donor NONOate was shifted to the right after preincubation of sGC with GTN. Furthermore at higher concentration, GTN significantly decreased the efficacy of the NO donor [120]. Inhibition of sGC, *via* haem or thiol oxidation, are likely inhibitory mechanisms for GTN, which is capable of rapid oxidation of deoxy-Hb to met-Hb [134], and oxidation of protein thiols [211,212]. Modification of cysteine residues of the β_1 subunit of sGC has been shown to inhibit NO-dependent activation of sGC [213]. Spectroscopic studies of sGC activation by GTN concluded that there was no evidence for a Fe^{2+}–sGC–NO complex, but that there was substantial evidence for Fe-haem oxidation [144].

Several heterocyclic small molecules are able to activate sGC in an NO-independent manner; for example YC-1, that is thought to bind to an allosteric site on sGC [214]. These activators often have modest efficacy, but substantially increase potency and efficacy of the poor sGC activator CO by inhibiting dissociation of the Fe^{2+}–sGC–CO complex [215,216]. YC-1 potentiated the activation of sGC by GTN (in the presence of cysteine) to the same activity levels seen for NO and true NO-donors [120]. The magnitude of the potentiation of GTN activity byYC-1 suggests that GTN activation of sGC could be subject to an endogenous allosteric potentiator. YC-1 has been shown to potentiate GTN-induced tissue relaxation in both naïve and tolerant aortic tissue [217]. This phenomenon may be general to nitrates since cGMP accumulation in PC12 cells in response to a hybrid nitrate drug was reported to be profoundly amplified by YC-1 [218].

ALDH2, a major NAD-dependent Phase II metabolic enzyme, essential for alcohol clearance *via* acetaldehyde dehydrogenation, contains 3 cysteines at the active site and is a candidate for sulfhydryl-dependent nitrate bioactivation. ALDH2 was reported in 1985 to mediate biotransformation of both GTN and ISDN [190], and in 1994 the inactivation of ALDH2 by ISDN and ISMN denitration was further studied [191,219]. ALDH2 activity in hepatic mitochondria shows significant regioselectivity towards 1,2-GDN formation from GTN, which is lost in mitochondria from tolerant tissue [152,155]. Therefore ALDH2 fulfils one of the traditional criteria for an enzyme mediating GTN bioactivation and contributing to nitrate tolerance. In support of this, cyanamide was shown to attenuate the increase in coronary blood flow and hypotension produced by GTN in anaesthesized dogs; an effect ascribed to ALDH2 inhibition [220,221]. Elegant studies by Stamler and co-workers have supported a role for ALDH2 in nitrate bioactivation, for example, aortic tissue from ALDH2 null mice was less responsive to relaxation induced by GTN ($\leq 1\ \mu M$), although relaxation induced by GTN ($>1\ \mu M$) and by ISDN was insensitive to the deletion of ALDH2 [222].

Nitrate tolerance can be simulated in animal models by continuous delivery of GTN, in preference to repetitive bolus administration, however, whether this completely models the human clinical condition is problematic [223]. In animal models of GTN tolerance, ALDH2 inhibitors induce a rightward shift in the GTN dose–response curve for tissue relaxation [155], however, rightward shifts were also observed in tolerant tissue that should not be susceptible to inhibitors if inhibition of ALDH2 is the cause of nitrate tolerance [152]. The last observation prompted the conclusion that ALDH2 is able to contribute to nitrate biotransformation, but there is no clear evidence to suggest that ALDH2 provides the primary pathway of bioactivation, nor does attenuation of ALDH2 represent a primary cause of nitrate tolerance.

It is useful to move discussion directly to human studies, since nitrate tolerance is a human clinical phenomenon. The ALDH2*2 allele has a prevalence up to 50% in certain Asian populations, notably Japanese and Chinese, resulting in compromised alcohol clearance and associated presentations. This Glu504Lys point mutation results in 6% enzyme activity in heterozygotes and 0% in homozygotes through perturbed NAD+ binding and Cys319 activation at the active site [224]. Subjects possessing this allele would be predicted to be poorly responsive to GTN and not to manifest symptoms of nitrate tolerance. In Chinese subjects genotyped for ALDH2, the presence of the allele was observed to decrease the responsiveness to sublingual GTN from 86% to 58% [225]. A second study measured forearm blood flow in response to GTN in a group of genotyped East Asians and additionally in a group of subjects administered the ALDH2 inhibitor disulfiram or placebo [226]. The results were internally consistent, in that the response to GTN was blunted by disulfiram and in the ALDH2*2 population, but by less than a factor of two. Both human studies concluded that several different enzymes are involved in the bioactivation of GTN in addition to ALDH2. It was noted that the literature did not indicate clinical abnormalities with respect to GTN in populations with high ALDH2*2 incidences, although compensatory mechanisms may be in place.

NITRATE TOLERANCE

In the clinical setting, tolerance to the antianginal and haemodynamic effects of nitrates is associated with continuous exposure to nitrates, and has led to the use of elliptical dosage regimens to provide a nitrate-free interval, during which recovery of sensitivity to the vasodilator effects of the nitrate can occur. A rebound phenomenon is clinically characterized, wherein increase in the frequency of anginal attacks or deterioration in exercise performance is observed. Tolerance to nitrates is widely acknowledged to be multifactorial in nature, but the underlying causes can be classified as blood-vessel dependent or independent. The blood-vessel-dependent hypothesis of tolerance focuses on the inability of vascular smooth muscle cells to adequately respond to nitrates, whereas the vessel-independent hypothesis relates to activation of neurohormonal counter-regulatory mechanisms, such as activation of the renin-angiotensin system [227]. The involvement of neurohormonal mechanisms is questioned by observations that tolerance can be induced *in vitro*, where the presence of circulating vasoconstrictor hormones is excluded, however, the *in vitro* tolerance models are unlikely to incorporate the complete physiology of clinical nitrate tolerance [228].

Attenuation of nitrate bioactivation, discussed above, represents the most popular proposition to explain the phenomenon of blood-vessel-dependent tolerance to nitrate [6]. Other mechanisms proposed include: (a) induction of oxidative stress by nitrate biotransformation, increased vascular oxidase activity, or disrupted endothelial function [229]; (b) upregulation of phosphodiesterase activity with the subsequent reduction of intracellular levels of cGMP [230,231]; (c) loss of sGC responsiveness to NO [232,233]; and, (d) intracellular depletion of thiol equivalents [234]. There are numerous studies that are not in concordance with the early thiol depletion rationale for tolerance induction [234–236]. Spontaneous NO donors (e.g. nitroprusside, nitrosothiols, and diazeniumdiolates) are not subject to tolerance nor to cross-tolerance towards nitrates, which has been used to argue for attenuated bioactivation as the sole cause of tolerance. This argument is based on the erroneous assumption that the only interaction of nitrates with biological systems is *via* release of NO. Numerous biochemical pathways can potentially contribute to nitrate tolerance and available data are consistent with tolerance being a multifactorial phenomenon that includes attenuation of one or more bioactivation pathway.

The research discussed above using $ALDH^{-/-}$ mice demonstrated a profound difference between responses to GTN *vs.* ISDN, and also suggested a high affinity and low affinity pathway for GTN [222]. In GTN-tolerant tissue, potency is reduced by 5–10-fold, but the potency of GTN still surpasses that of other nitrate vasodilators, such as ISDN, in non-tolerant tissue. This common observation also hints at multiple nitrate bioactivation pathways, with GTN utilizing a high potency or high affinity pathway, sensitive to tolerance, in addition to the low affinity pathway that is open to all other nitrates. Early research reporting biphasic concentration–response curves for GTN-induced relaxation supported a dual mechanism hypothesis for GTN, with high affinity and low affinity components [237–239]. However, it is confounding that ETN, which incorporates the GTN moiety in its structure, was reported to be 10-fold less potent than GTN, to manifest no biphasic activity; but to abolish the putative high affinity pathway for GTN [240]. Further studies categorized classical nitrates as high potency ALDH2 bioactivated (GTN, PETN, pentaerythritol trinitrate), or low potency ALDH2 insensitive (ISDN, ISMN, and pentaerythritol mono and dinitrates) [241]. PETN is a marginally more potent vasodilator than GTN, but unlike GTN was not reported to inhibit ALDH2 activity. These observations demonstrate that even amongst classical nitrates, the structural diversity leads to quite different biological properties.

SUMMARY

Nitrates are an important class of drug for cardiovascular therapy under continuous development. Hybrid nitrates and NO chimeras, respectively, release a non-nitrate drug molecule or contain an ancillary pharmacophore; both approaches are designed to amplify or synergize the therapeutic activity of the nitrate in specific disease indications. Current drug discovery of new nitrate drug classes for indications beyond the vasculature emphasizes the need to better understand nitrate metabolism, bioactivation, and bioactivity. There is an opportunity to increase the selectivity of nitrates for target cells and tissues exploiting different bioactivation apparatus and mechanisms, beyond the serendipitous selectivity of many nitrates for bioactivation in hypoxic tissue. Structural modifications of nitrates can be envisioned that target specific enzymes for bioactivation, for example, those upregulated in cells under pathophysiological conditions. The biological half-life of GTN is in the order of minutes and several NO-NSAID nitrates cannot be detected intact in target tissue or even in plasma, yet contemporary nitrate drugs are often described as releasing NO in a sustained fashion for several hours. This apparent dilemma is resolved if one considers nitrates to provide a sustained source of “NO bioactivity,” certainly of longer duration than that provided by true NO. The biological and clinical data suggest that there are multiple bioactivation pathways accessible to nitrates and multiple sites of action, and also that the mediators of bioactivity include but are not limited to NO itself. After 130 years of nitrate-based therapy of human diseases, the prognosis is excellent for further and extended use of nitrates in improving human health.

REFERENCES

1 Landmesser U, Engberding N, Bahlmann FH, Schaefer A, Wiencke A, Heineke A, Spiekermann S, Hilfiker-Kleiner D, Templin C, Kotlarz D, Mueller M, Fuchs M, Hornig B, Haller H, Drexler H. Statin-induced improvement of endothelial progenitor cell mobilization, myocardial neovascularization, left ventricular function, and survival after experimental myocardial infarction requires endothelial nitric oxide synthase. Circulation 2004; 110: 1933–1939.

2 Hayashi T, Rani PJ, Fukatsu A, Matsui-Hirai H, Osawa M, Miyazaki A, Tsunekawa T, Kano-Hayashi H, Iguchi A, Sumi D, Ignarro LJ. A new HMG-CoA reductase inhibitor, pitavastatin remarkably retards the progression of high cholesterol induced atherosclerosis in rabbits. Atherosclerosis 2004; 176: 255–263.

3 Laufs U, Gertz K, Huang P, Nickenig G, Bohm M, Dirnagl U, Endres M. Atorvastatin upregulates type III nitric oxide synthase in thrombocytes, decreases platelet activation, and protects from cerebral ischemia in normocholesterolemic mice. Stroke 2000; 31: 2442–2449.

4 Murrell W. Nitroglycerin as remedy for anina pectoris. Lancet 1879; 1: 80.

5 Brunton TL. On the use of nitrite of amyl in angina pectoris. Lancet 1867; 2: 97.

6 Bennett BM, McDonald BJ, Nigam R, Simon WC. Biotransformation of organic nitrates and vascular smooth muscle cell function. Trends Pharmacol. Sci. 1994; 15: 245–249.

7 Fung HL. Biochemical mechanism of nitroglycerin action and tolerance: is this old mystery solved? Annu. Rev. Pharmacol. Toxicol. 2004; 44: 67–85.

8 Thatcher GRJ, Nicolescu AC, Bennett BM, Toader V. Nitrates and NO release: Contemporary aspects in biological and medicinal chemistry. Free Radic. Biol. Med. 2004; 37: 1122–1143.

9 Thatcher GRJ, Weldon H. NO problem for nitroglycerin: organic nitrate chemistry and therapy. Chem. Soc. Rev. 1998; 27: 331–337.

10 Fretland DJ, Connor JR, Pitzele BS, Currie MG, Manning PT. Inhibition of nitric oxide synthase for therapy in inflammatory diseases. Curr. Pharm. Des. 1997; 3: 447–462.

11 Wallace JL, Del Soldato P, Cirino G. Nitric oxide-releasing non-steroidal anti-inflammatory drugs: a new generation of antithrombotics? Expert Opin. Invest. Drugs 1997; 6: 533–538.

12 Thatcher GR, Nicolescu AC, Bennett BM, Toader V. Nitrates and NO release: contemporary aspects in biological and medicinal chemistry. Free Radic. Biol. Med. 2004; 37: 1122–1143.

13 Taylor AL, Ziesche S, Yancy C, Carson P, D'Agostino Jr. R, Ferdinand K, Taylor M, Adams K, Sabolinski M, Worcel M, Cohn JN. Combination of isosorbide dinitrate and hydralazine in blacks with heart failure. N. Engl. J. Med. 2004; 351: 2049–2057.

14 Bolla M, Almirante N, Benedini F. Therapeutic potential of nitrate esters of commonly used drugs. Curr. Top Med. Chem. 2005; 5: 707–720.

15 Minamino T, Jiyoong K, Asakura M, Shintani Y, Asanuma H, Kitakaze, M. Rationale and design of a large-scale trial using nicorandil as an adjunct to percutaneous coronary intervention for ST-segment elevation acute myocardial infarction: Japan-Working groups of acute myocardial infarction for the reduction of Necrotic Damage by a K-ATP channel opener (J-WIND-KATP). Circ. J. 2004; 68: 101–106.

16 Taira N. Similarity and dissimilarity in the mode and mechanism of action between nicorandil and classical nitrates: an overview. J. Cardiovasc. Pharmacol. 1987; 10(Suppl 8): S1–S9.

17 Benedini F, Bertolini G, Gromo G, Mizrahi J, Sala A. The discovery of a new organic nitrate: an overview. J. Cardiovasc. Pharmacol. 1995; 26(Suppl 4): S1–S5.

18 Van Bortel LM, Spek JJ, Balkestein EJ, Sardina M, Struijker Boudier HA. Is it possible to develop drugs that act more selectively on large arteries? J. Hypertens. 1999; 17: 701–705.

19 Asada H, Nanjo T, Itoh T, Suzuki H, Kuriyama H. Effects of 3,4-dihydro-8-(2-hydroxy-3-isopropylaminopropoxy)-3-nitroxy-2H-1-bezopyran (K-351) on smooth muscle cells and neuromuscular transmission in guinea-pig vascular tissues. J. Pharmacol. Exp. Ther. 1982; 223: 560–572.

20 Okamura T, Kitamura Y, Uchiyama M, Toda M, Ayajiki K, Toda N. Canine retinal arterial and arteriolar dilatation induced by nipradilol, a possible glaucoma therapeutic. Pharmacology 1996; 53: 302–310.

21 Taguchi R, Shirakawa H, Yamaguchi T, Kume T, Katsuki H, Akaike A. Nitric oxide-mediated effect of nipradilol, an alpha- and beta-adrenergic blocker, on glutamate neurotoxicity in rat cortical cultures. Eur. J. Pharmacol. 2006; 535: 86–94.

22 Thatcher GRJ, Bennett BM, Dringtenberg HC, Reynolds JN. Novel nitrates as NO mimetics directed at Alzheimer's Disease. J. Alzheimer's Dis. 2004; 6: S75–S84.

23 Bennett BM, Reynolds JN, Prusky GT, Nelson LS, Douglas RM, Sutherland RJ, Thatcher GRJ. Cognitive deficits in rats after forebrain cholinergic depletion are reversed by a novel NO mimetic nitrate ester. Neuropsychopharmacol. 2006; Mar, Epub.

24 Clarke JL, Kastrati I, Johnstone L, Thatcher GRJ. Photochemical reactions of thiols with organic nitrate oxygen atom transfer. Can. J. Chem. 2006; 84: 709–719.

25 Lipton SA. Paradigm shift in NMDA receptor antagonist drug development: molecular mechanism of uncompetitive inhibition by memantine in the treatment of Alzheimer's disease and other neurologic disorders. J. Alzheimer's Dis. 2004; 6: S61–S74.

26 Thatcher GRJ, Bennett BM, Dringenberg HC, Reynolds JN. Novel nitrates as NO mimetics directed at Alzheimer's disease. J. Alzheimer's Dis. 2004; 6: S75–S84.

27 Reynolds JN, Bennett BM, Boegman RJ, Jhamandas K, Ratz JD, Zavorin SI, Scutaru D, Dumitrascu A, Thatcher GRJ. Neuroprotection against ischemic brain injury conferred by a novel nitrate ester. Bioorg. Med. Chem. Lett. 2002; 12: 2863–2866.

28 Clarke JL, Kastrati I, Johnston LJ, Thatcher GRJ. Photochemical reactions of thiols with organic nitrates, Oxygen atom transfer via a thionitrate. Can. J. Chem. 2006; 84: 709–719.

29 Bredt DS, Snyder SH. Nitric oxide mediates glutamate-linked enhancement of cGMP levels in the cerebellum. Proc. Natl. Acad. Sci. USA 1989; 86: 9030–9033.

30 Garthwaite J, Garthwaite G, Palmer RM, Moncada S. NMDA receptor activation induces nitric oxide synthesis from arginine in rat brain slices. Eur. J. Pharmacol. 1989; 172: 413–416.

31 Hanley MR, Iversen LL. Muscarinic cholinergic receptors in rat corpus striatum and regulation of guanosine cyclic 3′,5′-monophosphate. Mol. Pharmacol. 1978; 14: 246–255.

32 Wang X, Robinson PJ. Cyclic GMP-dependent protein kinase and cellular signaling in the nervous system. J. Neurochem. 1997; 68: 443–456.

33 Dawson TM, Dawson VL, Snyder SH. A novel neuronal messenger molecule in brain: the free radical, nitric oxide. Ann. Neurol. 1992; 32: 297–311.

34 Son H, Lu YF, Zhuo M, Arancio O, Kandel ER, Hawkins RD. The specific role of cGMP in hippocampal LTP. Learn Mem. 1998; 5: 231–245.

35 Lu YF, Kandel ER, Hawkins RD. Nitric oxide signaling contributes to late-phase LTP and CREB phosphorylation in the hippocampus. J. Neurosci. 1999; 19: 10250–10261.

36 Snyder EM, Nong Y, Almeida CG, Paul S, Moran T, Choi EY, Nairn AC, Salter MW, Lombroso PJ, Gouras GK, Greengard P. Regulation of NMDA receptor trafficking by amyloid-beta. Nat. Neurosci. 2005; 8: 1051–1058.

37 Puzzo D, Vitolo O, Trinchese F, Jacob JP, Palmeri A, Arancio O. Amyloid-beta peptide inhibits activation of the nitric oxide/cGMP/cAMP-responsive element-binding protein pathway during hippocampal synaptic plasticity. J. Neurosci. 2005; 25: 6887–6897.

38 Cleary J, Hittner JM, Semotuk M, Mantyh P, O'Hare E. Beta-amyloid(1-40) effects on behavior and memory. Brain Res. 1995; 682: 69–74.

39 Redish AD, Touretzky DS. The role of the hippocampus in solving the Morris water maze. Neural. Comput. 1998; 10: 73–111.

40 Silva AJ, Giese KP, Fedorov NB, Frankland PW, Kogan JH. Molecular, cellular, and neuroanatomical substrates of place learning. Neurobiol. Learn Mem. 1998; 70: 44–61.

41 Prusky GT, Douglas RM, Nelson L, Shabanpoor A, Sutherland RJ. Visual memory task for rats reveals an essential role for hippocampus and perirhinal cortex. Proc. Natl. Acad. Sci. USA 2004; 101: 5064–5068.

42 Nguyen T, Brunson D, Crespi CL, Penman BW, Wishnok JS, Tannenbaum SR. DNA damage and mutation in human cells exposed to nitric oxide in vitro. Proc. Natl. Acad. Sci. USA 1992; 89: 3030–3034.

43 Maragos CM, Andrews AW, Keefer LK, Elespuru RK. Mutagenicity of glyceryl trinitrate (nitroglycerin) in Salmonella typhimurium. Mutat. Res. 1993; 298: 187–195.

44 Routledge MN, Mirsky FJ, Wink DA, Keefer LK, Dipple A. Nitrite-induced mutations in a forward mutation assay: influence of nitrite concentration and pH. Mutat. Res. 1994; 322: 341–346.

45 Ali NA, Eubanks WS, Stamler JS, Gow AJLagoo-Deenadayalan SA, Villegas L, El-Moalem HE, Reynolds JD. A method to attenuate pneumoperitoneum-induced reductions in splanchnic blood flow. Ann. Surg. 2005; 241: 256–261.

46 Moya MP, Gow AJ, Califf RM, Goldberg RN, Stamler JS. Inhaled ethyl nitrite gas for persistent pulmonary hypertension of the newborn. Lancet 2002; 360: 141–143.

47 Kleschyov AL, Oelze M, Daiber A, Huang Y, Mollnau H, Schulz E, Sydow K, Fichtlscherer B, Mulsch A, Munzel T. Does nitric oxide mediate the vasodilator activity of nitroglycerin? Circ. Res. 2003; 93: e104–e112.

48 Kleschyov AL, Munzel T. Advanced spin trapping of vascular nitric oxide using colloid iron diethyldithiocarbamate. Methods Enzymol. 2002; 359: 42–51.

49 Singh RJ, Hogg N, Joseph J, Konorev E, Kalyanaraman B. The peroxynitrite generator, SIN-1, becomes a nitric oxide donor in the presence of electron acceptors. Arch. Biochem. Biophys. 1999; 361: 331–339.

50 Nunez C, Victor VM, Tur R, Alvarez-Barrientos A, Moncada S, Esplugues JV, D'Ocon P. Discrepancies between nitroglycerin and NO-releasing drugs on mitochondrial oxygen consumption, vasoactivity, and the release of NO. Circ. Res. 2005; 97: 1063–1069.

51 Kojima H, Hirotani M, Nakatsubo N, Kikuchi K, Urano Y, Higuchi T, Hirata Y, Nagano T. Bioimaging of nitric oxide with fluorescent indicators based on the rhodamine chromophore. Anal. Chem. 2001; 73: 1967–1973.

52 Schror K, Forster S, Woditsch I. On-line measurement of nitric oxide release from organic nitrates in the intact coronary circulation. Naunyn Schmiedebergs Arch. Pharmacol. 1991; 344: 240–246.

53 Liu X, Miller MJ, Joshi MS, Thomas DD, Lancaster Jr. JR. Accelerated reaction of nitric oxide with O_2 within the hydrophobic interior of biological membranes. Proc. Natl. Acad. Sci. USA 1998; 95: 2175–2179.

54 Hunter CJ, Dejam A, Blood AB, Shields H, Kim-Shapiro DB, Machado RF, Tarekegn S, Mulla N, Hopper AO, Schechter AN, Power GG, Gladwin MT. Inhaled nebulized nitrite is a hypoxia-sensitive NO-dependent selective pulmonary vasodilator. Nat. Med. 2004; 10: 1122–1127.

55 Pluta RM, Dejam A, Grimes G, Gladwin MT, Oldfield EH. Nitrite infusions to prevent delayed cerebral vasospasm in a primate model of subarachnoid hemorrhage. JAMA 2005; 293: 1477–1484.

56 O'Donnell VB, Eiserich JP, Chumley PH, Jablonsky MJ, Krishna NR, Kirk M, Barnes S, Darley-Usmar VM, Freeman BA. Nitration of unsaturated fatty acids by nitric oxide-derived reactive nitrogen species peroxynitrite, nitrous acid, nitrogen dioxide, and nitronium ion. Chem. Res. Toxicol. 1999; 12: 83–92.

57 O'Donnell VB, Eiserich JP, Bloodsworth A, Chumley PH, Kirk M, Barnes S, Darley-Usmar VM, Freeman BA. Nitration of unsaturated fatty acids by nitric oxide-derived reactive species. Methods Enzymol. 1999; 301: 454–470.

58 Rubbo H, Radi R, Trujillo M, Telleri R, Kalyanaraman B, Barnes S, Kirk M, Freeman BA. Nitric oxide regulation of superoxide and peroxynitrite-dependent lipid peroxidation. Formation of novel nitrogen-containing oxidized lipid derivatives. J. Biol. Chem. 1994; 269: 26066–26075.

59 Goss SP, Singh RJ, Hogg N, Kalyanaraman B. Reactions of *NO, *NO2 and peroxynitrite in membranes: physiological implications. Free Radic. Res. 1999; 31: 597–606.

60 Pryor WA, Lightsey JW, Church DF. Reaction of nitrogen dioxide with alkenes and polyunsaturated fatty acids: addition and hydrogen abstraction mechanism. J. Am. Chem. Soc. 1982; 104: 6685–6692.

61 Gallon AA, Pryor WA. The reaction of low levels of nitrogen dioxide with methyl linoleate in the presence and absence of oxygen. Lipids 1994; 29: 171–176.

62 Gallon AA, Pryor WA. The identification of the allylic nitrite and nitro derivatives of methyl linoleate and methyl linolenate by negative chemical ionization mass spectroscopy. Lipids 1993; 28: 125–133.

63 Pryor WA, Castle L, Church DF. Nitrosation of organic hydroperoxides by nitrogen dioxide/dinitrogen tetraoxide. J. Am. Chem. Soc. 1985; 107: 211–217.

64 Giamalva DH, Kenion GB, Church DF, Pryor WA. Rates and mechanism of reaction of nitrogen dioxide with alkenes in solutions. J. Am. Chem. Soc. 1987; 109: 7059–7063.

65 van der Vliet A, Eiserich JP, Halliwell B, Cross CE. Formation of reactive nitrogen species during peroxidase-catalyzed oxidation of nitrite. A potential additional mechanism of nitric oxide-dependent toxicity. J. Biol. Chem. 1997; 272: 7617–7625.

66 Sampson JB, Ye Y, Rosen H, Beckman JS. Myeloperoxidase and horseradish peroxidase catalyze tyrosine nitration in proteins from nitrite and hydrogen peroxide. Arch. Biochem. Biophys. 1998; 356: 207–213.

67 Rubbo H, Parthasarathy S, Barnes S, Kirk M, Kalyanaraman B, Freeman BA. Nitric oxide inhibition of lipoxygenase-dependent liposome and low-density lipoprotein oxidation: termination of radical chain propagation reactions and formation of nitrogen-containing oxidized lipid derivatives. Arch. Biochem. Biophys. 1995; 324: 15–25.

68 O'Donnell VB, Chumley PH, Hogg N, Bloodsworth A, Darley-Usmar VM, Freeman BA. Nitric oxide inhibition of lipid peroxidation: kinetics of reaction with lipid peroxyl radicals and comparison with alpha-tocopherol. Biochemistry 1997; 36: 15216–15223.
69 Rubbo H, Freeman BA. Nitric oxide regulation of lipid oxidation reactions: formation and analysis of nitrogen-containing oxidized lipid derivatives. Methods Enzymol. 1996; 269: 385–394.
70 Zavorin SI, Artz JD, Dumitrascu A, Nicolescu A, Scutaru D, Smith SV, Thatcher GR. Nitrate esters as nitric oxide donors: SS-nitrates. Org. Lett. 2001; 3: 1113–1116.
71 Nicolescu AC, Zavorin SI, Turro NJ, Reynolds JN, Thatcher GR. Inhibition of lipid peroxidation in synaptosomes and liposomes by nitrates and nitrites. Chem. Res. Toxicol. 2002; 15: 985–998.
72 Nicolescu AC, Reynolds JN, Barclay LR, Thatcher GRJ. Organic nitrites and NO: inhibition of lipid peroxidation and radical reactions. Chem. Res. Toxicol. 2004; 17: 185–196.
73 Schopfer FJ, Baker PR, Giles G, Chumley P, Batthyany C, Crawford J, Patel RP, Hogg N, Branchaud BP, Lancaster Jr. JR, Freeman BA. Fatty acid transduction of nitric oxide signaling: Nitrolinoleic acid is a hydrophobically-stabilized nitric oxide donor. J. Biol. Chem. 2005; 280: 19289–19297.
74 Nicolescu AC, Li Q, Brown L, Thatcher GR. Nitroxidation, nitration, and oxidation of a BODIPY fluorophore by RNOS and ROS. Nitric Oxide 2006; 15: 163–176.
75 Gooden DM, Chakrapani H, Toone EJ. C-nitroso compounds: synthesis, physicochemical properties and biological activities. Curr. Top Med. Chem. 2005; 5: 687–705.
76 Yoshida K, Kita Y. Hydroxyimine NO-donors; FK409 and derivatives. Curr. Top Med. Chem. 2005; 5: 675–685.
77 Csizmadia IG, Hayward LD. Photolysis of nitrate esters. II. Solution photolysis of aralkyl nitrate esters: kinetics, E.S.R. spectra, and photoproducts. Photochem. Photobiol. 1965; 4: 657–671.
78 Hayward LD, Kitchen RA, Livingstone DJ. Photolysis of nitrate esters. I. Photonitration of diphenylamine. Can. J. Chem. 1962; 40: 434–440.
79 Batelaan JG, Hageman HJ, Verbeek J. Photodecomposition of α(nitratomethyl)benzoin as studied by ESR. Tetrahedron Lett. 1987; 28: 2163–2166.
80 Turberg MP, Giolando DM, Tilt C, Soper T, Mason S, Davies M, Klingensmith P, Takacs GA. Atmospheric photochemistry of alkyl nitrates. J. Photochem. Photobiol. A. 1990; 51: 281–292.
81 Kuhn LP. The ionization of ethyl nitrate in sulfuric acid. J. Am. Chem. Soc. 1947; 69: 1974.
82 Honeyman J, Morgan JWW. Advances in Carbohydrate Chemistry; Academic Press Inc., New York, 1957, p. 117.
83 Baker JW, Easty DM. Hydrolysis of organic nitrates. Nature (Lond.) 1950; 166: 156.
84 Capellos C, Fisco WJ, Ribaudo C, Hogan VD, Campisi J, Murphy FX, Castorina TC, Rosenblatt DH. Basic hydrolysis of glyceryl nitrate esters. III. Trinitroglycerin. Int. J. Chem. Kinet. 1984; 16: 1027–1051.
85 Robertson RE, Koshy KM, Annessa A, Ong JN, Scott JMW, Blandamer MJ. Kinetics of solvolysis in water of four secondary alkyl nitrates. Can. J. Chem. 1982; 60: 1780–1785.
86 Baker JW, Heggs TG. Hydrolytic decomposition of esters of nitric acid. V. The effects of structural changes in aralphyl nitrates on the SN and ECO reactions. J. Chem. Soc. 1955; 616–630.
87 McKeown GG, Hayward LD. The action of pyridine on dulcitol hexanitrate. Can. J. Chem. 1955; 33: 1392–1398.
88 Anbar M, Dostrovsky I, Samuel D, Yoffe AD. Esters of inorganic oxyacids. I. The mechanism of hydrolysis of alkyl esters, O18 being used as a tracer, and its relation to other reactions in alkaline media. J. Chem. Soc. 1954; 3603–3611.
89 Merrow RT, Van Dolah RW. Reactions of nitrate esters. III. Evidence for nitrogen oxygen cleavage in reductions with hydrazine and alkaline hydrosulfides. J. Am. Chem. Soc. 1955; 77: 756–757.
90 Merrow RT. Reactions of nitrate esters. IV. Kinetics of hydrazinolysis. J. Am. Chem. Soc. 1956; 78: 1297–1300.
91 Hayward LD, Purves CB. Action of hydroxylamine on methyl α- and β-D-glucopyranoside tetranitrate in pyridine. Can. J. Chem. 1954; 32: 19–30.
92 Baker JW, Neale AJ. Hydrolytic decomposition of esters of nitric acid. IV. Acid hydrolysis, and the effects of change in the nucleophilic reagent on the SN and ECO reactions. J. Chem. Soc. 1955; 608–615.

93 Carlson T. Alkaline hydrolysis of alkyl nitrates in presence of hydrogen peroxide. Berichte der Deutschen Chemischem Geseuschaft 1908; 40: 4191–4194.

94 Merrow RT, Cristol SJ, Van Dolah RM. The reaction of butyl nitrate with alkaline hydrogen sulfides. J. Am. Chem. Soc. 1953; 75: 4259–4265.

95 Rheinboldt H, Mott F. Uber Thionitrate, I. (vorlaufige) Mitteil.: tert-Butyl-thionitrat. Berichte 1932; 65B: 1223.

96 Kim YH, Shinhama K, Fukushima D, Oae S. New selective oxidation of thiols to the corresponding thiosulfonates with dinitrogen tetroxide: one pot synthesis of thiosulfonates from thiols. Tetrahedron Lett. 1978; 14: 1211–1212.

97 Oae S, Shinhama K, Fujimori K, Kim YH. Physical properties and various reactions of thionitrates and related substances. Bull. Chem. Soc. Jpn. 1980; 53: 775–784.

98 Goto K, Hino Y, Kawashima T, Kaminaga M, Yano E, Yamamoto G, Takagi N, Nagase S. Synthesis and crystal structure of a stable S-nitrosothiol bearing a novel steric protection group and of the corresponding S-nitrothiol. Tetrahedron Lett. 2000; 41: 8479–8483.

99 Goto K, Hino Y, Takahashi Y, Kawashima T, Yamamoto G, Takagi N, Nagase S. Synthesis, structure, and reactions of the first stable aromatic S-nitrosothiol bearing a novel dendrimer-type steric protection group. Chem. Lett. 2001; 1204–1205.

100 Bartberger MD, Mannion JD, Powell SC, Stamler JS, Houk KN, Toone EJ. S-N dissociation energies of S-nitrosothiols: on the origins of nitrosothiol decomposition rates. J. Am. Chem. Soc. 2001; 123: 8868–8869.

101 Artz JD, Yang K, Lock J, Sanchez C, Bennett BM, Thatcher GRJ. Reactivity of thionitrate esters: putative intermediates in nitrovasodilator activity. J. Chem. Soc. Chem. Commun. 1996; 927–928.

102 Okazaki R, Goto K. Synthesis of highly reactive organosulfur compounds. Heteroatom Chem. 2002; 13: 414–418.

103 Doyle MP, LePoire DM, Pickering RA. Oxidation of hemoglobin and myoglobin by alkyl nitrites. Inhibition by oxygen. J. Biol. Chem. 1981; 256: 12399–12404.

104 Oh SMNYF, Williams DL. Nitrosation by alkyl nitrites. Part 7. Comparison with thionitrites: reactions with phenols. J. Chem. Soc. Perkin Trans. 1991; 2: 685–688.

105 Iglesias E, Garcia Rio L, Leis JR, Pena ME, Williams DLH. Evidence for concerted acid hydrolysis of alkyl nitrites. J. Chem. Soc. Perkin Trans. 1992; 2: 1673–1679.

106 Patel HMS, Williams DLH. Nitrosation by alkyl nitrites. Part 3. Reactions with cysteine in water in the pH range 6–13. J. Chem. Soc. Perkin Trans. 1989; 2: 339–341.

107 Williams DLH. Nitrosation; Cambridge University Press, Cambridge, New York, Melbourne, 1988, p. 214.

108 Kochi JK. Free radicals. In Reactive Intermediates in Organic Chemistry; (Olah GA, ed.), John Wiley & Sons, New York, 1973; p. 906.

109 Nussbaum AL, Robinson CH. Recent developments in the preparative photolysis of organic nitrites. Tetrahedron 1962; 17: 35–39.

110 Barton DHR, Hesse RH, Pechet MM, Smith LC. The mechanism of the Barton reaction. J. Chem. Soc. Perkin Trans. 1979; 1; 1159–1165.

111 Grossi L, Strazzari S. A new synthesis of alkyl nitrites: the reaction of alkyl alcohols with nitric oxide in organic solvents. J. Org. Chem. 1999; 64: 8076–8079.

112 Grossi L. The photolysis of cycloalkyl nitrite esters: the radical intermediates as studied by EPR. Tetrahedron 1997; 53: 6401–6410.

113 Grossi L. The photoinduced ring expansion of five membered ring nitrites: a 1,6-exo ring closure process of the intermediate 5-nitrosopentanoyl-type radical. Tetrahedron 1997; 53: 3205–3214.

114 Nicolescu AC, Reynolds JN, Barclay LR, Thatcher GR. Organic nitrites and NO: inhibition of lipid peroxidation and radical reactions. Chem. Res. Toxicol. 2004; 17: 185–196.

115 Artz JD, Thatcher GR. NO release from NO donors and nitrovasodilators: comparisons between oxyhemoglobin and potentiometric assays. Chem. Res. Toxicol. 1998; 11: 1393–1397.

116 Litwinienko G, Ingold KU. Abnormal solvent effects on hydrogen atom abstractions. 1. The reactions of phenols with 2,2-diphenyl-1-picrylhydrazyl (dpph*) in alcohols. J. Org. Chem. 2003; 68: 3433–3438.

117 Doyle MP, Pickering RA, da Conceicao J. Structural effects in alkyl nitrite oxidation of human hemoglobin. J. Biol. Chem. 1984; 259: 80–87.
118 Doel JJ, Godber BL, Goult TA, Eisenthal R, Harrison R. Reduction of organic nitrites to nitric oxide catalyzed by xanthine oxidase: possible role in metabolism of nitrovasodilators. Biochem. Biophys. Res. Commun. 2000; 270: 880–885.
119 Ratz JD, McGuire JJ, Anderson DJ, Bennett BM. Effects of the flavoprotein inhibitor, diphenyleneiodonium sulfate, on ex vivo organic nitrate tolerance in the rat. J. Pharmacol. Exp. Ther. 2000; 293: 569–577.
120 Artz JD, Toader V, Zavorin SI, Bennett BM, Thatcher GR. In vitro activation of soluble guanylyl cyclase and nitric oxide release: a comparison of NO donors and NO mimetics. Biochemistry 2001; 40: 9256–9264.
121 Ignarro LJ. Nitric oxide. A novel signal transduction mechanism for transcellular communication. Hypertension 1990; 16: 477–483.
122 Kroncke KD, Fehsel K, Schmidt T, Zenke FT, Dasting I, Wesener JR, Bettermann H, Breunig KD, Kolb-Bachofen V. Nitric oxide destroys zinc-sulfur clusters inducing zinc release from metallothionein and inhibition of the zinc finger-type yeast transcription activator LAC9. Biochem. Biophys. Res. Commun. 1994; 200: 1105–1110.
123 Ridnour LA, Thomas DD, Mancardi D, Espey MG, Miranda KM, Paolocci N, Feelisch M, Fukuto J, Wink DA. The chemistry of nitrosative stress induced by nitric oxide and reactive nitrogen oxide species. Putting perspective on stressful biological situations. Biol. Chem. 2004; 385: 1–10.
124 Becker K, Savvides SN, Keese M, Schirmer RH, Karplus PA. Enzyme inactivation through sulfhydryl oxidation by physiologic NO-carriers. Nat. Struct. Biol. 1998; 5: 267–271.
125 DeMaster EG: Quast BJ, Redfern B, Nagasawa HT. Reaction of nitric oxide with the free sulfhydryl group of human serum albumin yields a sulfenic acid and nitrous oxide. Biochemistry 1995; 34: 11494–11499.
126 Muscara MN, Lovren F, McKnight W, Dicay M, del Soldato P, Triggle CR, Wallace JL. Vasorelaxant effects of a nitric oxide-releasing aspirin derivative in normotensive and hypertensive rats. Br. J. Pharmacol. 2001; 133: 1314–1322.
127 Cuzzolin L, Adami A, Degan M, Crivellente F, Bonapace S, Minuz P, Benoni G. Effect of single and repeated doses of a new nitroderivative of acetylsalicylic acid on platelet TXA2 production in rats. Life Sci. 1996; 58: PL207–PL210.
128 Wallace JL, Muscara MN, McKnight W, Dicay M, Del Soldato P, Cirino G. In vivo antithrombotic effects of a nitric oxide-releasing aspirin derivative, NCX-4016. Thromb. Res. 1999; 93: 43–50.
129 Carini M, Aldini G, Orioli M, Maffei Facino R. In vitro metabolism of a nitroderivative of acetylsalicylic acid (NCX4016) by rat liver: LC and LC-MS studies. J. Pharm. Biomed. Anal. 2002; 29: 1061–1071.
130 Carini M, Aldini G, Stefani R, Orioli M, Facino RM. Nitrosylhemoglobin, an unequivocal index of nitric oxide release from nitroaspirin: in vitro and in vivo studies in the rat by ESR spectroscopy. J. Pharm. Biomed. Anal. 2001; 26: 509–518.
131 Lauer T, Preik M, Rassaf T, Strauer BE, Deussen A, Feelisch M, Kelm M. Plasma nitrite rather than nitrate reflects regional endothelial nitric oxide synthase activity but lacks intrinsic vasodilator action. Proc. Natl. Acad. Sci. USA 2001; 98: 12814–12819.
132 Bryan NS, Rassaf T, Maloney RE, Rodriguez CM, Saijo F, Rodriguez JR, Feelisch M. Cellular targets and mechanisms of nitros(yl)ation: an insight into their nature and kinetics in vivo. Proc. Natl. Acad. Sci. USA 2004; 101: 4308–4313.
133 Gladwin MT, Crawford JH, Patel RP. The biochemistry of nitric oxide, nitrite, and hemoglobin: role in blood flow regulation. Free Radic. Biol. Med. 2004; 36: 707–717.
134 Bennett BM, Kobus SM, Brien JF, Nakatsu K, Marks GS. Requirement for reduced, unliganded hemoprotein for the hemoglobin- and myoglobin-mediated biotransformation of glyceryl trinitrate. J. Pharmacol. Exp. Ther. 1986; 237: 629–635.
135 Prosperi C, Scali C, Pepeu G, Casamenti F. NO-flurbiprofen attenuates excitotoxin-induced brain inflammation, and releases nitric oxide in the brain. Jpn. J. Pharmacol. 2001; 86: 230–235.
136 Jantzen PT, Connor KE, DiCarlo G, Wenk GL, Wallace JL, Rojiani AM, Coppola D, Morgan D, Gordon MN. Microglial activation and β-amyloid deposit reduction caused by a nitric oxide-releasing

nonsteroidal anti-inflammatory drug in amyloid precursor protein plus presenilin-1 transgenic mice. J. Neurosc. 2002; 22: 2246–2254.

137 Aldini G, Carini M, Orioli M, Facino RM, Wenk GL. Metabolic profile of NO-flurbiprofen (HCT1026) in rat brain and plasma: a LC-MS study. Life Sci. 2002; 71: 1487–1500.

138 Williams JL, Nath N, Chen J, Hundley TR, Gao J, Kopelovich L, Kashfi K, Rigas B. Growth inhibition of human colon cancer cells by nitric oxide (NO)-donating aspirin is associated with cyclooxygenase-2 induction and β-catenin/T-cell factor signaling, NF-κB, and NOS2 inhibition: implications for chemoprevention. Cancer Res. 2003; 63: 7613–7618.

139 Williams JL, Kashfi K, Ouyang N, del Soldato P, Kopelovich L, Rigas B. NO-donating aspirin inhibits intestinal carcinogenesis in Min (APC(Min/+)) mice. Biochem. Biophys. Res. Commun. 2004; 313: 784–788.

140 Gao J, Liu X, Rigas B. Nitric oxide-donating aspirin induces apoptosis in human colon cancer cells through induction of oxidative stress. Proc. Natl. Acad. Sci. USA 2005; 102: 17207–17212.

141 Bratasz A, Weir NM, Parinandi NL, Zweier JL, Sridhar R, Ignarro LJ, Kuppusamy P. Reversal to cisplatin sensitivity in recurrent human ovarian cancer cells by NCX-4016, a nitro derivative of aspirin. Proc. Natl. Acad. Sci. USA 2006; 103: 3914–3919.

142 Spiegel A, Hundley TR, Chen J, Gao J, Ouyang N, Liu X, Go MF, Tsioulias GJ, Kashfi K, Rigas B. NO-donating aspirin inhibits both the expression and catalytic activity of inducible nitric oxide synthase in HT-29 human colon cancer cells. Biochem. Pharmacol. 2005; 70: 993–1000.

143 Kashfi K, Borgo S, Williams JL, Chen J, Gao J, Glekas A, Benedini F, Del Soldato P, Rigas B. Positional isomerism markedly affects the growth inhibition of colon cancer cells by NO-donating aspirin in vitro and in vivo. J. Pharmacol. Exp. Ther. 2004; 312: 978–988.

144 Artz JD, Schmidt B, McCracken JL, Marletta MA. Effects of nitroglycerin on soluble guanylate cyclase: implications for nitrate tolerance. J. Biol. Chem. 2002; 277: 18253–18256.

145 Murray J, Macartney D, Thatcher GR. Catalysis of NO production by a molybdoenzyme model. Org. Lett. 2001; 3: 3635–3638.

146 Kirby AJ. Effective molarities of intramolecular reactions. Adv. Phys. Org. Chem. 1980; 17: 183.

147 Page MI, Jencks WP. Entropic contributions to rate accelerations in enzymic and intramolecular reactions and the chelate effect. Proc. Natl. Acad. Sci. USA 1971; 68: 1678–1683.

148 Smith S, Dringenberg HC, Bennett BM, Thatcher GR, Reynolds JN. A novel nitrate ester reverses the cognitive impairment caused by scopolamine in the Morris water maze. Neuroreport 2000; 11: 3883–3886.

149 Ignarro LJ, Lippton H, Edwards JC, Baricos WH, Hyman AL, Kadowitz PJ, Gruetter CA. Mechanism of vascular smooth muscle relaxation by organic nitrates, nitrites, nitroprusside and nitric oxide: evidence for the involvement of S-nitrosothiols as active intermediates. J. Pharmacol. Exp. Ther. 1981; 218: 739–749.

150 Kowaluk EA, Chung SJ, Fung HL. Nitrite ion is not an active intermediate in the vascular metabolism of organic nitrates and organic nitrites to nitric oxide. Drug Metab. Dispos. 1993; 21: 967–969.

151 Bennett BM, Marks GS. How does nitroglycerin induce vascular smooth muscle relaxation? Trends Pharmacol. Sci. 1984; 5: 329–332.

152 DiFabio J, Ji Y, Vasiliou V, Thatcher GR, Bennett BM. Role of mitochondrial aldehyde dehydrogenase in nitrate tolerance. Mol. Pharmacol. 2003; 64: 1109–1116.

153 Romanin C, Kukovetz WR. Guanylate cyclase activation by organic nitrates is not mediated via nitrite. J. Mol. Cell. Cardiol. 1988; 20: 389–396.

154 Parks NJ, Krohn KJ, Mathis CA, Chasko JH, Geiger KR, Gregor ME, Peek NF. Nitrogen-13-labeled nitrite and nitrate: distribution and metabolism after intratracheal administration. Science 1981; 212: 58–60.

155 Chen Z, Zhang J, Stamler JS. Identification of the enzymatic mechanism of nitroglycerin bioactivation. Proc. Natl. Acad. Sci. USA 2002; 99: 8306–8311.

156 Samouilov A, Kuppusamy P, Zweier JL. Evaluation of the magnitude and rate of nitric oxide production from nitrite in biological systems. Arch. Biochem. Biophys. 1998; 357: 1–7.

157 Kozlov AV, Staniek K, Nohl H. Nitrite reductase activity is a novel function of mammalian mitochondria. FEBS Lett. 1999; 454: 127–130.

158 Brudvig GW, Stevens TH, Chan SI. Reactions of nitric oxide with cytochrome c oxidase. Biochem. 1980; 19: 5275–5285.

159 Benedini F, Bertolini G, Cereda R, Dona G, Gromo G, Levi S, Mizrahi J, Sala A. New anti-anginal nitro esters with reduced hypotensive activity. Synthesis and pharmacological evaluation of 3-[(nitrooxy)alkyl]-2H-1,3-benzoxazin-4(3H)-ones. J. Med. Chem. 1995; 38: 130–136.

160 Abrams J. Mechanisms of action of the organic nitrates in the treatment of myocardial ischemia. Am. J. Cardiol. 1992; 70: 30B–42B.

161 Giuffre A, Barone MC, Mastronicola D, D'Itri E, Sarti P, Brunori M. Reaction of nitric oxide with the turnover intermediates of cytochrome c oxidase: reaction pathway and functional effects. Biochemistry 2000; 39: 15446–15453.

162 Sarti P, Giuffre A, Barone MC, Forte E, Mastronicola D, Brunori M. Nitric oxide and cytochrome oxidase: reaction mechanisms from the enzyme to the cell. Free Radic. Biol. Med. 2003; 34: 509–520.

163 Green LC, Wagner DA, Glogowski J, Skipper PL, Wishnok JS, Tannenbaum SR. Analysis of nitrate, nitrite, and [15N]nitrate in biological fluids. Anal. Biochem. 1982; 126: 131–138.

164 Preik-Steinhoff H, Kelm M. Determination of nitrite in human blood by combination of a specific sample preparation with high-performance anion-exchange chromatography and electrochemical detection. J. Chromatogr. B. Biomed. Appl. 1996; 685: 348–352.

165 Zhang Y, Hogg N. The formation and stability of S-nitrosothiols in RAW 264.7 cells. Am. J. Physiol. Lung Cell Mol. Physiol. 2004; 287: L467–L474.

166 Godber BL, Doel JJ, Sapkota GP, Blake DR, Stevens CR, Eisenthal R, Harrison R. Reduction of nitrite to nitric oxide catalyzed by xanthine oxidoreductase. J. Biol. Chem. 2000; 275: 7757–7763.

167 Doyle MP, Herman JG, Dykstra RL. Autocatalytic oxidation of hemoglobin induced by nitrite: activation and chemical inhibition. J. Free Radic. Biol. Med. 1985; 1: 145–153.

168 Kowaluk EA, Fung HL. Vascular nitric oxide-generating activities for organic nitrites and organic nitrates are distinct. J. Pharmacol. Exp. Ther. 1991; 259: 519–525.

169 Yeates RA. Possible mechanisms of activation of soluble guanylate cyclase by organic nitrates. Arzneim.-Forsch./Drug Research 1992; 42: 1314–1317.

170 Marks GS, McLaughlin BE, Jimmo SL, Poklewska-Koziell M, Brien JF, Nakatsu K. Time-dependent increase in nitric oxide formation concurrent with vasodilation induced by sodium nitroprusside, 3-morpholinosydnonimine, and S-nitroso-N-acetylpenicillamine but not by glyceryl trinitrate. Drug Metab. Dispos. 1995; 23: 1248–1252.

171 Fung HL, Chung SJ, Bauer JA, Chong S, Kowaluk EA. Biochemical mechanism of organic nitrate action. Am. J. Cardiol. 1992; 70: 4B–10B.

172 Patel HMS, Williams DLH. Nitrosation by alkyl nitrites. Part 6. Thiolate nitrosation. J. Chem. Soc. Perkin Trans. 1990; 2: 37–42.

173 Buckell F, Hartry JD, Rajalingam U, Bennett BM, Whitney RA, Thatcher GRJ. Hydrolysis of nitrite esters: Putative intermediates in the biotransformation of organic nitrates. J. Chem. Soc. Perkin Trans. 1994; 2: 401–403.

174 Bauer JA, Nolan T, Fung HL. Vascular and hemodynamic differences between organic nitrates and nitrites. J. Pharmacol. Exp. Ther. 1997; 280; 326–331.

175 Zimmermann T, Leitold M, Yeates RA. Comparison of isobutyl nitrate and isobutyl nitrite: tolerance and cross-tolerance to glyceryl trinitrate. Eur. J. Pharmacol. 1991; 192: 181–184.

176 Kowaluk EA, Fung HL. Vascular nitric oxide-generating activities for organic nitrites and organic nitrates are distinct. J. Pharmacol. Exp. Ther. 1991; 259: 519–525.

177 Needleman P, Blehm DJ, Rotskoff KS. Relationship between glutathione-dependent denitration and the vasodilator effectiveness of organic nitrates. J. Pharmacol. Exp. Ther. 1969; 165: 286–288.

178 Yeates RA, Laufen H, Leitold M. The reaction between organic nitrates and sulfhydryl compounds. A possible model system for the activation of organic nitrates. Mol. Pharmacol. 1985; 28: 555–559.

179 Artz JD, Yang K, Lock J, Sanchez C, Bennett BM, Thatcher GRJ. Reactivity of thionitrate esters: putative intermediates in nitrovasodilator activity. Chem. Commun. 1996; 927–928.

180 Cameron DR, Borrajo AMP, Bennett BM, Thatcher GRJ. Organic nitrates, thionitrates, peroxynitrites, and nitric oxide: a molecular orbital study of the $RXNO_2$ <=> RXONO (X=O,S) rearrangement, a reaction of potential bilogical significance. Can. J. Chem. 1995; 73: 1627–1638.

181 Doel JJ, Godber BL, Eisenthal R, Harrison R. Reduction of organic nitrates catalysed by xanthine oxidoreductase under anaerobic conditions. Biochim. Biophys. Acta 2001; 1527: 81–87.

182 Meah Y, Brown BJ, Chakraborty S, Massey V. Old yellow enzyme: reduction of nitrate esters, glycerin trinitrate, and propylene 1,2-dinitrate. Proc. Natl. Acad. Sci. USA 2001; 98: 8560–8565.

183 Habig WH, Keen JH, Jakoby WB. Glutathione S-transferase in the formation of cyanide from organic thiocyanates and as an organic nitrate reductase. Biochem. Biophys. Res. Commun. 1975; 64: 501–506.

184 McDonald BJ, Bennett BM. Cytochrome P-450 mediated biotransformation of organic nitrates. Can. J. Physiol. Pharmacol. 1990; 68: 1552–1557.

185 McGuire JJ, Anderson DJ, McDonald BJ, Narayanasami R, Bennett BM. Inhibition of NADPH-cytochrome P450 reductase and glyceryl trinitrate biotransformation by diphenyleneiodonium sulfate. Biochem. Pharmacol. 1998; 56: 881–893.

186 Nigam R, Whiting T, Bennett BM. Effect of inhibitors of glutathione S-transferase on glyceryl trinitrate activity in isolated rat aorta. Can. J. Physiol. Pharmacol. 1993; 71: 179–184.

187 Nigam R, Anderson DJ, Lee SF, Bennett BM. Isoform-specific biotransformation of glyceryl trinitrate by rat aortic glutathione S-transferases. J. Pharmacol. Exp. Ther. 1996; 279: 1527–1534.

188 Tsuchida S, Maki T, Sato K. Purification and characterization of glutathione transferases with an activity toward nitroglycerin from human aorta and heart. Multiplicity of the human class Mu forms. J. Biol. Chem. 1990; 265: 7150–7157.

189 McDonald BJ, Bennett BM. Biotransformation of glyceryl trinitrate by rat aortic cytochrome P450. Biochem. Pharmacol. 1993; 45: 268–270.

190 Towell J, Garthwaite T, Wang R. Erythrocyte aldehyde dehydrogenase and disulfiram-like side effects of hypoglycemics and antianginals. Alcohol. Clin. Exp. Res. 1985; 9: 438–442.

191 Mukerjee N, Pietruszko R. Inactivation of human aldehyde dehydrogenase by isosorbide dinitrate. J. Biol. Chem. 1994; 269: 21664–21669.

192 Millar TM, Stevens CR, Benjamin N, Eisenthal R, Harrison R, Blake DR. Xanthine oxidoreductase catalyses the reduction of nitrates and nitrite to nitric oxide under hypoxic conditions. FEBS Lett. 1998; 427: 225–228.

193 Brien JF, McLaughlin BE, Breedon TH, Bennett BM, Nakatsu K, Marks GS. Biotransformation of glyceryl trinitrate occurs concurrently with relaxation of rabbit aorta. J. Pharmacol. Exp. Ther. 1986; 237: 608–614.

194 Fung HL, Poliszczuk R. Nitrosothiol and nitrate tolerance. Z. Kardiol. 1986; 75: 25–27.

195 McGuire JJ, Anderson DJ, Bennett BM. Inhibition of the biotransformation and pharmacological actions of glyceryl trinitrate by the flavoprotein inhibitor, diphenyleneiodonium sulfate. J. Pharmacol. Exp. Ther. 1994; 271: 708–714.

196 Sage PR, de la Lande IS, Stafford I, Bennett CL, Phillipov G, Stubberfield J, Horowitz JD. Nitroglycerin tolerance in human vessels: evidence for impaired nitroglycerin bioconversion. Circulation 2000; 102: 2810–2815.

197 Brien JF, McLaughlin BE, Kobus SM, Kawamoto JH, Nakatsu K, Marks GS. Mechanism of glyceryl trinitrate-induced vasodilation. I. Relationship between drug biotransformation, tissue cyclic GMP elevation and relaxation of rabbit aorta. J. Pharmacol. Exp. Ther. 1988; 244: 322–327.

198 Bennett BM, Schroder H, Hayward LD, Waldman SA, Murad F. Effect of in vitro organic nitrate tolerance on relaxation, cyclic GMP accumulation, and guanylate cyclase activation by glyceryl trinitrate and the enantiomers of isoidide dinitrate. Circ. Res. 1988; 63: 693–701.

199 Slack CJ, McLaughlin BE, Nakatsu K, Marks GS, Brien JF. Nitric oxide-induced vasodilation of organic nitrate-tolerant rabbit aorta. Can. J. Physiol. Pharmacol. 1988; 66: 1344–1346.

200 Ratz JD, Fraser AB, Rees-Milton KJ, Adams MA, Bennett BM. Endothelin receptor antagonism does not prevent the development of in vivo glyceryl trinitrate tolerance in the rat. J. Pharmacol. Exp. Ther. 2000; 295: 578–585.

201 Stewart DH, Hayward LD, Bennett BM. Differential biotransformation of the enantiomers of isoidide dinitrate in isolated rat aorta. Can. J. Physiol. Pharmacol. 1989; 67: 1403–1408.

202 Ratz JD, McGuire JJ, Bennett BM. Enantioselective inhibition of the biotransformation and pharmacological actions of isoidide dinitrate by diphenyleneiodonium sulphate. Br. J. Pharmacol. 1999; 126: 61–68.

203 Li H, Liu X, Cui H, Chen YR, Cardounel AJ, Zweier JL. Characterization of the mechanism of cytochrome P450 – P450 reductase mediated nitric oxide and nitrosothiol generation from organic nitrates. J. Biol. Chem. 2006; 281: 12546–12554.
204 Hille R. Molybdenum enzymes. Essays Biochem. 1999; 34: 125–137.
205 Harrison R. Structure and function of xanthine oxidoreductase: where are we now? Free Radic. Biol. Med. 2002; 33: 774–797.
206 Zhang Z, Naughton DP, Blake DR, Benjamin N, Stevens CR, Winyard PG, Symons MC, Harrison R. Human xanthine oxidase converts nitrite ions into nitric oxide (NO). Biochem. Soc. Trans. 1997; 25: 524S.
207 Li H, Cui H, Liu X, Zweier JL. Xanthine oxidase catalyzes anaerobic transformation of organic nitrates to nitric oxide and nitrosothiols: characterization of this mechanism and the link between organic nitrate and guanylyl cyclase activation. J. Biol. Chem. 2005; 280: 16594–16600.
208 Marks GS. The 1986 Upjohn award lecture. Interaction of chemicals with hemoproteins: implications for the mechanism of action of porphyrinogenic drugs and nitroglycerin. Can. J. Physiol. Pharmacol. 1987; 65: 1111–1119.
209 Koesling D, Friebe A. Soluble guanylyl cyclase: structure and regulation. Rev. Physiol. Biochem. Pharmacol. 1999; 135: 41–65.
210 Zhao Y, Brandish PE, Ballou DP, Marletta MA. A molecular basis for nitric oxide sensing by soluble guanylate cyclase. Proc. Natl. Acad. Sci. USA 1999; 96: 14753–14758.
211 Lee WI, Fung HL. Mechanism-based partial inactivation of glutathione S-transferases by nitroglycerin: tyrosine nitration vs sulfhydryl oxidation. Nitric Oxide 2003; 8: 103–110.
212 You KS, Benitez LV, McConachie WA, Allison WS. The conversion of glyceraldehyde-3-phosphate dehydrogenase to an acylphosphatase by trinitroglycerin and inactivation of this activity by azide and ascorbate. Biochim. Biophys. Acta 1975; 384: 317–330.
213 Friebe A, Wedel B, Harteneck C, Foerster J, Schultz G, Koesling D. Functions of conserved cysteines of soluble guanylyl cyclase. Biochemistry 1997; 36: 1194–1198.
214 Stasch JP, Becker EM, Alonso-Alija C, Apeler H, Dembowsky K, Feurer A, Gerzer R, Minuth T, Perzborn E, Pleiss U, Schroder H, Schroeder W, Stahl E, Steinke W, Straub A, Schramm M. NO-independent regulatory site on soluble guanylate cyclase. Nature 2001; 410: 212–215.
215 Schmidt K, Schrammel A, Koesling D, Mayer B. Molecular mechanisms involved in the synergistic activation of soluble guanylyl cyclase by YC-1 and nitric oxide in endothelial cells. Mol. Pharmacol. 2001; 59: 220–224.
216 Galle J, Zabel U, Hubner U, Hatzelmann A, Wagner B, Wanner C, Schmidt HH. Effects of the soluble guanylyl cyclase activator, YC-1, on vascular tone, cyclic GMP levels and phosphodiesterase activity. Br. J. Pharmacol. 1999; 127: 195–203.
217 Wu WP, Hao JX, Ongini E, Impagnatiello F, Presotto C, Wiesenfeld-Hallin Z, Xu XJ. A nitric oxide (NO)-releasing derivative of gabapentin, NCX 8001, alleviates neuropathic pain-like behavior after spinal cord and peripheral nerve injury. Br. J. Pharmacol. 2004; 141: 65–74.
218 Mulsch A, Bauersachs J, Schafer A, Stasch JP, Kast R, Busse R. Effect of YC-1, an NO-independent, superoxide-sensitive stimulator of soluble guanylyl cyclase, on smooth muscle responsiveness to nitrovasodilators. Br. J. Pharmacol. 1997; 120: 681–689.
219 Pietruszko R, Mukerjee N, Blatter EE, Lehmann T. Nitrate esters as inhibitors and substrates of aldehyde dehydrogenase. Adv. Exp. Med. Biol. 1995; 372: 25–34.
220 Zhang J, Chen Z, Cobb FR, Stamler JS. Role of mitochondrial aldehyde dehydrogenase in nitroglycerin-induced vasodilation of coronary and systemic vessels: an intact canine model. Circulation 2004; 110: 750–755.
221 DeMaster EG, Redfern B, Shirota FN, Nagasawa HT. Differential inhibition of rat tissue catalase by cyanamide. Biochem. Pharmacol. 1986; 35: 2081–2085.
222 Chen Z, Foster MW, Zhang J, Mao L, Rockman HA, Kawamoto T, Kitagawa K, Nakayama KI, Hess DT, Stamler JS. An essential role for mitochondrial aldehyde dehydrogenase in nitroglycerin bioactivation. Proc. Natl. Acad. Sci. USA 2005; 102: 12159–12164.
223 Wang EQ, Balthasar JP, Fung HL. Pharmacodynamics of in vivo nitroglycerin tolerance in normal conscious rats: effects of dose and dosing protocol. Pharm. Res. 2004; 21: 114–120.

224 Singh S, Fritze G, Fang BL, Harada S, Paik YK, EckeyR, Agarwal DP, Goedde HW. Inheritance of mitochondrial aldehyde dehydrogenase: genotyping in Chinese, Japanese and South Korean families reveals dominance of the mutant allele. Hum. Genet. 1989; 83: 119–121.
225 Li Y, Zhang D, Jin W, Shao C, Yan P, Xu C, Sheng H, Liu Y, Yu J, Xie Y, Zhao Y, Lu D, Nebert DW, Harrison DC, Huang W, Jin L. Mitochondrial aldehyde dehydrogenase-2 (ALDH2) Glu504Lys polymorphism contributes to the variation in efficacy of sublingual nitroglycerin. J. Clin. Invest. 2006; 116: 506–511.
226 Mackenzie IS, Maki-Petaja KM, McEniery CM, Bao YP, Wallace SM, Cheriyan J, Monteith S, Brown MJ, Wilkinson IB. Aldehyde dehydrogenase 2 plays a role in the bioactivation of nitroglycerin in humans. Arterioscler Thromb Vasc. Biol. 2005; 25, 1891–1895.
227 Munzel T, Kurz S, Heitzer T, Harrison DG. New insights into mechanisms underlying nitrate tolerance. Am. J. Cardiol. 1996; 77: 24C–30C.
228 Gori T, Parker JD. The puzzle of nitrate tolerance: pieces smaller than we thought? Circulation 2002; 106: 2404–2408.
229 Warnholtz A, Tsilimingas N, Wendt M, Munzel T. Mechanisms underlying nitrate-induced endothelial dysfunction: insight from experimental and clinical studies. Heart Fail. Rev. 2002; 7: 335–345.
230 Kim D, Rybalkin SD, Pi X, Wang Y, Zhang C, Munzel T, Beavo JA, Berk BC, Yan C. Upregulation of phosphodiesterase 1A1 expression is associated with the development of nitrate tolerance. Circulation 2001; 104: 2338–2343.
231 Macpherson JD, Gillespie TD, Dunkerley HA, Maurice DH, Bennett BM. Inhibition of phosphodiesterase 5 selectively reverses nitrate tolerance in the venous circulation. J. Pharmacol. Exp. Ther. 2006; 317: 188–195.
232 Molina CR, Andresen JW, Rapoport RM, Waldman S, Murad F. Effect of in vivo nitroglycerin therapy on endothelium-dependent and independent vascular relaxation and cyclic GMP accumulation in rat aorta. J. Cardiovasc. Pharmacol. 1987; 10: 371–378.
233 Schroder H, Leitman DC, Bennett BM, Waldman SA, Murad F. Glyceryl trinitrate-induced desensitization of guanylate cyclase in cultured rat lung fibroblasts. J. Pharmacol. Exp. Ther. 1988; 245: 413–418.
234 Needleman P, Johnson Jr. EM. Mechanism of tolerance development to organic nitrates. J. Pharmacol. Exp. Ther. 1973; 184: 709–715.
235 Fung HL, Bauer JA. Mechanisms of nitrate tolerance. Cardiovasc. Drugs Ther. 1994; 8: 489–499.
236 Haj-Yehia AI, Benet LZ. In vivo depletion of free thiols does not account for nitroglycerin-induced tolerance: a thiol-nitrate interaction hypothesis as an alternative explanation for nitroglycerin activity and tolerance. J. Pharmacol. Exp. Ther. 1996; 278: 1296–1305.
237 Ahlner J, Andersson RGG, Torfgard K, Axelsson KL. Organic nitrate esters: clinical use and mechanisms of actions. Pharmacol. Rev. 1991; 43: 351–409.
238 Malta E. Biphasic relaxant curves to glyceryl trinitrate in rat aortic rings. Evidence for two mechanisms of action. Naunyn Schmiedebergs Arch. Pharmacol. 1989; 339: 236–243.
239 Ahlner J, Axelsson KL, Ljusegren ME, Grundstrom N, Andersson RG. Demonstration of a high affinity component of glyceryl trinitrate induced vasodilatation in the bovine mesenteric artery. J. Cyclic Nucleotide. Protein Phosphor. Res. 1986; 11: 445–456.
240 Axelsson KL, Andersson C, Ahlner J, Magnusson B, Wikberg JE. Comparative in vitro study of a series of organic nitroesters: unique biphasic concentration-effect curves for glyceryl trinitrate in isolated bovine arterial smooth muscle and lack of stereoselectivity for some glyceryl trinitrate analogues. J. Cardiovasc. Pharmacol. 1992; 19: 953–957.
241 Daiber A, Oelze M, Coldewey M, Bachschmid M, Wenzel P, Sydow K, Wendt M, Kleschyov AL, Stalleicken D, Ullrich V, Mulsch A, Munzel T. Oxidative stress and mitochondrial aldehyde dehydrogenase activity: a comparison of pentaerythritol tetranitrate with other organic nitrates. Mol. Pharmacol. 2004; 66: 1372–1382.

PART V

Dithiocarbamate Iron Complexes: Implication for NO Studies

CHAPTER 18

Mononitrosyl-iron complexes with dithiocarbamate ligands: physico-chemical properties

Anatoly F. Vanin[1] and Ernst van Faassen[2]

[1] *Semenov Institute of Chemical Physics, Russian Academy of Sciences, Moscow, 119991, Russian Federation*
[2] *Debye Institute, Section Interface Physics, Ornstein Laboratory, Utrecht University, 3508 TA, Utrecht, The Netherlands*

INTRODUCTION

Iron–dithiocarbamate complexes are currently the most widely used spin traps for nitric oxide radicals, and are one of the very few techniques available for *in vivo* NO detection [1–6]. Although conventional spin traps like dimethyl-1-pyrroline N-oxide (DMPO) or 2-methyl-2-nitrosopropane (MNP) are capable to trap NO radicals [7,8], the NO-adducts are prone to subsequent conversion *via* complex pathways. Alternatively, the consumption of NO specific radical scavengers like paramagnetic carboxy-PTIO or carboxymethoxy-PTIO may be used to demonstrate the presence of NO radicals [7] in an indirect manner. Nitric oxide has high affinity for iron, and many different ferrous iron complexes have been shown to act as traps for free NO. The first example of NO spin trapping leading to paramagnetic nitrosyl-iron adducts seems to have been achieved by McDonald who reported [9] the formation of iron-nitrosyl complexes with phosphate and catechol ligands. As always in spin trapping, the stability of the nitrosyl adduct determines the sensitivity of the method. In this respect, the adducts of NO with iron–dithiocarbamate complexes have good stability in biological tissues and are fairly specific for free-NO radicals (see below).

The dithiocarbamates form a class of disulfur compounds that bind a broad range of metal ions and are used to clear aqueous solutions from traces of Cd, Co, Hg, Mn, Mo, Ni and Pb. They have particular affinity for Fe^{2+}, Fe^{3+} and Cu^{2+}. Besides, as spin traps in research, these compounds are applied industrially as metal chelators and in synthesis of polymers, in agriculture as pesticides (e.g. thiram = tetramethyl thiuram disulfide), seed disinfectants and soil fumigants. In medicine, they are applied as metal chelators and for treatment of alcoholism. For example, tetraethyl thiuram disulfide (disulfiram, teruram) induces alcohol

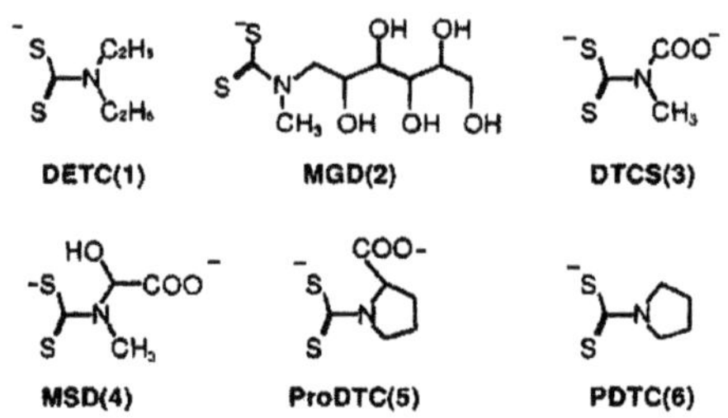

Fig. 1. Structures of the commonly used dithiocarbamate ligands. Diethyl-dithiocarbamate (DETC,1), *N*-methyl-D-glucamine dithiocarbamate (MGD,2), *N*-dithiocarboxy-sarcosine (DTCS,3), *N*-methyl-L-serine DTC (MSD,4), L-proline DTC (ProDTC, 5) and pyrroline DTC (PDTC, 6).

intolerance by inhibition of the acetaldehyde dehydrogenase enzyme. All dithiocarbamates share the characteristic disulfur motif that binds to the metal as a bidentate ligand (cf. Fig. 1). The nitrogen may be functionalized in various ways to modify the physico-chemical properties of the ensuing metal complex, in particular its solubility and lipophilicity. Some widely used dithiocarbamates are represented in Fig. 1.

Most of these bidentate ligands show cooperative binding where two ligands keep the metal ion in a planar complex with approximate C_{2v} symmetry, with free access to the axial coordination sites of the metal ion (Fig. 2).

The ligands themselves generally have good solubility in water, but the solubility of the metal–(ligand)$_2$ complex is strongly affected by the nature of the ligand. The polar tail of the MGD ligand makes the Fe–(MGD)$_2$ complex highly soluble in water (up to 2 mM at *RT* [10]) and it remains in the extracellular compartment as it does not penetrate the lipid cell membrane. The DETC ligands are very soluble in water themselves but produce a lipophilic complex of modest solubility (10^{-7} M at pH = 7 [11]) in water. Nevertheless, concentrations up to ca 50 μM can be achieved in aqueous solutions for some 20 min, with small black

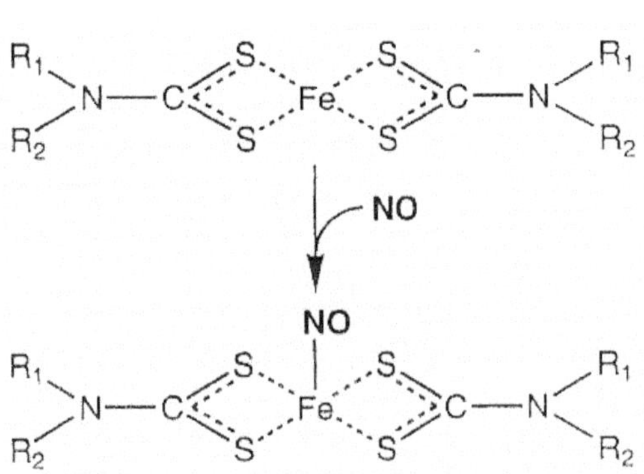

Fig. 2. Geometrical structure and trapping reaction of a typical iron–dithiocarbamate complex. The iron can be in ferric and ferrous state. The ferrous nitrosyl complex is paramagnetic.

particles becoming noticeable after longer duration). As a result, the Fe–$(DETC)_2$ complexes accumulate in the apolar lipid and protein fractions of biological samples.

The dithiocarbamate motif is susceptible to decomposition in aqueous solution, releasing the highly toxic carbon disulfide and amines in the process. The decomposition is accelerated by the presence of peroxides, acid halides and acid conditions. Therefore, in actual applications of this class of ligands, acidification should be avoided by adequate buffering at neutral or basic pH. The release of carbon disulfide is the mechanism behind the fungicidal and disinfectant properties observed in many dithiocarbamates (e.g. thiram). The risk of carbon disulfide toxicity is manifest in all dithiocarbamates. Individual dithiocarbamates act as inhibitors of certain enzymes. The inhibition of acetaldehyde dehydrogenase by disulfiram was already mentioned in connection with alcohol intolerance. Pyrrolidine dithiocarbamate (PDTC) is considered an antioxidant and inducer of apoptosis in lymphocytes and tumour cells [12,13]. The antioxidant property was related to electron donation under formation of thiuram disulfides, whereas the pro-apoptotic action required the presence of Cu^{2+} or Zn^{2+} [12]. The thiuram disulfides themselves have also been implicated in apoptosis *via* inhibition of nuclear factor NF-κB and p53 proteins [14]. In this context, it should be noted that administration of high doses of iron–dithiocarbamate complexes will also influence the NO status of tissues and cell cultures by acting as scavengers of the free-NO radicals. In certain cases, this phenomenon might be beneficial. For example, it has been noted that iron–dithiocarbamate complexes offer significant protection against the rejection of cardiac allografts following tissue transplantation [15]. For more information on this interesting phenomenon we refer the reader to Chapter 19 of this book.

LIGAND STRUCTURE OF IRON–DITHIOCARBAMATE COMPLEXES

The number n of dithiocarbamate ligands in the iron complex has been a subject of discussion. For Fe^{3+}–DETC in DMSO experiments are compatible with $n = 2$ or 3 [16,17]. The geometrical structure of this complex has not been determined, but the three-ligand structure is possibly stabilized by hydrophobic interactions between the tails of the ligands.

For MGD ligands in water it was found [17] that $n = 2$. The value of n for the $Fe^{3+}(MGD)_2$ complex was determined by using the optical absorption of the complexes in strong HEPES buffer (150 mM, pH 7.4) (Fig. 3) as a function of ligand concentration.

Fig. 3. confirms that the dithiocarbamate ligands are very good chelators for ferric iron, since half of the available iron ([Fe] = 0.1 mM) is sequestered as $Fe^{3+}(MGD)_2$ at a low concentration of ligand: [MGD] ~0.25 mM.

TRAPPING OF FREE NITRIC OXIDE BY IRON–DITHIOCARBAMATE COMPLEXES

The application of iron–dithiocarbamate complexes as spin trap is motivated by their high affinity to bind nitric oxide radicals at the axial position of the iron (Fig. 2). Many other ferric and ferrous iron complexes bind NO to form nitrosyl complexes as well, but the dithiocarbamate ligands provide the iron complex with high binding rates (Table 1) and good

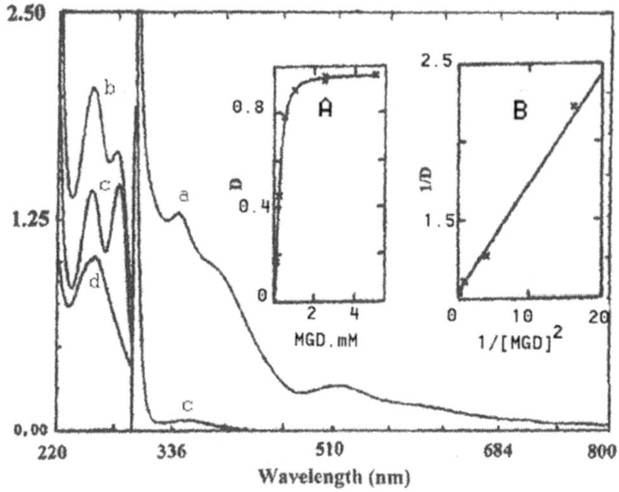

Fig. 3. UV/VIS optical spectra of the $Fe^{3+}(MGD)_2$ complex in HEPES (150 mM, pH 7.4, [Fe] = 0.1 mM, [MGD] = 1 mM). The optical pathlength of the cuvette is 1 cm (VIS) or 0.1 cm (UV). (Curve a) VIS spectrum of the $Fe^{3+}(MGD)_2$ complex with absorption bands at 314, 340 and 385 nm. Free MGD does not absorb in this wavelength region. (Curve b) The UV spectrum appears as superposition of spectra of 0.8 mM free MGD and 0.1 mM $Fe^{3+}(MGD)_2$. It has a pair of non-equal bands at 256 and 283 nm. (Curve c) Reference spectrum of 0.8 mM free MGD ligands with bands at 256 and 283 nm. (Curve d) Computed difference spectrum b–c. It shows the spectrum of 0.1 mM $Fe^{3+}(MGD)_2$ with a prominent band at 258 nm. The inset A shows the yield of $Fe^{3+}(MGD)_2$ as function of ligand concentration and monitored at 340 nm. Inset B shows the same data plotted as 1/D *vs.* $1/[MGD]^2$. The linear dependence shows that $n = 2$. (From Ref. [17].)

stability of the nitrosyl adduct. Typical trapping rates are of the order 10^6–10^8 $(Ms)^{-1}$, i.e. orders of magnitude faster than the trapping rate $(0.5–1.5) \cdot 10^{-4}$ $(Ms)^{-1}$ reported [18] for PTIO traps. Interestingly, ferric as well as ferrous iron dithiocarbamate complexes have high binding rates for NO (Table 1). Rapid binding and stability of the adduct are crucial for good NO traps. This adduct is a ferric or ferrous mononitrosyl-iron complex (MNIC) which will preserve the NO ligand indefinitely unless destroyed by oxidation or decomposition of the dithiocarbamate ligands.

Table 1 Trapping rates of free-NO radicals by various iron–dithiocarbamate complexes. ProDTC = L-proline dithiocarbamate; DTCS = *N*-(dithiocarboxy) sarcosine; MGD = *N*-methyl-D-glucamine dithiocarbamate

Complex	NO trapping rate $(Ms)^{-1}$	Solvent	Remarks	Method	Refs.
Fe^{2+}–ProDTC	$1.1 \cdot 10^8$	Phosphate buffer	*RT*, pH = 7.0	Photolysis of tetrahydro piridazin	[19]
Fe^{3+}–DTCS	$4.8 \cdot 10^8$	Phosphate buffer	*RT*, pH = 7.4	Pulse radiolysis of nitrite	[20]
Fe^{2+}–MGD	$1.2 \cdot 10^6$	Phosphate buffer	*RT*, pH = 7.4	Competition with oxyhemoglobin	[21]
Fe^{2+}–DTCS	$1.7 \cdot 10^6$	Phosphate buffer	*RT*, pH = 7.4	Competition with oxyhemoglobin	[21]

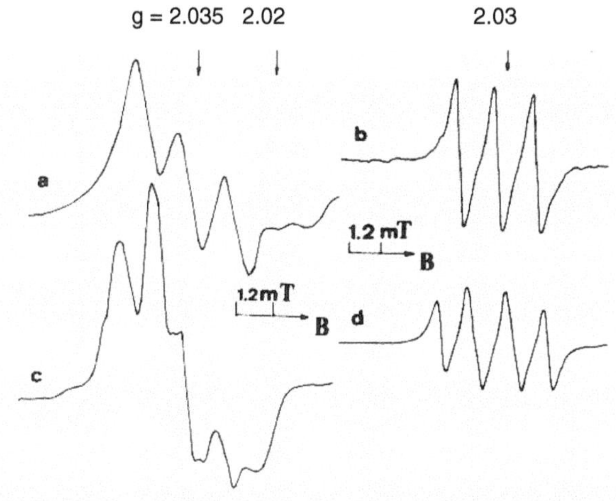

Fig. 4. Typical EPR spectra of paramagnetic NO–Fe^{2+}–$(MGD)_2$ complexes in solution or frozen state. The spectra show resolved hyperfine coupling with the nitrogen nucleus of the nitrosyl ligand and with the nuclear magnetic moment of the ^{57}Fe isotope ($I = 1/2$, abundance 2.2%). (Curve a) Complex with ^{56}Fe in frozen state. (Curve b) Complex with ^{56}Fe in solution at room temperature. (Curve c) Complex with ^{57}Fe in frozen state. (Curve d) Complex with ^{57}Fe in solution at room temperature [16].

NO is a π-binding ligand that binds with the nitrogen atom pointing towards the iron center. Most nitrosyl-iron complexes show a small but significant bending of the Fe-NO unit. NO would be a pure π-bonding ligand if the NO-axis were perfectly aligned in axial direction from iron, but bending of the Fe–N–O leads to the admixture of significant pseudo-σ bonding. For DETC ligands, the MNIC was reported to have a Fe–N–O bond angle of ca 174° [22]. The ferric nitrosyl complex NO–Fe^{3+}–$(DETC)_2$ is diamagnetic, but the ferrous form is paramagnetic ($S = 1/2$) and may be detected in solutions or frozen tissues with electron paramagnetic resonance (EPR) if the quantity of complexes exceeds the instrumental detection limit of a few picomoles. Typical spectra at room temperature and in frozen state are given in Fig. 4. This property has given iron–dithiocarbamates widespread application as a spin trap for NO *in vitro* and *in vivo*. As such, it still is one of the very few methods to trap and detect NO in living tissues. A typical application involves laboratory animals in which the trapping complexes assemble spontaneously endogeneously after separate injection of DETC ligands and iron. For the final MNIC yield it is irrelevant whether the iron be injected in ferric or ferrous state, as long as endogenous formation of iron precipitates is avoided. Given the low solubility of ferric iron, the iron is usually administered as a ferrous citrate complex. At physiological pH, citrate acts as a tridentate ligand giving good stability, solubility and some protection of the ferrous complex against oxidation. It allows distribution of the iron and DETC ligands throughout the animal *via* the blood circulation and the formation of adequate numbers of Fe–DETC traps in the various tissues like brain, kidney, heart or liver. Certain tissues like liver or brain are known to have particularly high MNIC yields. In most applications, the trapping proceeds *in vivo*, while the detection and quantification of adduct yields are done *ex vivo* on frozen samples at 77 K. Only recently, instrumental advances in EPR instrumentation have made possible the *in vivo* detection of ferrous MNIC under favorable conditions [5]. Typical experimental protocols for NO trapping in animals have been described in [3].

It should be mentioned that iron-dithiocarbamate traps are not fully specific for the presence of free NO. Paramagnetic MNIC adducts are also formed in the presence of *S*-nitrosothiols by transnitrosylation of the trap by the nitrosothiol. It was verified [23] that the nitric oxide is transferred to the iron as the neutral NO radical. However, the rate of these transfer reaction is orders of magnitude lower than the trapping rate of the highly mobile NO radical itself. Fe^{2+}–$(MGD)_2$ is transnitrosylated [24] by Cys-NO [$k \sim 30 \pm 5$ $(Ms)^{-1}$] and GSNO [$k \sim 3.0 \pm 0.3$ $(Ms)^{-1}$]. Ferric Fe^{3+}-$(MGD)_2$ is transnitrosylated by Cys-NO at a modest rate of 83 ± 3 $(Ms)^{-1}$ [23]. In the case of free NO, the trapping rates for iron–dithiocarbamate complexes in water are far higher in the order of 10^6–10^8 $(Ms)^{-1}$ (cf Table 1).

Dinitrosyliron complexes (DNIC) are also capable of transnitrosylating iron-dithiocarbamate traps [1] and may form endogenously in living tissues [25]. The transfer reactions from DNIC are discussed in Chapter 2 of this book. Finally, MNIC adducts are easily obtained in the presence of nitrite anions. In animal tissues, three mechanisms have been identified so far. The first pathway is acidic reduction of nitrite to free NO, which is trapped by the iron–dithiocarbamate. This pathway may be inhibited by preventing acidification. The second pathway involves enzymatic reduction of nitrite to NO. Dedicated nitrite reductases are known in plants, algae and cyanobacteria but are lacking in mammals. However, certain mammalian enzymes have shown some nitrite reductase activity under anoxia. This pathway is supplementary to their normal physiological function. A detailed discussion of these reactions is given in Chapters 14 and 15. As a third pathway, nitrite is known to induce formation of nitrosothiols in tissues. It was observed [26] that the nitrosothiol concentration in living tissues increases proportionally to exogenously administered nitrite, and these nitrosothiol intermediates will generate MNIC adducts. Additional pathways exist *in vitro*. For example, very high levels of ferrous dithiocarbamate complexes themselves were reported to release NO from nitrite even at physiological pH [27,28]. Fortunately, in actual spin trapping experiments such concentrations are unlikely to be met.

Besides these problems with specificity, users of Fe-dithiocarbamates should be aware of potential artifacts introduced by the method. First, the administration of exogenous iron could potentially initiate unwanted redox reactions, with Fenton reactions being the most harmful. The dithiocarbamate ligands are known inhibitors of the nuclear transcription factor kappa B (NFκB) which affects the induction of NOS enzymes. For example, DETC was reported [29] to inhibit the induction of NOS in murine bone marrow macrophages when exposed to LPS. The ligands are known to extract Cu^{2+} ions from the Cu–Zn superoxide dismutase enzyme, thereby disabling an important agent in the antioxidant defense mechanism in tissues [30]. The dithiocarbamates themselves were reported to act as reducing agents [27] when applied at high concentrations. *In vitro* studies [31] have shown that Fe^{3+}–DTCS and Fe^{3+}–MGD act as an inhibitor for neuronal NOS with $IC_{50} \sim 25$ μM and 10 μM, respectively. Finally, although the iron–dithiocarbamate complexes are generally known to have good stability in biological tissues, they may participate in redox reactions [4,32]. This holds for the traps as well as for the nitrosylated adduct. In water, the Fe^{2+}–MGD complex is oxidized by dioxygen with a fast rate [27] of 5×10^5 $(Ms)^{-1}$. Given the micromolar oxygen concentrations in normoxic tissues, this rapid oxidation rate makes that the traps exist in predominantly oxidized state (see below). In contrast, the nitrosylated NO–Fe^{2+}–MGD adduct is only slowly oxidized by dioxygen. The ferrous complex has a lifetime of nearly an hour even in the presence of millimolar oxygen concentrations. For example, only half of initial 1 mM NO–Fe^{2+}–MGD is oxidized to ferric state by bubbling with air for an hour at ambient temperature. Clearly, the presence of the nitrosyl ligands helps to maintain the complex in reduced state and this fact is very important for the success of iron-dithiocarbamates in NO trapping *in vivo* [17].

Paramagnetic adducts may be lost from observation with EPR *via* several other mechanisms: Halogen and nitrogen dioxide molecules may incorporate into ferrous MNIC complexes [33] and render the complex diamagnetic. This reaction may be reversed by reduction. Paramagnetism from ferrous MNIC may also be lost by reactions with peroxynitrite [34] and superoxide [4,21,34]. The reaction of MNIC-MGD and superoxide produces a diamagnetic iron-nitroso complex with a fast rate of 3×10^7 $(Ms)^{-1}$ (37°C, pH = 7.4). This shows that the NO moiety retains reactivity with superoxide even when liganded to the iron, in spite of significant transfer (ca 70%) of the unpaired electron density towards the iron. Still, this rate is two orders of magnitude below the rate of 7×10^9 $(Ms)^{-1}$ between free NO and superoxide. With Fe-dithiocarbamate traps usually in the millimolar range, we note that binding to the latter may provide a mechanism to protect free NO against superoxide in biological tissues. Superoxide levels in tissues are normally kept sufficiently low to avoid significant loss of paramagnetic MNIC, but caution should be exercised when NO is trapped under conditions where high levels of superoxide are expected. Loss of significant fractions of MNIC due to superoxide has been explicitly demonstrated in animal tissues [4]. In severe cases, the lost fraction may be estimated by the so-called ABC method [4]. This transformation to an EPR silent species can be partially reversed [34] by addition of reductants like ascorbate. The extent of the reversal depends on the time lapsed. Application of ascorbate within 1 min achieves complete restoration of the ferrous MNIC. After a delay of ca 10 min, only partial restoration of about 50% is achieved. It indicates that a reversible reaction with superoxide produces an intermediate EPR silent species that undergoes slow irreversible interconversion to a stable EPR silent end-product. This stable end-product was not identified but probably is some iron–nitrato complex.

The preceding applications describe NO trapping where the traps are distributed throughout the system. This is not always desirable and not necessary. In principle, solid surfaces of polymeric composites may be functionalized with iron-dithiocarbamate traps and used for detection of NO from liquid or gaseous phase without contamination of the system with traps [35,36]. BSA proteins may be loaded with Fe-DETC traps and infused into the bloodstream and applications of small colloidal Fe-DETC particles have been reported [37].

SPECTROSCOPIC PROPERTIES OF (NITROSYLATED) IRON–DITHIOCARBAMATE COMPLEXES

From the above discussion it is clear that applications of iron–dithiocarbamates (Fe-DTC, where DTC are for example MGD or DETC) as NO traps require the consideration of four different complexes: ferric and ferrous, with and without a nitrosyl ligand. These forms are listed in Scheme 1.

These four complexes are easily distinguished by differences in optical and EPR spectra. The optical spectra of Fe^{3+}-MGD and Fe^{2+}-MGD complexes (I and II complexes)

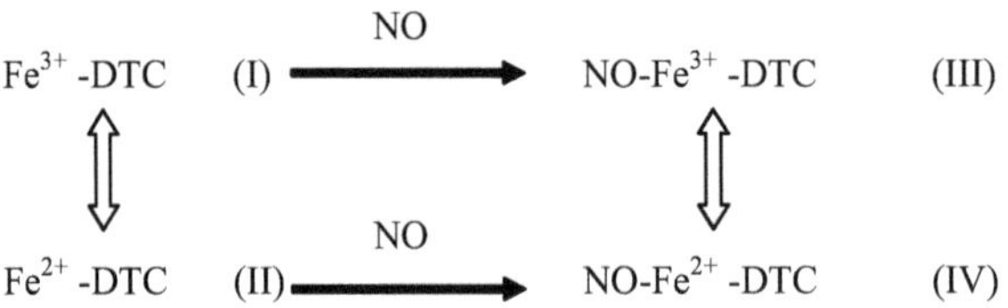

Scheme 1. The four possible Fe-DTC complexes considered here. Complex I is a high spin ($S = 5/2$) complex observable with EPR at $g = 4.3$. The ferric mononitrosyl-iron complex III is diamagnetic. The ferrous MNIC complex IV is paramagnetic ($S = 1/2$) and observable with EPR at $g = 2.03$. The vertical arrows represent reversible redox reactions. The nitrosylation reactions are irreversible as the nitrosyl ligand is very strongly bound to the iron.

or Fe-DETC complexes without or with a nitrosyl ligand (I–IV complexes) are given in Fig. 3 and Fig. 5, respectively. Optical absorption peaks and extinctions for the various Fe-MGD complexes are presented in Table 2. Depending on the dithiocarbamate ligand, the IR NO-stretch vibration in NO-Fe^{3+}-DTC falls in the range 1690–1710 cm^{-1} [38,39].

Often, the samples contain mixtures of I–IV complexes. In such cases, the optical spectra appear as superpositions and are difficult to analyze. The EPR spectra are much simpler in this respect since complexes II and III are EPR silent and the paramagnetic species I and IV are easily distinguished from differences in g-factor. Species I is a high spin $S = 5/2$ complex

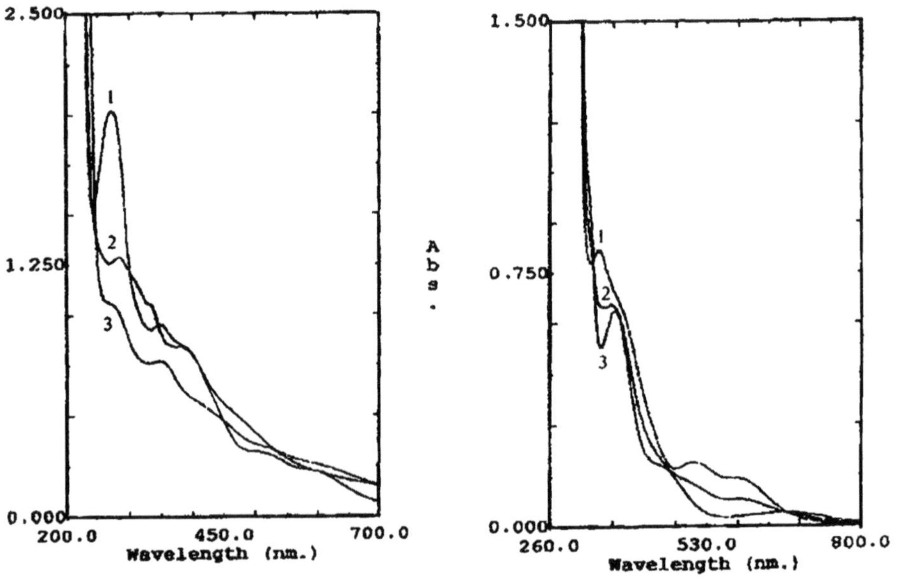

Fig. 5. Optical absorption spectra of iron-DETC complexes in solution. Complex II does not absorb in the visible region. (Curve 1) Fe^{3+}-DETC = complex I (orange brown). (Curve 2) NO- Fe^{3+}-DETC = complex III (yellow). (Curve 3) NO- Fe^{2+}-DETC = complex IV (bright green). Left panel: [Fe] = 50 μM, [DETC] = 200 μM in HEPES (150 mM, pH = 7.4) optical pathway 10 mm. These spectra appear superposed on a background due to light scattering from small solid particles. Right panel: [Fe] = 100 μM, [DETC] = 1 mM in DMSO-water mixture (3:1) optical pathway 5 mm. (From Ref. [43].)

Table 2 Optical absorption peaks and extinctions for the various Fe-MGD complexes. The extinction coefficients are given in $(Mcm)^{-1}$ and were compiled from [17]. The single peak of the yellow complex III appears as a poorly resolved shoulder on a broad smooth background (Fig. 5)

Complex	I	II	III	IV
Color	Orange brown	Uncolored	Yellow	Green
UV peaks	258 nm (52,000)	–		
VIS peaks	340 nm (12,700)	–	328 nm (12,500)	327 nm (10,400)
	385 nm (9600)			365 nm (6400)
	514 nm (2200)			440 nm (2050)
	591 nm (1005)			672 nm (670)

Table 3 EPR parameters of NO-Fe^{2+}-DETC crystals [22]

	x	y	z
Sample	Crystal	Crystal	Crystal
g-tensor	2.039	2.035	2.025
$A(^{14}N)$ (G)	13.4	12.1	15.5
$A(^{57}Fe)$ (G)	14	14	2

with prominent absorption near $g = 4.3$ [32], and the nitrosyl species IV has a characteristic spectrum near $g = 2.035$ (Fig. 4, Table 3).

ISOTOPIC SUBSTITUTIONS AFFECT THE EPR LINESHAPES OF THE MNIC ADDUCTS

Isotopic substitution provides a valuable technique to investigate the reaction mechanisms involving NO. Isotopic substitution of the natural $^{14}N(I = 1)$ by $^{15}N(I = 1/2)$ is commonly used to establish the identity of the substrate from which the ^{15}NO originates. A good example is described in Chapter 14 where nitrite is identified as the substrate for anoxic reduction by endothelial NOS. The difference in multiplicity of the hyperfine coupling is clearly reflected in the EPR lineshape (cf. Fig. 6). Substitution of iron affects the EPR spectrum as shown in Fig. 4 and is used to investigate the pool of endogenous iron in tissues. Applications are given later in this chapter. In a few cases [40], substitution of the oxygen on the nitrosyl ligand provided valuable information on the reaction pathways. The values of the various couplings are collected in Table 4 [40]. The wavefunction of the unpaired electron has clearly large overlap with the iron and nitrogen nuclei, whereas the interaction with the oxygen remains fairly small at a few Gauss.

After isotopic enrichment of the complexes with ^{57}Fe $(I = 1/2)$, Mossbauer spectroscopy may be used to investigate the ligand field surrounding the iron atom (cf e.g. [41]). In the absence of a ligand field, the transitions between nuclear ground state ($I = 1/2$, $g_n = 0.181$) and nuclear excited state ($I = 3/2$, $g_n = -0.106$ with quadrupole moment $Q = +0.4 \cdot 10^{-24}$ cm^2) are degenerate and the spectrum appears as a single line. If the nucleus is subject to a ligand field with symmetry lower than cubic or tetrahedral, the interaction between the quadrupole moment Q and the field gradient tensor lifts the degeneracy of the nuclear

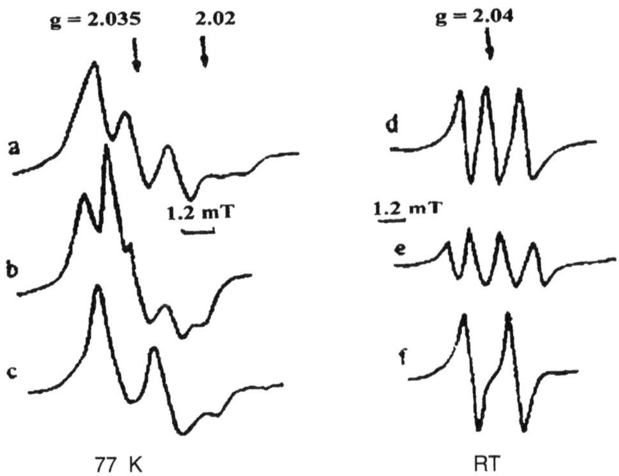

Fig. 6. EPR spectra of NO-Fe^{2+}-DETC in DMSO with various isotopes. The left column shows frozen solutions at 77 K. The right column shows liquid solutions at *RT*. (Curve a + d) with ^{56}Fe and ^{14}NO (natural abundance). (Curve b + e) with ^{57}Fe and ^{14}NO. (Curve c + f) with ^{56}Fe and ^{15}NO [16].

Table 4 Isotropic EPR parameters of NO-Fe^{2+}-MGD complexes in solution at room temperature. The hyperfine couplings are in Gauss. The isotropic g-factor is $g_{iso} = 2.040$ [40]

	$A(^{14}N)$	$A(^{15}N)$	$A(^{57}Fe)$	$A(^{17}O)$
Nuclear g_n	0.4038	−0.5664	0.1806	−0.7575
Nuclear spin	1	1/2	1/2	5/2
Coupling (G)	12.6	17.6	8.6	2.5
Refs.	[40]	[40]	[40]	[40]

excited state. In the absence of magnetic hyperfine interactions, the spectra appear as doublets. The spectral shape is characterized by two parameters, the quadrupole splitting ΔE_Q and the isomer shift δ. The latter measures the offset of the centroid of the Mossbauer spectrum and reflects the S-electron density at the nucleus. The Mossbauer spectrum of ferrous complex IV is shown in the lower panel of Fig. 7. It appears as a doublet spectrum with $\Delta E_Q =$ 0.87 mm/s and $\delta = 0.32$ mm/s. The ferric complex III appears as a doublet with $\Delta E_Q =$ 1.25 mm/s and $\delta = 0.22$ mm/s (upper panel) [32]. These parameters reflect the ligand field at the iron nucleus. This field is defined by the bidentate sulfur motif of the ligand, so that the various dithiocarbamate ligands have very similar Mössbauer parameters.

THE DETERMINANTS OF MNIC FORMATION IN TISSUES: REDOX STATE AND LIGANDS OF IRON

Iron-diethyldithiocarbamate (Fe-DETC) complexes are widely used as *in vivo* spin traps for nitric oxide (NO) [1,3,4] and this method has proven very successful by the formation of

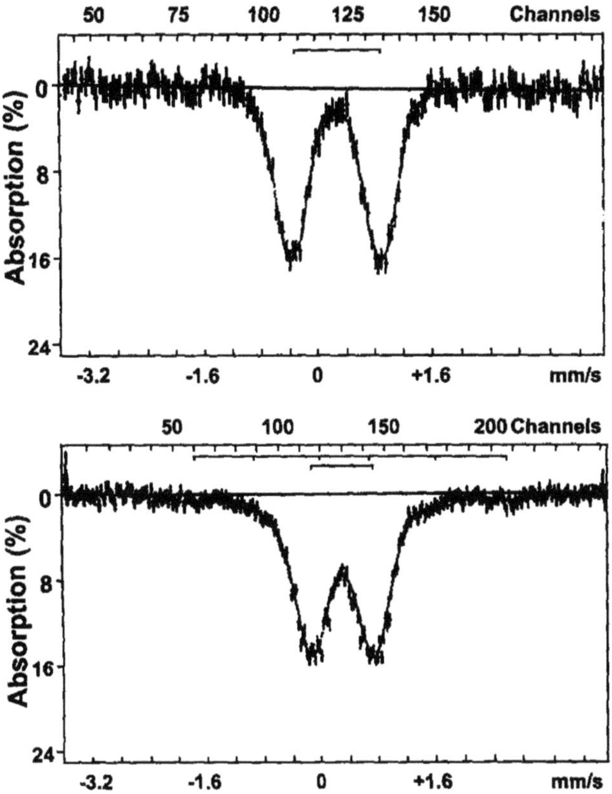

Fig. 7. Mossbauer spectra of $^{57}Fe^{2+}$-DETC in DMSO exposed to NO and then treated with NO_2 leading to the formation of NO-$^{57}Fe^{3+}$-DETC complex (top panel). Subsequent treatment with dithionite causes the formation of the mononitrosyl NO-$^{57}Fe^{2+}$-DETC complex (bottom panel) [32]. The spectra are taken at 80 K.

detectable quantities of paramagnetic ferrous MNIC adducts (complex IV) in many animal experiments.

Conventionally, the formation of complex IV is explained by a straightforward trapping of the NO radicals by diamagnetic ferrous Fe^{2+}-dithiocarbamate complexes. It is a reaction as in Fig. 2 with iron being in ferrous state. However, this conventional viewpoint is difficult to reconcile with the presence of micromolar levels of oxygen in living tissues except under conditions of hypoxia or ischemia. It also contradicts the experimental observation that the MNIC yields in animal tissues do not depend on whether the iron was originally added in ferrous or ferric form [42]. Our recent investigations [17,43] have revealed that the formation of ferrous MNIC proceeds *via* a much more complex mechanism whereby the MNIC forms only one observable compound in a complex reaction network where iron, dithiocarbamate or other ligands, nitrosothiols and NO are coupled *via* chemical and redox equilibria. Two basic ingredients are the charge state of the iron atom, and the choice of iron ligands. The former is determined by the redox state of the tissue, primarily by local oxygen levels and the glutathione GSH/GSSH ratio or by exogenous reductants like ascorbate or dithionite.

The second is largely determined by the presence of strongly binding ligands like citrate or phenantroline. Both ingredients may be modulated by interventions from outside and have significant effect on the MNIC yield in the tissue. We will first consider the redox state of the trapping complex and the reaction pathway which brings the mononitrosyl complexes to the ferrous paramagnetic state.

Frozen tissue samples usually show a prominent EPR absorption line at $g = 4.3$ from a combination of different endogenous ferric high spin complexes ($S = 5/2$). The intensity of this line is significantly enhanced by administration of exogenous ferric of ferrous iron, as in Fe-DETC trapping, for example. This EPR absorption at $g = 4.3$ is not specific for a given complex, since many different ferric complexes show such EPR absorption [43], for example, Fe^{3+}-citrate and Fe^{3+}-dithiocarbamates. The redox state of the iron–dithiocarbamate complexes was investigated [17,43] in male adult BALB mice that had been treated with lipopolysaccharide (LPS) four hours before the experiment in order to stimulate NO production. The basal concentrations of endogenous high-spin complexes were ca 10 μmol/kg wet tissue in various tissues. These results show that mammalian tissues maintain significant quantities of iron in ferric state, in spite of the ubiquitous presence of endogenous reductants like ascorbate or glutathione. This iron forms part of the non-transferrin-bound iron (NTBI) pool since the signal intensity was not diminished by perfusive removal of blood together with its transferrin content. After injection of DETC ligands together with exogenous ferrous sulfate, the EPR absorption near $g = 4.3$ was increased from 10 to 20 μmol/kg wet tissue, showing that a significant fraction of the iron was rapidly oxidized to ferric state. We conclude that endogenous iron as well as exogenous iron exists as a mixture of ferrous and ferric state.

REDUCTION WITH DITHIONITE ENHANCES MNIC YIELDS *IN VIVO*

This finding is highly relevant for NO trapping since both charge states of iron bind NO *in vitro*, and the nitrosylated adducts should be a mixture of ferric and ferrous complexes as well. This was confirmed by our experiments on animal tissues [43]. We studied the effect of *ex vivo* reduction of tissue extracts with sodium dithionite. Table 5 gives MNIC yields in various tissues of BALB mice that had been injected with LPS 4 h before the trapping experiment. The table shows that reduction of the tissue significantly enhances the yield of complex IV in all tissues by a factor of about 3–4.

These yields show that, contrary to the common assumption, the ferric complex I is the dominant NO trap in mouse tissues and cell cultures. The minority ferrous population of traps becomes observable with EPR after nitrosylation produces the paramagnetic complex IV. The ferric majority binds NO as the diamagnetic complex III and is observed only after *ex vivo* reduction with dithionite. It is important to note that a significant population of nitrosylated ferric complexes is formed and remains undetected in a conventional trapping experiment where tissue samples are analyzed without reduction. *Ex vivo* reduction then enhances the yields.

A similar increase in yield of complex IV by reduction was observed [43] in cultures of immortalized murine microvascular brain endothelial cells: When stimulated by calcium-ionophore, 7.5×10^6 BEND3 cells formed a total of 420 ± 60 pmol paramagnetic complex IV

Table 5 Effect of reduction with dithionite on the yields of paramagnetic complex IV in tissues of BALB mice pretreated with LPS 4 h prior to NO trapping [43]. The trapping proceeded for 10 or 30 min after injection of DETC intraperitonally and Fe citrate subcutaneously. For reduction, the tissues were incubated in HEPES (150 mM, pH = 7.4) for 30 min with solid sodium dithionite (10 mM final). The errors reflect the variation in MNIC yield between individual mice

Tissue preparation	MNIC-DETC yields (in nmol/g)			
	10 min incubation		30 min incubation	
	Unreduced	Reduced	Unreduced	Reduced
Liver	2 ± 1	12 ± 3	12 ± 4	33 ± 9
Kidney	0.5 ± 0.2	3 ± 1	3 ± 1	7 ± 3
Heart	0.3 ± 0.1	2 ± 1	0.8 ± 0.3	2 ± 1
Spleen	0.4 ± 0.2	1.0 ± 0.5	1.0 ± 0.5	2 ± 1
Lung	0.4 ± 0.2	1.0 ± 0.5	2 ± 1	4 ± 2

in a conventional trapping experiment (i.e. without reduction). The corresponding EPR spectrum is shown in curve (b) of Fig. 8. Subsequent reduction with dithionite raised the yield of complex IV fivefold to ca 2.0 ± 0.3 nmol (curve c). The yield remained below the EPR detection limit if the cells were preincubated with the NOS inhibitor L-NNA, clearly identifying NOS as the most significant source of NO in this experiment.

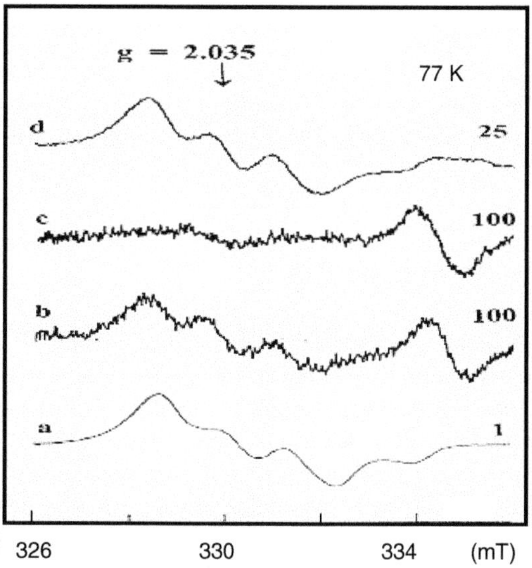

Fig. 8. EPR spectra at 77 K of MNIC adducts formed in cultured BEND3 cells after stimulation with Ca-ionophore. The aliquots contain about half of the lipid cell fraction of $7.5 \cdot 10^6$ cells. Scaling factors are indicated on the right-hand side. (Curve a) Reference spectrum of 25 nmol NO–Fe^{2+}–MGD complexes in frozen PBS buffer. (Curve b) MNIC signal after 15 min trapping in cell culture, unreduced. (Curve c) absence of MNIC signal in presence of 1 mM NOS-inhibitor L-NNA. (Curve d) sample (b) after reduction for 30 min with 10 mM dithionite [43].

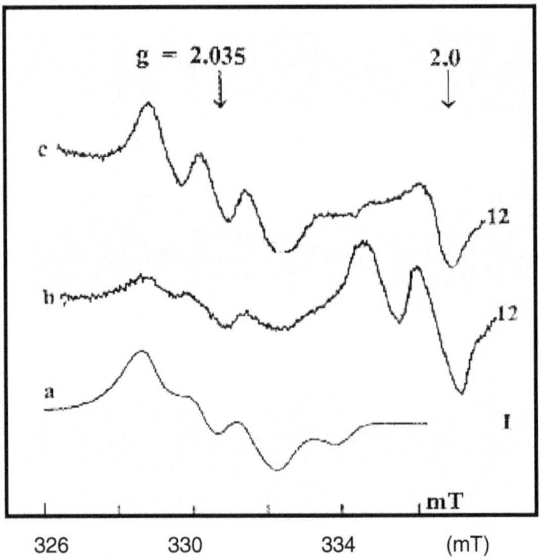

Fig. 9. EPR spectra at 77 K of MNIC adducts from 0.3 g spinach leaves in HEPES buffer (150 mM, pH = 7.4). Scaling factors are given on the right-hand side. (Curve a) reference spectrum of 25 nmol NO–Fe^{2+}–MGD complexes in frozen PBS buffer. (Curve b) MNIC signal after 4 h trapping in spinach, unreduced. (Curve c) sample (b) after reduction with 10 mM dithionite for 30 min [43].

A similar increase in yield was even observed [43] when Fe-DETC traps were used to detect the NO production from spinach leaves in strong HEPES buffer (pH = 7.4). A basal yield of 3.1 nmol MNIC per gram leaf tissue was increased to 10 nmol by reduction with dithionite (cf Fig. 9, curve c).

OTHER EFFECTS OF REDUCTION ON EPR SPECTRA

The reduction of tissue samples has beneficial effects beyond the enhancement of the MNIC yield. Fig. 10 gives representative EPR spectra from tissue samples of BALB mice.

In unreduced tissues, the EPR spectra appear as the superposition of complex IV (triplet at $g = 2.035$) and signals from other paramagnetic species like the hyperfine quartet of Cu^{2+}-DETC ($S = 1/2$, $I = 3/2$) commonly observed in animal tissues [3,44]. Other contaminants are a broad line attributed to Hb-NO ($g = 2.07$, 1.98), a free-radical signal near $g \sim 2.0$, a broad absorption from iron–sulfur proteins near $g \sim 1.94$ and the EPR signal at $g = 4.3$ from high spin ferric complexes described above.

In all cases, the yields of complex IV were sharply increased upon reduction with dithionite. The high-spin ferric complexes at $g = 4.3$ were diminished fourfold (data not shown). At the same time, the peak at $g = 2.01$ from the overlapping Cu^{2+}-DETC signal decreased significantly and this facilitated the quantification of the yields of complex IV. In many cases the Cu^{2+}-DETC signal is so strong that straightforward quantification of the MNIC yield is

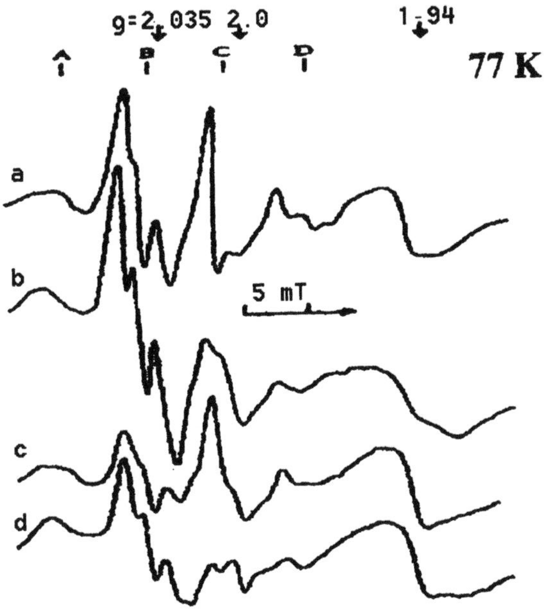

Fig. 10. EPR spectra of frozen tissue sections of BALB mice after 30 min of NO trapping. The mice had been injected with 2 mg/kg LPS 4 h prior to the experiment. The samples contain ca 1 g tissue in strong HEPES buffer (150 mM, pH = 7.4). (Curve a) Liver unreduced. (Curve b) The same liver after reduction with 10 mM dithionite. (Curve c) Four kidneys combined, unreduced. (Curve d) The same four kidneys after reduction with 10 mM dithionite. A,B,C,D are positions where hyperfine lines of the Cu^{2+}-DETC complex appear.

actually impossible without reduction. Fig. 11 shows EPR spectra from 180 mg brain of a young (2 week) spontaneously hypertensive rat. In this case, the yields are small because we trapped the basal NO production, without applying any stimulus like LPS. The strong Cu^{2+}-DETC signal prevents quantification of the MNIC yield from the unreduced spectrum (bottom). After reduction of the sample with dithionite, a total quantity of ca 250 ± 25 pmol MNIC was detected (top). The yield in this case was ca 1.4 nmol MNIC per gram wet tissue. Reduction also affects the EPR spectra of samples containing blood: In heart, spleen and lung of BALB mice, reduction enhanced the Hb-NO signal up to a factor two (data not shown).

PREVENTION OF REDUCTION OF ENDOGENOUS NITRITE BY BUFFERING

Application of strong reducing agents brings the risk of inadvertent release of NO from endogenous nitrite, in particular under acidic conditions. Therefore, we studied the formation of paramagnetic complexes IV in aqueous solutions containing 1.5 mM $NaNO_2$, 0.5 mM DETC and 0.05 mM ferrous sulfate. Significant quantities of complex IV were formed by adding 10 mM dithionite to unbuffered or weakly buffered solutions: After 5 min, the yields were 40 μM in distillate water and 10 μM in the mildly buffered PBS (10 mM, pH 7.4).

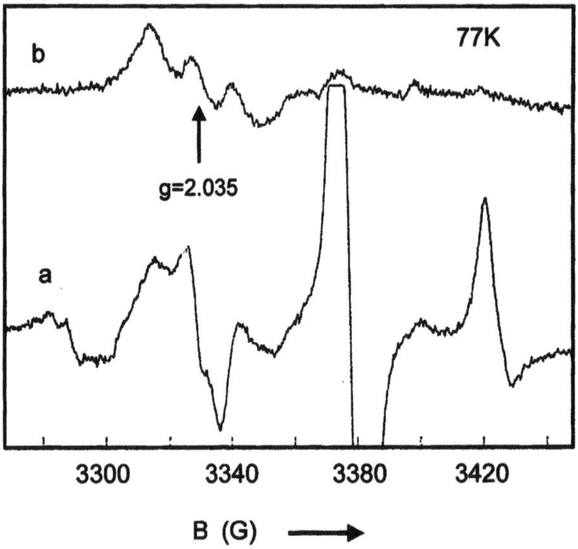

Fig. 11. EPR spectra in 180 mg brain of a young spontaneously hypertensive rat after 20 min trapping with Fe-DETC *in vivo*. (Curve a) Strong Cu^{2+}-DETC signal in unreduced sample. (Curve b) Same sample after reduction with dithionite reveals 250 pmol MNIC.

The yield was proportional to the initial nitrite concentration and caused by acidification of the weak 10 mM PBS buffer. The acidification is caused by the consumption of OH^- upon electron donation by dithionite:

$$S_2O_4^{2-} + 4OH^- \rightarrow 2SO_3^{2-} + 2H_2O + 2e^-$$

This acidification and the NO release could be prevented by adequate buffering: in 150 mM HEPES (pH 7.4) the yields of complex IV remained below 10 nM and 80 nM after addition of 10 mM and 100 mM dithionite, respectively.

Endogenous nitrite levels in biological systems are in the micromolar range for plasma and normoxic tissues [26,45,46] with higher levels up to 20 μM reported for vascular tissue (cf Table 5 later in the chapter). This is 2–3 orders of magnitude below the nitrite concentration considered above. Therefore, dithionite can be safely used for the reduction of complex III to IV if the buffering is adequate to prevent local acidification. In particular, the reduction does not significantly distort the yield of NO adducts by the artificial release of NO from nitrite. Direct proof of this assertion was obtained by reduction of tissue samples in strong HEPES. The presence of 100 μM exogenous nitrite did not affect the yield of complexes IV before or after reduction with 10 mM dithionite for 30 min. Interestingly, the reduction with dithionite did not require disruption of the structural integrity of the tissue, since fine grinding of the tissue prior to reduction did not affect the MNIC yield. However, the samples had been subject to freeze-thawing in strong buffer, which is disruptive for membranes and admits strong buffer into the intracellular space.

Reduction of biological samples also carries the risk of inadvertent release of NO from nitrosothiols. The latter are not by themselves reduced by mild reductants like ascorbate, but may release NO under catalytic action of reduced transition metal ions like Fe^{2+} or Cu^{1+}. The levels of these catalytic species are automatically kept very low by the presence of excess metal chelating dithiocarbamate ligands. However, reduction of tissue samples is known to release additional iron from the endogenous ferritin stores. The possibility of this artefact should be carefully considered in experiments with tissue samples.

DIFFERENT LIGANDS COMPETE WITH DETC FOR EXOGENOUS IRON IN TISSUES

The ligand structure of the iron is known to depend sensitively on the iron charge state and type of solvents [47]. In water, the solubility of Fe^{3+}-DETC in water is low, but the dithiocarbamate ligands are strongly bound because 50 μM of this complex remains intact in the presence of 20 mM competing citrate ligands, although the latter are known as a very strong ligand for Fe^{3+}-ions ($\log K = 11.4$ in water). However, the Fe-DETC bond is much weaker in non-hydrogen-bonding solvents like DMSO or DMF where the DETC ligand is easily replaced by 1 mM citrate [43]. This brings to mind the lipid compartment in cells and tissues. The lipid compartment has low polarity and competing endogenous ligands like citrate could compromise the formation of iron-DETC complexes, irrespective of whether this iron have endogenous or exogenous origin. Actual trapping experiments with laboratory animals employ mM DETC concentrations [3]. This quantity of DETC is probably insufficient for complete sequestration of the exogenous iron. A substantial fraction of the iron is bound by other ligands like citrate or phosphate. The presence of Fe^{3+}-complexes with non-DETC ligands was directly demonstrated by exposing homogenized liver tissue to very high doses of gaseous NO. Such excessive nitrosylation did not affect the EPR intensity at $g = 4.3$ although the copious formation of paramagnetic MNIC-DETC was detected as usual. It suggests that the high spin signal at $g = 4.3$ in tissues be dominated by Fe^{3+}-complexes with ligands other than DETC [43].

DIFFERENT LIGANDS COMPETE WITH DETC FOR EXOGENOUS IRON IN CULTURED ENDOTHELIAL CELLS

Experiments on BEND3 endothelial cell cultures support this viewpoint. After administration of a total of 100 nmol ferrous sulfate and 25 μmol DETC to a cell culture containing ca 7.5×10^6 BEND3 cells, we found a basal, unstimulated yield of 110 pmol of complex IV. The yield was increased to 400 pmol by stimulation of the cells with 5 μM calcium-ionophore. The number of Fe-DETC traps in the culture was of the order of 4 nmol, since a MNIC yield of 4 nmol was found after addition of 2.5 μmol/10 ml = 250 μM nitrite and reduction by dithionite. Therefore, the number of Fe-DETC traps (4 nmol) remains far below the total quantity of exogenous iron (100 nmol). It confirms that only a small fraction of the exogenous iron is bound to DETC ligands. By implication, the majority of iron is being bound to other ligand species. The process of ligand replacement is a dynamic equilibrium between the various ligands. Accordingly, the non-DETC majority act as a pool of iron to compensate the depletion of Fe-DETC complexes by nitrosylation. As mentioned before, the nitrosyl adducts of iron-DETC can be considered stable end products of the trapping reaction since they are very robust against ligand replacement by, for example, citrate.

LOOSELY BOUND IRON PARTICIPATES IN THE REDOX EQUILIBRIUM

We observed that the NTBI in tissues exists as a mixture of ferrous and ferric form. The ferric species is unambiguously attested by the EPR signal at $g = 4.3$. The ferrous fraction was manifest from the enhancement of $g = 4.3$ upon extraction of Fe-DETC complexes with DMF and subsequent oxidation by oxygen from air. The EPR intensity showed that the two fractions were of comparable magnitude in tissue extracts of mice and rats [43].

Exogenous iron also rapidly partitions into ferric and ferrous state. EPR showed that quantity of high spin ferric iron was increased by administration of ferrous-citrate complexes to the animals, due to oxidation by dioxygen in the liver tissue.

These experiments on animal tissues showed that the endogenous as well as exogenous iron finds itself as a mixture of ferrous and ferric redox states. EPR clearly showed that a significant fraction of iron in the tissue extracts exists in high spin ferric state. This finding is highly relevant for the trapping of NO, since the nitrosyl ligand binds to both redox state of iron. The presence of ferrous traps in tissues is evidenced by formation of complex IV. Trapping by ferric complexes is inferred from the large effect on yields by a wide range of interventions like reduction, perfusion with chelators and harvesting of complexes with DMF. Interestingly and contrary to common assumption, both ferric and ferrous fraction were seen to contribute to the yields of NO-trapping. Upon injection of DETC, ferrous iron, even when carrying strong ligands like citrate, forms ferrous NO-DETC complexes that bind NO to form paramagnetic MNIC-DETC (complexes IV) in animal tissues. This population is observed with EPR in a conventional NO trapping experiment. The ferric fraction forms ferric traps that bind NO as the ferric diamagnetic complex III. The latter may be transformed into paramagnetic state *ex vivo* by dithionite. Accordingly, the reduction enhances the yield of complex IV significantly in animal tissues (Table 5).

NO trapping in spinach leaves or endothelial cell culture was accomplished simply by addition of ferrous iron and DETC in the incubation medium [43]. The medium rapidly acquired a dark appearance due to rapid oxidation of ferrous DETC complexes to ferric state. With 2.5–5 mM concentration in spinach leaves and cell cultures, DETC is clearly in excess over endogenous ligands like citrate. So, endogenous and exogenous iron binds mainly with DETC to form complexes that rapidly oxidize to ferric state. Further experiments on spinach leaves support the hypothesis. If leaves were incubated with 5 mM DETC, bubbling with NO led to complete disappearance of the signal $g = 4.3$. The effect is characteristic of ferric-DETC and would not happen for ferric-citrate complexes. We conclude that incubation of leaves with DETC leads to the formation of significant quantities of complex I. Bubbling with NO forms diamagnetic complexes III that are observed with EPR only after reduction with dithionite. That is why the reductive increase in MNIC yield is more pronounced in leaves or cultured cells than in animal tissues. The reduction magnifies yields better if the iron carries DETC ligands.

ENDOGENOUS COMPOUNDS AS IMPOSTORS FOR TRUE NITRIC OXIDE

It should be noted that the beneficial effect of dithionite on adduct yield has been observed before [48,49] *in vitro* and in tissue preparations. However, inadequate buffering in these cases suggests that these early observations can be wholly or partially attributed to the artifact of reduction of nitrite. The *ex vivo* reduction of tissue samples requires careful consideration of possible artificial release of NO from nitrite, nitrate, nitrosothiols or other metabolites of NO. As shown above, the reduction of nitrite may be prevented by adequate buffering. Nitrate is not reduced unless exposed to far stronger reductants or nitrate reductases as may exist in certain bacteria.

Alternative compounds have been proposed as endogenous NO donors. First, dinitrosyl-iron complexes (DNIC) may form endogeneously in living tissues [50] and are capable of donating the NO moiety to iron-dithiocarbamate complexes in a transnitrosation reaction. Under the conditions used here, the tissue samples showed no EPR signals from paramagnetic DNIC, so that their concentration should be lower than the detection limit of ca 10 pmol/250 μl = 40 nM. This upper limit is negligible in comparison with the yields of complex IV in our samples. *S*-nitrosothiols may decay into NO and a thiyl radical by photolysis, thermolysis or by a reaction catalyzed by Fe^{2+} or Cu^{+}. Alternatively, they may transfer their NO moiety directly to Fe-DETC in a slow reaction. Cys-NO is known to transfer the NO moiety to Fe^{2+}-MGD with a rate of ca 30 ± 5 $(Ms)^{-1}$ (see above). The transfer rates to Fe^{2+}-DETC or Fe^{3+}-DETC in the lipid compartment have not been determined, but are expected to be slower. It was reported [26,51,52] that the tissue levels of *S*-nitrosothiols are in the 30–40 nM range in healthy animals with normal nitrite status. Higher levels were found in aortic tissues (~100 nM) and erythrocytes (~250 nM). At such concentrations, transnitrosylation by *S*-nitrosothiols remains a negligible source of NO for MNIC formation. Typical concentrations of NO-metabolites are collected in Table 6.

The nitroxyl anion NO^- has been proposed [53] as an alternative potential source of paramagnetic adducts by reacting directly with Fe^{3+}-dithiocarbamate complexes I into complex IV. In this proposal, complex III would be bypassed as a reaction intermediate and only appear at a later stage *via* the very slow re-oxidation of the paramagnetic complex IV

Table 6 Magnitude of various pools of NO-metabolites in Wistar rats (adapted from Ref. [26]). The rightmost column gives values from human plasma for reference

NO-metabolite	Aorta	Brain	Heart	Plasma	Erythrocytes	Human plasma¶
Nitrite (μM)	23 ± 9	1.7 ± 0.3	$0.80 \pm .08$	0.29 ± 0.05	0.68 ± 0.06	0.20 ± 0.02
Nitrate (μM)	49 ± 7	6.1 ± 1.1	5.9 ± 1.7	5.7 ± 0.6	10.2 ± 1.2	14.4 ± 1.7
S-nitroso moieties (nM)	96 ± 24	22 ± 6	13 ± 2	1.4 ± 0.5	246 ± 32	7.2 ± 1.1
N-nitroso moieties (nM)	19 ± 11	61 ± 8	14 ± 2	3.5 ± 0.4	95 ± 14	32.3 ± 5.0
Nitrosyl-heme (nM)	Below detection*	160 ± 30	15 ± 1	Below detection*	10.8 ± 1.8	n.d.

* Detection limit 1 nM.
¶ [52].

by oxygen. The known stability of complex IV against re-oxidation makes the nitroxyl hypothesis incompatible with our observation that the mononitrosyl complexes in biological materials appear as a mixture of ferric and ferrous forms. The nitroxyl hypothesis is also at odds with abundant observations of MNIC formation by trapping of NO in unambiguous neutral radical state (for example, when supplied by true NO donors, gaseous NO, or identified with NO electrodes).

THE MEANING OF THE FERRIC HIGH-SPIN SIGNAL AT $g = 4.3$

In conclusion, the EPR signal at $g = 4.3$ appears as a superposition of several non-heme high spin ferric iron complexes. A noticeable contribution from ferric DETC complexes was observed in cells and tissues only at excessive levels of DETC as reached in animal tissues perfused with 100 mM DETC or in cells and tissues incubated in the media with high concentration of DETC (a few mM). In actual *in vivo* trapping experiments in lab animals such levels are not achieved, and ferric DETC will represent only a minority fraction of total high spin ferric iron as observed with EPR.

After DETC is injected into animals together with ferrous-citrate complexes, the $g = 4.3$ signal in tissue extracts is dominated by endogenous ferric and exogenous ferric citrate complexes. However, quantification of the EPR signal shows that the majority of exogenous iron remains in ferrous state. Ferric as well as ferrous DETC have high affinity for NO, and are located in the apolar lipid compartment favored by the NO radical. Therefore, the mononitrosyl iron complexes are formed as a mixture of ferric and ferrous state, i.e. as a mixture of complex III and complex IV. Subsequent treatment of tissue preparations with dithionite leads to transformation of the complex III to complex IV increasing the yield of the latter in the preparations. As a bonus, the overlapping EPR signal from Cu^{2+}-DETC complexes is removed.

The solvent has a large influence on the rates of reduction of complex III to complex IV by excess dithiocarbamate ligands or mild reducing agents like ascorbate and thiols. Water and DMSO are both rather polar ($\varepsilon = 80$ *vs.* 46, respectively) but very different with respect to hydrogen bonding and Gutmann donor number. Reduction of NO-Fe^{3+}-DETC with ascorbate is slow in aqueous solutions, but is readily achieved in DMSO. It shows that the effect of such endogenous reductants on the redox balance depends on the local microenvironment and is different for the cytosolic and lipid compartments. The lipid compartment, where the DETC complexes accumulate, may occupy a small fraction of the total volume of tissues, but its properties are clearly determining the effect of the endogenous reductants on the formation of complex IV in conventional NO trapping experiments in animals or cultured cells.

CONCLUDING REMARKS

During the past fifteen years, iron dithiocarbamate complexes were proven to be a valuable tool for the detection and quantification of nitric oxide in biological systems. EPR detection of MNIC adducts has reasonable sensitivity in the sense that basal NO levels can now be detected in animal tissues and certain cell lines, without need for stimulation of the NO release.

The ambition to achieve NO imaging with MRI is driving the technical development at a rapid pace. The shape of the EPR spectrum is affected by isotopic substitution of the Fe, N and O constituents of the iron-nitrosyl moiety. Isotopic labelling provides a very valuable tool for elucidation of the reaction pathways that release NO *in vivo*. Additionally, DETC ligands have allowed us to study the nature and redox state of various endogenous iron pools in animal tissues. The iron appears to be a dynamic mixture of ferric and ferrous iron, and we found that loosely bound iron is the subject of competition between various endogenous ligands. Given the importance of loosely bound iron for human health we expect that the combination of dithiocarbamate ligands with EPR will provide an important experimental tool for future investigations into the role of iron in physiology.

REFERENCES

1 Vanin AF, Mordvintcev PI, Kleschyov AL. Appearance of nitric oxide in animal tissues. Stud. Biophys. 1984; 102: 135–143.

2 Mordvintcev P, Mulsch A, Busse R, Vanin A. On-line detection of nitric oxide formation in liquid aqeous phase by electron paramagnetic resonance spectroscopy. Anal. Biochem. 1991; 199: 142–146.

3 Ohnishi S. Measurement of NO using electron paramagnetic resonance. Meth. Mol. Biol. 1996; 100: 129–153.

4 Vanin AF, Huisman A, van Faassen E. Iron dithiocarbamates as spin trap for nitric oxide: pitfalls and successes. Meth. Enzymol. 2002; 359: 27–42.

5 Nagano T, Yoshimura T. Bioimaging of nitric oxide. Chem. Rev. 2002; 102: 1235–1269.

6 Yoshimura T, Kotake Y. Spin trapping of nitric oxide with the iron-dithiocarbamate complex: chemistry and biology. Antioxid. Redox Signal. 2004; 5: 639–647.

7 Henry YA, Guissani A, Ducastel B. Nitric oxide research from chemistry to biology: EPR spectroscopy of nitrosylated compounds. Austin, Texas, USA; Landes, RG, 1997, p. 294.

8 Rosen GM, Britigam BE, Halpern HJ, Pou S. Free radicals: Biology and detection by spin trapping. New York, Oxford, Oxford University Press, 1999, pp. 400–403.

9 MacDonald C, Philips W, Mower H. An electron spin resonance study of some complexes of iron, nitric oxide and anionic ligands. J. Am. Chem. Soc. 1965; 87: 3319–3326.

10 Shinobu L, Jones S, Jones M. Acta Pharm. Toxicol. 1984; 54: 189–194.

11 Bode H, Tusche H, Wahrhausen H. Zeit. Anal. Chem. 1962; 190: 48–60.

12 Erl W, Weber C, Hansson G. Pyrrolidine dithiocarbamate-induced apoptosis depends on cell type, density and the presence of Cu^{2+} and Zn^{2+}. Am. J. Physiol. Cell Physiol. 2000; 278: C1116–C1125.

13 Bach S, Williamson S, O'Dwyer S, Potten C. Watson A. Regional localization of p53-independent apoptosis determines toxicity to 5-fluorouracil and pyrrolidine dithiocarbamate in the murine gut. Br. J. Cancer 2006; 95: 35–41.

14 Liu G.-Y, Frank N, Bartsch H, Lin J.-K. Induction of apoptosis by thiuramdisulfides, the reactive metabolites of dithiocarbamates, through coordinative modulation of NF-κB, c-fos/c-jun and p53 proteins. Mol. Carcinog. 1998; 22: 235–246.

15 Pieper G, Nilakantan V, Hilton G, Halligan N, Felix C, Kampalath B, Khanna A, Roza A, Johnson C, Adams M. Mechanisms of the protective action of diethyldithiocarbamate-iron complex on acute cardiac allograft rejection. Am. J. Physiol. Heart Circ. Physiol. 2003; 284: H1542–H1551.

16 Vanin AF. Iron dithiocarbamate as spin trap for nitric oxide detection. Meth. Enzymol. 1998; 301 (Part C): 269–279.

17 Vanin A, Poltorakov A, Mikoyan V, Kubrina L, van Faassen E. Why iron-dithiocarbamates ensure detection of nitric oxide in cells and tissues, Nitric Oxide: Biol. Chem. 2006; 15: 295–311.

18 Woldman Y, Khramtsov V, Grigorev I, Kiriljuk I, Utepbergenov D. Spin trapping of nitric oxide by nitronylnitroxides: measurement of the activity of NO synthase from rat cerebellum. Biochem. Biophys. Res. Commun. 1994; 202: 195–203.

19 Paschenko SV, Khramtsov VV, Skatchkov MP, Plyusnin VF, Bassenge E. EPR and laser flash photolysis studies of the reaction of nitric oxide with water soluble NO trap Fe(II)-proline-dithiocarbamate complex. Biochim. Biophys. Res. Comm. 1996; 225: 577–584.

20 Fujii S, Kobayashi K, Tagawa S, Yoshimura T. Reaction of nitric oxide with the iron(III) complex of N-(dithiocarboxy)sarcosine: a new type of reductive nitrosylation involving iron(IV) as an intermediate. J. Chem. Soc. Dalton Trans. 2000; 3310–3315.

21 Pou S, Tsia P, Porasuphatana S, Halpern H, Chandramouli G, Barth E, Rosen G. Spin trapping of nitric oxide by ferro-chelates: kinetic and in-vivo pharmacokinetic studies. Biochim. Biophys. Acta 1999; 1427: 216–226.

22 Goodman B, Raynor J, Symons M. Electron spin resonance of Bis(N,N-diethylthiocarbamato) nitrosyl iron. J. Chem. Soc. A. 1969; 2572–2575.

23 Butler AR, Elkins-Daukes S, Parkin D, Williams DL. Direct NO group transfer from S-nitrosothiols to iron centers. Chem. Commun. (Camb.) 2001; 18: 1732–1733.

24 Vanin A, Papina A, Serezhenkov V, Koppenol W. The mechanism of S-nitrosothiol decomposition catalyzed by iron. Nitric Oxide 2004; 10: 60–73.

25 Mikoyan VD, Serezhenkov VA, Brazhnikova NV, Kubrina LN, Khachtryan GN, Vanin AF. Formation of paramagnetic nitrosyl complexes of nonheme iron in the animal organism with the participation of nitric oxide from exogenous and endogenous sources. Biophysics (Translated from Russian) 2004; 41: 110–116.

26 Bryan NS, Rassaf T, Maloney RE, Rodriguez CM, Saifo F, Rodriguez JR, Feelish M. Cellular targets and mechanisms of nitros(yl)ation: An insight into their nature and kinetics in vivo. Proc. Natl. Acad. Sci. USA 2004; 101: 4308–4313.

27 Tsuchiya K, Jiang J, Yoshizumi M, Tamaki T, Houchi H, Minakuchi K, Fukuzawa K, Mason R. Nitric oxide forming reactions of the water-soluble nitric oxide spin-trapping agent MGD. Free Rad. Biol. Med. 1999; 27: 347–355.

28 Tsuchiya K, Yoshizumi M, Houchi H, Mason R. Nitric oxide forming reactions between the iron-N-methyl-D-glucamine dithiocarbamate complex and nitrite. J. Biol. Chem. 2000; 275: 1551–1556.

29 Mulsch A, Schray-Utz B, Mordvintcev P, Hauschildt S, Busse R. Diethyldithiocarbamate inhibits induction of macrophage NO synthase. FEBS Lett. 1993; 321: 215–218.

30 Heikkila R, Cabbat F, Cohen G. In vivo inhibition of superoxide dismutase in mice by diethyldithiocarbamate. J. Biol. Chem. 1976; 251: 2182–2185.

31 Yoneyama H, Kosaka H, Ohnishi T, Kawazoe T, Mizoguchi K, Ichikawa Y. Reaction of neuronal nitric oxide synthase with the nitric oxide spin-trapping agent, iron complexed with N-dithiocarboxysarcosine. Eur. J. Biochem. 1999; 266: 771–777.

32 Vanin A, Liu X, Samouilov A, Stukan R, Zweier J. Redox properties of iron-dithiocarbamates and their nitrosyl derivatives: implications for their use as traps for nitric oxide in biological systems. Biochim. Biophys. Acta 2000; 1474: 365–377.

33 Ileperuma OA, Feltham RD. Iron–sulphur complexes of nitric oxide. Synthesis and exchange studies of $Fe(NO)(S_2CN(CH_3)_2)_2$. Crystal and molecules of cys-$Fe(NO)(NO_2)(S_2CN(CH_3)_2)_2$. Inorg. Chem. 1977; 16: 1876–1883.

34 Vanin A, Huisman A, Stroes E et al. Antioxidant capacity of mononitrosyl-iron-dithiocarbamate complexes: implications for NO trapping. Free Rad. Biol. Med. 2001; 30: 813–824.

35 Melo J, Biazotto J, Brunello C, Graeff C. Solid state nitric oxide sensor prepared by sol-gel entrapment of iron-dithiocarbamate in a siloxane matrix. J. Non-cryst. Solids 2004; 348: 235–239.

36 van Faassen E, Vanin A. NO trapping in biological systems with a functionalized zeolite network. Nitric Oxide: Biol. Chem. 2006; 15: 233–240.

37 Kleschyov A, Mollnau H, Oelze M, Meinertz T, Huang Y, Harrison D, Munzel T. Spin trapping of vascular nitric oxide using colloid Fe(II)-diethyldithiocarbamate. Biochem. Biophys. Res. Commun. 2000; 275: 672–677.

38 Fujii S, Yoshimura T, Kamada H. Nitric oxide trapping efficiencies of water soluble iron (III) complexes with dithiocarbamate derivatives. Chem. Lett. 1996; 785–786.

39 D'Autreaux B, Horner O, Oddou J-L et al. Spectroscopic description of the two nitrosyl-iron complexes responsible for Fur inhibition by nitric oxide, J. Am. Chem. Soc. 2004; 126: 6005–6016.

40 Kotake Y, Yanigawa T, Tanigawa M, Ueno I. Spin trapping isotopically labeled nitric oxide produced from [^{15}N] L-Arginine and [^{17}O] dioxygen by activated macrophage using a water soluble Fe dithiocarbamate complex. Free Rad. Res. 1995; 23: 287–295.

41 Palmer G. Electron Paramagnetic Resonance of metalloproteins. In Physical Methods in Bioinorganic Chemistry (Que L. ed.), University Science Books, Sausalito 2000.

42 Vanin A, Kubrina L, Kurbanov I, Mordvintcev P, Khrapova N, Galagan M, Matkhanov E. Iron as an inducer of the formation of nitric oxide in animal organisms. Biochemistry (Mosc.) 1989; 54: 1974–1979.

43 Vanin A, Bevers L, Mikoyan V, Poltorakov A, Kubrina L, van Faassen E. Reduction enhances yields of nitric oxide trapping by iron-diethyldithiocarbamate complexes in biological systems, Nitric Oxide: Biol. Chem. 2007; 16: 71–81.

44 Suzuki Y, Fujii S, Tominaga T, Yoshimura T, Kamada H. The origin of an EPR signal observed in dithiocarbamate loaded tissues: Copper (II)-dithiocarbamate complexes account for the narrow hyperfine lines. Biochim. Biophys. Acta 1997; 1335: 242–245.

45 Kelm M. Nitric oxide metabolism and breakdown. Biochim. Biophys. Acta 1999: 1411: 273–289.

46 Manoghan J, Cook K, Gara D, Crowther D. Determination of nitrite and nitrate in human serum. J. Chrom. 1997; A770: 143–149.

47 Vanin AF, Bevers LM, Slama-Schwok A, van Faassen EE. Nitric oxide synthase reduces nitrite to NO under anioxia. Cell. Mol. Life Sci. 2007, in press.

48 Mulsch A, Vanin A, Mordvintcev P, Hauschildt S, Busse R. NO accounts completely for the oxygenated nitrogen species generated by enzymatic L-arginine oxygenation. Biochem. J. 1992; 288: 597–603.

49 Tsuchiya K, Takasugi M, Minakuchi K, Fukuzawa K. Sensitive quantitation of nitric oxide by EPR spectroscopy. Free Rad. Biol. Med. 1996; 21: 733–737.

50 Vanin A, Kleschyov A. EPR studies and biological implications of nitrosyl nonheme complexes. In Nitric Oxide in Transplant Rejection and Anti-tumor Defense. (Lukiewicz S, Zweier J. eds.), Kluwer Academic Publishers, Amsterdam 1998.

51 Rodriguez J, Maloney R, Rassaf T, Bryan N, Feelisch M. Chemical nature of nitric oxide storage forms in rat vascular tissue. Proc. Natl. Acad. Sci. 2003; 100(1): 336–341.

52 Rassaf T, Bryan N, Kelm M, Feelisch M. Concomitant presence of N-nitroso and S-nitroso proteins in human plasma. Free Rad. Biol. Med. 2002; 33: 1590–1596.

53 Xia Y, Cardounel A, Vanin A, Zweier J. Electron paramagnetic resonance spectroscopy with MGD-iron complexes distinguishes nitric oxide and nitroxyl anion in a redox dependent manner: applications in identifying nitrogen monoxide products from nitric oxide synthase. Free Rad. Biol. Med. 2000; 29: 793–797.

CHAPTER 19

Protection against allograft rejection by iron–dithiocarbamate complexes

Galen M. Pieper

Division of Transplant Surgery and Free Radical Research Center, Medical College of Wisconsin, Milwaukee, WI, USA

INTRODUCTION

The expression of inducible nitric oxide synthase (iNOS) has been implicated in graft rejection. Despite caveats of certain experimental designs, the role of iNOS in cardiac rejection has been investigated by a broad range of pharmacological and gene deletion strategies. The pharmacological strategies concentrate on either inhibition of NOS enzyme activity or on scavenging and neutralizing nitric oxide (NO). Iron–dithiocarbamates are a class of compounds that scavenge NO *in vitro* and *in vivo*. Iron–dithiocarbamates appear to have unique attributes as well: besides scavenging NO they were found to have anti-inflammatory and immunosuppressant properties. This chapter reviews the experimental evidence for the role of NO in cardiac allograft rejection with special emphasis on the protective action of iron–dithiocarbamates.

Acute rejection of solid organ grafts involves a complex array of inflammatory mediators. Among leading candidates for these mediators are lymphokines, cytokines and NO derived from iNOS. Much of the work analyzing a role of iNOS in graft rejection has been performed in rodent models of cardiac transplant rejection. The clinical relevance of the upregulation of iNOS was shown in human cardiac transplants in which cardiac contractile dysfunction [1] and graft rejection [2] was associated with strong expression of iNOS.

ROLE OF iNOS IN CARDIAC REJECTION: EVIDENCE FROM PHARMACOLOGICAL APPROACHES

NO release from iNOS plays an important role in cardiac graft rejection. The evidence comes from several experimental studies performed in rodent models using pharmacological approaches. Most investigators in this area of research have utilized one of two criteria to determine benefits on graft rejection: graft survival time and/or histological scoring for rejection. A few investigators have examined contractile dysfunction assessed in *ex vivo* preparations such as isolated papillary muscles or isolated, perfused hearts. In our laboratory, we have extended functional analysis to *in situ* evaluation using sonomicrometry.

Original studies were conducted using non-specific, substrate-based, NOS inhibitors. Many studies indicated that treatment with these inhibitors extended the survival time of the grafts [3–5]. However, disparate findings were observed in which inhibitors left unchanged [6] or even decreased [7] graft survival time. One possible explanation for these disparate findings is the fact that the inhibitors used in the studies were not specific for inhibition of iNOS or that the doses used for the studies were too high, thereby, potentially inhibiting beneficial properties of constitutive NOS activity as well. Subsequent studies in our laboratory [8] and elsewhere [9] used inhibitors that were selective for the iNOS isoform only. These selective inhibitors consistently extended the survival time of the grafts or decreased histological rejection scores. In addition to substrate-based NOS inhibitors, one study using a pterin-based NOS inhibitor also showed significant improvement in graft survival time [10]. This finding suggests that NOS activity can be controlled at the pterin-binding domain of iNOS protein as well.

Taken collectively, the preponderance of experimental evidence confirms that treatment with iNOS inhibitors protects against rejection of cardiac allografts. Upon closer examination of the literature; however, it appears in many studies that the protective effect of non-selective and selective iNOS inhibitors on prolonging graft survival time or decreasing histological rejection scores occurred under conditions of incomplete inhibition of iNOS. A number of studies seemed to show a residual increase in NO production from upregulation of iNOS [9,11,12]. This phenomenon in fact was first reported in our studies using an iNOS selective inhibitor. Our experiments revealed that the protective action on graft survival time was lost if the inhibitor was administered at higher concentrations that ablated the increase in NO bioactivity. This observation suggests that NO may possess both detrimental and beneficial properties in allograft rejection. This state of affairs may explain the contradictory conclusions reported in various previous studies on graft survival. For histological rejection scores the situation appears simpler: our experiments with the selective iNOS inhibitor showed that histological rejection scores were consistently improved for all doses of inhibitor. This is in contradistinction to the nonconsistent effects on graft survival *per se* [8]. This surprising but important finding indicates that the two common criteria for determining transplant rejection have significantly different etiologies. This consideration should be kept in mind when the two criteria are used to determine the beneficial effects of drugs or treatments of cardiac transplant rejection.

It is also important to consider that the application of NOS inhibitors may change iNOS expression as a secondary effect. In fact, this has been shown using aminoguanidine [5] and certain inhibitors of iNOS dimerization [9]. This side effect greatly complicates the interpretation of experiments on the role of NO in rejection of cardiac allografts. Fortunately, this problem does not occur for all inhibitors. For example, this property was not found in our own experiments using the iNOS inhibitor, N^6-(1-iminoethyl)-L-lysine (denoted as L-NIL) [8].

ROLE OF iNOS IN CARDIAC REJECTION: EVIDENCE FROM GENE DELETION STRATEGIES

Controlled deletion of certain genes is a well established and powerful tool to investigate the role of particular genes in the etiology of various pathologies. Such tools have already been applied to acute cardiac allograft rejection relating to the iNOS gene. As with pharmacological approaches, the experimental studies produced contradictory findings as well. An initial report indicated a decrease in rejection based upon histological criteria [13]. In this study, graft survival per se as an endpoint had not been examined. The conclusion of this research group was not confirmed by a second study [14]. In this second study, the investigators observed no change in either histological rejection or graft survival time between BALB/c mouse donor hearts transplanted into iNOS wild-type and iNOS-/- recipients nor between iNOS wild-type and iNOS-/- donor hearts transplanted into BALB/c recipient mice. The lack of differences in either rejection or graft survival between wild-type and iNOS-/- mice in the second study was corroborated by a third study [15].

The experimental design of this third study provided some important insights that could help explain the divergent observations using the gene deletion approach. A significant finding was that iNOS deletion strategies applied to either the donor or the recipient did not prevent iNOS upregulation within the cardiac allograft. This finding had not been understood previously. The primary reason for this unexpected finding in the heterotopic cardiac transplant model is that the upregulation of iNOS was caused by two different sources. One source is from inflammatory cells that infiltrate the graft at early stages but which are derived from the recipient of the transplant. The second source is from parenchymal cells (i.e. cardiac myocytes and endothelial cells) of the donor graft. Both contributions appear to be significant in the problem of allograft rejection. This poses a significant limitation for the traditional gene deletion strategy against upregulation of iNOS. The consequences are potentially problematic in the heterotopic cardiac transplant model where the upregulation of iNOS is not eliminated but rather only the distribution of iNOS changes.

Hypothetically, it is interesting to speculate that the localization of iNOS expression could account for the fact that iNOS deletion strategy did not alter either rejection or graft survival in the cardiac allograft transplantation model. This hypothesis would provide a plausible explanation for the independent finding in which the proliferation of splenocytes stimulated *ex vivo* by alloantigen was found to be higher in splenocytes isolated from iNOS-/- *vs.* wild-type mice [16]. Thus, any benefit of ablated iNOS expression within the tissue parenchymal per se could be potentially offset by compensatory increases in activation and proliferation of inflammatory cells.

EFFECTS OF IRON–DITHIOCARBAMATES ON GRAFT SURVIVAL AND REJECTION

Iron dithiocarbamate complexes may easily bind free NO radicals to form a stable mononitrosyl iron complex. Therefore, iron–dithiocarbamates may be used to scavenge free NO. In a rat model of acute cardiac transplant rejection in the absence of any immunosuppressant therapy, daily post-operative intraperitoneal injections with Fe^{2+}-diethyldithiocarbamate caused a dose-dependent increase in graft survival time (Fig. 1, top panel) [17]. This effect approaches the magnitude of protection seen with low-dose cyclosporine, a common agent used for immunosuppression following transplantation. To determine if efficacy was unique to Fe^{2+}-diethyldithiocarbamate (Fe-DETC), we also showed a similar level of protection using the same daily dose of the water-soluble derivative, Fe^{2+}-*N*-methyl-D-glucamine-dithiocarbamate (Fe-MGD) (Fig. 1). The two types of complexes have very different physical properties as the first is hydrophobic and preferentially partitions into the low-polarity membrane and protein compartments. The second, Fe-MGD remains dissolved in the aqueous compartments (cytosol, plasma, urine etc.). Oral administration of the precursor agent *N*-methyl-D-glucamine-dithiocarbamate at the dose of 2.5 and 5 mg/ml in drinking water also prolonged graft survival time. This latter finding indicated that the protective activity on cardiac allograft survival occurs *via* different routes of administration and may not require pre-bound Fe^{2+} for efficacy (Fig. 1, bottom panel).

In addition to the assessment of graft survival time, some agents were also examined in our laboratory for histological rejection. Histological rejection scores were elevated in untreated allografts *vs.* isograft controls. Rejection was significantly decreased at post-operative days 5 and 6 using these agents (Fig. 2). Interestingly, decreased rejection scores could be achieved by administration of either Fe^{2+}-*N*-methyl-D-glucamine-dithiocarbamate or iron-free, *N*-methyl-D-glucamine-dithiocarbamate. This suggests that there may be sufficient quantity of loosely bound endogenous iron available inside the inflamed tissue during graft rejection to bind the *N*-methyl-D-glucamine-dithiocarbamate and serve as a neutralizer of NO bioactivity. Our work has clearly shown the benefits of daily treatment with iron–dithiocarbamates against cardiac allograft rejection. The complexes significantly improved graft survival. We are not aware of any other reports in the literature where these iron–dithiocarbamates were applied to other models of transplant rejection.

EVIDENCE THAT IRON–DITHIOCARBAMATES ACT *IN VIVO* TO SCAVENGE NITRIC OXIDE

In cell-free systems, there is ample evidence that iron–dithiocarbamates scavenge NO in solution. Iron complexes with a variety of dithiocarbamate ligands have been used to detect increases in NO in a wide range of experimental conditions (e.g. septic shock, ischemia-reperfusion injury). We note that the NO trapping takes place *in vivo* over an interval of 20–30 min before sacrifice of the lab animal for analysis. The nitrosylated NO–Fe–dithiocarbamate complexes are subsequently detected *ex vivo* in tissue samples using electron spin resonance spectroscopy. The principal of this activity is that NO rapidly binds to ferric or ferrous Fe-diethyldithiocarbamate complexes. The resulting mononitrosyl complex is stable in tissues in frozen state. The ferrous mononitrosyl complexes are paramagnetic and amenable to detection by EPR spectroscopy. When measured under liquid nitrogen temperatures, the detection of mononitrosyl iron complexes consists of an EPR signal giving a characteristic three-line spectra ($g_{\perp} = 2.035$ with 12.5 G splitting). The properties of the ferric and ferrous species and details of the detection technique were discussed in Chapter 18.

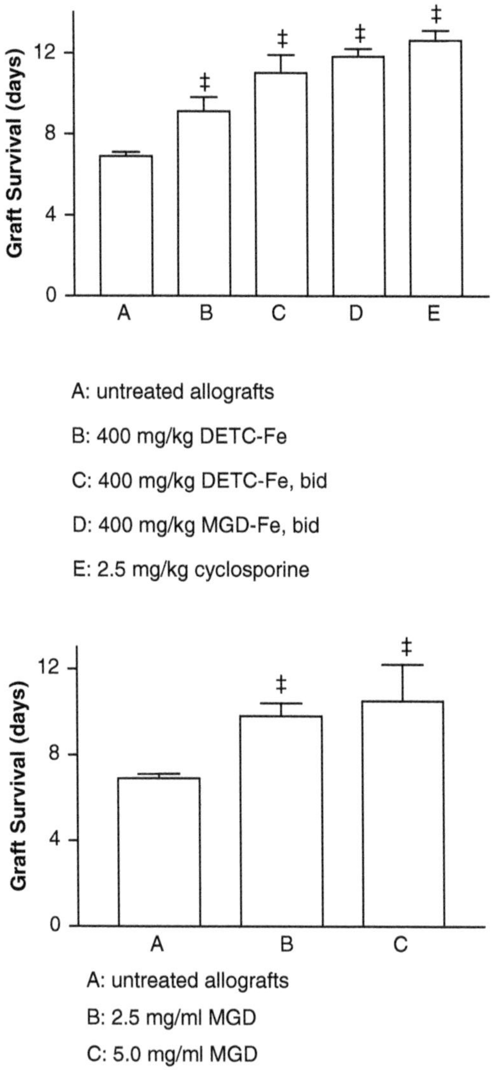

Fig. 1. Upper Panel: Effects of treatment with iron–dithiocarbamate derivatives *vs.* low-dose cyclosporine on cardiac allograft survival time. Fe^{2+}-diethyldithiocarbamate, DETC-Fe; Fe^{2+}-*N*-methyl-D-glucamine-dithiocarbamate, MGD-Fe. $^{\ddagger}P < 0.01$ *vs.* untreated allografts. Lower Panel: Effects of oral administration of iron-free methyl-D-glucamine-dithiocarbamate (MGD) on cardiac allograft survival time. $^{\ddagger}P < 0.01$ *vs.* untreated allografts.

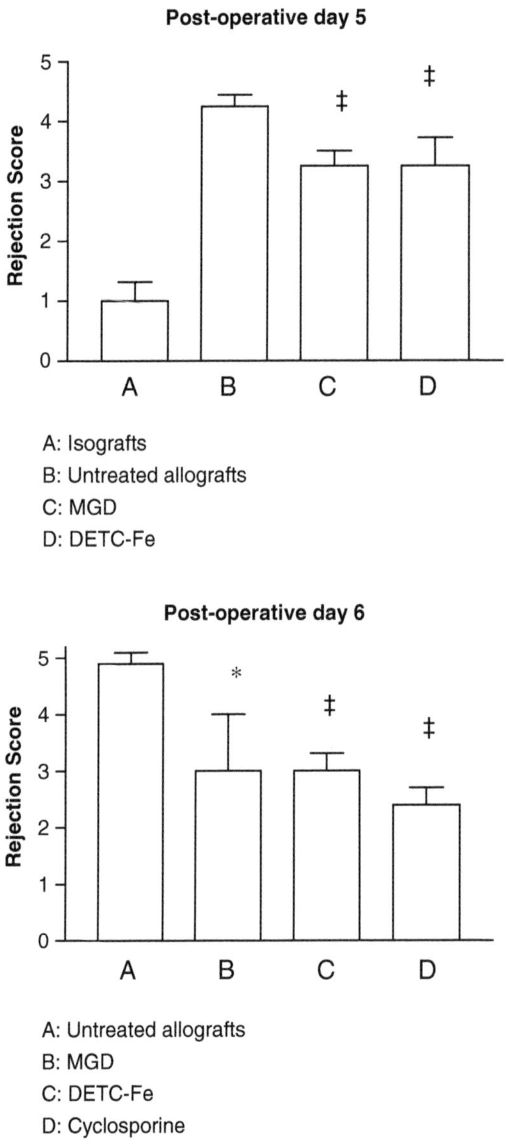

Fig. 2. Effects of iron-free methyl-D-glucamine-dithiocarbamate (MGD) and Fe^{2+}-diethyldithiocarbamate (400 mg/kg injected twice daily) and low-dose cyclosporine (2.5 mg/kg injected daily) on histological rejection scores at post-operative day 5 (upper panel) and day 6 (lower panel). $^{\ddagger}P < 0.01$ *vs.* untreated allografts and $^{*}P < 0.05$.

Akizuki et al. [18] were the first research group to document NO formation within graft tissue using EPR spectroscopy. This group used a rat model for cardiac transplantation in which ACI donors hearts were transplanted into Lewis recipients. In this study, a single injection of Fe^{2+}-diethyldithiocarbamate was followed about 30 min later by saline perfusion to remov contamination of NO trapped in circulating blood. NO was then detected unambiguously by EPR analysis of tissue samples at 110 K. The EPR spectra showed the presence of a three-line spectra characteristic of the paramagnetic species, mononitrosyl-Fe^{2+}-diethyldithiocarbamate complex.

We have investigated [17,19] a different rat strain model for cardiac rejection where hearts were transplanted from Wistar–Furth to Lewis rats. We expanded on this initial observation by documenting NO formation within cardiac allograft tissue per se. The NO was detected using either Fe^{2+}-diethyldithiocarbamate or other derivatives [17,19]. We also observed that the yield of NO trapping within cardiac allograft tissue by a single injection of Fe^{2+}-diethyldithiocarbamate was canceled if NO production had been inhibited by previous administration of a selective inhibitor of iNOS [17]. This is illustrated in Fig. 3 which indicates the formation of the three-line EPR spectra from NO-Fe^{2+}-DETC. This signal is superimposed on a background signal of nitrosylated ferrous heme ascribed to nitrosylated myoglobin. Formation of NO-Fe^{2+}-DETC was prevented by co-administration of the iNOS inhibitor, L-NIL. This was the first study to determine that the NO trapped actually derived from enzymatic production *via* iNOS. This finding is potentially important because it refutes speculations in other models that trapping of NO might arise from NOS-independent pathways [20,21]. Such alternative pathways are discussed in Chapters 14 and 15.

Peripheral measurements of NO metabolites, nitrate plus nitrite, have also been traditionally used to determine NO levels in biological systems. Experiments have confirmed that allograft rejection indeed significantly raises the plasma concentrations of NO metabolites. We have shown that this increase seen in rejecting allografts is inhibited by treatment with

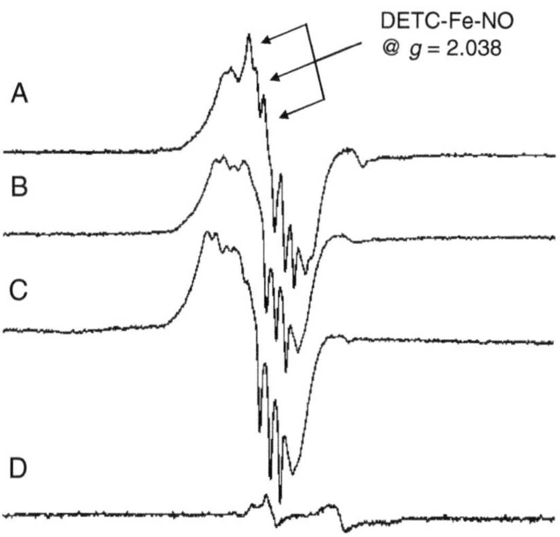

Fig. 3. EPR spectra of rat cardiac allograft tissue. Spectra A: Spectrum showing intragraft trapping of NO within cardiac allografts denoted by the DETC-Fe-NO complex at $g = 2.038$ after a single pulse injection of DETC-Fe at 30 min prior to freezing sample for EPR analysis. Spectrum B: Background spectrum of an untreated allograft showing the signals arising from endogenous MNIC. Spectrum C: Blockade of trapping of DETC-Fe-NO by a single treatment of iNOS inhibitor, L-NIL, prior to injection with DETC-Fe. Spectrum D: Spectrum of isograft control. Spectra are analyzed at the same time under identical instrument settings.

Fe^{2+}-diethyldithiocarbamate. This gives secondary evidence that this agent acts *in vivo* as a NO scavenger.

Collectively, these studies demonstrate that iron–dithiocarbamates fulfill two independent roles in allograft rejection. First, a therapeutic role by acting as scavenger of NO. Second, they are very useful for diagnostic purposes. This may sound preposterous given that the lab animals in the above experiments were all sacrificed in order to extract their tissues for EPR detection. This is not a fundamental restriction, however. The use of water-soluble Fe-MGD complexes allows extraction of NO adducts from body fluids like blood or urine, so that sacrifice of the animal is not necessary for NO detection. In these cases, however, the spatial selectivity for given tissue sections (like heart or kidney) is lost. This area of study has been fully developed to date.

IRON–DITHIOCARBAMATES FOR QUANTIFYING NITRIC OXIDE LEVELS

The trapping of NO and evaluation by EPR spectroscopy for diagnosis of increased levels in NO in biological systems is highlighted above. In principal, this property can potentially be extended to the actual quantification of NO levels. A limitation of this procedure was highlighted in early studies indicating the potential oxidation of Fe^{2+} to Fe^{3+} under aerobic conditions in physiological settings producing the diamagnetic NO-Fe^{3+}-dithiocarbamate species which are visible by EPR spectroscopy [22,23]. Thus, this EPR technique might considerably underestimate NO levels and limit its value for quantitative analysis of NO levels in tissue. This limitation seems to be circumvented by appropriate modifications including exogenous reducing agents [22,24,25]. Details of the protocols are given in Chapter 18. Thus, it is possible that this may be a more direct, alternative approach to quantifying NO levels in cardiac grafts than the NO electrode technique previously described [26].

Reports using EPR techniques for quantifying NO content in transplanted organs are extraordinarily rare. To our knowledge there has been only one published report and this was a study performed in an experimental rat renal transplant model [27].

EFFECT OF IRON–DITHIOCARBAMATES TO INHIBIT TARGETS OF NITRIC OXIDE

The precise mechanism of how NO plays a role in cardiac allograft rejection remains an area of ongoing research. It has been known for quite some time that NO has an avid affinity for metalloproteins particularly those involved in mitochondrial respiration. Hence, interaction with such biomolecules has been implicated in cell injury. Among the array of intracellular metalloproteins are heme- and non-heme containing proteins.

Lancaster et al. [28] were the first research group to report the formation of mononitrosyl iron complexes (MNIC) in both packed red blood cell samples and within cardiac tissue after cardiac transplantation in rats. The MNIC formed were attributed to nitrosylation of heme proteins, presumably, hemoglobin and myoglobin, respectively. In our laboratory, we have documented that the formation of MNIC within rejecting cardiac allografts, but not in

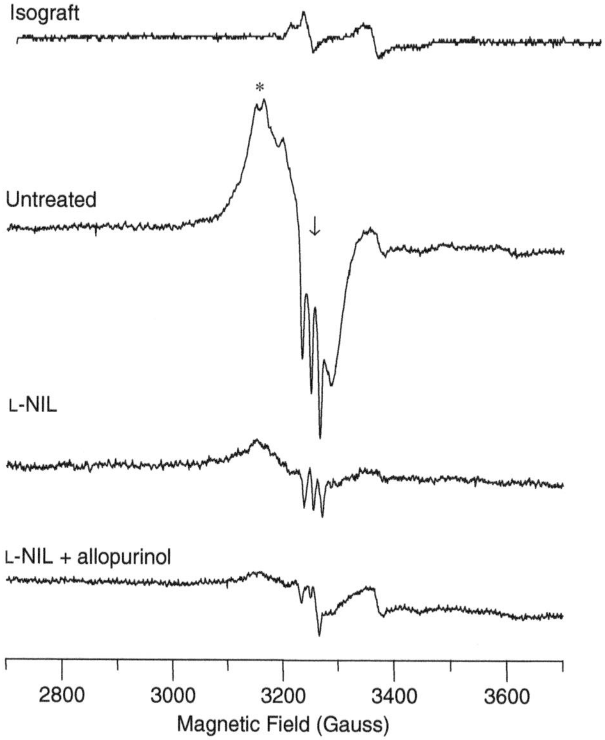

Fig. 4. EPR evidence showing the formation of mononitrosyl iron complexes (MNIC) in untreated cardiac allografts *vs.* isograft controls is predominantly inhibited by chronic treatment of allograft recipients with the iNOS inhibitor, L-NIL. The EPR signal is shown by the broad signal at $g = 2.08$ (shown by asterisk, *) and the triplet hyperfine component at $g = 2.014$ (shown centered by the arrow, ↓). There is only a small portion of L-NIL-resistant MNIC formation which is inhibited by the xanthine oxidase inhibitor, allopurinol. Spectra are analyzed at the same time under identical instrument settings.

isograft controls nor in native hearts of allograft recipients (not shown), is attributed primarily to enzymatically-derived NO *via* iNOS within cardiac allografts (Fig. 4). This conclusion is based upon the observation that EPR signals for these complexes are mostly prevented by post-operative treatment with a selective inhibitor of iNOS, L-NIL. This inhibitor inhibits the enzymatic activity of iNOS but does not affect the expression of iNOS protein. The finding in this study is important because of evidence in other models that NO can be derived from NOS enzyme independent pathways. Indeed, a xanthine oxidase-dependent reduction of endogenous nitrate or nitrite has been shown to be a significant source of NOS-independent NO formation in anoxic/ischemic tissue models [20,21]. Subsequent studies in our model show that concomitant treatment with iNOS inhibitor plus allopurinol, an inhibitor of xanthine oxidase, fully inhibited MNIC formation in cardiac allografts. Thus, we conclude that there is potentially only a very small contribution of xanthine-oxidase catalyzed NO formation to MNIC formation.

In our laboratory, various agents have been shown to decrease the levels of MNIC formation in rejecting cardiac allografts. These same agents have also been shown to display protective action by prolonging graft survival and decreasing histological rejection. While these studies show that interaction with heme protein is a major pathway of processing NO in cardiac allograft rejection, the biological significance of the formation of MNIC in this pathological state is not determined. Whether or not MNIC formation per se is a causative factor in rejection has not been ascertained to date.

Fe-S cluster proteins are known to bind NO and inhibit enzymatic activity in the case of aconitase, a key mitochondrial protein. In this context, we have shown that aconitase activity is lower in rejecting cardiac allografts. This lowering is ameliorated by agents such as cyclosporine that inhibit iNOS expression or L-NIL, an inhibitor of iNOS enzyme activity [29]. Here we show that treatment of allografts with Fe^{2+}-diethyldithiocarbamate also protect against the loss of aconitase activity. The effects of Fe^{2+}-diethyldithiocarbamate are very similar to those seen with L-NIL (Fig. 5). This is consistent with the hypothesis that NO or a downstream product of NO mediates inhibition of myocardial aconitase activity.

Considering that NO has this property of interacting with metals, iron-containing compounds such as the iron–dithiocarbamate class of agents have been developed as tools to trap NO in biological tissue for detection by biophysical techniques such as electron paramagnetic resonance (EPR) spectroscopy [30]. This is especially true for Fe^{2+}-dithiocarbamates *vs.* Fe^{3+}-dithiocarbamates that have lower affinity for NO. As such competition for binding NO by these agents may also be expected to have therapeutic efficacy in limiting the interaction

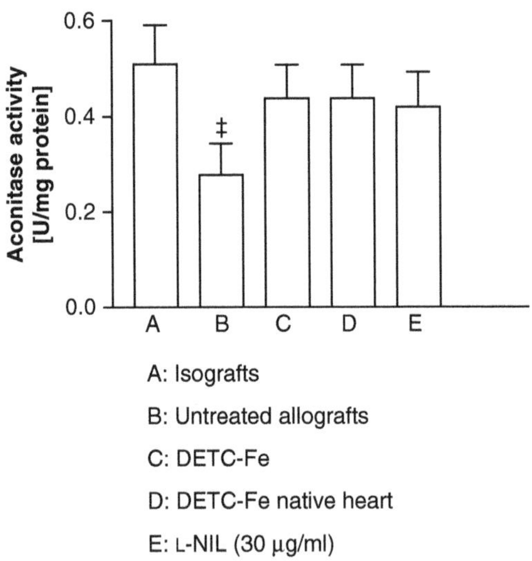

Fig. 5. Decrease in aconitase enzyme activity in cardiac allografts is reversed by treatment with Fe^{2+}-diethyldithiocarbamate. Aconitase activity is restored to levels seen in the native heart of allograft recipients treated with Fe^{2+}-diethyldithiocarbamate and similar to levels restored by alternative treatment with the selective inhibitor of iNOS, L-iminoethyllysine (L-NIL). $^{‡}P < 0.01$ *vs.* isograft controls.

of intracellular NO or NO-derived species such as peroxynitrite with target molecules within biological systems during inflammatory states including allograft rejection.

ANTI-INFLAMMATORY ACTIONS OF DITHIOCARBAMATES

Fe^{2+}-diethyldithiocarbamates have undisputed merits as scavengers of NO. As such they provide protection against rejection of cardiac allografts. However, it appears that they have other properties as well when used in the *in vivo* setting. Indeed, we have shown that Fe^{2+}-diethyldithiocarbamate inhibited activation of redox-activated transcription factors, nuclear factor κB and activator protein-1 [17]. This could explain its action to partially inhibit iNOS gene expression as well as inflammatory cytokine gene expression such as interferon-γ. These are properties that are shared with traditional antioxidant treatments such as dimethylthiourea [31] and the spin-trap agent, α-phenyl-*N-tert*-butylnitrone [32]. The similarity suggests that iron–dithiocarbamate treatments may have antioxidant efficacy as well.

An interesting concept is that iron–dithiocarbamate preparations are never traditionally constructed in a 1:1 ratio of iron:dithiocarbamate. Thus it is possible that the actions on inhibiting activation of redox-sensitive transcription factors might reflect antioxidant properties of iron-free dithiocarbamates. The antioxidant could result from scavenging of iron or by scavenging other reactive oxygen species [33,34]. An example of this is the protective action of pyrollidine dithiocarbamate. This compound inhibits activation of nuclear factor κB and prolongs the survival time of grafts. These effects were shown in early experiments conducted in our laboratory [35]. Subsequently, we found prolonged graft survival using ruthenium (III) polyaminocarboxylate complexes. These have an entirely different chemical structure from dithiocarbamates. This suggests that the protection was achieved primarily *via* the NO scavenging action [36]. The findings of protection with the ruthenium-metal-based NO scavenger, however, does not exclude the antioxidant contributions to protection by iron–dithiocarbamate derivatives.

EFFECTS ON LYMPHOCYTE ACTIVATION AND PROLIFERATION

The mechanisms of the benefits of *in vivo* treatment of cardiac allografts with iron–dithiocarbamate complexes is not fully explored. An outstanding issue is the possible multi-site action of iron–dithiocarbamates. Of interest is the possibility that these agents might have direct effects as immunosuppressant agents to inhibit T-lymphocyte activation and proliferation which occur *in vivo* during the process of organ rejection. Studies conducted in our laboratory give preliminary insight to support this role.

In one portion of the study, the pro-drug, NOX700, caused a concentration-dependent inhibition of cell proliferation in splenocytes stimulated *ex vivo* with concanavalin A [19]. In a second portion of the study, NOX700 enhanced the effects of cyclosporine A to inhibit concanavalin-A-induced proliferation. In addition to these findings, splenocytes isolated from animals treated with NOX700 also displayed decreased expression of interferon-γ in response to alloimmune activation *ex vivo* in a mixed lymphocyte response assay. Furthermore, expression of cyclin D3 was also inhibited. These studies illustrate the latent action of *in vivo* treatment with NOX700 to make cells refractory *ex vivo* to activation and proliferation in response to alloimmune stimuli. Thus, it is intriguing to speculate on how this finding might explain our original findings of the induction of long-term graft acceptance by chronic treatment and removal of these types of agents given in combination with cyclosporine [37].

These findings may have some potential bearing on the potential immunosuppressive properties of iron–dithiocarbamates. However, we know of no studies which have yet specifically examined iron–dithiocarbamates per se as immunomodulators.

THE ROLE OF NITRIC OXIDE IN LYMPHOCYTE PROLIFERATION

NO per se also has significant immunosuppressant activity [38–43]. This complicates a better understanding of the role of dithiocarbamates on cell proliferation. This action is very complex and contradictory findings in the literature show that the phenomenon is incompletely understood [44]. An interesting finding is one report indicating that the immunosuppressive activity of NO on lymphocyte proliferation was inhibited by superoxide production [45]. Again, this highlights that there is much information that is not fully understood in this area of research.

Related to this discussion is the possibility that dithiocarbamate agents may have direct antioxidant properties or that certain dithiocarbamate derivatives only partially neutralize NO levels. All these considerations have the potential to explain how these agents act to inhibit lymphocyte proliferation. Implicit in all these studies is that a better understanding of the amount of NO produced *in vivo* and the environment of that production is needed in this area of research to delineate how and when NO acts as an immunosuppressant agent *vs.* a cytotoxic agent.

OTHER ACTIONS OF IRON–DITHIOCARBAMATES INVOLVING IMMUNOSUPPRESSION

Several studies conducted in our laboratory revealed that simultaneous administration of dithiocarbamates and low-doses of cyclosporine act synergetically to improve graft survival and benefits to histological rejection [19,35]. Classically, it has been understood that immunosuppressive agents act *via* limiting IL-2 production as a mechanism to promote immunosuppression. More recently, it has been appreciated that transforming growth factor-β (TGF-β) is implicated in the action of a range of clinical immunosuppressant drugs that find clinical use to suppress tissue rejection in human organ transplantation. Examples of such drugs are cyclosporine, tacrolimus and rapamycin [46–48].

To examine a possible role of TGF-β, we measured serum TGF-β levels from rat cardiac graft recipients using ELISA. Our analysis revealed increases in serum TGF-β levels if allograft recipients were treated with Fe^{2+}-diethyldithiocarbamate. Treatment with Fe^{2+}-diethyldithiocarbamate caused a nearly 5-fold increase in serum TGF-β (Fig. 6). An analogous increase was obtained by the oral pro-drug, dithiocarbamate derivative, NOX700. Although the mechanism behind this phenomenon remains unclear at this moment, it appears

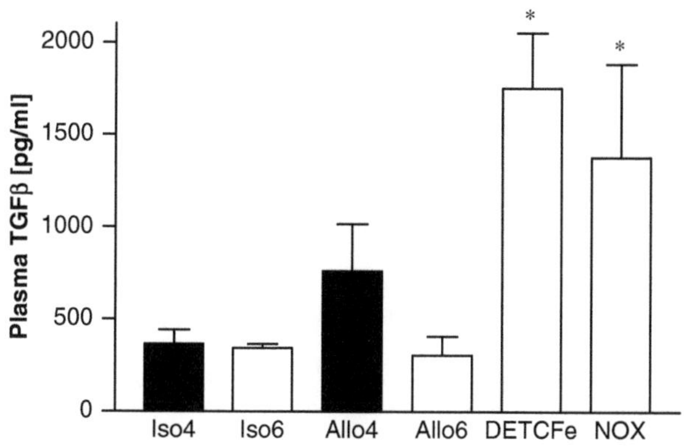

Fig. 6. Blood levels of the immunosuppressant mediator, TGF-β are increased at post-operative day 6 by treatment with Fe^{2+}-diethyldithiocarbamate (DETC-Fe) and derivative, NOX700 *vs.* isografts (Iso) and untreated allografts (Allo) at post-operative days 4 and 6. $^{*}P < 0.05$ *vs.* allografts and isografts at post-operative day 6 (Iso6, Allo6, respectively).

beyond doubt that dithiocarbamates have significant potential for immunosuppression as needed in clinical tissue transplants.

CONCLUSION

In conclusion, this paper has described the protective action of iron–dithiocarbamates in cardiac allograft rejection. It is clear that the action of this type of compound is complex and may involve not only NO scavenging but also antioxidant, anti-inflammatory and immune modulation properties. The mechanisms remain complex and not fully understood.

ACKNOWLEDGMENTS

The author acknowledges the valuable contributions of the microsurgery provided by Ms. Gail Hilton. This work was supported by NIH grants HL64637 and HL078937 to Dr. Galen M. Pieper and by NIH grant EB001981 for the Electron Paramagnetic Resonance Center at the Medical College of Wisconsin. The author also recognizes the in-kind support for surgical research facilities at the VA Medical Center, Milwaukee, Wisconsin.

REFERENCES

1 Lewis NP, Tsao PS, Rickenbacher PR, Xue C, Johns RA, Haywood GA, von der Leyen H, Trindale PT, Cooke JP, Hunt SA, Billingham ME, Valantine HA, Fowler MB. Induction of nitric oxide synthase in the human cardiac allograft is associated with contractile dysfunction of the left ventricle. Circulation 1996; 93: 720–729.

2 Szabolcs MJ, Ravalli S, Minanov O, Sciaacca RR, Michler RE, Cannon PJ. Apoptosis and increased expression of inducible nitric oxide synthase in human allograft rejection. Transplantation 1998; 65: 804–812.
3 Worrall NK, Lazenby WD, Misko TP, Lin TS, Rodi CP, Manning PT, Tilton RG, Williamson JR, Ferguson Jr. TB. Modulation of in vivo alloreactivity by inhibition of inducible nitric oxide synthase. J. Exp. Med. 1995; 181: 63–70.
4 Menon SG, Zhao L, Xu S, Samlowski WE, Shelby J, McGregor J, Barry WH. Relative importance of cytotoxic T lymphocytes and nitric oxide-dependent cytotoxicity in contractile dysfunction of rejecting murine cardiac allografts. Transplantation 1998; 66: 413–419.
5 Takahishi W, Suzuki J, Izawa A, Takayama K, Yamazaki S, Isobe M. Inducible nitric oxide-mediated myocardial apoptosis contributes to graft failure during acute cardiac allograft rejection in mice. Jap. Heart J. 2000; 41: 493–506.
6 Bastian NR, Xu S, Shao XL, Shelby J, Granger DL, Hibbs Jr. JR. N^{ω}-monomethyl-L-arginine inhibits nitric oxide production in murine cardiac allografts but does not affect graft survival. Biochim. Biophys. Acta 1994; 1226: 225–231.
7 Paul LC, Myllärniemi M, Muzaffar S, Benediktsson H. Nitric oxide synthase inhibition if associated with decreased survival of cardiac allografts in the rat. Transplantation 1996; 62: 1193–1195.
8 Pieper GM, Nilakantan V, Hilton G, Zhou X, Khanna AK, Halligan NLN, Felix CC, Kampalath B, Griffith OW, Hayward MA, Roza AM, Adams MB. Variable efficacy of N^6-(1-iminoethyl)-L-lysine in acute cardiac transplant rejection. Am. J. Physiol. Heart Circ. Physiol. 2004; 286: H525–H534.
9 Szabolcs MJ, Sun J, Ma N, Albala A, Sciacca RR, Philips GB, Parkinson J, Edwards N, Cannon PJ. Effects of selective inhibitors of nitric oxide synthase-2 dimerization on acute cardiac allograft rejection. Circulation 2002; 106: 2392–2396.
10 Brandacher G, Zou Y, Obrist P, Steurer W, Werner-Felmayer G, Margreiter R, Werner ER. The 4-amino analogue of tetrahydrobiopterin efficiently prolongs murine cardiac allograft survival. J. Heart Lung Transplant. 2001; 20: 747–748.
11 Worrall NK, Misko TP, Sullivan PM, Hui JJ, Ferguson Jr. TB. Inhibition of inducible nitric oxide synthase attenuates established acute cardiac allograft rejection. Ann. Thorac. Surg. 1996; 62: 378–385.
12 Worrall NK, Pyo PT, Botney MD, Misko TP, Sullivan PM, Alexander DG, Lazenby WD, Ferguson TB. Inflammatory cell-derived NO modulates cardiac allograft contractile and electrophysiological function. Am. J. Physiol. Heart Circ. Physiol. 1997; 273: H28–H37.
13 Koglin J, Granville DJ, Glysing-Jensen T, Mudgett JS, Carthy CM, McManus BM, Russell ME. Attenuated acute cardiac rejection in NOS2-/- recipients correlates with reduced apoptosis. Circulation 1999; 99: 836–842.
14 Mannon RB, Roberts K, Ruiz P, Laubach V, Coffman TM. Inducible nitric oxide synthase promotes cytokine expression in cardiac allografts but is not required for efficient rejection. J. Heart Lung Transplant. 1999; 18: 819–827.
15 Szabolcs MJ, Ma N, Athar E, Zhong J, Ming M, Sciacca RR, Husemann J, Albala A, Cannon PJ. Acute cardiac allograft rejection in nitric oxide synthase $2^{-/-}$ and nitric oxide $2^{+/+}$ mice. Effects of cellular chimeras on myocardial inflammation and cardiomyocyte damage and apoptosis. Circulation 2001; 103: 2514–3520.
16 Casey JJ, Wei XQ, Orr DJ, Gracie JA, Huang FP, Bolton EM, Liew FY, Bradley JA. Skin allograft rejection in mice lacking inducible nitric oxide synthase. Transplantation 1997; 64: 589–593.
17 Pieper GM, Nilakantan V, Hilton G, Halligan NLN, Felix CC, Kampalath B, Khanna AK, Roza AM, Johnson CP, Adams MB. Mechanisms of the protective action of diethyldithiocarbamate-iron complex on acute cardiac allograft rejection. Am. J. Physiol. Heart Circ. Physiol. 2003; 284: H1542–H1551.
18 Akizuki E, Akaike T, Okamoto S, Fujii S, Yamaguchi Y, Ogawa M, Maeda H. Role of nitric oxide and superoxide in acute cardiac allograft rejection in rats. Proc. Soc. Exp. Biol. Med. 2000; 225: 151–159.
19 Pieper GM, Khanna AK, Kampalath BN, Felix CC, Hilton G, Johnson CP, Adams MB, Roza AM. Inhibition of nitrosylation, nitration, lymphocyte proliferation and gene expression in acute and delayed cardiac allograft rejection by an orally active dithiocarbamate. J. Cardiovasc. Pharmacol. 2004; 43: 522–530.

20 Li H, Samouilov A, Liu X, Zweier JL. Characterization of the magnitude and kinetics of xanthine oxidase-catalyzed nitrite reduction. Evaluation of its role in nitric oxide generation in anoxic tissues. J. Biol. Chem. 2001; 276: 24482–24489.

21 Li H, Samouilov A, Liu X, Zweier JL. Characterization of the magnitude and kinetics of xanthine oxidase-catalyzed nitrate reduction: evaluation of its role in nitrite and nitric oxide generation in anoxic tissue. Biochemistry 2003; 42: 1150–1159.

22 Vanin VF, Liu X, Samouilov A, Stukan RA, Zweier JL. Redox properties of iron-dithiocarbamates and their nitrosyl derivatives: implications for their use as traps of nitric oxide in biological systems. Biochim. Biophys. Acta 2000; 1474: 365–377.

23 Xia Y, Cardounel AJ, Vanin AF, Zweier JL. Electron paramagnetic resonance spectroscopy with N-methyl-D-glucamine dithiocarbamate iron complexes distinguishes nitric oxide and nitroxyl anion in a redox-dependent manner: applications in identifying nitrogen monoxide products from nitric oxide synthase. Free Radical Biol. Med. 2000; 29: 793–797.

24 Tsuchiya K, Takasugi M, Minakuchi K, Fukuzawa K. Sensitive quantitation of nitric oxide by EPR spectroscopy. Free Radical Biol. Med. 1996; 21: 733–737.

25 Vanin AF, Poltorakov AP, Mikoyan VD, Kubrina LN, van Faassen E. Why iron-dithiocarbamates ensure detection of nitric oxide in cells and tissues. Nitric Oxide 2006; 15: 295–311.

26 Joshi MS, Lancaster Jr. JR, Liu X, Ferguson Jr. TB. *In situ* measurement of nitric oxide production in cardiac isografts and rejecting allografts by an electrochemical method. Nitric Oxide. Biol. Chem. 2001; 5: 561–565.

27 Huisman A, Vos I, Van Faassen EE, Joles JA, Gröne H-J, Martasek P, Van Zonneveld A-J, Vanin AF, Rabelink TJ. Anti-inflammatory effects of tetrahydrobiopterin on early rejection in renal allografts: modulation of inducible nitric oxide synthase. FASEB J. 2002; 16: 1135–1137.

28 Lancaster Jr. JR, Langrehr JM, Bergonia HA, Murase N, Simmons RL, Hoffman RA. EPR detection of heme and nonheme iron-containing protein nitrosylation by nitric oxide during rejection of rat heart allograft. J. Biol. Chem. 1992; 267: 10994–10998.

29 Pieper GM, Halligan NLN, Hilton G, Konorev EA, Felix CC, Roza AM, Adams MB, Griffith OW. Non-heme iron protein: a potential target of nitric oxide in acute cardiac allograft rejection. Proc. Natl. Acad. Sci. USA 2003; 100: 3125–3130.

30 Lai CS, Komarov AM. Dithiocarbamate spin traps for *in vivo* detection of nitric oxide production in mice. In Bioradicals Detected by ESR Spectroscopy, (Hhya-Nishiguchi H, Packer L, eds.), Birkhäuser Verlag, Basel, 1995, pp. 164–171.

31 Pieper GM, Olds C, Hilton G, Lindholm PF, Adams MB, Roza AM. Antioxidant treatment inhibits activation of myocardial nuclear factor κB and inhibits nitrosylation of myocardial heme protein in cardiac transplant rejection. Antioxid. Redox Signal 2001; 3: 81–88.

32 Pieper GM, Nilakantan V, Zhou X, Khanna AK, Johnson CP, Roza AM, Adams MB, Hilton G, Felix CC. Treatment with α-phenyl-*N-tert*-butynitrone, a free radical-trapping agent, abrogates inflammatory cytokine gene expression during alloimmune activation in cardiac allografts. J. Pharmacol. Exp. Thera. 2005; 312: 774–779.

33 Mankhetkorn S, Abedinzadeh Z, Houee-Levin C. Antioxidant action of sodium diethyldithiocarbamate: reaction with hydrogen peroxide and superoxide radical. Free Radical Biol. Med. 1994; 17: 517–527.

34 Liu J, Shigenaga MK, Yan LJ, Mori A, Ames BN. Antioxidant activity of diethyldithiocarbamate. Free Radical Res. 1996; 24: 461–472.

35 Cooper M, Lindholm P, Pieper G, Seibel R, Moore G, Nakanishi A, Dembny K, Komorowski R, Johnson C, Adams M, Roza A. Myocardial nuclear factor-κB activity and nitric oxide production in rejecting cardiac allografts. Transplantation 1998; 66: 838–844.

36 Pieper GM, Roza AM, Adams MB, Hilton G, Johnson M, Felix CC, Kampalath B, Darkes M, Wanggui Y, Cameron B, Fricker SP. A ruthenium (III) polyaminocarboxylate complex, a novel nitric oxide scavenger, enhances graft survival and decreases nitrosylated heme protein in models of acute and delayed cardiac transplant rejection. J. Cardiovasc. Pharmacol. 2002; 39: 441–448.

37 Roza AM, Cooper M, Pieper G, Hilton G, Dembny K, Lai CS, Lindholm P, Komorowski R, Felix C, Johnson C, Adams M. NOX 100, a nitric oxide scavenger, enhances cardiac allograft survival and promotes long-term graft acceptance. Transplantation 2000; 69: 227–231.

38 Hoffman RA, Langrehr JM, Wren SM, Dull KE, Ildstad ST, McCarthy SA, Simmons RL. Characterization of the immunosuppressive effects of nitric oxide in graft vs. host disease. J. Immunol. 1993; 151: 1508–1518.

39 MacMicking J, Xie QW, Nathan C. Nitric oxide and macrophage function. Ann. Rev. Immunol. 1997; 15: 323–350.

40 Hoffman RA, Ford HR. Nitric oxide regulation of lymphocyte function. In Nitric Oxide and Inflammation, (Salveminie D, Billiar TR, Vodovotz Y, eds.), Birkhäuser Verlag, Basel, Switzerland, 2001, pp. 131–143.

41 Allione A, Bernabei P, Bosticardo M, Ariottie S, Forni G, Novelli F. Nitric oxide suppresses human T lymphocyte proliferation through IFN-γ-dependent and IFN-γ-independent induction of apoptosis. J. Immunol. 1999; 163: 4182–4191.

42 Mahidhara RS, Hoffman RA, Huang S, Wolf-Johnston A, Vodovotz Y, Simmons RL, Billiar TR. Nitric oxide-mediated inhibition of caspase-dependent T lymphocyte proliferation. J. Leuk. Biol. 2003; 74: 403–411.

43 Van der Veen RC, Dietlin TA, Pen L, Gray JD. Nitric oxide inhibits the proliferation of T-helper 1 and 2 lymphocytes without reduction in cytokine secretion. Cell. Immunol. 1999; 193: 194–201.

44 Macphail Se, Gibney CA, Brooks BM, Booth CG, Flanagan BF, Coleman JW. Nitric oxide regulation of human peripheral blood mononuclear cells: critical time dependence and selectivity for cytokine versus chemokine expression. J. Immunol. 2003; 171: 4809–4815.

45 Van der Veen RC, Dietlin TA, Hofman FM, Pen L, Segal BH, Holland SM. Superoxide prevents nitric oxide-mediated suppression of helper T lymphocytes: decreased autoimmune encephalomyelitis in nicotinamide adenine dinucleotide phosphate oxidase knockout mice. J. Immunol. 2000; 164: 5177–5183.

46 Khanna AK, Cairns VR, Becker CG, Hosenpud JD. Transforming growth factor (TGF)β and anti-TGF-β antibody abrogates the in vivo effects of cyclosporine: demonstration of a direct role of TGF-β in immunosuppression and nephrotoxicity of cyclosporine. Transplantation 1999; 67: 882–889.

47 Khanna AK. Mechanism of the combination immunosuppressive effects of rapamycin with either cyclosporine or tacrolimus. Transplantation 2000; 70: 690–694.

48 Khanna AK. The immunosuppressive agent tacrolimus induces p21WAF/CIP1WAF1/CIP1 via TGF-β secretion. Biochem. Biophys. Res. Commun. 2003; 302: 266–272.

INDEX

E

M

N

O

P

Q

R

S

www.ingramcontent.com/pod-product-compliance
Ingram Content Group UK Ltd.
Pitfield, Milton Keynes, MK11 3LW, UK
UKHW020229290926
15470UKWH00001B/430

9 780444 522368